Masonry Skills
Fifth Edition

Richard T. Kreh, Sr.

THOMSON

DELMAR LEARNING ™ Australia Canada Mexico Singapore Spain United Kingdom United States

THOMSON

DELMAR LEARNING

Masonry Skills, 5th Edition
Richard T. Kreh, Sr.

Executive Director:
Alar Elken

Executive Editor:
Sandy Clark

Acquisitions Editor:
Mark Huth

Development:
Dawn Daugherty

Executive Marketing Manager:
Maura Theriault

Channel Manager:
Fair Huntoon

Marketing Coordinator:
Brian McGrath

Executive Production Manager:
Mary Ellen Black

Production Editor:
Ruth Fisher

Library of Congress Cataloging-in-Publication Data:
Kreh, R. T.
   Masonry skills / Richard T. Kreh, Sr.—5th ed.
      p. cm.
   Includes index.
   ISBN 0-7668-5936-3—
   ISBN 0-7668-5937-1 (ig)
   1. Masonry. I. Title.
TH5313 .K73 2004
693'.1—dc21   2002031396

NOTICE TO THE READER

Contents

SECTION 11 UNDERSTANDING AND READING CONSTRUCTION DRAWINGS

SECTION 12 MASONRY AS A CAREER AND MASONRY PROJECTS

Preface

Ever since MASONRY SKILLS was first published, it has been considered the best book available for really explaining and showing how good masonry work is performed on the job. The major reasons for this are that it is one of the few masonry books in print and that it is written by a master bricklayer and teacher of the trade. The author has spent his entire life working as a journeyman mason and masonry instructor in the public school system of Maryland.

It is also one of the few books that has been reviewed and approved by the various construction institutes around the nation that are the recognized authorities in the masonry field. This revision has been carefully redone by the author to point out new developments and techniques in the masonry trade and add new, relevant, and important information necessary to train apprentice masons in their goals of becoming journeymen masons.

Masons are highly skilled craftsmen who use specialized equipment and tools to mix mortar, bond materials together to form patterns, and lay out and construct masonry walls and other important features of buildings. In this day and age, this requires both technical as well as on-the-job skill training to be successful.

Once you have learned and perfected your trade, it becomes a part of you, a part that no one can ever take away. The pride that one knows, that we can do good work in our chosen field, is very satisfying and also gives us a purpose in our life. To be really good at what you do, you must like it! This completely revised edition of MASONRY SKILLS provides the necessary materials to guide the student, apprentice, or reader through a secondary or vocational-technical program in the study of masonry basics. It will act as a handbook and guide for the apprentice and as a reference for the journeyman mason in their everyday work. It will also greatly benefit the serious amateur masons who want to do work of their own.

This revised edition of MASONRY SKILLS retains the major textbook features that made the other editions a trade standard. These include:

- performance objectives that explain to the student and instructor what information is to be learned through careful study of the material.
- an easy-to-follow format, high in illustrations and hands-on-application.
- progressive skill objectives in each unit project that stress specific skills each student or apprentice should have learned up to that point in study.

- safety features and comments integrated throughout the text where they are needed.
- safety introduction at the beginning of the book.
- latest updated tables and information on brick sizes and nomenclature from the Brick Industry Association.
- manufacturing methods for brick and concrete block units.
- information on the latest development in masonry tools and equipment including laser transits.
- information in the mortar unit on repointing mortar joints in historical buildings, as well as tips on how to add color and tint to match the original mortar.
- how to operate the masonry saw.
- unit reviews are included where applicable, and math check points to assist the reader to apply the information in estimating materials and labor for masonry jobs.
- up-to-date information on OSHA scaffolding safety standards and requirements.
- a large number of student masonry projects at the end of selected units, including some new ones from previous editions, that guide the reader through actual construction in a step-by-step procedure format.
- all units are written in a trade competency format with objectives listed at the beginning of each unit.
- achievement review questions following each unit that allow readers to evaluate their understanding and progress, section summaries followed by questions which help the reader to gauge his or her understanding of each section or block of related materials.
- the latest information on chimney and fireplace design current in the masonry trade today.
- a detailed section on arches, and how they are built, including revised arch projects for students to build in the shop.
- a plan with complete instructions about how to build your own corner pole guide for laying brick or concrete block in a shop situation.
- a totally new Unit on Glass Block Masonry with many photographs and details of current construction practices reflecting the latest advances and techniques in installing glass block. This unit was reviewed and approved by Pittsburgh Corning Corporation, Glass Block Division, the leading manufacturer of glass block products in the world, and it meets the current standards for installing glass block products
- a unit containing an additional assortment of interesting, challenging masonry projects for the students to build in order to test their skills in a school shop setting.
- some new selected tips and techniques from the author with photographs for masonry projects that will enhance and beautify your home.
- An instructor's guide that includes lesson plans for each unit, answers to review questions, pre-tests for the materials in the units to point the student toward the important information to be learned and a comprehensive final test for the instructor or reader, to measure their mastery of the materials presented.
- an important unit on masonry work as a career that has been revised; it includes apprenticeship standards and requirements.

- technical review of text materials by field experts, including the Brick Industry Association, National Concrete Masonry Association, National Lime Association, and Pittsburgh Corning Corporation, Glass Block Division.
- this revision was aided by a survey and review from a number of leading masonry instructors around the nation asking for their input and what should be added or changed in this revision. A number of these comments and suggestions were used in the revision of this edition by the author.

RELATED SKILLS

The apprentice mason, in addition to attending 144 hours per year of related classroom instruction in the technical areas, should be proficient in the following skills:

- proper care and use of the tools
- study of mortar and mixing mortar
- building of foundations and parging
- trowel skills
- laying of brick, concrete block, and other masonry units
- safe, accurate cutting of masonry materials and walling in and around openings
- laying brick to create arches
- laying out and building fireplaces and chimneys
- proper cleaning of masonry work
- observing good safety practices and practicing accident prevention at all times
- able to read and use the different mason's scale rules
- be able to lay out and construct the various types of masonry walls and projects mentioned in the text
- be familiar with and understand how to form, place, and finish concrete footings and slabs
- reading a basic set of plans and carrying them out in the layout and construction of masonry work

At the completion of a masonry apprenticeship program, the apprentice becomes a journeyman mason and is entitled to full prevailing scale wages determined by the area of the country he or she works in. Interested journeyman masons who want to advance and better themselves may continue their studies in programs offered by technical schools, union schools or community colleges, which may lead to becoming a masonry foreman or eventually operating their own masonry contracting business. Because new techniques are being constantly developed, masons should continually read trade literature and professional journals, attend related trade organizations or workshops, and study new methods and products as they are developed.

ABOUT THE AUTHOR

The author of MASONRY SKILLS, Richard T. Kreh Sr., was most recently an instructor of masonry and building trades at Linganore High School, near Frederick, Maryland. He is a member of a number of professional organizations, including the National Educational Association, Maryland State Teachers Association, the Iota Lambda Sigma Fraternity of Professional Technical Educators, and the National Association of Home and Workshop Writers. Mr. Kreh is also a Past President of the Masonry Instructors of Maryland.

Mr. Kreh has earned numerous awards, such as the Distinguished Professional Award from Iota Lambda Sigma Fraternity, National Brick Masonry Instructors Association, Professional Masonry Contractors Association, Meritorious Service Award from the Masonry Instructors of Maryland, and the James L. Reid Award, which is given to the instructor who has made the most outstanding and praiseworthy contribution to the masonry profession in the State of Maryland.

Other Accomplishments

Mr. Kreh is a nationally and internationally renowned author of masonry books, such as Delmar Learning's MASONRY SKILLS, ADVANCED MASONRY SKILLS, and SAFETY FOR MASONS, which are distributed in eight different countries by International Thomson Publishing. Mr. Kreh has published numerous other materials, including DCA (Display Corporation of America) Masonry Transparencies Series Sets, which were designed to correlate with *Masonry Skills* and *Advanced Masonry Skills*. Mr. Kreh has also published BRICKLAYING AND STONE MASONRY and MASONRY PROJECTS AND TECHNIQUES for Popular Science Book Club. Mr. Kreh writes feature articles for The Taunton Press Inc. publications, including FINE HOMEBUILDING MAGAZINE, Fine Home Building Q & A Page, and he is the author of BUILDING WITH MASONRY, BRICK, BLOCK AND CONCRETE for the Taunton Press Inc. Book Club. Mr. Kreh is also a contributor to The Taunton Press Inc. book, BUILDER'S LIBRARY. Mr. Kreh has also served as a Consulting Editor for Time-Life Books, ADVANCED MASONRY publication, has written freelance articles for FINE GARDENING MAGAZINE, MASONRY CONSTRUCTION MAGAZINE, and SATURDAY EVENING POST MAGAZINE.

Mr. Kreh served as a consultant for ORTHO PUBLICATIONS (owned by Chevron Chemical Company) and also was a consultant for the United States Department of Labor in the rewriting of Standards for National Masonry Job Title. Mr. Kreh is also a professional trade photographer, a member of the NPS (Nikon Professional Services) and has supplied photographs for a number of other authors to use in their books. In addition to all of his other accomplishments, Mr. Kreh is a Master Mason by trade, consultant to architects and builders on historic renovation of older buildings and incorporates over 45 years of experience and knowledge in the masonry trade. His books, MASONRY SKILLS, ADVANCED MASONRY SKILLS, and SAFETY FOR MASONS, published by Delmar/Thomson Learning, have been the best-selling masonry books in the United States in their field for many years.

ACKNOWLEDGMENTS

The author wishes to extend special thanks for their assistance in the preparation of this revision to the Brick Industry Association; Pittsburgh Corning Corporation, Glass Block Product Division; Robert DeGusipe, Marketing Manager for PCC; Nicholas T. Loomis, P.E., Senior Systems Engineer for PCC; Mike Fraser, Vice President and General Manager of Stabila Levels; Dennis Grabler, Director of Technical Publications, National Concrete Masonry Association; Larry Schaffert, President of Schaffert Construction; George Moehrle Masonry; and Carl A. Bongiovanni, President of Bon Tool Co.

Reviewers

The author also wishes to express his appreciation to the following reviewers for their comments and recommendations:

Robert Fleurent
Upper Cape Cod Regional Voc-Tech
Bourne, MA

David McCosby
New Castle School of Trades
Pulaski, PA

Mark Tanner
Fulton-Montgomery Career & Technical Center
Johnstown, NY

Marty Swan
Lincoln Land Community College
Taylorville, IL

RECOMMENDED ADDITIONAL DELMAR LEARNING BOOKS

Safety for Masons, Kreh   ISBN 0827316682
Practical Problems in Math for Masons, 2E, Ball   ISBN 0827312830
Understanding Construction Drawings, Huth   ISBN 0766815803

UNIT 1
Safety on the Job

─────────────────── OBJECTIVES ───────────────────

After studying this unit, the student will be able to

■ explain why good safety practices on the job are essential.

■ describe the correct dress and safety measures to be utilized when on the job.

■ list some of the more common hazards presented on jobsites and how to avoid them.

Nothing should be more important to masons than their own personal safety and the prevention of accidents on the job. Besides the physical handicap which results from injury, the worker may be deterred from work for a period of time. Loss of time costs the mason and the employer money.

Many accidents could be avoided if common sense were practiced and shortcuts were not taken. The quickest way to complete a job is not always the best way. Other reasons for injuries on the job include lack of interest and attention to the job at hand, tiredness, and careless, slipshod methods. Sometimes, there is a tendency for workers to become careless on the job after they have learned the basics of the job very well. It is imperative that the mason guard against this attitude. Safety requirements should be explained to the mason by the contractor when starting work. It is then the responsibility of the mason to observe and obey these regulations.

Besides the physical condition and the mental attitude of the worker, injury and accidents may result from unsafe tools and equipment and improper clothing and safety devices. Construction work is one of the most hazardous of all occupations because one must work with power tools, on scaffolds, and under and around moving equipment. For this reason, the federal law OSHA (Occupational Safety and Health Act) was passed to guarantee the worker a safe working environment. It should be remembered that OSHA only sets minimum standards and more strict state requirements may be in effect in many areas of the country. It also should be remembered that the OSHA law was

passed because of the high accident rate due to careless methods and lack of regard for the safety of the worker. It is to everyone's benefit to practice good safety rules at all times.

THE IMPORTANCE OF PROPER DRESS

Clothing and Shoes

Clothing should be kept in good repair at all times when working, Figure 1-1. Torn coveralls or work pants may snag on materials or tools, thereby causing a fall. Shirt sleeves should be either buttoned or rolled up and shirttails tucked into the pants. Since pant cuffs may also catch on scaffolding or ladders when climbing, the mason should turn the cuffs down or wear pants without cuffs.

Short pants cut off above the knee may be comfortable in warm weather but offer no protection to the legs when working around protruding objects such as wall ties and wire. Also, pieces of masonry units and welding sparks may cause injury to exposed legs. On many construction sites, pants cut above the knee are prohibited by the supervisor or safety engineer.

The mason should be careful to protect the feet, as they may be struck and crushed by heavy weights. Steel-toed shoes which are capable of bearing heavy blows are available and are a recommended safety measure. Many masons resist wearing steel-toed shoes because of the extra weight, and because workers sometimes forget they are wearing them and injure themselves by kicking an object and jamming the toes against the hard surface of the shoe.

1

Fig. 1-1 Mason in safety gear operating a saw. Notice the hard hat, safety glasses, rubber gloves, and coveralls.

Fig. 1-2 Mason wearing proper safety headgear. Notice that the hard hat is set slightly off the head.

Working shoes should be equipped with sturdy heels so that the feet are properly supported and to reduce the chance of a turned ankle. Excessively thin soles on shoes should be avoided, as they can cause sore feet and fatigue.

Many firms offer safety programs which provide safety shoes to employees at a reduced price.

Protective Headgear

Protective headgear constructed of a sturdy material such as fiberglass are known as *hard hats*. Hard hats are required on all construction jobsites by federal and state law. Lightweight plastic hats, commonly known as *bump hats*, afford little protection and do not conform to regulations. Approved OSHA hard hats are recommended for use on all construction jobsites.

The hard hat should be raised slightly off the worker's head so that if a blow is encountered, the hard hat will cushion the effect, acting much as a shock absorber does in a car, Figure 1-2. It is extremely dangerous practice to remove the lacing from inside the hat, Figure 1-3, as the hat would be in direct contact

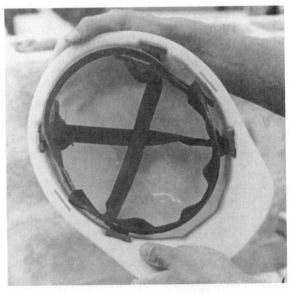

Fig. 1-3 Hard hat is supported on head by lacing to cushion a blow to the head.

with the skull and could cause serious, if not fatal, injury.

Eye Protection

The masonry saw presents a hazard, as particles may fly off during the cutting process and enter the eyes. Eye protection in the form of goggles or shatterproof glasses should, therefore, always be worn when cutting masonry materials, Figure 1-4. Although goggles may be sprayed with water and partially obstruct sight when operating a wet saw, flying chips present a greater danger and should be worn when any saw is in operation. They also help considerably to prevent grit and dirt from entering the eyes on windy days. If dirt, grit, or other foreign matter enters the eye and cannot be removed easily, the mason should see an eye doctor as soon as possible.

Fig. 1-4 Mason operating saw with proper eye protection.

Safety and face protection should meet the requirements of American National Standards Institute Z87.1-1968 as stated in the OSHA Federal Regulations. Safety glasses worn by masons should always include protective side shields to protect the corners of the eye. People who wear glasses to correct their vision can obtain matching prescription safety glass lenses through their doctor.

Hearing Protection

Hearing protection should meet the requirements of American National Standards Institute S-1926.52 as stated in the OSHA Federal Regulations. Protective measures should be taken when the noise exposure level exceeds 90 decibels. When noise levels vary over two or more periods, their combined effect should be considered, rather than the individual level of each. The only way noise exposure levels can be accurately determined on the job is by having a safety engineer conduct a test. It is the engineer's responsibility to inform the contractor or workman on the job that there is a danger present and to take protective action. Devices to protect the ear, such as ear muffs or ear plugs, should be fitted to suit each individual's needs. Just stuffing plain cotton in the ear does not give acceptable protection. The best rule to follow is that if you have any questions about or discomfort from excessive noise, wear ear protection.

Working Gloves

While restricting the sense of feeling to some degree, work gloves protect the hands from rough materials or when cutting masonry work. Work is more efficient in cold weather when gloves are worn. Wear a glove light enough so that it is not clumsy and difficult to flex the fingers.

Hair

It should be remembered that exceptionally long hair presents a very dangerous hazard if it is allowed to hang loose when working around running machinery and protruding objects. Tuck the hair well under the hard hat or tie it back to prevent possible injury.

ARRANGING MATERIALS SAFELY IN THE WORK AREA

Materials in the general working area and on the stockpile should be stacked safely. Masonry units which are not stacked properly and bound are very likely to be upset, possibly causing injury.

The accepted method on most jobs is to reverse the direction of every other course when stacking materials so that they are secure, Figure 1-5. The mason should attempt to keep the pile neat and vertically in line to eliminate the possibility of snagging the clothing. The pile should not be so high that the mason does not have easy access to the material. Materials stacked too high on the pile not only pose a safety hazard, but may drastically curtail the mason's production.

Mortar pans should be placed back from walls approximately 2′ to allow proper working space. Pans should have approximately 5′ between them lengthwise to allow stacking of masonry units between each pan. Tools should be stored underneath the mortar pans out of the paths of workers.

Workers should never run on the job or jump over trenches or stacks of materials. In cold weather, be es-pecially cautious of walkways and runways into buildings as they may be covered with ice and may be very slippery.

SAFETY NETS

One of the most important life-saving devices to be developed in recent years is the safety net, Figure 1-6. During the two years that the 110-story Sears Tower was being built in Chicago, Illinois, nine construction workers fell into elevator shafts and could have been killed or seriously injured. But in each case, the worker was caught in a safety net hung across the shaft. Most of them returned to work immediately.

Construction workers in general feel much safer when working at heights if they know a net is in place underneath. Since it is costly to erect the nets, they usually are used only on big jobs.

The original nets were made of manila rope, but now synthetic fibers are used, as they are stronger, lighter in weight, and more resistant to wear and weather elements. One of the most popular types is made from polypropylene and has more than twice the strength of manila rope.

Fig. 1-5 Stacked materials can be secured by reversing every other course on the pile.

Fig. 1-6 Steel framework building protected by safety nets.

USING TOOLS SAFELY

Tools are the masons' means of earning a living. They must be kept in good repair if the mason is to be productive and a safe worker. Poor tools can cause injuries and they become potential weapons. Worn and defective tools should be replaced or repaired immediately.

Hammers and chisels must be kept sharp and tempered to the proper degree of hardness for the materials being cut. The best procedure is to take the tools to a local blacksmith, if one is available, so that the steel can be heated and tempered when being sharpened. Excessive grinding of a cutting edge results in a defective tool. A sharp tool is a safe tool.

Mushroomed or Burred Chisels

After prolonged use, the head of a chisel becomes flared out from striking with a steel hammer, Figure 1-7. In these cases, the chisel has become *mushroomed* or *burred*. This is a dangerous condition, since the metal curled over the edges eventually comes off; and

Fig. 1-7 Burred chisel heads, a potential cause of injury, should be avoided.

invariably, the pieces of steel become lodged in the body. These steel pieces are very difficult to remove and may cause injury and loss of time at work.

To prevent this from happening, grind off the chisel head on a grinding wheel to remove the burred metal. While doing this, cool the tool periodically with water to prevent it from becoming too hot and, therefore, losing the temper of the steel.

> **Caution:** Heavy gloves should always be worn when cutting with or grinding a chisel to protect the hands from possible blows and particles of steel.

Electrical Equipment

Grounding Electrical Tools. All electrical tools should be grounded. Grounded plugs have three prongs. The third prong acts as a grounding device for the electricity feeding into the piece of equipment. If it were not present, the electricity might pass through the worker's body in the event of an electrical problem. Most new plugs on tools act as grounding devices.

> **Caution:** All receptacles should be properly grounded.

Electrical Extension Cords. A simple act such as the unplugging of an electrical extension cord can be dangerous if not done correctly, Figures 1-8A and 1-8B. Heavy-duty cords are also costly to replace if damaged through careless handling and neglect. The mason should never stand in water while unplugging a cord or yank the cord carelessly. When there is an electrical problem on the job, a trained electrician should always be called.

Using Ladders Safely

The ladder is an essential piece of equipment for the mason. Practical safety rules should be practiced whenever using ladders.

Be sure that the ladder is in a good workable condition, with no split rungs or sides. If a cracked rung is

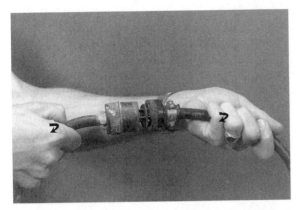

Fig. 1-8A Incorrect method of unplugging electrical cord. Notice the mason's grip on the wire and the downward pull. This is a potential work hazard.

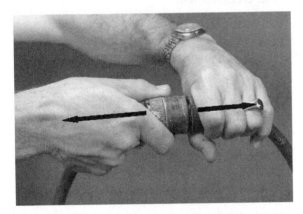

Fig. 1-8B Correct method of unplugging cord. The mason is gripping the plug firmly and pulling the plug in a straight, outward direction.

found, remove the defective rung with a hammer to warn others of the hazard. Report the unsafe condition immediately so that repairs can be made.

Do not erect a ladder in a place where a passing vehicle or passerby might collide with it. Always place the ladder on a solid, nonskid surface. When using a ladder to reach another walking surface, be certain that the ladder extends at least 3′ above the point where you expect to step off of it. Be sure that the ladder is firmly attached to a strong support. The areas on which the mason is stepping should be free from clutter and should offer some type of grip for the hands. Rung clamps must always be engaged before ascending the

ladder, or the top section of the ladder will slip, possibly causing injury.

Aluminum ladders, while stronger, are lighter in weight. For this reason, they skid more easily than heavy wood ladders. Aluminum, an excellent conductor of electricity, must not touch or lie against electrical wire.

If tools are to be carried while climbing a ladder, shove the plumb rule through the handle of the tool bag and sling it over the shoulder, Figure 1-9. This frees both hands for climbing.

At no time should the mason stand on the highest rung of the ladder to work. This practice is extremely dangerous. Where there is a danger of slipping on concrete floors, another worker should hold the bottom of the ladder steady. Nonskid feet are also available for ladders.

Ragged-Edged Mortar Pans

The rough treatment given mortar pans, such as being beaten to remove mortar, soon takes its toll on the pans, in the form of sharp, torn metal edges. This can

Fig. 1-9 Proper way to ascend a ladder with tool bag on shoulder.

be very dangerous since the mason's hand is placed below the top edge when mortar is taken from the pan with the trowel.

When mortar pans are in this condition, they should be taken out of use and repaired immediately. If they are beyond repair, they should be replaced. Metal pans can be welded, but plastic ones must eventually be discarded.

Removing Nails from Lumber

When wooden braces which contain nails are removed, either remove the nails at once or bend them over into the wood. Never throw lumber with nails which have the sharp end facing up on the ground or floor. A common accident on jobsites is stepping on nails, but this can be prevented through simple safety practices. If a nail punctures the skin, see a doctor immediately for treatment of the wound.

The sharp end of nails should never be exposed underneath scaffolding, since a person passing underneath must bend over and, therefore, will not see them. Remove or bend the nails over.

Overhead Objects

The danger derived from objects being dropped from overhead cannot be overemphasized. This danger may come from two major sources. First, workers located at a high level pose a threat, as they may drop tools or materials on the workers below. Protective measures can be taken to prevent this, such as wooden planking laid on framework or heavy wire mesh. The best course is to avoid working directly underneath other workers, but this is not always possible. Always be on the alert for falling objects when working in this type of situation.

The second source of danger from overhead objects originates on high-rises or tall buildings, where cranes are necessary for lifting and placing various materials on the structure. Workers must exercise extreme caution in these situations. For example, sometimes steel bar joists which support floors must be fastened to a crane by means of a steel hook and cable so that they can be moved, Figure 1-10. It is entirely possible that the joists may become loosened from the grip as they are being

Fig. 1-10 Steel bar joists being lifted by crane up on building. No worker should ever be under the load.

transported to the desired level, endangering the workers below.

> **Caution:** Under no circumstances should any workers be located underneath cranes when they are transporting heavy equipment. As a rule, the best practice is to suspend work in the immediate area while materials are being moved by the crane.

Protective Safety Rod Caps

An important safety improvement for masons working around standing steel reinforcement is the use of rigid plastic safety caps on the ends of the steel rods, Figure 1-11. This protects workers from being injured on the sharp edges of the rods. There have been many accidents and injuries over the years caused by workers falling against or onto the ends of the steel rods and the caps are now required by federal OSHA Regulations.

Fig. 1-11 A protective safety cap on the end of a steel reinforcement rod.

EXERCISING CARE ON SCAFFOLDING

Remember the general rules concerning safety on scaffolding, and always apply these rules when on the job.

- Inspect every scaffold to be sure it is built correctly before working on it. It should be set on a firm base, preferably a wooden plank (called a *mud sill).*
- Spaces or gaps left near the wall are potential hazards through which the mason could fall and be injured. A substantial thickness of plywood or boards should cover the hole and be nailed down on both sides.
- When more than one section of scaffolding is used, the danger becomes greater because of the increased height. Attach all braces and tighten thumbscrews. Lap boards far enough over so that they are secure. Guardrails and toe boards should be in place when the height demands them.

- Tie scaffolding to the structure if more than two sections are used or if it is more than 12' in height. Use a strong wire that can be cut off later. Fill the holes with mortar.
- Tie ladders to the scaffold so that they won't become dislodged.
- Place stacks of materials over the scaffolding frames.
- Use good, strong planking for the flooring of the scaffold.
- Practical joking cannot be tolerated when working on scaffolding.

LIFTING SAFELY

Many masons suffer unnecessary, painful injuries such as back strain and hernias due to poor lifting practices. Good lifting practices are primarily the application of good common sense, Figure 1-12. The following are suggestions to be followed when lifting.

- Remove all scraps of materials from the working area and clear the area to which the object is to be moved. Be sure that the area is free from holes and uneven footing.
- Wear steel-toed safety shoes if possible.
- Get extra help for large or heavy objects. Never attempt to lift a weight when it is questionable as to whether you are able to lift it or not.
- Study the object to determine the best way to pick up and grip the object.
- Place the feet close to the object, and apply a good grip.
- Bend the knees and keep the back straight when lifting.
- Keep the load close to the body.

WHEN AN INJURY OCCURS

Any injury serious enough to require medical attention should be reported to the supervisor or instructor immediately. There are two important reasons for this: quick action and first aid may stop the injury from becoming more serious, and workers' compensation insurance requires that injuries be reported as soon as

CORRECT

INCORRECT

Fig. 1-12 Correct and incorrect methods of lifting. The worker on the left bends her legs as she lifts the block, thereby reducing unnecessary strain. The worker on the right is lifting with her back, not her arms.

possible after they occur. Failure to report accidents may result in the loss of rights or benefits.

As a rule, the worker should not attempt to treat the injured person if the injury is serious. However, if a person is bleeding severely, immediate action must be taken. It should be remembered, though, that improper treatment may do more harm than good.

So that work is safe, profitable, and satisfying, be constantly alert to hazards on construction work and practice good safety rules at all times.

HAZARDOUS MATERIALS

There are a large number and a variety of hazardous materials used on construction jobs that workers may come in contact with in the course of their daily work. Two of the hazardous materials most frequently encountered by masons in their specific work are cements and masonry cleaners. OSHA (the Occupational Safety and Health Administration) is the federal agency in charge of enforcing these regulations on the job site.

Material Safety Data Sheets

Chemical manufacturers and importers must obtain or develop a Materials Safety Data Sheet for each hazardous chemical they produce or distribute. All employers must have a Material Safety Data Sheet on the job site for each hazardous material they use on that job. Each materials sheet must be in English and must contain important safety information for the particular product.

The employer must also ensure that the sheets are readily accessible to all employees in the work area during each work shift. Materials Safety Data Sheets can be kept in any format, including operating procedures, and may be designed to cover groups of hazardous chemicals in a work area where it may be more appropriate to address the hazards of a process rather than individual hazards.

In addition to listing chemical composition and the hazards associated with the particular product, the Material Safety Data Sheet should also list the safety protective equipment that needs to be worn by a worker using the material and the emergency and first aid procedures that should be enacted if a problem arises.

Employers must also, according to the law, conduct safety training for their employees regarding hazardous chemicals or products in the work area at the time of the employee's initial assignment and also whenever a new hazard is introduced into the work area. Training sessions for all employees are required and must include important information such as location of the Materials Safety Data Sheet on the job site, methods to detect the presence or release of hazardous chemicals in the work area, the physical and health hazards of the chemical in the work area, measures employees can take to protect themselves from the hazard, and the procedure for prompt reporting of any hazardous situation in the work area to a supervisor or the employer. Periodically, regular safety and training meetings must be held to keep all employees updated and informed of hazards in their work areas. These meetings must be carefully recorded and the records should be readily available if an OSHA official needs to examine them.

For further information, consult the 1995 OSHA SAFETY AND HEALTH STANDARDS FOR CONSTRUCTION 29 CFR 1926, Subpart 1926.59 Hazard Communications. This is available from The United States Government Printing Office.

OSHA REVISES FALL-PROTECTION RULES

The Secretary of Labor announced revised OSHA standards for protection from falls, which are the leading cause of worker fatalities in the construction business. Although the construction industry's 5 million workers constitutes slightly more than 5% of the total U.S. work force, the industry accounts for 17% of all job-related fatalities annually; about 21% of these fatalities are the result of a fall. According to OSHA officials, these new standards are necessary because the existing rules are unclear, do not cover all construction activities, set varying heights, and are difficult to enforce.

The new standards were published in the Federal Register on August 9, 1994, and the final rules became effective February 6, 1995. The revised standards:

- Set a uniform threshold height of 6 feet for providing consistent protection. Fall protection can generally be provided through the use of guardrail, safety net, or personal fall protection.
- Prohibit the use of body belts as part of a personal fall-arrest system as of January 1, 1998. In addition, only locking-type snap-hooks will be permitted for use in personal fall-arrest and positioning systems as of that date.
- Allow employers to choose from various options to provide fall protection. For example, when working on a low-sloped roof with protected sides, workers could be protected by guard-rails, safety nets, personal fall-arrest systems, or warning lines and safety monitoring systems.
- Provide sample fall-protection plans, which can be modified for site-specific conditions.

One provision of the final rules applies especially to masonry work. Masons who engage in overhand bricklaying 6 feet or more above the lower level must be protected from falling by a guard-rail, safety net, or personal fall-arrest system. Overhand bricklaying occurs, by OSHA's definition, when a mason leans over a wall to lay units or tool mortar joints on the opposite side of the wall from where he or she is standing.

Although the employer is responsible for creating and maintaining a safe work place, it is the responsibility of the workers to observe all safety regulations as they are the ones who suffer if injured on the job. Employers will not tolerate workers who continually violate safe work practices and will discharge them as soon as possible. This is because any serious accident on the job is investigated by not only the State Safety and Health Administrations, but also can involve OSHA. The fines are extremely large and the employer's insurance will either be canceled or will increase dramatically. This in turn seriously affects the employer's profit margin and ability to stay in business.

ACHIEVEMENT REVIEW

Answer the following questions with a short statement.

1. Describe a properly dressed mason ready for work.
2. State two reasons why good work shoes are important to the mason while working.
3. Give a reason why it is not desirable to wear work pants with cuffs while working.
4. Although the wearing of steel-toed shoes is recommended, there is a danger if care is not exercised. What is this danger?
5. Removing the lacing from inside a hard hat can be very dangerous. Why is this true?
6. Describe two situations when the wearing of safety glasses or goggles is required.
7. What is a *burred head* on a chisel and what action should be taken to correct this situation?
8. Leaving a brick hammer lying in the work path can cause a leg to be painfully bruised. Describe how this may happen.
9. Describe the proper method of unplugging an electrical extension cord.
10. What is the purpose of a third prong on the end of the cord that is plugged into the receptacle?
11. The mason is preparing to climb a ladder set on a concrete floor. The ladder does not have nonskid feet. What is the proper procedure to follow to prevent an accident?
12. Describe the safe method of stacking bricks. Use a sketch or drawing to explain more fully.
13. What is the proper procedure to follow when erecting the first sections of scaffolding to properly secure it?
14. Accidents must be reported to the supervisor as soon as possible. Why is this necessary?

SECTION ONE
DEVELOPMENT AND MANUFACTURE OF BRICK AND CONCRETE BLOCK MASONRY UNITS

UNIT 2
Development of Clay and Shale Brick

OBJECTIVES

After studying this unit, the student will be able to

- describe the development of brick.
- discuss how brick is used in a modular building system.
- list some of the characteristics of the modern brick.

THE DEVELOPMENT OF BRICK

Brick is one of the oldest manufactured building materials. Recent excavations have uncovered remains of brick walls dating back 6000 years. Considering their age, the walls were in surprisingly good condition.

Greek historians of the fifth century B.C. relate accounts of the splendid wonders of the city Babylon. These include striking descriptions of immense walls and temples, many of which were built of brick. In Dashur, Egypt, two ancient pyramids of sun-dried brick still stand as monuments to the craft of bricklaying.

Highly respected in early civilizations, masons not only laid bricks but made them as well. Kings often gave their support to masons by having a royal seal molded into bricks when they were made. This practice continues today as some brick manufacturers mold the name of their company into the bottoms of the bricks.

The recessed panel in the bottom of bricks manufactured today is called a *frog,* Figure 2-1. It has several purposes:

- To lock the mortar into the depression for a better bond
- To create special effects when the brick is laid on its side, as in decorative walls
- To save on the cost of materials

Fig. 2-1 Example of two types of cored brick and a recessed frogged brick.

Bricks also sometimes have holes in the bottom. These holes help provide a better bond. Such bricks are called *cored bricks,* Figure 2-1.

One of the earliest types of brick and one still used in many countries today is a clay or shale sun-dried brick called *adobe brick.* These bricks contain straw for greater strength, just as steel rods reinforce modern concrete.

After the sun-dried adobe brick had been in use for some time, a discovery was made. It was found that a brick subjected to fire in a closed area such as a *kiln,* or oven, for a definite period of time became very hard and highly fire-resistant. The fired brick resisted erosion far better than unfired bricks. Some of the bricks were coated with a thick enamel or glaze. The glazes were commonly red, yellow, green, or a combination of these colors. When subjected to heat in the kiln, the color hardened and developed a glass-like finish. Some of these glazed bricks, recovered from old buildings, still retain their original color after 2000 years. Glazed bricks are made today but have limited use. This is because they are costly to manufacture.

Brickmaking and bricklaying were regarded by many of the old world craftspersons as secret processes. To keep them secret and confined to their own groups, masons banded together in organizations called *guilds.* These specialized guilds were the forerunners of modern unions.

In 1666, a great fire changed London, England from a city of wooden buildings to a city of brick construction. The manufacture of brick attained a high degree of excellence and dominated the building field in this period of history.

Early records indicate that the first bricks manufactured in the United States were made in Virginia in 1611 and in Massachusetts in 1629. The bricks were made by hand using very simple methods and tools. Many of the bricks used in construction in the early American settlements were brought from England as ballast in sailing ships. Some of these bricks can still be found in the foundations and walls of the remaining original houses in the eastern part of the United States, Figure 2-2.

The invention of the steam engine in 1760, and the subsequent Industrial Revolution, brought a change from manual labor to the use of power-driven machinery to make bricks. This change started the true development of the brick industry in America. The first brickmaking machine was patented in 1800.

THE MODERN BRICK

The term *brick* as used today denotes a solid masonry rectangular unit formed in a plastic state from clay and shale and burned in a kiln. The United States Federal Trade Commission has ruled that no product made from materials other than clay or shale can be called brick. The exception to this is if the name includes the material from which the unit is manufactured, such as cinder brick, sand lime brick, or concrete brick.

Raw Materials

Clay and shale are the principal materials used to make bricks. Usually concentrated in large deposits, these materials are found all over the world.

Clay is a natural product formed by the weathering of rocks. Shale is made in very much the same way and from the same material. However, shale is compressed into layers in the ground. Shale is very dense and is more difficult to remove from the ground than clay. As a result, shale is a more costly raw material.

Two or more kinds of clay and shale may be mixed together to obtain a material having the proper consis-

Fig. 2-2 Early brick row houses that have been restored to their original beauty in an urban renewal project.

tency and composition. Good raw material is the backbone of the brick industry. The following are several forms of clay. They have a similar chemical composition but different physical characteristics.

Surface clays are found near the surface of the earth. They may be offshoots of old deposits or the result of more recent weathering of rocks. *Shales* are clays that have been formed by natural conditions under high pressure until they resemble slate. *Fire clays* are mined from a greater depth than are the other clays. They have fire-resistant qualities (ordinary brick is also fire resistant). They contain fewer impurities and have more uniform chemical and physical properties than shales or surface clays.

Although surface clays and fire clays differ in physical structure from shale, the three types of clay are chemically similar. All three are made of silica and alumina with varying amounts of metallic oxides and other impurities.

Metallic oxides act as fluxes and promote fusion at lower temperatures. The amount of iron, magnesium, and calcium oxides in the clays influences the color of the finished product. The material from each deposit of clay and shale has chemical characteristics which may be uniform for that deposit but may differ from the characteristics of material in other deposits. The changes in characteristics from deposit to deposit are due to differences in the relative amounts of the chemical components. As a result, brick made from the material in one deposit will have one set of characteristics for color, finish, and texture. Brick made from material in another location may look different because the chemical composition of the material varies slightly from that at the first location. In addition, all clay and shale do not react in the same manner to processing methods.

Brick Classification

Depending on its use, brick can be classified by one of several specifications. The American Society for Testing Materials (ASTM) publishes the most widely accepted standards on brick. The Canadian Standards Association also publishes standards for certain brick. Most of the model codes in the United States reference ASTM standards. Many of the requirements in the brick standards aid in predicting the durability of the brick in actual use. These predictors and other requirements in ASTM standards are not infallible. All ASTM standards are reviewed and updated periodically.

Depending on its use, brick can be classified by one of a number of specifications or designations, as listed below:

Type of Brick Unit	*ASTM Designation*	*CSA Designation*
Building brick	C 62	
Facing brick	C 216	A 82.1
Hollow brick	C 652	A 82.8
Paving brick	C 902	
Ceramic glazed brick	C 126	
Thin brick veneer units	C 1088	
Sewer and manhole brick	C 32	
Chemical resistant brick	C 279	
Industrial floor brick	C 410	

Following is a brief explanation of each type of brick.

Building Brick.   Building brick is intended for use in both structural and non-structural masonry where appearance is *not* a requirement. Building brick is typically used as a backing material that is not exposed and represents a very small amount of the brick made today.

Facing Brick (more commonly known as face brick). Facing brick is intended for use in both structural and non-structural masonry where appearance is a necessary requirement. The great majority of all brick made today is classified as facing brick according to the Brick Industry Association.

Hollow Brick.   Hollow Brick is identical to facing brick but has a larger core area. Most hollow brick is used in the same application as facing brick. Hollow brick with a large core is used in walls that are reinforced with steel and grout and is placed in those units where building brick, facing brick, or some hollow brick would be difficult to use.

Paving Brick.   Paving brick is intended to be used as a paving material to support pedestrian or light vehicular traffic.

Ceramic Glazed Brick.   Ceramic glazed brick is brick with a ceramic glaze that is fused to the body and used as facing brick or some other solid masonry unit.

Thin Brick.   Thin brick veneer units are fired clay units with normal face dimensions but with a reduced thickness; they may be adhered to a surface to appear as full thickness brick masonry.

Sewer and Manhole Brick.   Sewer and manhole brick are intended for use in drainage structures for the conveyance of sewage, industrial waste, storm water, and in related structures such as manholes and catch basins.

Chemical resistant and industrial floor brick are only used in very specialized situations.

The color and texture of brick must meet the specifications of the range number established for the variety of brick being made. The *range number,* an identifying number or letter, means the blend, texture, and color as-

signed by manufacturers to each of their products. The bricks made for a range number must conform to a sample brick produced earlier for the same coded numbers or letters. The user of the brick is then assured that bricks of the same range number will always match the original sample selected for the job. If it becomes necessary to build an addition to a structure, a match of the brick can be obtained by consulting the range number.

How Brick Fits into the Modular Building System

The modern brick is made for use in the *modular grid system* of building. The main reason for making a brick on a modular grid is for economy purposes. Standards for modular dimensions have been approved by the American Standards Association for all building materials. These dimensions are based upon a $4''$ unit of measure called the *module*. This module is used as a basis for the grid system, which must be used when two or more different materials are to be used in a construction job. Any building construction in which the size of the building materials used is based on the $4''$ grid system is called *modular design.*

Most masonry materials will tie and level off together at a height of $16''$ vertically ($16''$ is a multiple of the $4''$ grid system). For example, 2 *courses* (layers) of block for a wall including the mortar joint will equal $16''$ vertically. Six courses of standard brick in mortar will also equal $16''$ vertically. As a result, the wall can be tied together at $16''$ intervals or at multiples of $16''$. Masons should learn early in their work how various building materials tie together on the job.

In modular design, the *nominal* dimension of a masonry unit (such as a brick or a block) means the specified or manufactured dimension plus the thickness of the mortar joint to be used. That is, the brick size is designed so that when the size of the mortar joint is added to any of the brick dimensions (thickness, height, and length), the sum will equal a multiple of the $4''$ grid. For example, a modular brick whose nominal length is $8''$ will have a manufactured dimension of $7\frac{1}{2}''$ if it is designed to be laid with a $\frac{1}{2}''$ mortar joint. The same $8''$ brick will have a manufactured dimension of $7\frac{5}{8}''$ if it is designed to be laid with a $\frac{3}{8}''$ joint.

Mortar Joints

Mortar joints play an important part in the modular system of building, as they are an overall part of the height of each course of brick. The beginning mason should become familiar with the meaning of the terms that describe mortar joints. The two basic types of mortar joints are shown in Figure 2-3.

Masons form mortar joints by using the trowel to place mortar between and under the brick being laid. The two most basic joints are the *head joint* and the *bed joint.*

BRICK SIZES AND NOMENCLATURE

Brick are manufactured in many sizes and have been called by different names. This assortment of sizes and names can lead to confusion for designers and specifiers. Recent efforts led jointly by the Brick Industry Association and the National Association of Brick Distributors have led to the development of standard nomenclature for brick, which represents roughly 90 percent of all sizes currently manufactured in the United States. The term nomenclature, as used here, refers to description and names used for parts of a brick.

Important information in this Unit lists the sizes of brick generally available in the United States and pre-sents the standard nomenclature for brick sizes. The differences between nominal, specified, and actual dimensions are explained. Guidance is given on the recommended order in which brick dimensions should be listed. Vertical and horizontal coursing tables are also presented to aid the reader.

Brick sizes have varied a lot over the centuries but have always been small enough that a mason could easily pick them up and lay them in a wall. Because brick are a relatively small sized material, they tend to give large expanses of wall a visually pleasing appearance. The use of oversize brick alters the scale of the masonry units in relation to the wall. Because people perceive what the size of a standard brick should be, the use of oversize brick makes a wall appear smaller than it really is.

Over the years, new sizes have been developed to meet specific design, production, or construction needs. New types of construction have required new sizes, such as hollow units for reinforced masonry and larger units for increased economy. Hollow units do have varying coring patterns (interior webs and cells) but typically are bigger than standard or modular sizes and have larger cell areas to allow installing vertical steel reinforcement rods and concrete grout to make the wall stronger. Brick with larger face dimensions allow the bricklayer to lay more square feet of wall per

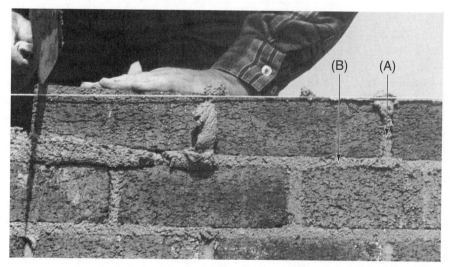

Fig. 2-3 Two basic types of mortar joints are (A) the head joint and (B) the bed joint.

day. Such larger units, when compared with standard or modular brick, may increase the number of brick laid per day by over fifty percent. One needs to remember that as a brick gets larger and heavier, the number of brick a person can lay will decrease due to the mason becoming fatigued from the extra weight. This is the same problem encountered when laying concrete block. The heavier and bigger a block is, the more the productivity declines. In many senses, it is a

trade off that has to be studied before deciding on which size brick the designer will specify in a brick building. Another drawback with using larger units is that filling the mortar head joints requires more skill.

Until recently, a given brick size may have been known by several names due to regional variations. The standard nomenclature for brick sizes is presented in Figure 2-4. These terms were developed by a cooperative process involving companies across the United

MODULAR BRICK SIZES								
Unit Designation	Nominal Dimensions, in.			Joint Thickness[2], in.	Specified Dimensions[3], in.			Vertical Coursing
	w	h	l		w	h	l	
Modular	4	2⅔	8	⅜	3⅝	2¼	7⅝	3C = 8 in.
				½	3½	2¼	7½	
Engineer Modular	4	3⅕	8	⅜	3⅝	2¾	7⅝	5C = 16 in.
				½	3½	2¹³⁄₁₆	7½	
Closure Modular	4	4	8	⅜	3⅝	3⅝	7⅝	1C = 4 in.
				½	3½	3½	7½	
Roman	4	2	12	⅜	3⅝	1⅝	11⅝	2C = 4 in.
				½	3½	1½	11½	
Norman	4	2⅔	12	⅜	3⅝	2¼	11⅝	3C = 8 in.
				½	3½	2¼	11½	
Engineer Norman	4	3⅕	12	⅜	3⅝	2¾	11⅝	5C = 16 in.
				½	3½	2¹³⁄₁₆	11½	
Utility	4	4	12	⅜	3⅝	3⅝	11⅝	1C = 4 in.
				½	3½	3½	11½	
NON-MODULAR BRICK SIZES								
Standard				⅜	3⅝	2¼	8	3C = 8 in.
				½	3½	2¼	8	
Engineer Standard				⅜	3⅝	2¾	8	5C = 16 in.
				½	3½	2¹³⁄₁₆	8	
Closure Standard				⅜	3⅝	3⅝	8	1C = 4 in.
				½	3½	3½	8	
King				⅜	3	2¾	9⅝	5C = 16 in.
					3	2⅝	9⅝	
Queen				⅜	3	2¾	8	5C = 16 in.

1 in. = 25.4 mm; 1 ft = 0.3 m
[2]Common joint sizes used with length and width dimensions. Joint thicknesses of bed joints vary based on vertical coursing and specified unit height.
[3]Specified dimensions may vary within this range from manufacturer to manufacturer.

Fig. 2-4 Standard nomenclature for brick sizes.

States. These standard terms describing brick are strongly recommended when ordering or specifying brick for a job.

Figure 2-5 lists other brick sizes that are produced by a limited number of manufacturers. Because clay is such a flexible material, manufacturers can make many different sizes. Also, modular and non-modular sizes are illustrated in Figures 2-5 and 2-6, respectively. The coring patterns shown in these figures are for illustrative purposes only, as they can vary with different requirements and manufacturers.

Brick Dimensions

Brick are identified by three dimensions: width, height, and length. Height and length are sometimes called face dimensions because these are the dimensions that show when a brick is laid in the wall as we normally see it. Specifications and purchase orders should list

dimensions in this standard order of width first, followed by height, then length.

The specified dimensions are the anticipated manufacturers' dimensions. These dimensions are the ones that should be stated in purchase orders when ordering from the supplier. Specified dimensions are used by the structural engineer in the rational design of brick masonry. In non-modular construction, only the specified dimensions should be used. Figures 2-4 and 2-5 provide the specified and nominal dimensions where applicable. The actual dimensions of a brick signify the size it is when manufactured. Actual dimensions may vary slightly from a specified size due to shrinkage or expansion from the burning and cooling process in the kilns. This is to be expected, but they must fall within the range of sizes defined by the specified dimensions, plus or minus the specified dimensional tolerances. Dimensional tolerances are found listed in ASTM standard specifications for brick, such as

MODULAR BRICK SIZES							
Nominal Dimensions, in.			Joint Thickness[2], in.	Specified Dimensions[3], in.			Vertical Coursing
w	h	l		w	h	l	
4	6	8	$\frac{3}{8}$ $\frac{1}{2}$	$3\frac{5}{8}$ $3\frac{1}{2}$	$5\frac{5}{8}$ $5\frac{1}{2}$	$7\frac{5}{8}$ $7\frac{1}{2}$	2C = 12 in.
4	8	8	$\frac{3}{8}$ $\frac{1}{2}$	$3\frac{5}{8}$ $3\frac{1}{2}$	$7\frac{5}{8}$ $7\frac{1}{2}$	$7\frac{5}{8}$ $7\frac{1}{2}$	1C = 8 in.
6	$3\frac{1}{5}$	12	$\frac{3}{8}$ $\frac{1}{2}$	$5\frac{5}{8}$ $5\frac{1}{2}$	$2\frac{3}{4}$ $2\frac{13}{16}$	$11\frac{5}{8}$ $11\frac{1}{2}$	5C = 16 in.
6	4	12	$\frac{3}{8}$ $\frac{1}{2}$	$5\frac{5}{8}$ $5\frac{1}{2}$	$3\frac{5}{8}$ $3\frac{1}{2}$	$11\frac{5}{8}$ $11\frac{1}{2}$	1C = 4 in.
8	4	12	$\frac{3}{8}$ $\frac{1}{2}$	$7\frac{5}{8}$ $7\frac{1}{2}$	$3\frac{5}{8}$ $3\frac{1}{2}$	$11\frac{5}{8}$ $11\frac{1}{2}$	1C = 4 in.
8	4	16	$\frac{3}{8}$ $\frac{1}{2}$	$7\frac{5}{8}$ $7\frac{1}{2}$	$3\frac{5}{8}$ $3\frac{1}{2}$	$15\frac{5}{8}$ $15\frac{1}{2}$	1C = 4 in.
NON-MODULAR BRICK SIZES							
			$\frac{3}{8}$	3 3	$2\frac{3}{4}$ $2\frac{5}{8}$	$8\frac{5}{8}$ $8\frac{5}{8}$	5C = 16 in.

1 in. = 25.4 mm; 1 ft = 0.3 m
[2]Common joint sizes used with length and width dimensions. Joint thicknesses of bed joints vary based on vertical coursing and specified unit height.
[3]Specified dimensions may vary within this range from manufacturer to manufacturer.

Fig. 2-5 Other brick sizes.

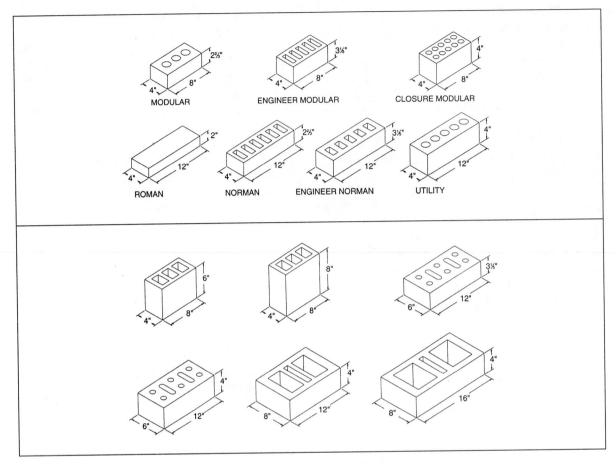

Modular brick sizes (nominal dimensions).

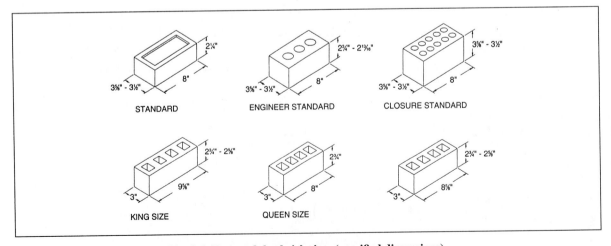

Fig. 2-6 Non-modular brick sizes (specified dimensions).

ASTM C 216 Standard Specification for Facing Brick or in project literature.

Coursing

Although nominal dimensions are given only for modular brick, it should be noted that the heights of both modular and non-modular brick can be the same. The reason for this is that when modular brick were first introduced, brick manufacturers were faced with the problem of supplying matching brick to existing non-modular construction. For the sake of appearance, most designers required that the vertical coursing of modular brick match the existing non-modular brick. Therefore, all bricks are really modular in height. The vertical coursing given in Figures 2-4 and 2-5 reflect this fact. Figure 2-7 provides vertical dimensions based on the modular vertical coursing given in Figures 2-4 and 2-5. For example, brick with heights which course vertically 2 courses to 4 inches (2C = 4 in) such as Roman size brick, should use column 1 of Figure 2-7. The dimensions given in Figure 2-7 include typical mortar joints of 3/8″ to 1/2″. The actual mortar joint size can be determined from the vertical coursing information and the specified unit size. For example, when coursing out with a modular height brick, the mortar bed joint is ever so slightly larger than 3/8″ and slightly less than 1/2″, so that 3 courses of brick and mortar will equal a 8″ module. For most brick sizes, the mortar bed joint will not be exactly 3/8″ nor 1/2″. Figure 2-7 is applicable to both modular and non-modular brick. In this table brick are assumed to be positioned or laid as stretchers or headers. Horizontal coursing information is shown in Figure 2-8. Figure 2-8 includes coursing for both modular and non-modular brick.

No. of Courses	Vertical Coursing of Unit			
	2C = 4 in.	3C = 8 in.	5C = 16 in.	1C = 4 in.
1	0' - 2"	0' - 2⅔"	0' - 3⅕"	0' - 4"
2	0' - 4"	0' - 5⅓"	0' - 6⅖"	0' - 8"
3	0' - 6"	0' - 8"	0' - 9⅗"	1' - 0"
4	0' - 8"	0' - 10⅔"	1' - 0⅘"	1' - 4"
5	0' - 10"	1' - 1⅓"	1' - 4"	1' - 8"
6	1' - 0"	1' - 4"	1' - 7⅕"	2' - 0"
7	1' - 2"	1' - 6⅔"	1' - 10⅖"	2' - 4"
8	1' - 4"	1' - 9⅓"	2' - 1⅗"	2' - 8"
9	1' - 6"	2' - 0"	2' - 4⅘"	3' - 0"
10	1' - 8"	2' - 2⅔"	2' - 8"	3' - 4"
11	1' - 10"	2' - 5⅓"	2' - 11⅕"	3' - 8"
12	2' - 0"	2' - 8"	3' - 2⅖"	4' - 0"
13	2' - 2"	2' - 10⅔"	3' - 5⅗"	4' - 4"
14	2' - 4"	3' - 1⅓"	3' - 8⅘"	4' - 8"
15	2' - 6"	3' - 4"	4' - 0"	5' - 0"
16	2' - 8"	3' - 6⅔"	4' - 3⅕"	5' - 4"
17	2' - 10"	3' - 9⅓"	4' - 6⅖"	5' - 8"
18	3' - 0"	4' - 0"	4' - 9⅗"	6' - 0"
19	3' - 2"	4' - 2⅔"	5' - 0⅘"	6' - 4"
20	3' - 4"	4' - 5⅓"	5' - 4"	6' - 8"
21	3' - 6"	4' - 8"	5' - 7⅕"	7' - 0"
22	3' - 8"	4' - 10⅔"	5' - 10⅖"	7' - 4"
23	3' - 10"	5' - 1⅓"	6' - 1⅗"	7' - 8"
24	4' - 0"	5' - 4"	6' - 4⅘"	8' - 0"
25	4' - 2"	5' - 6⅔"	6' - 8"	8' - 4"
26	4' - 4"	5' - 9⅓"	6' - 11⅕"	8' - 8"
27	4' - 6"	6' - 0"	7' - 2⅖"	9' - 0"
28	4' - 8"	6' - 2⅔"	7' - 5⅗"	9' - 4"
29	4' - 10"	6' - 5⅓"	7' - 8⅘"	9' - 8"
30	5' - 0"	6' - 8"	8' - 0"	10' - 0"
31	5' - 2"	6' - 10⅔"	8' - 3⅕"	10' - 4"
32	5' - 4"	7' - 1⅓"	8' - 6⅖"	10' - 8"
33	5' - 6"	7' - 4"	8' - 9⅗"	11' - 0"
34	5' - 8"	7' - 6⅔"	9' - 0⅘"	11' - 4"
35	5' - 10"	7' - 9⅓"	9' - 4"	11' - 8"
36	6' - 0"	8' - 0"	9' - 7⅕"	12' - 0"
37	6' - 2"	8' - 2⅔"	9' - 10⅖"	12' - 4"
38	6' - 4"	8' - 5⅓"	10' - 1⅗"	12' - 8"
39	6' - 6"	8' - 8"	10' - 4⅘"	13' - 0"
40	6' - 8"	8' - 10⅔"	10' - 8"	13' - 4"
41	6' - 10"	9' - 1⅓"	10' - 11⅕"	13' - 8"
42	7' - 0"	9' - 4"	11' - 2⅖"	14' - 0"
43	7' - 2"	9' - 6⅔"	11' - 5⅗"	14' - 4"
44	7' - 4"	9' - 9⅓"	11' - 8⅘"	14' - 8"
45	7' - 6"	10' - 0"	12' - 0"	15' - 0"
46	7' - 8"	10' - 2⅔"	12' - 3⅕"	15' - 4"
47	7' - 10"	10' - 5⅓"	12' - 6⅖"	15' - 8"
48	8' - 0"	10' - 8"	12' - 9⅗"	16' - 0"
49	8' - 2"	10' - 10⅔"	13' - 0⅘"	16' - 4"
50	8' - 4"	11' - 1⅓"	13' - 4"	16' - 8"
100	16' - 8"	22' - 2⅔"	26' - 8"	33' - 4"

1 in. = 25.4 mm; 1 ft = 0.3 m
Brick positioned in wall as stretchers or headers.

Fig. 2-7 Vertical coursing.

Number of Units	Unit Length					
	Nominal Dimensions, in.		Specified Dimensions, in.			
	8	12	8		8⅝	9⅝
			½ in. jt.	⅜ in. jt.	⅜ in. jt.	⅜ in. jt.
1	0' - 8"	1' - 0"	0' - 8½"	0' - 8⅜"	0' - 9"	0' - 10"
2	1' - 4"	2' - 0"	1' - 5"	1' - 4¾"	1' - 6"	1' - 8"
3	2' - 0"	3' - 0"	2' - 1½"	2' - 1⅛"	2' - 3"	2' - 6"
4	2' - 8"	4' - 0"	2' - 10"	2' - 9½"	3' - 0"	3' - 4"
5	3' - 4"	5' - 0"	3' - 6½"	3' - 5⅞"	3' - 9"	4' - 2"
6	4' - 0"	6' - 0"	4' - 3"	4' - 2¼"	4' - 6"	5' - 0"
7	4' - 8"	7' - 0"	4' - 11½"	4' - 10⅝"	5' - 3"	5' - 10"
8	5' - 4"	8' - 0"	5' - 8"	5' - 7"	6' - 0"	6' - 8"
9	6' - 0"	9' - 0"	6' - 4½"	6' - 3⅜"	6' - 9"	7' - 6"
10	6' - 8"	10' - 0"	7' - 1"	6' - 11¾"	7' - 6"	8' - 4"
11	7' - 4"	11' - 0"	7' - 9½"	7' - 8⅛"	8' - 3"	9' - 2"
12	8' - 0"	12' - 0"	8' - 6"	8' -4½"	9' - 0"	10' - 0"
13	8' - 8"	13' - 0"	9' - 2½"	9' - 0⅞"	9' - 9"	10' - 10"
14	9' - 4"	14' - 0"	9' - 11"	9' - 9¼"	10' - 6"	11' - 8"
15	10' - 0"	15' - 0"	10' - 7½"	10' - 5⅝"	11' - 3"	12' - 6"
16	10' - 8"	16' - 0"	11' - 4"	11' - 2"	12' - 0"	13' - 4"
17	11' - 4"	17' - 0"	12' - 0½"	11' - 10⅜"	12' - 9"	14' - 2"
18	12' - 0"	18' - 0"	12' - 9"	12' - 6¾"	13' - 6"	15' - 0"
19	12' - 8"	19' - 0"	13' - 5½"	13' - 3⅛"	14' - 3"	15' - 10"
20	13' - 4"	20' - 0"	14' - 2"	13' - 11½"	15' - 0"	16' - 8"
21	14' - 0"	21' - 0"	14' - 10½"	14' - 7⅞"	15' - 9"	17' - 6"
22	14' - 8"	22' - 0"	15' - 7"	15' - 4¼"	16' - 6"	18' - 4"
23	15' - 4"	23' - 0"	16' - 3½"	16' - 0⅝"	17' - 3"	19' - 2"
24	16' - 0"	24' - 0"	17' - 0"	16' - 9"	18' - 0"	20' - 0"
25	16' - 8"	25' - 0"	17' - 8½"	17' - 5⅜"	18' - 9"	20' - 10"
26	17' - 4"	26' - 0"	18' - 5"	18' - 1¾"	19' - 6"	21' - 8"
27	18' - 0"	27' - 0"	19' - 1½"	18' - 10⅛"	20' - 3"	22' - 6"
28	18' - 8"	28' - 0"	19' - 10"	19' - 6½"	21' - 0"	23' - 4"
29	19' - 4"	29' - 0"	20' - 6½"	20' - 2⅞"	21' - 9"	24' - 2"
30	20' - 0"	30' - 0"	21' - 3"	20' - 11¼"	22' - 6"	25' - 0"
31	20' - 8"	31' - 0"	21' - 11½"	21' - 7⅝"	23' - 3"	25' - 10"
32	21' - 4"	32' - 0"	22' - 8"	22' - 4"	24' - 0"	26' - 8"
33	22' - 0"	33' - 0"	23' - 4½"	23' - 0⅜"	24' - 9"	27' - 6"
34	22' - 8"	34' - 0"	24' - 1"	23' - 8¾"	25' - 6"	28' - 4"
35	23' - 4"	35' - 0"	24' - 9½"	24' - 5⅛"	26' - 3"	29' - 2"
36	24' - 0"	36' - 0"	25' - 6"	25' - 1½"	27' - 0"	30' - 0"
37	24' - 8"	37' - 0"	26' - 2½"	25' - 9⅞"	27' - 9"	30' - 10"
38	25' - 4"	38' - 0"	26' - 11"	26' - 6¼"	28' - 6"	31' - 8"
39	26' - 0"	39' - 0"	27' - 7½"	27' - 2⅝"	29' - 3"	32' - 6"
40	26' - 8"	40' - 0"	28' - 4"	27' - 11"	30' - 0"	33' - 4"
41	27' - 4"	41' - 0"	29' - 0½"	28' - 7⅜"	30' - 9"	34' - 2"
42	28' - 0"	42' - 0"	29' - 9"	29' - 3¾"	31' - 6"	35' - 0"
43	28' - 8"	43' - 0"	30' - 5½"	30' - 0⅛"	32' - 3"	35' - 10"
44	29' - 4"	44' - 0"	31' - 2"	30' - 8½"	33' - 0"	36' - 8"
45	30' - 0"	45' - 0"	31' - 10½"	31' - 4⅞"	33' - 9"	37' - 6"
46	30' - 8"	46' - 0"	32' - 7"	32' - 1¼"	34' - 6"	38' - 4"
47	31' - 4"	47' - 0"	33' - 3½"	32' - 9⅝"	35' - 3"	39' - 2"
48	32' - 0"	48' - 0"	34' - 0"	33' - 6"	36' - 0"	40' - 0"
49	32' - 8"	49' - 0"	34' - 8½"	34' - 2⅜"	36' - 9"	40' - 10"
50	33' - 4"	50' - 0"	35' - 5"	34' - 10¾"	37' - 6"	41' - 8"
100	66' - 8"	100' - 0"	70' - 10"	69' - 9½"	75' - 0"	83' - 4"

1 in. = 25.4 mm; 1 ft = 0.3 m

Fig. 2-8 Horizontal coursing.

ACHIEVEMENT REVIEW

Select the best answer from the choices offered to complete the statement or answer the question. List your choice by letter identification.

1. Many ancient bricks had an unusual feature called a frog. Which of the following describes a frog?
   a. A very hard red brick burned in the center of the kiln
   b. Brick with a recessed panel
   c. Brick with holes to lock the mortar in
   d. Glazed brick made for decorative work

2. One of the earliest types of brick was called
   a. a cinder brick.
   b. a salmon brick.
   c. a sand lime brick.
   d. an adobe brick.

3. Many years ago, masons formed into groups to guard their trade secrets and manufacturing processes. What term was applied to these groups?
   a. Unions
   b. Guilds
   c. Craftspeople
   d. Manufacturers

4. Bricks are
   a. made from cement and clay material.
   b. formed in a plastic state from clay and shale and then burned in a kiln.
   c. formed from a sand and lime mixture.

5. Clay is a natural product that is formed by
   a. earth and sand mixed together.
   b. weathering of rocks.
   c. decomposed vegetation and minerals.
   d. cement and sand mixture pressed into a mold.

6. Fire clays have a high resistance to heat and therefore can be used in fireplaces and smokestacks. When the raw material is taken from the ground, it is always found
   a. near the surface.
   b. on hillsides and in outcroppings of ground.
   c. deeper in the ground than other clays.
   d. on the surface.

7. The majority of brick made and sold today is
   a. hollow brick.
   b. face brick.
   c. building brick.
   d. sewer and manhole brick.

8. The brick manufacturer's range number indicates
   a. the hardness of the brick.
   b. the size of the brick.
   c. the blend, texture, and color of the brick.
   d. the degree of water resistance the brick shows under controlled test conditions.

9. The modular system of measurement for the building industry is based on the module. The module unit measures
   a. 16″.
   b. 8″.
   c. 12″.
   d. 4″.

10. The principal reason for making a brick on a modular grid is for
    a. beauty and design.
    b. economy.
    c. ease in manufacturing.

MATH CHECKPOINT

1. If a standard brick weighs $4\frac{1}{2}$ pounds, what would the exact weight of a cube of 500 bricks be?

2. If three courses high of standard brick laid in a standard $\frac{3}{8}''$ mortar bed joint equal $8''$, how many courses of brick would it take to equal a wall $4'$ high?

3. If each course of Roman brick, including the mortar bed joint, equals $2''$ in height, how many courses would there be at a $6'$ height?

4. A true modular brick length is $7\frac{5}{8}''$. When a $\frac{3}{8}''$ mortar head joint is added, it rounds off $8''$. Applying this information, how many bricks would it take to lay out a wall $20'$ in length?

UNIT 3
Manufacture of Brick

─────────────── OBJECTIVES ───────────────

After studying this unit, the student will be able to

- ■ list the various steps in modern brick making.
- ■ describe briefly each of the basic brick production steps listed.
- ■ describe various kilns used in brick making.

INTRODUCTION

Technological developments during the last century have helped to make the manufacture of brick a very efficient and productive process. More complete knowledge of the characteristics of the raw material, improved kiln designs, controlled heat in the kilns, and extensive mechanization have all played important roles in modernizing brick manufacturing.

There is such a tremendous demand for bricks that they are used faster than they can be manufactured. The modern brick plant meets the challenge of increasing production, while retaining a high quality for the final product, by using computerized manufacturing methods and highly skilled workers.

Basically, bricks are made by mixing a specified amount of water with finely ground clay or shale or a combination of both. The mixture is then formed into the desired shape, predried, and burned in a kiln for a predetermined time under carefully controlled conditions.

While the basic steps of brick manufacturing are standard throughout the industry, each brick-making plant has minor variations to these steps due to local conditions. For example, one brick plant may be near the source of the raw material, while another plant may have to transport the material from a distant source. These two plants will have different ways of obtaining and stocking their raw materials. If possible, the plant should be built on the site where the raw material is obtained.

STEPS IN THE MANUFACTURING PROCESS

The following are the six major steps in the manufacture of brick:

1. Mining the raw clay or shale from the ground
2. Preparing the raw materials for use
3. Forming the raw materials into bricks
4. Predrying the bricks for burning
5. Burning the brick in the kiln under controlled heat
6. Storing and shipping

All other operations stem from these six major steps. Figure 3-1A shows, in schematic form, the basic steps in the manufacture of bricks.

Mining the Material

The removal of the raw material from the ground is called *mining*. Power equipment mines surface clay and shale in open pits. Power shovels or bulldozers work the raw material to the conveyor belts. The conveyor belts then transport the material to the brick plant's storage area, where it is deposited on a pile, Figure 3-1B. If the raw material is not mined near the site of the brick plant, trucks or railways bring the material to the storage piles.

Enough raw material is stored to ensure plant operations for several days in the event that bad weather halts mining or shipping operations. Several storage areas are provided so that the clay and shale can be

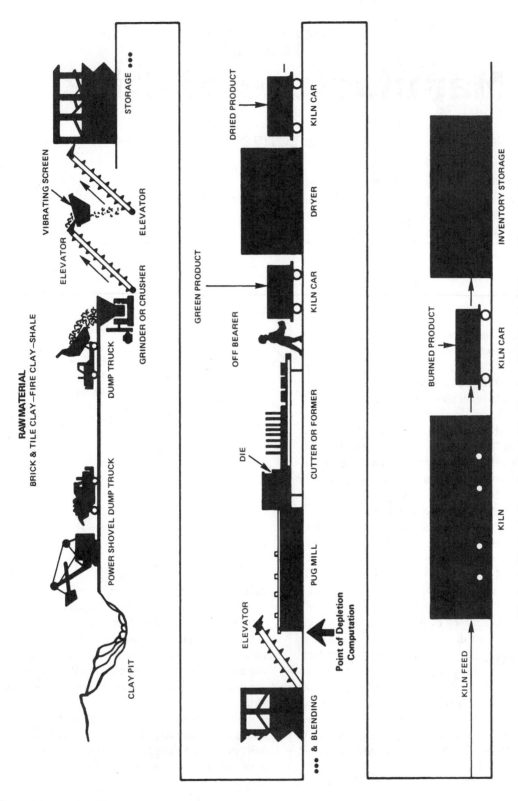

RAW MATERIAL
BRICK & TILE CLAY—FIRE CLAY—SHALE

CLAY PIT

POWER SHOVEL DUMP TRUCK

DUMP TRUCK

GRINDER OR CRUSHER

ELEVATOR

VIBRATING SCREEN

ELEVATOR

STORAGE •••

••• & BLENDING

ELEVATOR

Point of Depletion Computation

PUG MILL

DIE

CUTTER OR FORMER

OFF BEARER

GREEN PRODUCT

KILN CAR

DRYER

DRIED PRODUCT

KILN CAR

KILN FEED

KILN

BURNED PRODUCT

KILN CAR

INVENTORY STORAGE

Fig. 3-1A The basic flow of materials in the manufacture of brick in a modern plant.

26

Fig. 3-1B Raw clay and shale pile ready to be ground for use.

blended to yield material with a better composition. Blending may not be necessary if the shale alone is of sufficient quality that it can be used by itself. Blending produces more uniform raw material, helps control the color of the finished product, and permits some control over providing raw material suitable for manufacturing a given type of brick unit.

Preparing the Material

If the raw material is in large lumps, it may be crushed before it is placed on the storage pile. Crushing breaks up the large pieces and removes the stones. Next, 4-ton to 8-ton grinding wheels revolving in a circular pan grind and mix the material. It then passes through an inclined, vibrating screen which controls the particle sizes. The finely ground material is taken by a conveyor belt to the site where it is formed into single bricks.

Forming the Clay or Shale into a Brick Shape

Three methods of forming are used in the production of bricks: the *stiff-mud process,* the *soft-mud process,* and the *dry-press process.*

Stiff-Mud Process. The most frequently used process at present is the stiff-mud process. It produces a hard, dense brick. A greater volume of bricks can be manu-

factured by this method to meet the growing demands of the construction industry.

The first step in the stiff-mud process is to add water to the raw material to make a plastic, workable mass suitable for molding. This is called *tempering.* The mixing is done in a machine called a *pug mill.* The pug mill has a mixing chamber which contains one or two revolving shafts which thoroughly mix the raw material and a measured amount of water.

After thorough mixing, the tempered clay goes through a machine which removes air bubbles, giving the clay increased workability and plasticity. In addition, a brick made from clay without air bubbles will have greater strength.

The clay is next forced through an opening called a *die,* a process much like toothpaste being forced from a tube. The long, formed ribbon of brick being extruded through the die is called the *column,* Figure 3-2. Texture may be applied to the face of the brick as it leaves the die, if so desired. As the column moves away from the die, it is cut into lengths which are either the height *(side cut)* or length *(end cut)* of the brick. The cutting is done automatically by a large, circular wire cutter which cuts each brick to the same size, Figure 3-3.

Upon leaving the wire-cutting machine, the bricks move to a conveyor for inspection. Bricks passing

Fig. 3-2 A pug mill showing the column of brick after being forced through the die.

inspection are placed on the dryer cars. Imperfect bricks are returned to the pug mill for reprocessing. The soft bricks that have not yet been burned hard in the kiln are called *green bricks.*

Soft-Mud Process. This is the oldest way of making brick and was used before brickmaking machines were developed. Automated machinery is now used in this process, Figure 3-4.

The soft-mud process is suited for clays which contain too much natural water for the stiff-mud process. The clay is mixed with twice as much water as in the stiff-mud process and is pressed into wooden molds. The molds are lubricated with sand or water so the clay does not stick to the mold. When sand is used to lubricate the molds, the bricks are *sand-struck* and have a sandy finish. When water is used, the bricks are *water-struck* and have a very smooth finish.

Dry-Press Process. This method is particularly adaptable for clay of low plasticity. In this process, the clay is mixed with a small amount of water and then forced into steel molds under very high pressure. This method

Fig. 3-4 A soft-mud machine in operation. The soft-mud mixture is placed in wooden molds, formed, removed from the molds, and placed in the kiln. Note that the workers are wearing hard hats to comply with safety rules concerning work near machinery.

is seldom used today due to the high cost and low demand for pressed brick.

Predrying Brick Before Burning in the Kiln

To prevent cracking, excess moisture must be removed from the bricks before they are burned in the kiln. Years ago, bricks were allowed to dry in the open air before they were placed in the kiln. The modern method is to predry the bricks in the forward section of the kiln using the waste heat from the hot section of the kiln. The heat and humidity must be regulated carefully to prevent sudden changes in the temperature. Sudden changes in temperature will cause excessive cracking and deformation of the brick. The drying time in the kiln is greatly reduced from that of the open-air process. In other words, with the brick being dried in the kiln, the weather is not a factor affecting the moisture content of the brick. The bricks are placed in the drying chamber on special rolling steel cars. They are left there for a predetermined drying time before they move into the hot section of the kiln, Figure 3-5. The temperature in the drying area is 100 to 400 degrees Fahrenheit ($^\circ$F).

Fig. 3-3 Bricks are automatically cut to size by a wire cutter.

Fig. 3-5 Green brick entering the drying chamber. Note the heavy steel doors that close the drying chamber off. The brick can also be seen on the car that will take them on their journey through the kiln.

Burning the Brick in the Brick Kiln

The burning process is one of the most specialized steps in the manufacture of brick. Kilns have changed drastically over the years. Since brickmaking emerged as an industry, three kilns have come into use: the *scove kiln,* the *beehive kiln,* and the *tunnel kiln.*

The Scove Kiln. The scove kiln, Figure 3-6, was one of the first types to be used to burn bricks. Scove kilns are now obsolete. The unburned bricks were stacked in piles inside the kiln with air spaces left between the individual bricks to permit the heat to reach all surfaces. The kiln was then plastered with mortar to lock in the heat during the burning process. Openings were left in the bottom walls of the kiln where hardwood fires provided heat for the kiln (in later years, gas and oil were used). This method was not very efficient. The bricks nearest the fires were burned hard, while those near the top of the kiln were soft and could be used only for filler walls or interior construction.

Fig. 3-6 A scove kiln. This type of kiln was one of the first used to burn bricks. Note the openings at the base of the kiln where fires were burned to provide the heat for burning the bricks.

The Beehive Kiln. An improvement upon the scove kiln was the beehive kiln, Figure 3-7. The beehive kiln is a round brick structure wrapped with steel bands which control the expansion caused by the heat in the kiln. The bricks to be burned are stacked in the kiln with narrow spaces between them so the heat will pass completely around them. The kiln is sealed by walling the doors shut with brick and mortar. The heat may be applied either from the bottom or the top. When these kilns first came into use, wood and coke were the main sources of heat. Gradually, these fuels were replaced by gas and oil because they could be controlled and regulated more efficiently. The beehive kiln requires at least one week to burn the brick. As a result, this type of kiln yields a limited amount of bricks.

The Tunnel Kiln. The most modern kiln in use today is the tunnel kiln, which is built of bricks and lined on the inside with firebricks, Figure 3-8. The bricks to be burned are stacked on flat cars which move very slowly through the long, narrow kiln. The cars move from the predrying section of the kiln into the burning section.

The average time for the unit to pass completely through the kiln is about 36 hours. This is controlled by computers. The number of bricks on a car is rather small compared to the number of bricks stacked in the

Fig. 3-7 A beehive kiln. Kilns of this type are still in limited use. Each kiln is wound with steel bands to control expansion. The piping for each kiln supplies the gas or oil fuel used to burn the bricks.

older types of kilns. The cars, however, pass through the long tunnel kiln continuously and the bricks all receive the same heat treatment, resulting in a more uniform product.

The heat in the kiln is supplied by gas or oil. The heat gradually increases as the bricks pass from the kiln inlet to the zone in the center of the kiln. This is where the greatest temperature is reached, an average of 1950°F. As the bricks move from the center to the outlet, the temperature drops and the bricks cool slowly. Cooling the bricks slowly eliminates cracking, pitting, and other problems which are due to a rapid reduction of temperature, Figure 3-9.

The rate of temperature change in the kiln depends on the raw materials being used. Tunnel kilns are equipped with recording instruments which provide a constant check on the temperatures in the kiln.

While the bricks are still in the kiln, they can be given a treatment called flashing. *Flashing* means that the amount of oxygen used in the burning is reduced.

Fig. 3-8 A modern tunnel kiln. The large ductwork on the top of the kiln is used to recycle the hot gases to the preheating chamber of the kiln.

The flame changes from red to blue. Flashing lowers the temperature of the kiln which, in turn, causes the bricks to take on varying shades of color. The type of clay or shale in the brick determines how the brick reacts to flashing.

Storing and Shipping

The process of removing the bricks from the kiln is called *drawing*. The bricks leave the kiln on cars which are placed in a holding area until the bricks are cool. When the bricks are cool enough, they are banded with metal strips to hold them together for shipment and handling. The banded cubes of brick contain approximately 500 standard bricks. The number may vary slightly according to the variation in brick sizes, Figure 3-10.

After the bricks are banded, they are moved to a storage yard and are ready for delivery to a job. During the off-seasons of the year (cold weather), the stock piles in the storage yards are built up to prepare for the construction season ahead, Figure 3-11.

Fig. 3-10 Bricks coming through the banding machine. This machine places bands around the bricks to form cubes of approximately 500 bricks.

Fig. 3-9 Burned brick at the end of their journey through the kiln. The intense heat of the kiln can be seen in the background.

Fig. 3-11 Brick cubes being stored for future use. The banding holds the bricks together and simplifies handling.

It is true that modern masons no longer manufacture the brick they use as their forerunners did many years ago. However, an understanding of how bricks are manufactured and the materials of which they are made will help masons make better use of the material.

The modern brick is a superior product due to careful quality control during manufacturing. Much of the guesswork and trial-and-error methods of the past have been replaced by scientific processes and computerized manufacturing.

ACHIEVEMENT REVIEW

The column on the left contains a statement associated with brick making. The column on the right lists terms. Select the correct term from the right-hand list and match it with the proper statement on the left.

1. Coloring brick by reducing the amount of oxygen in kiln
2. Forcing the clay through a die into the shape of a brick
3. Forming the brick in a wooden mold
4. A long brick structure used to burn brick using the most modern controls
5. Clay mixed with only a small amount of water and formed under high pressure in steel molds
6. Round brick kiln surrounded by steel bands
7. The mass of clay being extruded through a die
8. The taking or obtaining of the raw materials from the ground
9. Product made from finely ground shale or clay and burned
10. The process of removing the bricks from the kiln, allowing them to cool off, and then unloading them from the cars
11. The most commonly used process for making bricks
12. A machine that mixes raw material with water to form a workable mass
13. Trade name given the soft condition of brick before burning in the brick kiln
14. Preparatory treatment of the green brick to remove excess moisture before burning
15. Early type of brick kiln, which has openings at the bottom of the kiln in which to build the fires; the most inefficient of the three kilns discussed in the unit

a. Forming
b. Drawing
c. End cut
d. Tunnel kiln
e. Column
f. Sand-struck
g. Flashing
h. Brick
i. Scove kiln
j. Dry-press process
k. Soft-mud process
l. Mining
m. Beehive kiln
n. Stiff-mud process
o. Humidity
p. Green brick
q. Predrying
r. Pug mill

UNIT 4
Development of Concrete Block

OBJECTIVES

After studying this unit, the student will be able to

■ describe the early methods of concrete block manufacturing.

■ explain the use of lightweight aggregates used in block.

■ list some of the uses of concrete block in construction.

EARLY DEVELOPMENT OF CONCRETE BLOCK

In the early nineteenth century, two American masons named Foster and Van Derburgh sought a method of making a precast building block which would be larger than the brick in use at that time. After a great deal of experimenting, the two masons made a discovery. They combined powdered quick lime and moist sand and placed the mixture in a mold under pressure. The masons found that through this process, a usable building block could be made, provided it was hardened or cured with steam. The natural mechanical heat generated by the pressure formed a silicate of lime. This silicate of lime cemented the block together.

Foster and Van Derburgh realized, however, that their solid block was very heavy and difficult for masons to handle. This was because the block had no hand or finger holes. This block was patented in the early nineteenth century. However, it never enjoyed great popularity.

The next major advance in concrete block making occurred in England in 1850. An English mason named Joseph Gibbs decided to make a hollow block which would be considerably lighter and easier to handle than Foster and Van Derburgh's block. Gibbs also cast his block in a mold, but the mold had dividers, so that the finished block had hollow *cells*.

Gibbs intended masons to lay these hollow blocks in the wall and then fill the hollow cells with cement to strengthen the wall. Although masons didn't always follow that procedure, Gibbs' block remains the forerunner of the hollow concrete block we know today.

BLOCK-MAKING MACHINES

The first American block-making machine for commercial use was patented by Harold S. Palmer in 1900. The machine made hollow block. Later machines used the same method of hand-tamping the mixture of cement and aggregate into a mold, Figure 4-1. Using a machine of this type, two workers could make about 80 blocks a day. This hand-tamping process was used to make block from 1904 to 1914. The next 10 years, however, saw a change to power-tamping machines. In power-tamping machines, bars or rods were lifted by mechanical action and pulled down by gravity. The bars compressed the mix into the mold. This operation was controlled by a machine operator. By 1924, the automatic-tamping machine could produce 3000 blocks a day. Block could now be produced in great enough volume to be competitive with other masonry materials. This marked the real beginning of the modern block industry.

The automatic-tamping machine was replaced by the vibration machine around 1938. The vibration machine produced a block with more uniform texture, sharper corners, and denser material composition. Production increased tremendously due to the vibration machine. Today, 15,000 blocks can be made on a single machine in a 10-hour day. New developments and

This is how concrete block was made in the very early days—and for a good many years thereafter in backyards and basements. Pictured here is the process involved in turning out one concrete block on a model of the early hand-tamp machines. In (1) the workman fills the machine with cement and aggregate, hand-mixed on the ground. In (2) he hand tamps the mixture, in (3) discharges the finished block and in (4) carries it to a crude pallet for curing.

Fig. 4-1 Early hand-tamp method of making concrete block.

research change this figure each year as methods and machinery are improved.

Since the first hand-tamping machine was used, automation has changed the block industry from a backyard business to the largest masonry products business in the country.

LIGHTWEIGHT AGGREGATES

Aggregates are sand and pea gravel or some inorganic substitute particles which are mixed with portland ce-

ment to form concrete block. About 2000 years ago, Romans used *pumice,* a light, porous volcanic rock. Pumice is still used by some manufacturers to make lightweight concrete block. A problem related to the use of pumice is its high absorption rate. This high rate makes it difficult to control the amount of water in the mix.

In 1913, a Pennsylvania bricklayer, Francis J. Straub, invented a system of making block out of waste cinders. The product, known as *cinder block,* was lightweight and was able to hold nails. Cinder block was sold for about the same price as competing products. Since the block was lighter in weight than block made with sand and gravel, masons could lay them with less physical effort, thereby enabling a faster work rate. The savings that it offered made it very popular as a building material. Cinder block was a big factor in the development and growing use of concrete block in the East, since lightweight pumice was only available in the western United States.

Other lightweight aggregates now in use include expanded fly ash, expanded shale, expanded clay, expanded slag, and expanded slate. Different companies give brand names to their own products. The trend in the block industry is to produce more lightweight concrete blocks as the demand increases.

CHARACTERISTICS AND PROPERTIES OF CONCRETE BLOCK

Durability

Durability is the ability of a product to withstand long, hard use. Concrete block is a very durable building material. Block made over 50 years ago is still in good condition. Block made today is even better than that made years ago due to better materials and improved methods of manufacturing.

Strength

Strength is another important characteristic of block. Not all block has the same degree of strength. Concrete block for heavy loads may be either solid or semisolid, depending on where it is used. Block is used in partition walls in a building where the structural frame is concrete or steel. This block is usually of low compressive strength and supports only its own weight. It is meant to provide privacy and fire protection at an economical cost. Generally, but not necessarily, heavier block has greater strength than lightweight block. Block made from stone dust is relatively heavy and is generally used for foundations. Building codes and federal and state regulations refer to ASTM standards for strength standards. Concrete block must meet ASTM standards before it may be used as a building material.

Standard Specifications for Load-Bearing Concrete Masonry Units (ASTM C 90) is the most frequently referenced standard for concrete masonry units throughout the industry. It includes minimum face shell and web thickness for different sizes of concrete masonry units. Overall unit dimensions (height, width, or length) are permitted to vary by $+ \frac{1}{8}''$ (3.2 mm) from the dimensions specified by the manufacturer. ASTM C 90 also defines the difference between hollow and solid masonry units. In 1990, the basis for compressive strength was changed from gross area to average net area of the unit. This change was mandated by an increasing use of engineered masonry design which uses net area strength as a basis for allowable stress.

Incorporation of net area strength permitted a change in the scope of ASTM C 90 to include both hollow and solid units and eliminated the need for a separate standard for solid units. Thus, the old standard ASTM C 145 has been discontinued.

Compressive strength based on net cross-sectional area is required to be at 1700 psi for individual units and 1900 psi for the average of three units. Higher compressive strengths may be specified where required by design. Local suppliers determine availability of higher compressive strength blocks.

Fire Resistance

Fire resistance is another important factor to consider in the selection of concrete block for a building material. Tests have shown that all concrete block is highly effective in resisting fire and heat. In the test, heat is applied for a predetermined amount of time. The number of hours that the concrete block resists damage and the passage of heat is the fire-resistance rating. Because of its fire resistance, block is used to enclose steel columns and other less resistant materials in construction jobs.

Sound Resistance and Absorption

The sound resistance of concrete block is measured in terms of sound transmission class. *Sound transmission class* is the measure of the loss of sound between two points. Generally, concrete block resists the passage of sound from one room to another very effectively. A heavy, dense block effectively blocks more sound than a lightweight block. The ability of a block to reduce sound which starts inside the room is known as *sound absorption*. Generally, an open-textured block absorbs sound more efficiently than a smooth wall. On some construction jobs, the hollow cores of the block are filled with a sound-absorbing material such as sand, plaster, Styrofoam, or an aggregate. In addition, the cores often contain insulation to reduce energy costs for heating and cooling the building.

BLOCK SIZES

Like bricks, concrete blocks are generally made using the modular system. In most modern building construction, blocks are solid or hollow rectangular shapes. Block sizes are expressed as nominal sizes. This means each dimension includes the mortar joint, which is $\frac{3}{8}''$. For example, an $8'' \times 8'' \times 16''$ block has an actual measurement of $7\frac{5}{8}''$ (height) $\times 7\frac{5}{8}''$ (width) $\times 15\frac{5}{8}''$ (length). The addition of the $\frac{3}{8}''$ mortar joints gives the block a nominal dimension of $8'' \times 8'' \times 16''$. Figure 4-2 shows actual and nominal dimensions of an $8'' \times 8'' \times 16''$ block. A *solid block* is defined by the ASTM as a unit in which the core (or hollow) area is no more than 25% of the total cross-sectional area. A hollow block has a core area of at least 25% of its total cross-sectional area. Concrete block may contain two or three hollow cells, depending on the job specifications determined by the architect.

Concrete blocks are produced in many different lengths and widths. The proper dimensions depend upon the requirements of the particular job. In addition, manufacturers produce concrete blocks for special uses, such as for solar screens or for decoration, that vary from normal dimensions. Figure 4-3 shows the various sizes of common concrete masonry units.

BLOCK SHAPES

Block units are made in a variety of patterns and shapes to suit modern construction. In some cases, the name of a shape describes its function in construction such as sash block, pilaster units, control joint block, etc. In other cases, the design itself determines the name, such

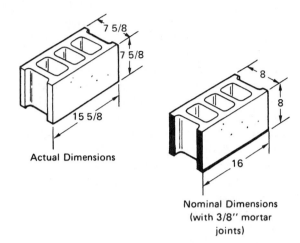

Actual Dimensions

Nominal Dimensions (with 3/8'' mortar joints)

Fig. 4-2 Actual and nominal dimensions of 8'' block.

as bull-nose unit, solid-top block, concrete brick, etc. While terminology has not been formally standardized, the shape names given here are fairly well established by common usage—and some variations are peculiar to certain localities. In most cases, terminology is self-explanatory. Shape design can be more readily understood by referring to Figure 4-4. The drawings are those of units generally available. It is impossible to show all shapes made by all manufacturers. This is because some units are dropped and new ones added.

Full Size Units		Supplementary	
Height	Length	Heights	Length
2 2/3	8	4	6, 4
3	8	2	6, 4
4	12		10, 8, 6, 4
5 1/3	12	2 2/3, 4	10, 8, 6, 4
6	12	2, 4	10, 8, 6, 4
8	12	4	10, 8, 6, 4
8	16	4	12, 8, 4

NOTE: Supplementary sizes listed are the sizes required for complete 4-inch flexibility. Such supplementary sizes as are required for a particular job may be cut on the job or furnished by the manufacturer. Maximum economy in construction is achieved when walls are planned and designed on the basis of the 4-inch module.

Fig. 4-3 Concrete masonry nominal dimensions.

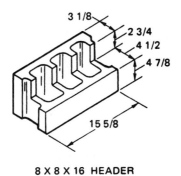

8 X 8 X 16 HEADER

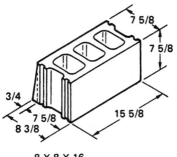

8 X 8 X 16
COLONIAL SIDING REGULAR

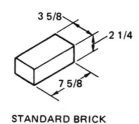

8 X 8 X 16 DOUBLE SHADOWAL

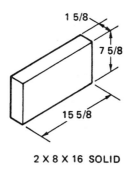

2 X 8 X 16 SOLID

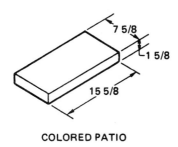

COLORED PATIO

STANDARD BRICK

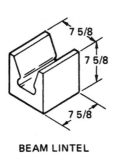

BEAM LINTEL

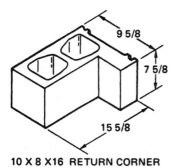

10 X 8 X16 RETURN CORNER

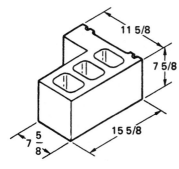

12 X 8 X 16 RETURN CORNER

Fig. 4-4 Typical concrete block shapes.

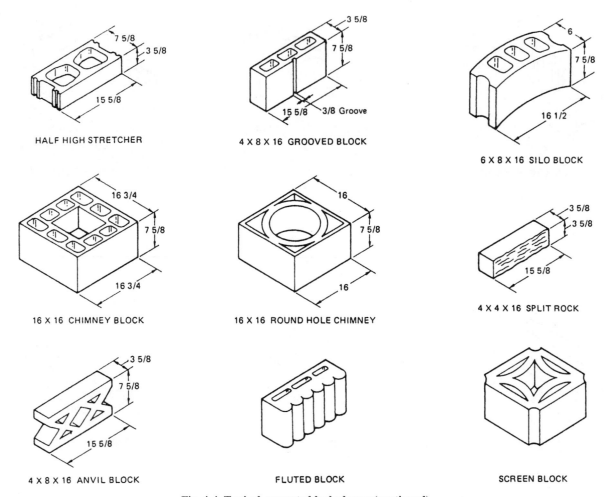

HALF HIGH STRETCHER

4 X 8 X 16 GROOVED BLOCK

6 X 8 X 16 SILO BLOCK

16 X 16 CHIMNEY BLOCK

16 X 16 ROUND HOLE CHIMNEY

4 X 4 X 16 SPLIT ROCK

4 X 8 X 16 ANVIL BLOCK

FLUTED BLOCK

SCREEN BLOCK

Fig. 4-4 Typical concrete block shapes (continued).

Most block manufacturers have up-to-date catalogs showing the shapes and sizes they stock.

Surface Finishes and Design

The finished appearance can be greatly enhanced or varied with the size, shape, color, mortar, bond pattern and surface finish of the unit or block. Ribs, flutes, projections, recesses, or special effects can be created by the mold in which the blocks are made. The split-faced units are molded with two units face-to-face. Then the units are mechanically split apart after being removed from the mold, leaving the split face with a rough texture resembling natural stone. These are classified in the industry as an architectural concrete masonry unit and are very popular in building large stores or commercial buildings. Figure 4-5 is an example of a split-face block.

The mason can also arrange concrete blocks to create a shadow effect. Figure 4-6 shows how recessing and projecting half blocks in a stack bond accomplishes this. Another increasingly popular architectural

Fig. 4-5 Architectural split-face block resembles natural stone.

Fig. 4-7 Splitrock used for decorative purposes.

Fig. 4-6 Concrete block wall showing how recessing and projecting of block creates a shadow effect.

Fig. 4-8 Closeup of 4″ fluted block.

concrete split block unit is solid with a white or gray stone aggregate which presents the appearance of natural stone. In some areas this is known as splitrock even though it is a block product. Figure 4-7 shows splitrock block in a wall pattern.

Fluted block, which have projecting vertical ribs, are also very popular for commercial construction. Figure 4-8 gives a closeup view of fluted block.

Figure 4-9 shows an example of a concrete block masonry unit retaining wall and steps. There are over

700 different sizes, shapes, and types of concrete masonry units available today from various manufacturers. The demand for new products by architects and designers to fill new requirements in construction will push this figure even higher.

Fig. 4-9 Example of concrete block masonry unit retaining wall and steps.

ACHIEVEMENT REVIEW

Select the best answer to each question from the choices offered. List your choice by letter identification.

1. The solid block is difficult for the mason to work with because
   a. the block is too long.
   b. the block is very fragile.
   c. it does not contain hand or finger holes.
   d. it is too big to handle.

2. Cinder block, an important factor in the development of concrete block, was a popular building material because it was
   a. lightweight.
   b. available in more colors than other materials.
   c. insulated.
   d. much bigger than other materials.

3. Production increase in the concrete block industry since 1900 is mainly due to
   a. the automatic-tamping and vibration machines.
   b. improved cement for the mix.
   c. better management.
   d. highly trained workers.

4. Pumice, used in early times to make masonry units, is still used today in modern block making. One problem with pumice is
   a. it is too dry to use economically in the mix.
   b. it is very heavy.
   c. the amount of water in the mixture is difficult to control.
   d. it is too coarse to pass through a machine.

5. In 1913, a Pennsylvania bricklayer named Francis J. Straub devised a new system of making concrete block. The block was called
   a. slag block.   c. cinder block.
   b. cement block.   d. pumice block.

6. In masonry, the term *durability* is used in connection with concrete block. To which of the following does the term *durability* refer?
   a. Color   c. Weight of the block
   b. Type of aggregate used   d. Ability to withstand long, hard use

7. The term *sound transmission class* is used when discussing building with concrete block. It refers to the ability of concrete block to
   a. reduce the noise in a room.
   b. block the sound from other rooms.
   c. measure the amount of noise in a room.
   d. increase the sound in an area.

8. The concrete masonry business has converted to standard sizes in manufacturing to avoid confusion among builders and architects. This system is called
   a. standardizing.   c. nominal.
   b. modular.   d. systematic.

9. To define a hollow block, the ASTM requires that the core area comprise a certain percentage of the block. The percentage required is
   a. 50%.   c. 25%.
   b. 75%.   d. 15%.

10. The development of concrete block has brought on the design and construction of many buildings which would have previously been built with wooden frames. The main reason for this is
    a. availability of the material.   c. durability and economy.
    b. improved color arrangements.   d. shapes and textures.

MATH CHECKPOINT

1. If a concrete block is 16″ long including the mortar head joint, how many blocks would it take to lay out a wall 40′ long?

2. If one concrete block, including the mortar joint, equals 8″ in height, how many courses would it take to build a wall 8′ high?

3. If each 8″ × 8″ × 16″ concrete block weighs 35 pounds, how much would a stack of 100 blocks weigh?

4. If 15,000 concrete blocks can be made on a single machine in a 10-hour day, how many blocks can be made in 50 hours?

UNIT 5
Manufacture of Concrete Block Masonry Units

—————————————— OBJECTIVES ——————————————

After studying this unit, the student will be able to

■ describe the steps in concrete block manufacturing

■ explain how concrete blocks are cured with steam.

■ describe safety practices in relation to the manufacturing of concrete block.

The change from the hand method of forming concrete block to the modern, fully-automated block plant is a fairly recent development. Concrete block has become an economical building material due mainly to power-driven machinery and automatic control.

BLOCK PRODUCTION

Block plants may differ in size and types of equipment and methods used, but the basic operation is the same throughout the industry. The operation includes the following:

1. Obtaining and storing the raw materials
2. Batching and mixing the materials
3. Molding the blocks
4. Curing the blocks
5. Cubing and storing the blocks
6. Delivering the units to the job

Raw Materials

The raw materials primarily used to manufacture blocks are crushed stone, expanded shale (lightweight aggregate), portland cement, water, and additives. Some types of raw materials may vary in individual geographical areas due to availability, but all have to be able to produce a concrete masonry unit that conforms to ASTM requirements. They are delivered to the plant by truck or railroad cars. A large supply of materials

must be stockpiled on hand at all times to allow the plant to operate 24 hours a day to meet customer demands.

To make one $8'' \times 8'' \times 16''$ concrete block, it requires approximately 19 lbs. of crushed stone, 7.5 lbs. of expanded shale (lightweight aggregate), 3.1 lbs. of portland cement, and 12% to 14% water, most of which will evaporate. Additives are added to improve surface texture and green strength, form sharper edges and corners on block, reduce cycle time of forming blocks, etc. As shown in Figure 5-1 raw materials, crushed stone and expanded shale are stored in stockpiles.

A conveyor belt, fed by gravity from a storage bin above, transports the raw materials to storage areas over the batching and mixing area. Figures 5-2 and 5-3 show this procedure.

Batching and Mixing

The raw materials are released by gravity into a bin directly above the scales. A computer scale (such as the one shown in Figure 5-4) weighs and measures out enough materials for one batch. A computerized automatic batching controller (shown in Figure 5-5) releases the proper amount of dry raw materials into the mixer, guaranteeing that each batch will be the same, resulting in a uniform block every time. A controlled amount of water and additives are then fed into the mixer and each batch is mixed for 6 to 8 minutes. After

43

Fig. 5-1 Raw materials, crushed stone, and expanded shale stockpiled at plant.

Fig. 5-2 Storage bin feeds raw materials to conveyor belt.

Fig. 5-3 Conveyor transporting raw materials to bins over batching area.

Fig. 5-4 Scales used to weigh and measure one batch of raw materials.

Fig. 5-5 Automatic batching controller.

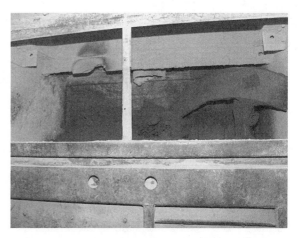

Fig. 5-6 Mixing raw materials.

Molding

The concrete mix is forced into the molds by vibration and pressure. The molding machine operates at 8 cycles per minute. Dies are changed in the machine to make different sized blocks when needed. When forming $8'' \times 8'' \times 16''$ blocks, five are made each cycle. The number of larger blocks will be fewer per cycle. Continuous operation over a 48-hour period at peak production can produce over 46,000 blocks. Most blocks made today have two cells instead of the older design of three cells. To simplify cutting halves, every fifth block made has a pre-formed slot in the center of the block that allows it to be cut on the job easily. Otherwise, they would have to be sawed or special halves would have to be made. Figure 5-7 shows a block with half slot. The block emerges from the machine on a belt or rollers. As the block leaves the machine (see Figure 5-8), a brush rolls across the top, removing any loose particles which may have collected.

Curing

The green (uncured) concrete blocks are transferred by computer control from the block machine to automated metal carts that travel on railroad tracks. These carts travel to the kiln area, where kiln doors open and the block is moved into the kiln (see Figure 5-9), also on a track in the floor. The blocks are spaced on individual

the batch is mixed, it is transported by conveyor belt above and into the block machine for the forming and molding of concrete blocks. Figure 5-6 shows the mixing of the raw materials.

Fig. 5-7 Center slot formed in concrete block to make it easier to cut halves.

shelves so there is an even circulation of heat in the kiln throughout the blocks. These carts can operate automatically and do not need a manual operator on the machine even though one is shown in the photo.

Curing is the process by which a concrete block is permanently hardened. The basic system of curing concrete block today is the low pressure or atmospheric pressure steam process. The kiln is normally built of insulated concrete blocks and is equipped with metal doors on one end which close off the heated area as the blocks are cured by steam. The curing process begins with the "presetting" of the block, which requires an ambient kiln temperature of 75 to 90 degrees. This takes about one hour. During this time, the cement is hydrating (combining with water) and starting to harden. The heat is gradually increased to a maximum of approximately 175 degrees, and the block is allowed to season or soak at this temperature for 24 hours. The goal is for the ambient temperature not to increase more than 5 degrees in relation to the internal temperature of the block. Using this method, a concrete block can be produced that has a net area of 1900 psi, which meets ASTM C 90 specifications. The heat in the kilns is continuously monitored and controlled by recording instruments to achieve an effective curing process.

Fig. 5-8 Green uncured block leaving molding machine.

Fig. 5-9 Computer-operated kiln cart moving blocks into kiln area.

Cubing and Storing

After the steam curing stage in the kiln, the blocks are removed, again by a computerized cart, and transported to the cubing area (see Figure 5-10).

The blocks are assembled in cubes (either semiautomatically or automatically) in any arrangement required to fit individual customers' needs. The blocks are assembled in cubes usually comprising six layers, Figure 5-11.

The cubes of blocks are moved by forklift tractors from the cubing area to the storage yard, where they are stacked four high (see Figure 5-12). They are allowed to open-air cure for another 28 days before being delivered to the customer or builder. Depending on weather conditions, this storage curing time could be extended so that they meet ASTM requirements.

Delivery to the Job

Special trucks deliver the block to the job site. These trucks (see Figure 5-13) are equipped with a hydraulic-operated forklift which is controlled by the driver. The boom of the forklift has a drop reach of about 12 feet and is able to swing in any direction. The truck driver can usually place the block in any desired area around the site. This is a real-labor saving practice compared to the old method of manually unloading blocks, and is very popular with customers and builders.

Fig. 5-10 Steam cured blocks being moved from kiln to cubing area.

Fig. 5-11 Concrete block leaving the automatic cubing machine.

Quality Control and Safety

Quality control of the product is required of all block manufacturers. The quality control area laboratory tests each production run of raw materials daily to make sure they are the same and meet ASTM requirements for the block being made.

Compressive strength and size variation tests are performed as soon as the blocks are removed from the kilns. Samples for a 7- and 28-day test are recorded. The result is a quality concrete block and a high degree of customer satisfaction with the many blocks that are produced. Figure 5-14 shows a sample display of typical concrete blocks manufactured at a block plant.

According to federal and state laws, it is the responsibility of each manufacturer to operate and maintain a safe place of employment. Workers should wear

Fig. 5-12 Concrete blocks stored on yard.

clothing suitable for the work being performed and steel-toed shoes to minimize any injury caused by job hazards. Hard hats are required, as well as ear plugs for high-noise areas. Guards on equipment must be in place when operating, and the workplace must be free of objects lying around that may cause an accident. Most plants employ a safety engineer who supervises inspections of manufacturing operations. Regular safety meetings are held with all employees to discuss how to improve safety at the work place. Most plants post the dates of accident-free workdays as a matter of public record and take great pride in a safe operation.

Fig. 5-13 Hydraulic-operated forklift block truck.

Fig. 5-14 Sample display of typical concrete blocks manufactured at a block plant.

ACHIEVEMENT REVIEW

The following questions refer to the six major steps in manufacturing concrete block. Write the correct answer to each.

1. How do raw materials feed into the batcher from the storage area on the roof of the plant?
2. The materials for concrete block mix are measured by what device?
3. By what two principles is concrete mix packed into the machine molds?
4. What are the two main methods by which concrete block is cured?
5. Using the method of curing mentioned in this unit, what is the psi rating that can be achieved for a concrete block?
6. What is the ASTM specification number for concrete block?
7. What is the advantage in having a forklift device mounted on trucks delivering concrete blocks?

SUMMARY, SECTION 1

- The biggest change in the manufacture of bricks is the burning process—from the older method of baking in the sun to burning in a kiln under a controlled temperature.
- Clay and shale are the principal materials in fired bricks.
- Modern modular bricks are constructed so that they may be used in the modular grid system of building (based on the 4″ module).
- The introduction of computers and highly trained workers into the masonry industry has resulted in a great increase in the manufacture of bricks.
- The method most often used to prepare bricks for burning in the kiln is known as the stiff-mud process.
- The tunnel kiln has replaced most of the older types of kilns, as it is highly efficient and therefore allows mass production of brick units.
- The properties of bricks are directly affected by the raw materials contained and the manufacturing process to which the bricks are subjected.
- Bricks can be given texture during the manufacturing process by special machine processes and, as they are laid in the wall, by treatment of mortar joints.
- The burning of brick units in the kiln will cause some differences in size to occur, but all bricks must conform to measurements established by the masonry industry and the government.
- The way in which bricks are laid in mortar is critical. Properly done, it results in a strong bond and efficient work.
- Outstanding characteristics and properties of bricks as a building product include color, durability, fireproofing quality, and flexibility.
- The increased production of concrete block is credited primarily to the invention and use of the vibration machine.
- Concrete masonry units are made according to the modular grid system.
- The success of the concrete masonry unit is due largely to the fact that the larger units allow the structure to be built more quickly and economically.

- Concrete block is made primarily of portland cement, aggregates, and certain additives.
- The development of lightweight aggregates with no strength loss has greatly speeded the process of laying units.

SUMMARY ACHIEVEMENT REVIEW, SECTION 1

Complete each of the following statements referring to material found in Section 1.

1. Many of the bricks made in ancient times have a recessed panel in which seals and inscriptions were imprinted. This recess is called a _____.

2. The basic raw materials used in the manufacture of bricks are _____.

3. The three major methods of preparing bricks for burning in the kiln are the _____ , _____ , and _____.

4. The most often used of the three methods mentioned above is the _____.

5. Reducing the amount of oxygen in the kiln to cause coloring of the bricks is called _____.

6. The production of concrete block increased tremendously, and a block with a more uniform texture was achieved by the invention and use of the _____.

7. Francis J. Straub, a Pennsylvania mason, invented a block in 1913 that was lightweight and could receive nails without breaking. This invention, which greatly accelerated the concrete block industry, was called _____.

8. The size of concrete block, including the mortar joints, is sometimes expressed as $8'' \times 8'' \times 16''$. This is known as the _____ size of the unit.

9. During the first hour of curing concrete block in a steam kiln, the temperature remains at 75° to 90°. This process is known as _____.

10. The process of stacking concrete masonry units on wooden pallets, each containing the same number of blocks, in preparation for delivery to the job is known as _____.

SECTION TWO
TOOLS AND EQUIPMENT

UNIT 6
Basic Tools of the Trade

OBJECTIVES

After studying this unit, the student will be able to

- describe the basic tools used in masonry.
- select good-quality tools.
- list safety measures to be practiced while handling tools.

The quality of the work masons do depends to a great extent on the condition and quality of their tools. Poorly made, dirty tools may be a factor in poor work. With this in mind, masons should select their tools with care.

The standard set of tools used by masons is fairly small compared to that of some other trades. A modest sum will buy a good-quality set of standard tools. Masons should purchase tools which are made by a well-known company when possible.

TROWELS

The brick trowel, used to cut, spread, and handle mortar, is the most important tool of the mason. The ability to use the trowel correctly and efficiently demands time and practice. The two major styles of trowels made for brick masons are the London pattern and the Philadelphia pattern. Trowels may vary in width and length, but they still are considered to be one of these two patterns.

Trowels are made from highly tempered select steel. There are six parts of a trowel, Figure 6-1. Handles may be wood, plastic, or leather, but wood is preferred. Notice that the wooden handle in the illustration has a metal band at the point where it attaches to the

shank. The band, known as a *ferrule,* prevents the handle from splitting. The handle tapers where it attaches to the shank to hold it securely in place.

The *London pattern* has either a wide or narrow heel, called a *diamond* heel. Many masons prefer the narrow London pattern for laying brick and the wide London pattern for laying concrete block or handling mortar for stonework. However, either pattern can be used depending on the preference of the individual bricklayer. The *Philadelphia pattern* has a square-shaped heel and will hold more mortar than the London patterns. Figure 6-2 shows the two different patterns of trowels discussed here.

The *pointing trowel,* shown in Figure 6-3, is a miniature model of the standard mason's trowel. It must be of the same quality and constructed of the same materials to withstand hard use. It is used in places where a large trowel cannot fit or be easily maneuvered.

Selecting a Trowel

A good trowel should possess certain characteristics. Only high-quality materials should be used in the construction of this important tool. If the steel used is of

53

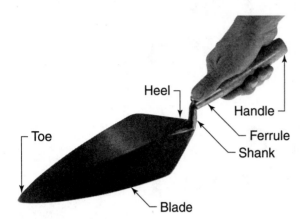

Fig. 6-1 Trowel in the narrow London pattern. On a well-made trowel, the weld between the blade and the shank is hardly visible. The diamond-shaped heel is a feature of a trowel in the London pattern.

good quality, the trowel will ring when tapped against a hard object. The best-quality steel will make the longest ring. Flexibility of the blade is very important (see Figure 6-4). The bending of the blade during the

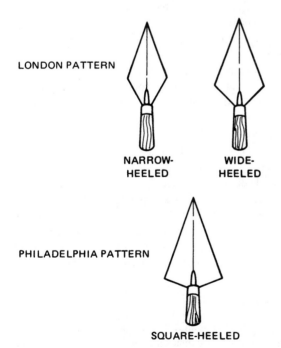

Fig. 6-2 Three types of brick trowels.

Fig. 6-3 Pointing trowel.

process of spreading mortar and cutting brick reduces tension on the mason's wrist.

A good trowel should be lightweight and well balanced to be handled easily (see Figure 6-5). The angle at which the handle is set in the trowel, known as the *set* of the trowel, is very important. If the set is too low or too high, the mason's wrist will be strained. The set of the trowel also aids in keeping the mason's

Fig. 6-4 Testing the flex of the blade. It should have some flex or it could break.

Fig. 6-6 Standard mason's hammer.

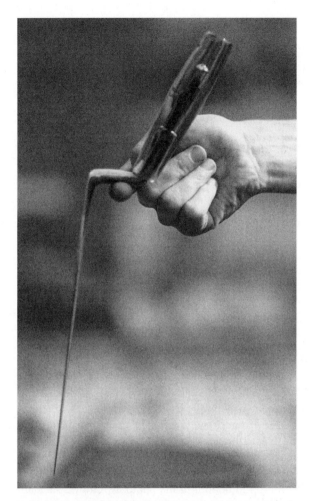

Fig. 6-5 Testing the balance of a trowel. It should balance on the forefinger if made correctly.

hand out of the mortar. The length of a trowel depends on the individual's preference and how much weight the mason can comfortably lift with a trowel. The standard length is 11″ to 11½″. If possible, masons should handle the trowel in the store before purchasing it to be certain that it is the proper one for them.

HAMMERS

Hammers used in bricklaying and concrete block construction are usually of the same design, shown in Figure 6-6. However, hammers used to cut concrete block sometimes have wider blades. The weight of the hammer may vary according to the mason's preference. Since the hammer is used to cut hard materials, it must be made of high-quality steel. The square end on the head of the hammer is used for cutting masonry materials, striking chisels, hammering nails, and tooling certain mortar joints. The other end of the hammer is wedge shaped. It forms a chisel peen for cutting and dressing masonry materials. The handle may be constructed of wood, fiberglass, or steel. Most masons prefer the wooden handle.

> **Caution:** If the wooden handle of a hammer becomes loose, replace the worn handle immediately. The hammer should be sharpened by a blacksmith since the temper may be lost if it is ground on a wheel.

When selecting a hammer, masons should choose the style and weight which suits them best. A good hammer of a medium weight (about 18 oz to 24 oz) is suitable for most masonry work.

The Tile Hammer

The tile hammer, Figure 6-7, resembles the brick hammer except that it is much smaller and lighter (approximately 9 oz). It is used to cut tile or to make a thin cut on bricks or block. The mason should not drive nails, strike chisels, or do other heavy work with a tile hammer.

Fig. 6-7 Tile hammer.

THE PLUMB RULE OR LEVEL

The plumb rule or level is used to establish a *plumb* line (aligned vertically with the surface of the earth) and *level* line (aligned horizontally with the surface of the earth), Figure 6-8. Plumb rules (known in the trade as *plumbrule*) may be constructed of seasoned hardwood, various metals, or a combination of both. They are made as lightweight as possible without sacrificing strength, since the plumb rule must be able to withstand fairly rough treatment.

Levels have a shape similar to a ruler and are equipped with vials enclosed in high-strength plastic or acrylic glass. Inside each vial is a bubble of air suspended in a liquid such as alcohol, light oil, or a liquid that will not be affected by extremes of low or high temperatures. The term *spirit level* indicates that one of these substances is being used and not water. When a bubble is located exactly between the two marks or bands on the vial, the object is either plumb or level, depending on if you are checking it vertically or horizontally.

In recent years, advancing technology has developed many important improvements that have been implemented in levels, both wood and metal ones. One good example of this is the Stabila Company, 332 Industrial Drive, South Elgin, Illinois 60177, 1-800-869-7460, which manufactures high-quality aluminum milled frame levels that are very comfortable to handle even in cold weather, have smooth edges, and have unbreakable acrylic vials that are filled with hydrocarbons fluids. Stabila guarantees that the levels will not break, leak, or fog up for life and that they will withstand cold or high temperatures. They are very strong, durable, will not warp, and will withstand a lot of rough usage. Their e-mail address is StabilaUSA@aol.com.

For many years masons carried only levels of two different sizes in their tool kits, a standard 48″ level and a 24″ one. Now, a variety of sizes is available to fit different situations and needs. The Stabila Company makes an assortment of excellent-quality levels in various sizes, as shown in Figure 6-9. They are arranged from top to bottom in the photo as follows: standard 48-inch level, 42-inch level, 24-inch level, 16-inch

Fig. 6-8 Plumbing a corner with a plumb rule.

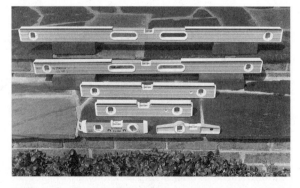

Fig. 6-9 Variety of different size levels. (Courtesy Stabila Inc.)

level, 10/25 cm level with a protractor vial for measuring and duplicating angles from 1 degree through 89 degrees and a standard 10/cm torpedo level. In addition Stabila also makes an 18 inch level that is not shown in the photograph.

In addition to the Stabila Company, there are many other excellent level companies with products available for masons who prefer the feel of a wood level. One of these companies I buy from and recommend highly is Bon Tool Company, 4430 Gibsonia Road, Gibsonia, PA 15044-7963. They will be happy to send you a free tool catalog on request or you can contact them at Website www.bontool.com.

Selecting a Plumb Rule

Plumb rules are tested in the factory, but many times are broken during shipping. Therefore, levels should always be checked before purchasing them. Hold the plumb rule against another in different positions to see if the bubbles match. Reverse one rule and check the bubbles again. The plumb rule is the most expensive of the basic tools and should be treated as such. Handling the plumb rule with care will give many years of service.

FOLDING MASON'S SCALE RULES

Masons use three different types of folding scale rules to evenly space and divide courses of masonry units including the mortar bed joints. The *course counter brick spacing rule*, shown at the top in Figure 6-10, is used to lay out and space standard brick courses to heights that are not modular in nature. The term *modular* is based on a scale rule that is multiples of the 4-inch grid called the *module,* which is a standard used in construction work. For example, this scale rule can be used to lay out even courses of brick to the top of a door or window frame that is 1″ higher than true modular spacing. There are ten different brick spacing scales on one side, and inches, 8ths, and 16ths on the opposite side, which is a standard 6-foot rule. An important feature of a course counter rule is that it shows how many courses of brick there are in a particular scale, which are numbered in red, as they would be laid up in mortar to a 6-foot height or width. A course counter brick spacing rule is also invaluable when laying out brick rowlock window sills.

The scale rule that is second from the top in Figure 6-10 is a *modular spacing rule*, which is based on the module of 4″. Six different scales represent the different sizes of modular masonry units including the mortar joints. The opposite side again contains the standard 6-foot rule divided into inches.

The scale rule that is third from the top in Figure 6-10 is relatively new on the scene and was developed as a result of the increased demand for and use of oversize high brick. This scale rule is called an *oversize brick spacing rule.* It has eleven different scales for "queen or king" size brick spacing ranging from $2\frac{7}{8}″$ to $3\frac{1}{2}″$ on the scale side and the standard 6-foot rule division on the other side. The scale side is marked with letters A thru K and red numerals to show the number of courses. It is essential for a working mason to own all three of these different scale rules to perform her or his work.

The *course counter brick spacing rule* and the *modular spacing rule*, as shown in Figure 6-11, are also

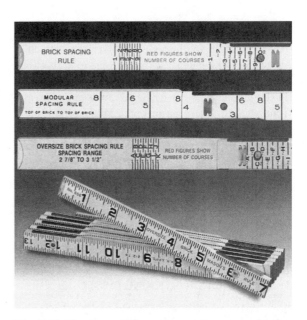

Fig. 6-10 The three different types of masonry folding scale rules that are used to lay out and evenly space and divide courses of masonry units. Top to bottom of photo: course counter brick spacing rule, modular spacing rule, and oversize brick spacing rule. The rule at the very bottom shows the reverse side of all scale rules in regular 16ths and inches. (Courtesy Bon Tool Company)

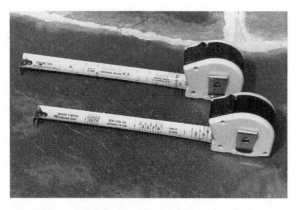

Fig. 6-11 The course counter brick spacing scale and the modular spacing scale are also available in steel tapes. (Courtesy Bon Tool Company)

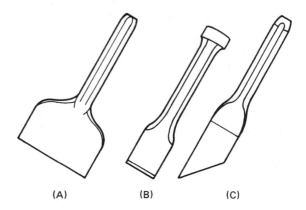

(A)　　　　　(B)　　　　　(C)

Fig. 6-12 Three chisels: (A) brick set chisel, (B) mason chisel, and (C) plugging or joint chisel.

available in steel tapes. Figure 6-10 and Figure 6-11 are courtesy of Bon Tool Company.

Selecting Rules

When buying a folding rule, purchase one which is heavy duty since it will last longer. Brass joints between the sections of a wood rule make the best joints. A brass tip should be located on each end of the rule to keep it from wearing. Many masons prefer wooden rules. Oil the joints of the rule occasionally with a light oil or paste wax and wipe off the excess. If it is not oiled, the rule may break due to dirt and mortar in the joints.

> **Caution:** Always open the rule with your fingers in the middle of the rule. Avoid opening with the hand on the end, which may pinch your fingers.

THE BRICK SET OR BLOCKING CHISEL

In everyday masonry work, bricks and blocks are cut with a hammer. However, when a very clean, sharp edge is required, a *brick set chisel* (also known as *blocking chisel* or *bolster chisel*) is used, Figure 6-12A.

The width of the blade is usually the same as the width of a standard-sized brick. Because of this, the mason is able to make a straight cut with one blow of

the hammer on the chisel. Chisel weights vary to meet the needs of individual masons.

The blade of the brick set chisel has a beveled edge on one side and a straightedge on the other. When cutting with the brick set chisel, be sure that the straight edge of the blade faces the finished cut. The purpose of the beveled edge is to cut off protruding parts of the brick on an angle, thereby allowing the head joints to fit neatly together.

The cutting edge of the blocking chisel must be perfectly straight or the bricks may be chipped and ruined.

> **Caution:** The steel striking end of the brick set will become burred after constant use. Grind off the burred metal to avoid the hazard of flying steel.

STANDARD MASON'S CHISEL

The standard mason's cutting chisel, Figure 6-12B, is designed for cutting block, brick, and veined stones. It has a narrow, strong, tempered blade which gives a neat, clean cut. This chisel is often used to cut out masonry work and make repairs. The enlarged head of the chisel provides more striking area for the hammer.

The Bon Tool Company has developed an excellent protective grip for a mason's chisel that really helps prevent painful injuries to your hand if you happen to miss the chisel head. The plastic grip fits tightly around the chisel handle and is grooved for comfort. It is available in a 3″- or 4″-wide blade depending on your requirements, Figure 6-13.

PLUGGING OR JOINT CHISEL

The *plugging* or *joint chisel,* Figure 6-12C, has a tapered blade for cleaning mortar joints or for removing a brick or block from a wall.

LINES

A *mason's line* is used to lay out walls and masonry materials over 4′ in length since plumb rules are only 4′ long and, therefore, cannot serve as a guide. The line is used as a guide for building straight, true walls.

For many years, cotton and linen line was used to construct masonry walls. However, the mason's line must be able to stretch without breaking, and cotton line has little stretch. Cotton line also deteriorates quickly if it is wet or exposed to weather for any period of time.

Most of these problems have been solved with the development of nylon line, Figure 6-14. Nylon line is long lasting and is not affected by moisture. It also resists the effects of mortar very well and tolerates a great deal of strain without breaking. Twisted nylon withstands well over 100 lb of pressure, while braided line may test as high as 170 lb. Braided line is slightly higher in cost than twisted line. However, braided line is smaller in diameter than twisted line and, therefore, does not tend to sag as much when pulled tight. Pulling the line tight without breaking it may at first be difficult for the mason and will require practice.

FASTENING THE LINE

Pins and nails, line blocks, and *trigs* are several fasteners that attach the line to the wall, Figure 6-15.

Pins and Nails

Most masons use a line pin and nails to fasten line to the structure. The *line pin* is a steel pin about 4″ in length which tapers to a point. The better pins are made of

Fig. 6-13 Brick chisel with a protective grip on it. (Courtesy Bon Tool Company)

Fig. 6-14 Braided nylon line.

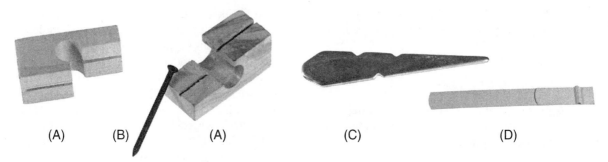

(A)   (B)   (A)   (C)   (D)

Fig. 6-15 Various fasteners used on line by the mason include (A) wooden blocks,
(B) the tempered nail, (C) the steel pin, and (D) the metal trig.

tempered steel. First, the end of the nail is driven into the wall and the line wrapped around it. The line pin is driven into the wall on the opposite end at the top of the course of the material being installed. The line is pulled tight and wrapped around the pin (see Figure 6-16).

The Line Block

Another method of fastening the line to the corner for use as a guide is with line blocks. *Line blocks* (see Figure 6-17) are L-shaped blocks made of wood, plastic, or metal which have a slit in the center of the block. Wooden blocks are preferred because they grip the corner securely. The line is drawn tight through the slit in

the block and held by tension against the finished corner. The block on the other end of the line is fastened in the same manner.

The advantage in using the line block is that the block can be moved up the corner for each succeeding course without measuring or tightening the line again. Also, line blocks do not leave holes in the wall.

> **Caution:** Since the line block projects slightly from the corner of the wall, exercise extra care when working with it. Dislodging the block and line can cause severe injury to anyone in its path. Line pins are considered safer than line blocks since they are driven more securely into the wall.

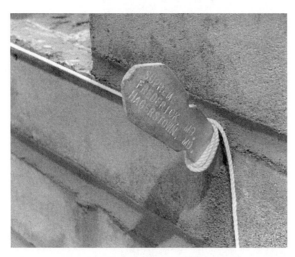

Fig. 6-16 Line held tight, pulled, and wrapped around a steel line pin.

Fig. 6-17 Line block with line attached to end of brick lead.

The Trig

Even though nylon line may be pulled tight on a long wall, the line will sag due to the weight of the line itself. To overcome this problem, a metal fastener known as a *trig* (or *twig*) holds the line in position. A trig can also be made from a small piece of line looped around the main line and fastened to the top of the brick. It may be necessary to use several trigs to control the line on a long wall.

There are other ways of fastening and controlling line, but the basic methods are pins and nails, blocks, and trigs. Normally, these fasteners are not purchased by the mason. Suppliers of masonry materials are usually glad to provide them free of charge.

JOINTERS

A *jointer* (also known as a *striking tool* or *striking iron*) is used to finish or tool vertical or horizontal mortar joints. Tooling is necessary to waterproof and beautify joints. Jointers are classified according to the specific joint which they are used to form, such as the concave joint or the V-joint, Figure 6-18. The type of joint which the mason forms depends mainly on the architectural design and specifications of the structure.

V-Jointer

The V-jointer is made from a V-shaped piece of angled steel, Figure 6-19. When passed through mortar, the

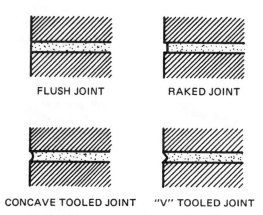

Fig. 6-18 Common mortar joints.

FLUSH JOINT RAKED JOINT

CONCAVE TOOLED JOINT "V" TOOLED JOINT

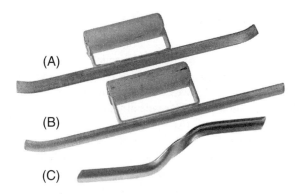

Fig. 6-19 Jointers commonly used by masons include (A) the V-jointer and (B and C) the long and short convex jointers.

angled edge of the V-jointer forms a V-shaped groove. This jointer can be obtained in the short or long styles. As with the convex jointer, the mason will find that the longer style is more effective. A solid, straight piece of wood may be used to obtain the same results as the metal V-jointer, but would not retain its straight, sharp edge very long.

Convex Jointer

The most commonly used jointers in various parts of the country are the *convex jointer* and the *V-jointer*. The convex jointer is made of a long, rounded piece of steel which forms a depressed, rounded indentation (often called a *round* or *concave* joint) in mortar. The convex jointer can be purchased in a pocket size or in longer lengths with a wooden handle. This type, *sled runner convex jointer,* is preferred over the short one. The short and long convex jointers are available in different widths to accommodate the various thicknesses of mortar joints being formed, Figure 6-19B and C.

Note: Remember that a *convex* jointer forms a *concave* joint.

Grapevine Jointer

The grapevine jointer has a raised bead of steel in the center which causes an indented line to form in the joint. This jointer is used mainly on Early American brickwork. The grapevine jointer is available only in the shorter size.

Rake Out Jointer

Originally, joints were raked out by a nail which was driven into a piece of wood and dragged across the mortar. This very inaccurate tool was replaced by the *rake out jointer.* A fairly recent development is a rake out jointer with wheels. This type of jointer has a screw, allowing the mason to set the tempered nail to the desired depth, Figure 6-20. The advantage of the skate wheel rake out jointer is the speed with which it may be operated, and the very neat, straight, hole-free joint that it forms (see Figure 6-21).

Slicker

A *slicker,* Figure 6-22, is a flat piece of steel used to smooth mortar and form flush, flat joints. It is often an S-shaped tool, either end of which may be used for striking. It is available in various sizes (usually with different sizes on each end) to form different mortar joints. The slicker is invaluable in working with stone and when paving. The slicker is also used as a pointing tool in tight places or in corners of buildings where a neat, clean joint is required. *Pointing* is another term for finishing joints.

MASON'S BRUSH

The finishing touch on masonry work includes brushing it with a good-quality, medium-stiff brush, Figure

Fig. 6-21 Raking a mortar joint out with a skate wheel rake out jointer.

6-23A. Most masons prefer the *stove* brush, which is equipped with a long handle. The long handle keeps the fingers away from the work. An old floor brush sawed in half works well, too.

In addition to brushing down work, the brush may also be used for brushing off footings and cleaning the work area, since a neat, clean work area is necessary to prevent accidents.

Fig. 6-20 A rake out jointer equipped with wheels. The screw located between the wheels allows the mason to set the desired depth of the nail.

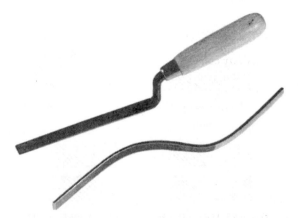

Fig. 6-22 Two styles of slickers. The slicker that does not have a handle may be used on two sizes of mortar joints.

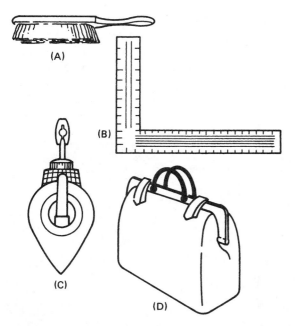

Fig. 6-23 Additional tools include (A) mason's brush, (B) steel square, (C) chalk box, and (D) mason's tool bag.

STEEL SQUARE

A *steel square* is a measuring tool that has at least two straight edges and one right angle, Figure 6-23B. A mason should carry a small square (about 12″ long) so that he can square *jambs* (sides of door and window openings) and returns. Large, 2′ squares are also used many times by the mason. The square can also be used to measure smaller materials.

CHALK BOX

A *chalk box* consists of a metal or plastic case with a good cotton line (usually in 50′ lengths) inside on a spool, Figure 6-23C. There is a hole in one end of the chalk box case through which the line is pulled. As the line is pulled from the case, it passes through finely ground chalk, usually blue or red in color. The chalk box helps the mason establish straight, accurate lines in layout work. The chalk box should be a standard item in every mason's tool kit.

Be careful that the line never gets wet, since the line will deteriorate, the chalk will get soggy, and the metal box will rust. However, if the line does get wet, un-

wind it from its spool and dry it out thoroughly before using it again. Replacements for worn line may be purchased in hardware stores.

TOOL BAG

Every mason should have a tool bag so that all tools are kept together and are always within reach. The average-sized tool bag measures 14″ to 18″ across and is equipped with an inside pocket in which small items are stored (see Figure 6-23D). Some tool bags may be converted to shoulder bags so that the mason's hands are free to grip scaffolds and ladders when climbing. The plumb rule is pushed through the handle when the hands must be free. Metal tool boxes are sometimes used. However, they are not preferred since the boxes tend to rust when wet and may be bulky and difficult to handle.

GROUT BAG

In recent years, the *grout bag* has become an important addition to the basic set of a mason's tools, Figure 6-24. A grout bag is cone shaped and is made of a vinyl or poly material that has a tip on the end with a hole in it.

Fig. 6-24 Grout bags made of vinyl or poly materials are available in either large or small sizes. (Courtesy Bon Tool Company)

The bag is filled with mortar mixed to the proper stiffness that will allow it to be forced out through the tip by squeezing the bag, so that the mortar joints are filled without excessively smearing the adjoining surface. It is a very efficient and fast method of filling mortar joints between masonry units, especially in pointing brick paving, re-pointing old mortar joints, or repairing cracks in masonry walls. Grout bags are available in large and small sizes; the choice of size depends on how much weight you can comfortably support with your arms.

Remember, spend time and effort when selecting tools (see Figures 6-25 and 6-26). Good-quality tools and equipment will pay off in the service you will receive from them.

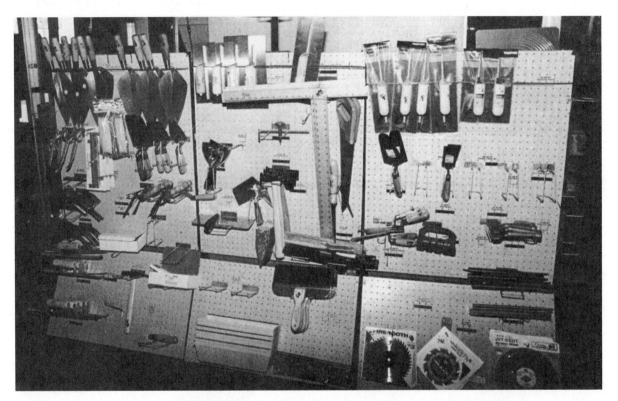

Fig. 6-25 Many building supply companies stock complete assortments of brand name masonry tools.

Fig. 6-26 Approved safety eyeglasses or goggles should be a part of any mason's tool set.

ACHIEVEMENT REVIEW

A. Selecting a set of tools is an important task for the beginning mason. Quality and handling features are the main factors to be considered in selecting basic tools, Figure 6-27. Listed are some of the basic tools discussed in this unit. For each tool, list the outstanding features you would look for when purchasing a tool set.

1. Trowel
2. Brick hammer
3. Brick set
4. Plumb rule
5. Rule
6. Striking tool
7. Mason's line

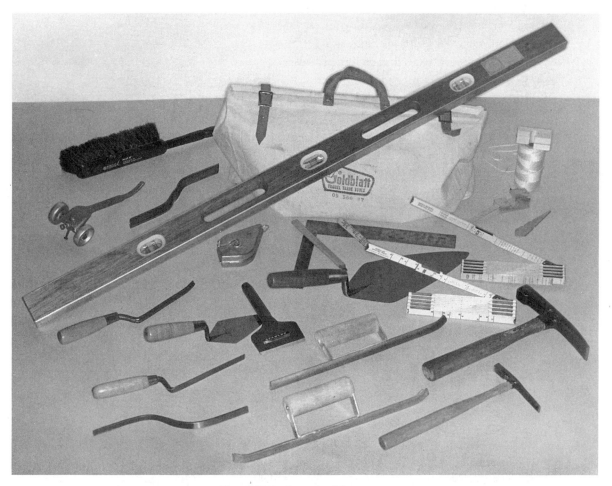

Fig. 6-27 Basic tools of the mason.

B. The column on the left contains a statement associated with mason's tools. The column on the right lists terms. Select the correct term from the right-hand list and match it with the proper statement on the left.

1. Used to form a depressed, rounded indentation in a mortar joint
2. Angle of the handle on a trowel
3. Flat tool used to smooth and form flat mortar joints
4. Tool equipped with wheels
5. L-shaped block used to hold line tight
6. Measurement tool based on the 4″ grid
7. Tool used for cutting brick accurately
8. Metal or plastic box filled with chalk and line; used for lines
9. Jointer with a raised bead of steel in center
10. Metal fastener used to keep line from sagging
11. Horizontal to the surface of the earth
12. Trowel pattern with a diamond-shaped heel
13. Vertical to the surface of the earth
14. Tool used to establish level and plumb points in masonry work

a. Philadelphia pattern
b. Plumb
c. Rake out jointer
d. Slicker
e. Modular rule
f. Grapevine jointer
g. Convex jointer
h. Chalk box
i. Trig
j. Level
k. Brick set chisel
l. London pattern
m. Line block
n. Set of the trowel
o. Plumb rule

MATH CHECKPOINT—TOOL SET

Determine the total cost of the following set of tools:

1.
48″ level	$48.50
Bricklayers trowel	$33.95
Pointing trowel	$ 8.65
Brick spacing rule	$15.75
Brick modular rule	$15.75
Skate wheel rake out jointer	$ 8.50
Slicker jointer	$ 2.90
Grapevine jointer	$ 7.95
Convex sled runner	$ 8.85
Regular convex jointer	$ 7.60
V sled runner	$ 8.85
Bricklayers brush	$ 6.90
Braided nylon line 500 ft.	$10.75
2 pair line blocks	$ 3.00
6 steel line pins	$ 3.95
Brick set chisel	$12.00
Canvas toolbag	$58.00
Brick hammer	$26.00
Tile hammer	$20.00

2 ft. square	$20.00
Chalk line and box	$10.00
1 doz. carpenters pencils	$ 6.00
100 ft. steel tape	$40.00

2. Add 5% sales tax to the total cost of the tools.

3. Assuming that the tools are paid for within 7 days, deduct 8% from the total cost of the tools.

4. Tools will be ordered from a mail-order supplier. Add $20.00 to the total cost of the tools to cover freight charges. This will be after sales tax has been added and discount has been subtracted and is the final cost of tools delivered to you.

UNIT 7
Learning to Use the Basic Tools

OBJECTIVES

After studying this unit, the student will be able to

- describe how to use the basic tools of the masonry trade.
- demonstrate the correct methods of using the basic tools.
- explain why both hands should be used when laying bricks.
- demonstrate good safety practices when using the tools.

The importance of learning to use the basic masonry tools correctly cannot be overemphasized. To a large extent, ability depends on how well a mason handles and uses the basic tools. While the masonry student must perfect skills with all masonry tools, the trowel is used more than any other tool. Therefore, more skill and practice are required to use it correctly.

As students begin working with the masonry tools, they will make mistakes. When mistakes are made, a demonstration of the correct technique or procedure by the masonry instructor or another mason will give the students guidance. This guidance will enable the students to continue their practice until they have mastered each problem area.

Students should try to keep work neat while they are learning and should not try to develop a great deal of speed. Students will find that as they become more skillful in the use of tools, speed will increase. After a time, they will develop a rhythm or a gait in work movements. A *gait* is the speed at which the student is most productive. To attain this rhythm, masons must eliminate unnecessary movements as they work.

At the end of this unit, there is a progress checklist of the basic skills used in masonry work. The students should check their work with this list. Also, with the help of the instructor, the students should determine their strong points and weak points. Students should

then work on the weak points until they have mastered the basic skills given in the list.

SAFETY NOTES

There are safe ways to use tools to prevent injury. Do not harm yourself and your fellow workers through carelessness.

- Always cut masonry materials away from a fellow worker's face.
- Always replace mortar cautiously in the mortar pan or on the mortar board. Do not throw it so that it splashes.
- Keep tools out of the paths of other people working on the job.
- Always hand tools to another person, never throw them.
- Use good tool habits and safety practices as tool skills are learned.

SPREADING MORTAR WITH THE MASON'S TROWEL

A great deal of the mason's working time is spent either spreading mortar or handling mortar with the trowel. There are several different ways of using the trowel to cut mortar free from the mortar pile. This unit will illustrate two accepted methods of doing this.

The Clock Method of Picking Up Mortar

The first method of picking up mortar is called the *clock method*. It is called the clock method because the movements of the mason are related to the location of the numbers on the face of a clock.

1. Using mortar mixed to the proper stiffness, fill the mortar pan. Mortar which is too stiff or too soft will not spread evenly. Place the mortar pan about 2′ from the wall to allow working space. It is recommended that the student use a mortar pan with the clock method. A mortar board can be used, but the pan will probably be easier to use.

2. Grasp the handle of the trowel between the thumb and first finger with the thumb resting well forward on the top of the handle. Figure 7-1A shows the correct way to hold the trowel. Note that the thumb should not drape over the handle and down the shank, Figure 7-1B. In this position, the thumb will be in the mortar. Prolonged contact with the mortar can cause irritation or cracking of the skin. The thumb stabilizes the trowel and guides its movements from side to side. The thumb also helps support

the wrist when lifting mortar from the board or pan. As shown in Figure 7-1A, the second, third, fourth, and fifth fingers are wrapped around the handle of the trowel. The muscles of the hand, wrist, and forearm must remain relaxed so that the trowel can be handled freely.

3. Figures 7-2A through 7-2D show the method of removing mortar from the pan. A clock face painted on a wooden frame is set over the pan so that the student can become used to the trowel positions. In Figure 7-2A, the trowel is in the starting position at 6:30 o'clock. The trowel is held at an angle so that as it is moved, it will cut the desired mortar away from the rest of the mortar.

4. The trowel is now moved upward to the 12:00 o'clock position, Figure 7-2B.

5. Next, the trowel is moved down to the 5:30 o'clock position, Figure 7-2C. Do not turn the trowel over and do not lift the trowel out of the mortar pan while making the final downward movement. If done correctly, the mortar is cut free and resembles a long church steeple. If the mortar looks like a wide piece of pie, it has not been cut at the correct angle. Therefore, it cannot be picked up properly.

6. Now, without removing the trowel from the mortar, slide the trowel over until it is face up.

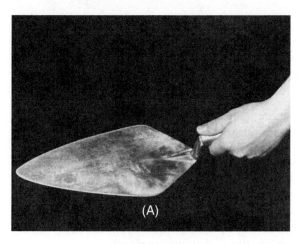

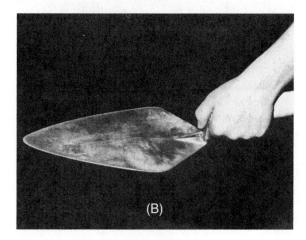

Fig. 7-1 Holding the trowel: (A) Correct way to hold the trowel (note that the thumb remains on the handle), and (B) incorrect way to hold the trowel.

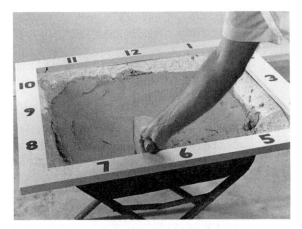

Fig. 7-2A Starting position at 6:30 o'clock.

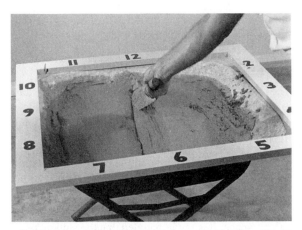

Fig. 7-2B Trowel moved upward at an angle to 12:00 o'clock position.

Fig. 7-2C Movement of trowel downward to 5:30 o'clock position.

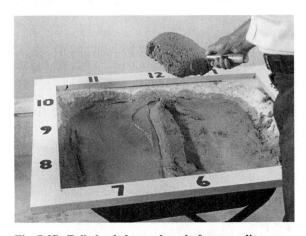

Fig. 7-2D Fully loaded trowel ready for spreading.

Then firmly push the trowel toward the 12:00 o'clock position. Finally, lift the trowel from the mortar at the end of the stroke. If the entire operation has been done correctly, the trowel will be fully loaded with a tapered section of mortar. The position of the mortar on the trowel is very important if the spreading is to be accomplished successfully. Figure 7-2D shows a fully loaded trowel and the mortar pan from which the mortar was removed.

To prevent the mortar from falling off the trowel, a slight snap of the wrist will set the mortar on the trowel.

This is one of the most important steps to master in spreading mortar. Do not give too much of a snap or the mortar will become dislodged and it will be necessary to start over again. Experienced masons can set the mortar so smoothly that one has to observe very carefully to see them do it.

The process of applying the mortar to a wall is called *stringing* the mortar. Approach the wall (in this case a practice $2'' \times 4''$) and keep the elbow of the right arm slightly away from the body so that the movements of the trowel hand are not restricted.

Figure 7-3 shows the mortar ready to be spread. Use a sweeping motion to spread the mortar on the wall.

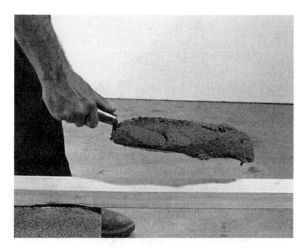

Fig. 7-3 Trowel with load of mortar ready for spreading.

Keep the trowel in the center of the wall for the length of the spread. In Figure 7-4, note the angle at which the trowel is held while the mortar is spread. Try to spread the mortar along the length of at least 2 bricks. Beginning students often try to obtain too long a spread. This results in an uneven bed of mortar on which it will be impossible to lay bricks without filling in the spaces. Students should repeat this spreading operation over and over until they have mastered the operation. This oper-

ation should be repeated until they can spread the correct amount of mortar with one motion of the trowel with little or no mortar thrown over the edge of the bricks. The ideal amount of mortar to be spread with one trowel motion equals the length of 3 to 4 bricks.

By holding the trowel face up at a slight angle, Figure 7-5, excess mortar hanging over the face of the wall can be cut off. The flat angle at which the trowel blade is held enables the mason to catch any mortar as it is cut off, thus preventing the excess mortar from smearing the wall. The excess mortar can be used to fill in any spaces in the bed joint or it may be returned to the mortar pan.

> **Caution:** Mortar splashed in an eye is very painful and can be avoided by wearing eye protection. Immediately flush eyes with water if this happens. To avoid accidents, mortar should not be thrown in the pan. Replace it carefully.

The next step in spreading the mortar is to furrow it, Figure 7-6. Hold the trowel at a 45° angle. Make a shallow furrow in the center of the bed joint by tapping with the point of the trowel along the length of the mortar spread. A shallow furrow in the bed joint allows enough movement of the mortar to permit the mason to

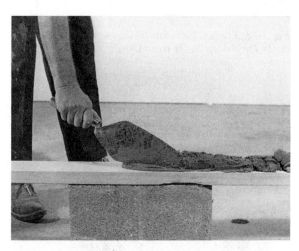

Fig. 7-4 Spreading the mortar on the wall. Note the location of the trowel in the center of the board and the flat back side of the trowel turned away from the mason.

Fig. 7-5 Using the trowel to cut off the excess mortar. Note the flat angle at which the trowel is held. The mason can catch the mortar so it does not fall on the floor.

Fig. 7-6 Furrowing the mortar with the point of the trowel. Note the angle at which the trowel is held.

Fig. 7-7A Cutting the mortar loose from the main pile with a downward slicing motion of the trowel. The smaller pile of mortar is then pushed to the edge of the mortar board with a rolling motion.

adjust the brick to its proper position. Avoid deeply furrowed joints. A deep furrow may expose the building unit below and may eventually cause a leak in the wall.

After furrowing, the excess mortar again is cut off. This mortar may be returned to the mortar pan or it may be used for the head joint on the brick to be laid.

The Cupping Method of Picking Up Mortar

The second method of picking up mortar is called *cupping* the mortar. This method is recommended when a mortar board is used.

Fig. 7-7B The mortar is shaped into a long, tapered form ready for pickup on the trowel.

1. In the cupping method, the amount of mortar to be picked up with the trowel is separated from the main pile on the mortar board. The mason slices down with the trowel and then pulls the mortar away from the main pile, Figure 7-7A.
2. The smaller pile of mortar is pulled and rolled to the edge of the mortar board.
3. At the edge of the board, the trowel works the mortar until it is shaped into a long, tapered form for convenient pickup, Figure 7-7B.
4. The trowel is then slid under the smaller, shaped pile of mortar. The trowel is held flat against the mortar board and then is pushed forward under the mortar. At the same time a quick, upward motion of the trowel lifts the mortar free from

Fig. 7-7C The trowel is moved under the mortar pile and is lifted to break the mortar free of the board. Note the tapered shape of the mortar on the trowel.

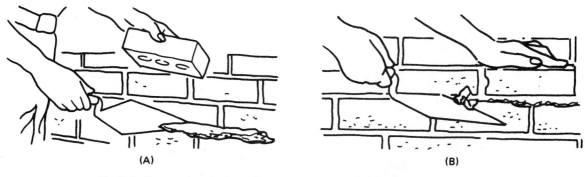

Fig. 7-8 Correct thumb placement: (A) Thumb in this position will not interfere with the trowel cutting off the mortar, and (B) thumb in this position will not interfere with the line as the brick is placed on the wall.

the mortar board at the completion of the stroke, Figure 7-7C.

From this point on, the spreading operation is the same as that shown in the previous clock method.

Regardless of which pickup method is selected, the speed with which students lay bricks depends on their ability to pick up and spread the mortar properly.

LAYING THE BRICKS

There is a correct way to hold a brick when laying it on the mortar bed. The masons's thumb should curl down over the top edge of the brick and be held slightly away from the face of the brick. With the thumb held in this manner, the line can fit between the thumb and the top edge of the brick. As a result, the mason does not cause the line to swing or bounce (called *crowding the line*) and other masons working on the same line do not have to wait for the line to stop swinging. Figures 7-8A and B show the correct thumb placement for laying the brick and cutting off the excess mortar.

Using Both Hands for Efficiency

The speed of bricklaying can be improved and the physical strain decreased if the operation is done with as few motions as possible. A common mistake made by many masons is to use only one hand to pick up and spread the mortar and place the bricks. Thus, one hand does all the work while the other remains idle. Masonry students should practice using both hands, Figure 7-9.

They will find that not only does the work go faster, but they will experience less fatigue.

For example, when working with *brick veneer* (brick that carries only its own weight and is used mainly as a facing material), the student should pick up bricks with one hand while the other hand picks up the mortar with the trowel, Figure 7-9. In addition, when working on a *solid wall* (brick backed by some other masonry material or another thickness of brick), a practice called *walling* the brick is recommended, Figure 7-10. In this procedure, both hands are used to pick up bricks and stack them on the finished back section of the wall. The bricks are stacked before any mortar is spread on the front wall. This eliminates a great deal of bending over to reach each brick on the brick pile.

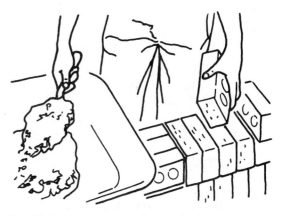

Fig. 7-9 Use both hands to spread mortar and lay brick.

Fig. 7-10 To wall brick, use both hands to place bricks on the completed back section of a solid wall before beginning work on the front wall section.

Another time-saving and labor-saving technique is shown in Figure 7-11. After the bricks are stacked on the completed back section of the wall, the mason picks up one brick and lays it in the mortar bed. While one hand cuts off the excess mortar and places it on the head

Fig. 7-11 Cutting off mortar with one hand while the other hand picks up the next brick to be laid.

of the brick just laid (to form the head joint), the mason should pick up another brick with the other hand to have it ready to lay as soon as the one hand completes the head joint.

Making the Head Joint

The proper way to form a head joint is shown in Figure 7-12. In Figure 7-12A, the first brick is placed on the bed joint. In Figure 7-12B, the brick is pressed into position on the mortar and the excess mortar is cut off with the trowel. The second brick is picked up and the mortar joint is applied to the end of the brick, Figure 7-12C. Finally, the second brick is pressed into position in Figure 7-12D. Note the way in which the mason's palm presses across the ends of both bricks. This step enables the mason to determine when the brick is in a level position. The mason then repeats the procedure and continues adding bricks to the course.

LEVELING AND PLUMBING THE COURSE

A course of bricks must be leveled and plumbed if a wall is to be constructed properly.

Leveling Bricks

After 6 bricks have been laid, they must be leveled using the following procedure. Remove any excess mortar on the top of the bricks before placing the level (plumb rule) on the wall. The level should be placed lengthwise in the center of the bricks to be checked. Tap down any bricks that are high in relation to the level, Figure 7-13. If the bricks are low in relation to the level, they must be picked up and relaid in a fresh bed of mortar until they are level.

When tapping down high bricks, keep the trowel blade parallel to the course of bricks and in the center of the bricks. By doing this, splattered mortar around the edges and chipped brick faces will be avoided. A hammer may be used to tap down a brick that is difficult to handle with the trowel. However, the trowel is usually sufficient to tap a brick into the proper position. Do not tap on the level itself to tap down high bricks. The level is a delicate tool and should be treated as such.

(A)

(B)

(C)

(D)

Fig. 7-12 The proper way to form a head joint.

Fig. 7-13 Leveling bricks by tapping them down with the blade of the trowel. Note the correct position of the trowel, parallel with the course of bricks and back from the edge of the bricks.

Plumbing the Bricks

Once the bricks have been adjusted so that they are all level, they must be plumbed. Hold the level in a vertical position against the end of the last brick laid, Figure 7-14. Tap the brick with the trowel as shown to adjust the brick face either in or out. Do this until the bubble in the glass vial of the level rests between the two lines. As the bricks are being plumbed, do not exert a great deal of pressure to force the level against the bricks. If too much force is used, the bricks may spring out of plumb when the level is removed. Next, move the level to the end of the first brick laid and repeat the plumbing process. Now that the first and last bricks have been plumbed, hold the level in a horizontal position against the top outside edge of the bricks, Figure 7-15. Tap the bricks either forward or back until they are all aligned against the level. Be careful not to move or change the original plumb points while the remaining bricks are being aligned.

Because bricks often have an uneven or curled top edge, the level must be held at this edge of the brick to assure that the edge is aligned. True alignment must be

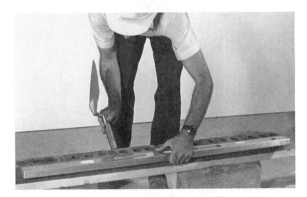

Fig. 7-15 To align the front faces of the remaining bricks, the mason places the level horizontally along the front upper edges of the bricks and taps the bricks until they are aligned. The level must be kept even with the top of each end brick while the mason adjusts the middle bricks.

achieved at the top edge to avoid uneven humps in the finished wall.

After the plumbing process is complete, place the level out of the way. Hang the level on a nail using the small hole in the top section of the level, or place it upright in one of the holes of a concrete block. Never place the level on the scaffold or on the ground near the work area as it will surely be broken.

CHECKING THE HEIGHT WITH THE MODULAR RULE

Of the three types of mason's rules listed in Unit 6, the modular rule, the course counter spacing rule, and the oversize rule, only the modular rule will be covered at this point. The correct spacing height for 1 course of standard brickwork equals the number 6 on the modular rule, indicating that there are 6 courses to every 16″ of vertical height. Since this value includes the mortar joints, the mason does not have to allow extra height for the mortar.

After the bricks are laid in the wall, unfold the modular rule and place it on the base used for the mortar and bricks. Hold the rule vertically, Figure 7-16. Be sure the end of the rule is flat on the base so that the reading is accurate. The top of the course of bricks should be even with the number 6 on the rule.

Fig. 7-14 Plumbing the end bricks with the level.

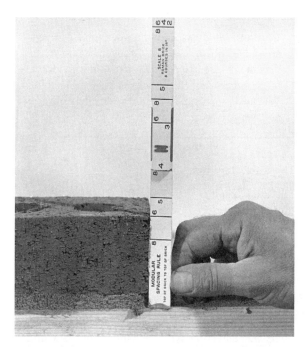

Fig. 7-16 Mason checking the height of a brick course with a modular spacing rule. Note that the top of the brick is even with number 6 of the rule. This is the correct value for standard brick spacing.

Note: If more than 1 course is being laid, always set the modular rule on the top of the first course to check the height as the base may have been irregular and a large joint may have been used to level the first course.

STRIKING THE WORK

To determine when the mortar is ready for striking or tooling, the student should press a thumb firmly into the mortar joint. If an impression is made and the mortar does not stick to the thumb, the mortar is ready to strike, Figure 7-17. A sled runner convex striking tool may be used to tool the joint, Figure 7-18. It is recommended that masonry students use the concave joint in their work. This is the joint used most often today. The tool should be slightly larger than the mortar joint to achieve the correct tool impression. Always strike the head joints first and then the bed joints. This order is followed because the bed joint should be straight and unbroken from one end of the wall to the other. If the

Fig. 7-17 Testing the mortar to determine if it is ready for striking. If the thumb impression remains, the mortar is ready.

head joints are struck last, there will be mortar ridges evenly spaced on the bed joints. These ridges will detract from the appearance of the joints. Masonry students should practice striking first the head joint and then the bed joints of their work. Most contractors require construction projects to show the smooth appearance resulting when the bed joint is struck last.

If a joint is struck while the mortar is too soft, the joint will be wet and runny. Also, the imprint of the striking tool will not be as sharp and neat as it should

Fig. 7-18 Striking the mortar joint with a sled runner tool. Note that the thumb is placed on the top outside edge of the handle so it does not scrape on the bricks.

be. However, if the joint is struck when the mortar is too hard, a black mark will appear on the surface of the joint. This is called *burning the joint*. Striking the joints when the mortar is too hard will also result in unnecessary fatigue for masons. Masons must test the joints repeatedly for the correct hardness so that they can strike them at the proper time.

BRUSHING THE WALL

Brushing is the final step in any masonry work and should be done before the mortar sets. A medium-soft brush should be used to brush the wall after the mortar is struck. In general, if the striking is done at the correct time, the brushing of the wall can be done immediately after. The masonry student should first brush all of the head joints vertically to remove any excess mortar. The brushing is completed by brushing lightly across the bed joints, Figure 7-19. The mortar joint may be restruck after brushing to obtain a sharp, neat joint without imperfections.

CUTTING BRICKS

Masons often must cut bricks to size to fit the construction job. Bricks can be cut with the brick hammer, brick set chisel, or trowel.

Caution: Students should not wrap their fingers around the brick set near the brick to be cut. When the cut is made, the downward pressure can force their fingers into the brick, causing abrasions. Whenever bricks must be cut, wear safety glasses or other eye protection.

The Brick Hammer

The brick hammer is the most frequently used cutting tool. To cut the brick with the brick hammer, hold the brick in one hand. Students must remember not to place their fingers and thumb on the face side of the brick where the cut will be made, Figure 7-20. Mark the desired cut on the face side of the brick either with a pencil or by scoring the brick with the hammer. Next, strike the brick with light blows of the hammer blade along the line marked to score the brick. Turn the brick on its side and strike with the hammer in line with the

Fig. 7-20 Cutting brick with the brick hammer. Note that the mason's fingers and thumb are not placed on the side of the brick being scored.

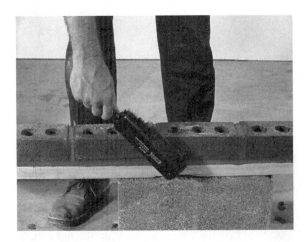

Fig. 7-19 Brushing the wall.

scoring just completed on the front of the brick. Repeat this procedure until the cutting line is marked on all four sides of the brick. Then strike the face of the brick with a hard, quick blow of the hammer blade. The brick should break cleanly along the lines scored. If the brick does not break the first time, restrike it gently until it has weakened enough to break. Scoring the brick causes it to break more evenly.

Striking an unscored brick usually results in an unacceptable break or shatters the brick. The cut made with the brick hammer is not a perfect cut. However, it will be accurate enough for use in the average construction job.

The Brick Set Chisel

When an accurate, straight cut is required, the brick set chisel is used. The mason can save time on the job by marking and cutting to size all of the bricks required at the edges of windows, around doors, or for window sills.

First, mark the cutting line on the face side of each brick using a pencil and a straightedge. Then place the brick, marked face up, on a wooden plank or on soft dirt. The bricks must never be placed on a hard surface such as concrete. This type of surface cannot give when the brick is struck with the brick set chisel. As a result, the brick may not cut true or it may shatter. The chisel of the brick set must be held vertically with the flat side of the blade facing the direction of the finished cut, Figure 7-21. The blade is beveled, so when the chisel is struck with the hammer, the blade travels downward at a slight angle and cuts off any particles of brick that may be protruding. As a result, the mason can place the cut brick in the wall immediately and does not have to do more chipping or dressing.

Cutting with the Trowel

The brick trowel is not recommended for most cutting, but may be used on medium-hard bricks when the hammer is not readily available. Masonry students should use the hammer and the brick set chisel as much as possible for cutting bricks until they have mastered cutting with these tools. Then they can learn how to cut with the trowel, which is a common practice on the job.

Fig. 7-21 Cutting brick with the brick set. The flat side of the chisel must face the finished cut being made. Keep fingers above the cutting edge of the chisel to avoid injuries.

As shown in Figure 7-22, the brick to be cut is held in one hand. Keep the fingers and thumb away from the side of the brick to be cut. Strike the brick with a quick, sharp blow at the point where the break is desired. To cut the brick with the trowel, a sharp snap of the wrist is required. A slow, swinging blow will not cut a brick. If the brick does not cut easily, use a brick hammer.

> **Caution:** Always hold the trowel downward when cutting bricks. Never cut bricks over the mortar pan or near other workers. When the weather is extremely cold, bricks should not be cut with the trowel as the blade of the trowel may break.

Fig. 7-22 Cutting brick with the trowel. Keep the fingers of the hand holding the brick well under the brick to avoid injuries from the blade.

PROGRESS CHECK

At this point, masonry students should have acquired certain skills. It is important to practice these skills. Students should demonstrate for their instructor each of the following skills. Any skills that are not mastered should be practiced until they can be demonstrated correctly.

- Spread mortar with the trowel for a length of at least 2 bricks.
- Apply the head joints on a brick.
- Level a course of bricks.
- Plumb a course of bricks.
- Check the brickwork for height with a modular rule.
- Strike the mortar joints with a concave tool.
- Brush the wall and restrike the joint.
- Cut bricks using the hammer, the brick set, and the trowel.

ACHIEVEMENT REVIEW

Select the best answer to complete each statement from the choices offered. List your choice by letter identification.

1. The mortar pan is placed away from the wall to allow for working space. The correct distance is
   a. 3′.
   b. 2′.
   c. 1½′.
   d. 5′.

2. The most efficient method of cutting mortar from the mortar pan with the trowel is called the
   a. cupping method.
   b. rolling method.
   c. clock method.
   d. scooping method.

3. If the mortar in the mortar pan is cut correctly with the trowel, it resembles
   a. a slice of pie.
   b. a circle.
   c. a church steeple.
   d. a square.

4. One of the most important steps in spreading mortar is the slight snap of the wrist after picking the mortar from the pan. The purpose of this movement is to
   a. shake the excess mortar off the trowel.
   b. distribute the mortar evenly on the trowel.
   c. set the mortar on the trowel to prevent it from falling off.
   d. shape the mortar for easier spreading.

5. The ideal amount of mortar spread with the trowel should equal the length of
   a. 1 to 2 bricks.
   b. 3 to 4 bricks.
   c. 5 to 6 bricks.
   d. 7 to 8 bricks.

6. Masons can improve their efficiency by
   a. learning to use both hands at once.
   b. using a larger trowel.
   c. spreading the mortar across the entire wall before laying any bricks.

7. The term *furrowed* is used when spreading mortar with the trowel. Which of the following best describes the term?
   a. Furrowing is the process of cutting mortar from the pan.
   b. A furrow is a slight indentation made in the center of the mortar joint with the point of the trowel.

    c. Furrowing is the spreading of the mortar on the wall.

    d. The trimming of the excess mortar from the wall after it is spread with the trowel is called furrowing.

8. The mason can determine the correct time to strike the mortar joint by using a simple test. Which of the following best describes the test?

    a. Measuring the rate of drying with a watch

    b. Pressing the thumb in the mortar of a bed joint

    c. Checking the mortar with the point of the trowel

    d. Light tapping on the face of the wall to see if the bricks have set correctly

9. A common expression in masonry work is *burning the joint.* This is defined as

    a. drying the mortar joint with heat.

    b. striking the mortar joint when it is too soft.

    c. striking the mortar joint when it is too hard, causing black marks on the joint.

    d. special effects created by raking out the joint.

10. The height of a standard course of bricks is indicated on a modular rule by the number

    a. 2.   c. 4.

    b. 5.   d. 6.

11. After the course of bricks is laid in mortar, leveled, and plumbed, it must be aligned horizontally with the level. This is done by

    a. holding the level across the top outside edges of the bricks and adjusting the bricks against the level.

    b. holding the level across the center of the course of bricks and adjusting the bricks until they are in line with the level.

    c. holding the level across the bottom of the course and adjusting the bricks to the level.

12. When bricks must be cut accurately, the best cutting tool to use is the

    a. brick hammer.   c. trowel.

    b. brick set chisel.   d. mortar hoe.

PROJECT 1: SPREADING MORTAR

OBJECTIVE

- The masonry student will be able to pick up and spread mortar for a length of from 1 to 3 bricks with the brick trowel.

EQUIPMENT, TOOLS, AND SUPPLIES

Mortar board or mortar pan and stand	Two 4″ concrete blocks
Standard dirt shovel	Mortar hoe
Brick trowel (not to exceed 11″ in length)	Bucket of clean water for tempering
2″ × 4″ lumber, at least 5′ long	Mortar

SUGGESTIONS

- Have your instructor check that you are holding the trowel correctly.
- Spread the mortar a short distance in the beginning. As you become more skillful, you will be able to spread the mortar for a length of 3 or 4 bricks.
- Do not furrow too deeply with the point of the trowel. Do not grip the trowel too tightly.
- Practice spreading mortar until it can be done smoothly and with ease.

PROCEDURE

Note: Mixed mortar will be provided by the instructor for this project.

1. Place the mortar pan or board on a mortar stand. (The stand may be metal or may consist of several concrete blocks placed under the mortar pan or board.)
2. Place the two 4″ blocks in the work area and place the 2″ × 4″ piece of lumber on top of the blocks. The 2″ × 4″ should be at least 2′ away from the mortar pan to allow working room.
3. Wet the mortar pan or board with some water before placing mortar in it. This will prevent the mortar from drying out too quickly.
4. Fill the mortar pan with mortar provided by your instructor.
5. Cut the mortar from the pan with the trowel as described in the unit and spread the mortar on the 2″ × 4″.
6. Furrow the mortar with the point of the trowel. Cut off the excess mortar and return the excess to the pan.
7. Temper the mortar with water as needed to keep it in a workable condition. If the mortar is kept rounded in the pan at all times, it requires less tempering.
8. Clean all tools with water immediately after use.

PROJECT 2: LAYING 6 BRICKS ON THE 2″ × 4″

OBJECTIVE

- The student will be able to use the basic tools in laying bricks without a line as a guide, Figure 7-23.

EQUIPMENT, TOOLS, AND SUPPLIES

Bucket of clean water	One 8″ block (use as level holder)
Mortar (from instructor)	Brick trowel
6 bricks	Plumb rule (level)
Mortar pan or board	Mason's modular rule
Standard dirt shovel	Convex jointer or V-jointer
2″ × 4″ lumber, at least 6′ long	Brick hammer
Mortar hoe	Brush
Two 4″ blocks	

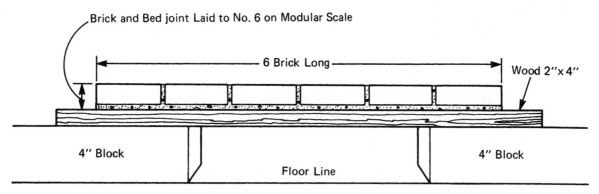

Fig. 7-23 Laying six bricks on a wooden $2'' \times 4''$.

SUGGESTIONS

- Stand close to the $2'' \times 4''$ and sight along the front edge of the bricks to determine if the wall is plumb.
- Press the bricks into position on the bed joint firmly. Do not use too much side motion. Do not tap the brick into position.
- When applying the head joint, do not smear the face of the brick.
- When laying the bricks, rest the palm of one hand half on the brick just laid and half on the brick being laid. This procedure helps the mason to develop the feel of level bricks.
- Keep the plumb rule in the center of the wall when leveling.
- In the plumbing operation, hold the plumb rule at the bottom of the brick and tap the brick true to the plumb rule.
- Never tap the plumb rule itself. When not in use, always place the plumb rule upright in one of the holes of a block.
- Never use a $4''$ block as a plumb rule holder. This block is not wide enough and the level will fall out easily. Any of the larger size blocks will work well.
- If a $2'' \times 4''$ is not available, a chalk line can be struck on the floor and the same procedure followed. A $2'' \times 4''$ is preferred because it is the same width as a standard brick wall.

PROCEDURE

1. Wet the mortar pan or board and fill it with mortar.
2. Set up the $2'' \times 4''$ as in Project 1.
3. Spread enough mortar on the $2'' \times 4''$ for 6 bricks.
4. Lay the 6 bricks on the mortar bed.
5. Level the bricks with the plumb rule.
6. Plumb the ends of the course and then straightedge the bricks with the level.
7. Check the height of the brickwork with the modular rule to No. 6.

8. When the mortar is thumbprint hard, strike the joints with the convex jointer.
9. Brush the brickwork, removing all excess dirt and mortar.
10. Clean all tools after use.

PROJECT 3: CUTTING BRICKS WITH THE HAMMER, BRICK SET CHISEL, AND TROWEL

OBJECTIVE

- The student will be able to cut bricks with the hammer, the brick set chisel, and the trowel. With practice, the quality of the cut will be such that the cut brick can be used in the finished wall.

EQUIPMENT, TOOLS, AND SUPPLIES

6 bricks	Trowel
Pencil	A piece of lumber at least 2′ long
Brick hammer	Safety glasses or goggles
Brick set chisel	

SUGGESTIONS

- Wear safety glasses or goggles to protect eyes when cutting.
- Keep fingers off the top edge of the brick where the cut is to be made to avoid injury.
- Have your instructor check the various cuts to see if they are of sufficient quality to use in the wall.

PROCEDURE

A. Cutting with the Hammer
 1. Mark the brick where it is to be cut.
 2. Score the brick on all four sides with the hammer.
 3. Strike the brick on the face side to complete the cut.
B. Cutting with the Brick Set Chisel
 1. Place the brick on the piece of lumber.
 2. Mark the desired cut with the pencil.
 3. Cut the brick on the pencil mark with the brick set chisel as explained in this unit.
C. Cutting with the Trowel
 1. Mark another brick with a pencil or scratch a mark on the brick with the trowel.
 2. Hold the brick down and away from the body. Strike the brick with a quick, sharp blow of the trowel. A second blow may be needed if the brick does not fracture the first time.

UNIT 8
Related Equipment

——————————————— OBJECTIVES ———————————————

After studying this unit, the student will be able to

■ describe important hand-powered equipment used in masonry work.

■ describe important power-driven equipment used in masonry work.

■ maintain power equipment and use it safely.

For masonry work to be done efficiently, both manual and power equipment are necessary. Most of the extremely heavy work in masonry is now done by using power-driven machinery. Work that does not require as much strength is done with manual tools.

MANUAL TOOLS

Brick Tongs

Brick tongs (or *brick carriers*), Figure 8-1, are used to manually move or place bricks near the mason. The ends of the tongs are equipped with a special flange so that the bricks are not damaged when they are picked up. They are adjustable to fit different-sized bricks. As a rule, 10 bricks may be picked up at one time. The tongs in the figure have a curved rubber handle which enables the mason to use the tongs with the least amount of discomfort. There are several types of brick tongs in use today. They are all built to withstand heavy use.

Wheelbarrows

There are two types of hand-pushed wheelbarrows used in masonry construction. One type, known as a *contractor's wheelbarrow,* is made of heavy-duty steel and is equipped with wooden handles and an air-filled rubber tire, Figure 8-2. Mortar or masonry materials can be transported directly to the work area with this wheelbarrow. The contractor's wheelbarrow is used for this reason instead of other types of wheelbarrows because it has a deeper bed.

Fig. 8-1 Moving units with the brick tongs. The weight of the tongs shown is about 4 pounds.

Fig. 8-2 Contractor's wheelbarrow.

The second type of wheelbarrow, equipped with a wooden body, is used for transporting bricks, blocks, or other materials to the work area, Figure 8-3. This wheelbarrow differs from the contractor's wheelbarrow in that it has no sides so that the material can be unloaded from any direction.

Mortar Box

Mortar is mixed in a *steel mortar box*. The box is leakproof and has smooth, welded edges for easy mixing, Figure 8-4. Mortar boxes are available in different lengths, widths, and depths. An average-sized box measures about $32'' \times 60''$ and weighs approximately 57 lb.

Mixing Tools

Several tools are used to mix mortar, Figure 8-5. Two different types of shovels are used in the mixing process. To proportion the ingredients in the mortar box, a standard dirt shovel is usually used. After the mortar has been mixed, it is transferred from the box to the wheelbarrow with a square-bladed shovel. The square blade allows the mason to scrape all corners of the mortar box, thereby picking up more mortar and cutting down handling time. Both shovels should be of a good-quality forged steel so that they can withstand hard use.

A large hoe is used to mix the mortar. The mortar hoe has two holes in the blade, centered on each side of the point at which the wooden handle is attached. The holes reduce the amount of energy exerted to pull and push the hoe through the mix and help break up ingredients in

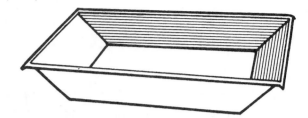

Fig. 8-4 Steel mortar mixing box.

the mix. After mixing, all mortar should be cleaned immediately from mixing tools with water.

The Mortar Pan, Stand, and Board

The *mortar pan* and *mortar board* are used to hold the mortar after it has been mixed.

The metal mortar pan, Figure 8-6, is approximately $30''$ square at the top, tapering to approximately $16'' \times 16''$ at the bottom. It is about $7''$ deep. Many masons and contractors prefer the mortar pan to the board for two reasons. It holds more mortar than the board and

Fig. 8-5 Mixing tools shown (left and right) are the standard dirt shovel, the mortar hoe, and the square-bladed shovel.

Fig. 8-3 Wooden body wheelbarrow.

Fig. 8-6 Mortar pan and stand.

does not deteriorate as quickly as the mortar board since it is made of steel or rigid plastic.

The mortar pan also fits on the mortar stand more securely. The mortar stand holds the pan at a convenient height, reducing excess strain and fatigue. The mortar stand consists of a folding tubular steel frame held together by a set of steel hinges, Figure 8-6.

Mortar boards, Figure 8-7, should measure about 30″ square. They can be made from any small lengths of lumber found around the job. The best selection of material for a mortar board, however, is a piece of ¾″ exterior plywood with 2″ × 4″ wooden runners on the bottom.

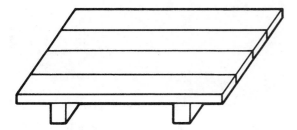

Fig. 8-7 Wood mortar board.

Mortar boards are limited in that they are not extremely strong and, since the mortar board has no sides, only a limited amount of mortar can be placed on the board. The dry wood also rapidly absorbs moisture from the mortar, making it necessary to temper the mortar with water more often than mortar which is mixed in a steel pan.

The Rope and Pulley

A heavy, reinforced steel pulley wheel and a strong rope, Figure 8-8, are used to hoist materials onto the scaffold. The pulley is hooked to a steel brace on the scaffolding. The mason pulls the rope over the wheel, raising the materials to the working level. The rope and pulley are usually used on small jobs or in construction where costs must be kept to a minimum.

Corner Pole

A *corner pole* (or *masonry guide*) is used when the mason wishes to lay bricks in a wall without building corners, Figure 8-9A. This means great savings and faster construction to the contractor. Most commercial corner poles are made of metal and have line blocks or

Fig. 8-8 Boom, rope, and pulley wheel.

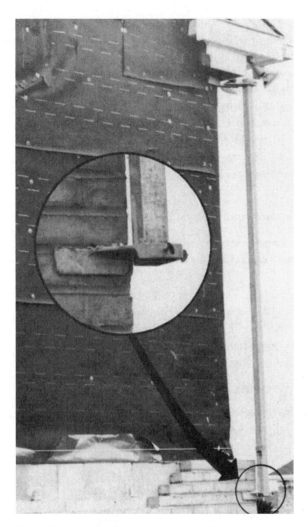

Fig. 8-9A Corner pole used in construction.

Fig. 8-9B Metal corner pole is attached and braced plumb to frame of house. Line blocks and line are attached to the pole at the scale marks.

Fig. 8-9C Corner poles set in place on a brick veneer house.

line carriers that can be attached to the pole. The corner pole has masonry units marked on it. In this way, the line can be moved to the desired position.

Corner poles may be erected differently, according to the type of pole used. All corner poles must be braced and plumbed, Figure 8-9B. They must also be located at a predetermined distance from the finished masonry work to enable the mason to work freely. The corner pole should be checked constantly to be certain that it remains plumb and tight in its bracing.

The mason can usually accomplish jobs much faster by using corner poles, Figure 8-9C. It has been estimated that on the average, masons lay at least 5 bricks in the time it takes them to lay, level, and plumb 1 brick on a corner. With a little experience, it should not take over 10 to 15 minutes to erect a corner pole. Except for occasionally checking a doorway or window jamb, the mason can lay bricks without using the plumb rule.

There are many places on a job, however, where the corner pole is not practical to use. It is not recommended for very small sections of wall even though it may be used to build the longer sections of that same wall.

Line Extenders

There are some very useful accessories available that can be attached to a corner pole in order to increase the efficiency of a mason when laying brick or concrete block. One of these that I really like is the metal *line extender*, Figure 8-10A. It attaches and clamps to the corner pole to allow a level line to be attached for backup work and face brick to be run alternately, Figure 8-10B. It is are an ideal way to establish the second masonry wall line when doing the interior backup of a brick and block composite wall. It can be used for all commercial solid masonry walls, for residential brick and backup, or for brick to grade foundations. It is sold in pairs and is available from the Bon Tool Company

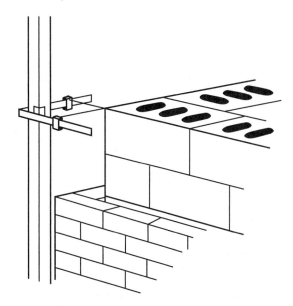

Fig. 8-10B Illustration of how line extenders are used as a guide for laying up a brick and block composite wall. (Courtesy Bon Tool Company)

through their catalog or by logging on their Website at www.bontool.com.

Chimney Brackets

There is also a handy metal *chimney bracket attachment*, Figure 8-11A, sold by Bon Tool Company that fastens to a metal corner pole, Figure 8-11B, which will hold plumb mason guide-lines to brick up a chimney of any width up to 25″ or 36″ deep. It eliminates plumbing and leveling. It will also improve work quality and speed, assuring you of square corners, plumb

Fig. 8-10A Line extenders. (Courtesy Bon Tool Company)

Fig. 8-11A Metal chimney bracket. (Courtesy Bon Tool Company)

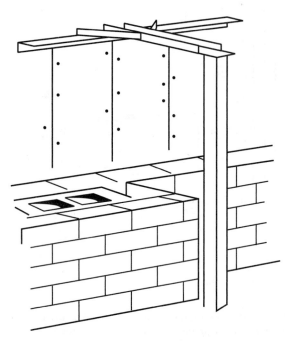

Fig. 8-11B Illustration of how chimney bracket attaches to a corner pole with plumb lines. (Courtesy Bon Tool Company)

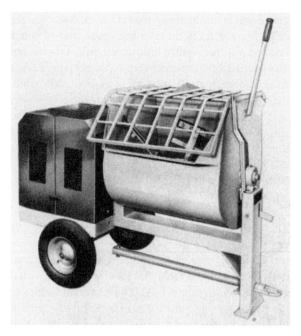

Fig. 8-12 Gasoline-powered mortar mixer.

walls, and uniform mortar joints. Chimney brackets are sold in pairs in the 25″ or 36″ size.

POWER-DRIVEN EQUIPMENT

Mortar Mixer

Mortar mixers may be powered by gasoline engines or electric motors, Figure 8-12. The gasoline-powered mixer is preferred since electricity is not always available on the job. The mortar is mixed by a turning shaft with attached blades which revolve through the mix. The mixer has a large dumping handle which allows the mason to empty the drum at the completion of the mixing process. Mixers are available in several different sizes. A typical mixer will mix about 4 cubic feet (cu ft) of mortar at one time.

The mixer should be washed immediately after use so that the mortar does not accumulate and ruin the mixing drum. Never beat the mixer drum with a hammer or other heavy equipment.

Masonry Saw

When a very precise cut is necessary or when material cannot be cut with a hand tool, the *masonry saw* is used (see Figures 8-13 and 8-14). Many architects specify that all masonry material on a job be cut with the masonry saw. This is because the saw does not weaken or fracture the material, with a straighter cut being the result.

The saw can be operated dry or with water. If it is used dry, a blade made of silicon carbide is usually used. If operated with water, an industrial diamond-chip blade is used. Diamond blades should never be used for dry cutting or they will burn and be ruined (see Unit 9, *Cutting with the Masonry Saw*).

> **Caution:** Safety goggles are required when cutting with a wet or dry saw, since serious eye damage could result if a chip should fly into the operator's eye. The operator should also have adequate ventilation and a dust mask or respirator if the material is cut dry.

Fig. 8-13 Cutting concrete with the masonry saw.

Hydraulic Masonry Splitter

Hydraulic splitters are available in a variety of sizes and models. Hydraulic force exerted by a pump presses the masonry unit against a cutting blade. It is often practical to use one of these cutters instead of a masonry saw.

Figure 8-15 shows a portable splitter used for stone, splitblock, face brick, and firebrick. It is especially popular with contractors because it is both portable and simple to operate.

Forklift Tractor

The *forklift tractor,* Figure 8-16, enables the mason to move and position materials mechanically. Mortar can be hoisted in a special steel cart or mortar box and placed automatically in the immediate work area.

High forklift tractors (see Figure 8-17) have revolutionized lifting and placing masonry materials on scaffolds and building floors, and are a big labor-saver. Most have the advantages of tilting forks, self-leveling, full hydraulic controls, high flotation tires, and the ability to operate on almost any job surface. They are especially adaptable in mud, which is common on most

Fig. 8-14 Portable masonry saw. (Courtesy Felker Company)

Fig. 8-15 Portable hydraulic masonry splitter.

Fig. 8-16 Forklift hoisting mortar box onto scaffold.

jobs. They are to be used strictly for transporting and lifting materials. Strict insurance and safety regulations forbid anyone but the operator to be on them.

Elevator

Elevators are used to carry the material to the correct level so that it can be delivered to the mason, Figure 8-18. To protect the worker from injury, safety rails and a heavy steel mesh screen are required around the elevator platform. The elevator is operated by the person who is loading or unloading from either the top or ground level by the use of weights. This particular type of elevator is preferred by many contractors since it does not require a full-time operator. These elevators are meant to carry materials only. Construction workers should never attempt to ride in them.

Fig. 8-17 High forklift tractors can place materials on high scaffolding.

Fig. 8-18 Material elevators are used to deliver supplies to the mason.

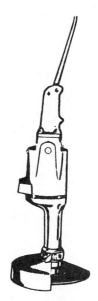

Fig. 8-19 Tuckpointer's grinder.

Electric Tuckpointer's Grinder

Tuckpointing is the replacing of old mortar joints with new mortar to repair the joint. In this way, the mason restores old masonry work. The term *tuck* is used since the mortar is tucked into the joints in the process with a steel tool such as a slicker. The *tuckpointer's grinder* is a hand-operated machine designed specifically for grinding out joints in the mortar bed and head joints of a masonry wall, Figure 8-19. It is equipped with shatterproof blades and a safety guard on that part of the machine which faces the mason.

Caution: Always wear safety goggles while operating the grinder to protect your eyes from flying debris.

Laser Transit

There has been a big improvement in transit levels in recent years with the development of the Laser Transit, which establishes true level and vertical lines or points in construction work. It is especially useful in laying out and establishing benchmarks on masonry jobs, Figure 8-20. A particular model that I like and had the opportunity to field test on a job is the *Self Leveling, Rotating Laser System*, Figure 8-21, manufactured and marketed by a German company, Stabila Levels, 332 Industrial Drive, South Elgin, Illinois 60177; call toll-free 1-800-869-7460.

One of the major advantages of this transit is that, once it is set up and leveled, one person can use it with the aid of a hand-held remote control to shoot level and horizontal points on the job site. It is very inexpensive to operate, as it is powered by regular alkaline batteries described in the manual. With the labor savings of a one-person operation, it pays for itself after a period of use. It is particularly useful for remodeling jobs, as it will even shoot level points through the glass windows of a building when checking existing floor levels for house additions and it is good for interior and exterior use.

Fig. 8-20 Workman adjusting the measurement rod until the desired height matches the electronic center point on the Roto Laser Receiver target. Once the two are the same, a level point can be marked on the existing wall or on a benchmark rod to serve as a reference point.

Self Leveling, Rotating Laser System with Remote Control #05100

- One man operation. Push a button and precise horizontal and vertical planes are established.
- Sensor blinks and rotating stops when self leveling range has been exceeded.
- Use the integrated adjustment knob to switch from the horizontal to the vertical plane.
- Move the laser beam with the remote control.
- Select from a Dot, 3 rotating speeds or 3 scanning line lengths.
- Receiver locates the level line outdoors or at long range 235'.
- Kit includes: Remote control, receiver, receiver holder, wall bracket, laser glasses, 5/8" x 11" mounting stud and case.

Fig. 8-21 Self Leveling Rotating Laser with important features listed. (Courtesy Stabila Level Co.)

It works on the principle of a self leveling rotating laser beam that is projected to the Roto Laser Receiver Target attached to a measuring rod. The target rod is either raised or lowered until it matches the target level point, which in turn indicates the level point or height desired. The correct height or mark can then be transferred onto the existing building or a rod, providing a level bench mark to work by.

ACHIEVEMENT REVIEW

Select the best answer from the choices offered to complete the statement or answer the question. List your answer by letter identification.

1. Brick tongs are used for
   a. cutting bricks.
   b. tooling mortar joints.
   c. carrying bricks.
   d. measuring materials.

2. The contractor's wheelbarrow is usually used instead of other types of wheelbarrows to transport mortar because
   a. it has a deeper bed.
   b. it is rustproof.
   c. it is less expensive.

3. The mortar box has smooth, welded joints and edges because
   a. they prevent the box from leaking.
   b. extra strength is provided.
   c. the mixing tools do not catch on the edge.

4. Most commercial corner poles are made of
   a. plastic.
   b. metal.
   c. wood.

5. The mortar hoe has two holes in the blade. Which of the following is not a valid reason for the holes being in the hoe?
   a. The holes greatly reduce the effort necessary to push and pull the hoe through the mix.
   b. The holes force air into the mix.
   c. The holes break up lumps in the material

6. The mortar stand is used for
   a. supporting the mortar pan at the correct working height.
   b. transporting the mortar to the working area.
   c. mixing the mortar.
   d. storing the dry mortar before mixing.

7. The equipment used to raise material onto the scaffold is the
   a. wheelbarrow.
   b. brick tongs.
   c. corner pole.
   d. rope and pulley.

8. When masonry material is cut with a wet masonry saw, the best blade to use is the
   a. diamond blade.
   b. silicon carbide blade.
   c. steel blade.
   d. fiberglass blade.

9. Measuring ingredients for mortar should be done with a
   a. square shovel with a V-shaped handle.
   b. regular, standard dirt shovel.
   c. 5-gal bucket.

MATH CHECKPOINT

1. A contractor buys a new mortar mixer that costs $990.00, a grinder for $225.00, a masonry saw for $1,200.00, and six mortar pans at $25.00 each. What is the total cost of the bill?

2. If the dealer decides to give a 12% discount from the total bill, what would the revised bill be?

3. A contractor needs the use of a materials elevator. He decides to rent one rather than buy it. The cost of renting one would be $150.00 per day. What would the cost be if the contractor rented it for ten days?

4. In addition to the rental fee in problem 3, a security deposit of $200.00 was required to assure that the elevator would be returned in good condition. When the elevator was returned it was found that a tire costing $45.00; bolts in the rear assembly, $16.00; and a gas cap costing $3.50 were missing and had to be deducted from the deposit. What was the total amount of money the contractor received back after the deductions were made?

UNIT 9
Cutting with the Masonry Saw

―――――――――――――― OBJECTIVES ――――――――――――――

After studying this unit, the student will be able to

- explain the two main methods of cutting: wet and dry.
- list the important safety practices to be followed when operating the masonry saw.
- make some of the more common cuts described in the unit.

The masonry saw is introduced at this point in the text because it will be needed for cutting masonry units around electrical and mechanical fixtures and when building fireplaces, arches, and complicated brick panel walls. Operating the masonry saw safely and correctly is one of the skills that all masons have to perform as part of their trade.

When laying masonry work around electrical and mechanical fixtures, usually the masonry unit must be cut rather thin to enclose it in the wall. This cannot always be done with a hammer or chisel because the shocking power may break the unit. To prevent breaking the masonry unit, architects on many jobs require that all masonry cuts be made with a saw.

Although it is true that on some jobs laborers operate the saw, the mason who is laying the unit in the wall should operate it. The mason has a better understanding of the cutting needs and therefore can be more exact in making the cuts.

Cutting with the masonry saw is faster, neater, and less wasteful than cutting with a chisel or hammer. On a well-run job, cuts are made ahead of the masons and stocked in the work area. This greatly speeds up the work because the masons do not have to wait for wet blocks or bricks to dry, or make cuts with the hammer or chisel. It also decreases the possibility of the bond shifting back and forth on the wall since all of the starting pieces are the same size. The mason must adjust the mortar joints so they work bond in the wall. Cutting

with the masonry saw, therefore, is economical and is an essential part of the masonry business.

DESCRIPTION OF THE MASONRY SAW

All brands of masonry saws have the same basic construction. If they are different, it is mainly in the quality of materials and devices built into the saws to improve performance. The parts of a masonry saw are shown in Figure 9-1.

A full-size stationary or portable saw can be purchased. The full-size saw can be made more movable by buying a wheel arrangement that can be lowered if necessary so the saw can be rolled from place to place. This requires the services of a truck and several people to move the saw from one job location to another. The portable saw, however, can be loaded into the trunk of a car and transported more easily.

Select the type of saw that fits the needs of the job. If much cutting is required, the full-size saw should be used. In recent years, one of the major improvements incorporated in new models is the electric pump. The pump forces water up from the pan onto the blade for use in wet cutting with a diamond blade. The adjustable cutting head operates by turning a crank to either raise or lower the blade.

Masonry saws can be run on either 115/120 volts or 220/440 volts. They will operate more efficiently on the higher voltage. Running on higher voltage, they have

97

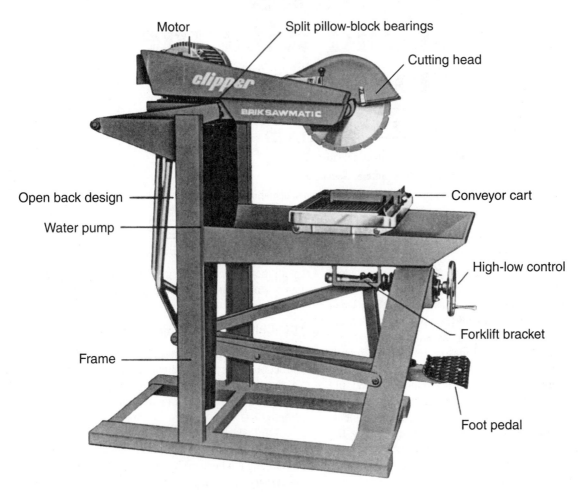

Fig. 9-1 Masonry saw.

less chance to bog down or overheat, and will probably last longer.

The motor is protected with a thermal overload switch for use if the machine does overheat. Once the motor cools, the reset button can be pushed, and the motor will run again, unless it has been damaged.

METHODS OF CUTTING WITH MASONRY SAWS

There are two ways of cutting with masonry saws: with water to cool the blade (*wet cutting,*) or with a dry shatter-resistant blade (*dry cutting*).

Wet Cutting

A wet saw blade has industrial diamond chips fused to the blade. As the blade cuts through the masonry unit, water is continuously pumped against the cutting edges and sides. This prevents the diamonds from overheating and burning up. If a diamond blade is run without water to cool it, the blade will be ruined in a very short time.

Caution: Eye and ear protection must be worn at all times when operating the masonry saw.

At the close of the workday, the saw operator should remove the blade from the saw and lock it in the toolshed or trailer. Be very careful not to bump or ram the cutting edge against a hard object since it can be damaged, making the blade useless for cutting.

With proper care, the diamond blade withstands long, hard wear and lasts a long time. Diamond blades are, however, expensive. They can be purchased in different diameters depending on the depth of cut desired. The average all-around size used is the 14-inch blade.

Diamond blades can also be purchased to suit the hardness of the materials being cut. When ordering blades be sure to specify the type of materials to be cut and the diameter of the blade. The edge of the diamond blade is made in segments (see Figure 9-2A) to help it run cooler and prolong its life.

Rapid advances have been made recently in diamond blade technology. A diamond blade properly used on well designed and maintained saws can provide the lowest cost per cut of all the methods now used for cutting masonry materials. However, if the same blade is put on a poorly maintained piece of equipment, and used by a poorly trained operator, this results in a much greater cost of cutting.

Dry Cutting

Cutting without the aid of water requires an abrasive type of blade (Figure 9-2B). Shatterproof abrasive ma-terials such as silicon carbide are bonded with a reinforcing mesh built into the blade to prevent it from breaking. Resins are added to help bond the blade and cutting abrasives together.

Dry cutting blades are also available to cut different hardnesses of masonry materials. If the blades are given proper care, they should not break.

> **Caution:** Dry cutting blades that are not reinforced should not be used; they can shatter, injuring the operator. Read the instructions carefully when ordering saw blades.

ADVANTAGES AND DISADVANTAGES OF CUTTING WITH WET BLADES

Wet cutting allows thinner, more delicate cuts with the saw because the diamond blade is sharper than the abrasive blade and less heat builds up. Wet cutting is also much faster than cutting with a dry blade. The diamond blade, although it does cost more, lasts longer than any other type of masonry saw blade with proper care and cutting practices. Therefore, the diamond blade is actually more economical to use.

The major disadvantage of cutting with the wet saw is that the masonry unit becomes soaked with water and is difficult to lay in the mortar bed. The wet cut does not set up as does a dry cut. The proper practice is to

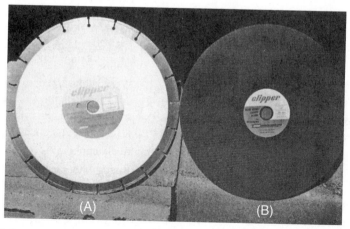

Fig. 9-2 (A) Diamond blade for wet cutting, (B) abrasive blade for dry cutting.

cut the units far enough in advance that they can dry out before being laid. This of course requires some planning ahead.

There are sometimes problems with the wet saw in cold weather. If it remains outside, the saw can freeze up. It is also uncomfortable for the person operating the saw and this can cause reduced work output. The water can be heated or the saw set up in a heated enclosure.

ADVANTAGES AND DISADVANTAGES OF CUTTING WITH A DRY BLADE

There are two important advantages of the dry saw. The masonry units set up more quickly in the mortar bed joint, and there is no weeping or smearing of the joints.

The disadvantage of dry cutting is that the units cannot be cut as thin as with the wet blade. Also, dry cutting is a much slower process. The heat generated from the dry blade can cause the masonry unit to blow out on the side or crack if the cut is too thin.

A major problem for the mason is the large amount of dust caused by the sawing process. A dust mask or respirator must always be worn when operating a dry saw. A strong exhaust fan can be placed in front of the saw to draw the dust away from the operator. The dust should be blown into a duct or tube and vented to the outside. This is for the safety of all workers inside the building.

Dry cut diamond blades are now available that perform well.

SAFETY PRACTICES TO FOLLOW WHEN OPERATING THE MASONRY SAW

- Clean safety eye protection often (see Figure 9-3).
- No horseplay.
- Wear ear protection and an approved hard hat to protect the head from any flying particles.
- Have the electrician ground the saw frame to a steel rod in the floor or ground to prevent electrical shock.
- Stand on a wooden platform or board to prevent grounding of the operator.
- Never stand in water when operating the saw.

Fig. 9-3 Safety eye protection should be cleaned whenever needed.

- Make sure the blade is on the saw correctly and tightened before starting the saw.
- Do not wear loose clothing.
- Do not wear rings that project out from the fingers.
- Wear shatterproof eye protection.
- Never ground or touch a metal object to another nearby object when running the saw because an electrical shock can result.
- Do not operate the saw if wires are frayed (worn) or bare.
- Never start the saw unless the masonry unit to be cut is clear of the blade.
- Do not cut close to the fingers. Back up the cut with a waste piece of masonry unit, Figures 9-4 and 9-5.
- Do not use the saw blade to dress chisel edges or points.
- Wear rubber outer wear to prevent getting wet and to reduce the chance of getting an electrical shock.
- Never force the saw blade to cut faster than it will freely cut.
- Keep blade clean to prevent material being cut from rocking back and forth.
- Make sure the unit to be cut is laid square and flat, free of chips, on the saw table so it cannot bind or kick back at the operator.

Fig. 9-4 Pushing the rolling conveyor table with brick onto saw blade.

- Use the miter gauge on the saw table when cutting angles, Figure 9-6.
- When operating a dry saw, face the saw so the wind carries the dust away from you.
- Secure an exhaust fan and place it in front of the saw to help carry away the dust.
- Wear a dust mask or respirator when running the dry saw.

Fig. 9-5 Holding a thin cut with waste pieces to prevent getting the hand too close to the blade.

Fig. 9-6 Using the miter gauge to hold brick securely when cutting angle pieces.

- Never mark a unit on the saw table while the blade is running.
- The conveyor cart that the masonry unit lies on should be rolled back and forth as the blade is applied to the cut. This lessens the possibility of binding the blade in the cut.
- If the saw blade binds in a cut, do not attempt to hold or grab it. Let it go.
- Do not cut a cracked masonry unit.
- Do not operate the saw when you are ill.

> **Caution:** Accidents happen to experienced operators, too, if they become careless after doing the cutting for a long period of time. Observe all safety rules and regulations and report any accidents immediately to the instructor, foreman, or employer.

MAKING CUTS WITH THE SAW

Before actually cutting a masonry unit, review the previous safety rules.

Although there are many cuts that can be made with the saw, there are certain basic cuts that are required on almost all masonry jobs. Masonry cuts are usually the same whether they are block, brick, tile, or other materials. The following cuts are made using concrete blocks.

One Cell Out.   Sometimes there is a pipe or wire in the wall that the block cannot be slipped over by using the hollow cell. If this is the case, a web has to be cut out with the saw.

If only the end web of the block is to be cut out, it is called one cell out. It is confusing to use the term *cell* when ordering cuts from the saw operator. Therefore, to save time, some masons have devised a set of terms to describe the number of cells to be cut out of a block.

If the end of one cell is to be cut out, it is called a quarterback. A quarterback cut is shown in Figure 9-7.

Two Cells Out.   When it is necessary to cut out the cross webs of two cells, it is called a halfback, Figure 9-8.

Fig. 9-8 Halfback cut block. Notice that two cells are cut out.

Three Cells Out.   If all of the cross webs must be cut leaving only the end of the block holding it together, it is called a fullback, Figure 9-9.

A comparison of the three cuts is shown in Figure 9-10. If a concrete block is being used that has only

Fig. 9-7 Quarterback cut block. Notice that the end cell is cut out of the block.

Fig. 9-9 Fullback cut block. Notice that all of the webs have been cut but the last one.

Fig. 9-10 Comparison of all three cut blocks described. Six-inch blocks are shown.

two cells, the cuts are called halfbacks and fullbacks; there is no quarterback due to the cell arrangement.

CUTTING AROUND ELECTRICAL BOXES

Electrical boxes installed flush with the face of the masonry wall must be cut back neatly into the block. Only the wall plates that the electrician fastens on later will hide the edges.

First, mark the area to be cut out with a crayon or soapstone. Then, using the saw, cut out as close as possible to the marked area. As a last step, chip out the remaining particles with a small tile hammer to allow the block to fit over the electrical box neatly. Figure 9-11 shows a standard receptacle box installed in a block lead.

SLABS

Standard size concrete block slabs can be ordered from the concrete block company. These slabs measure the same height and length of a standard block and are $1\frac{5}{8}$ inches thick. However, often when building around fixtures on the job, a slab thinner than normal is needed. A cut concrete block slab is shown in Figure 9-12. If cut wet, be sure to allow the slab time to dry before laying if possible. If time is not allowed, it will be difficult to hold the block in place.

RIPS

Concrete blocks that are cut across the face side lengthwise (horizontally) are called *rips*. They can be of

Fig. 9-11 Electric receptacle box cut into a concrete block corner. Notice the entire box is enclosed in the wall.

Fig. 9-12 Cut concrete block slab.

different sizes depending on the need. If the block is cut exactly in half lengthwise, it is sometimes called a *half-high* rip. Rip blocks are often used as the starting course off the floor. They are also used when there is a structural tile base, and a filler block piece is needed to

reach a height of eight inches for normal coursing. Other places where rip blocks are used are against the bottom of steel and concrete beams when sealing off a masonry wall, and under and over window frames.

In Figure 9-13 the mason is shown cutting a rip block that would be very difficult to cut with a mason's hammer. Note the proper dress and eye protection worn by the operator.

BOND BEAM

Sometimes the specifications require a course of concrete blocks to be laid across an opening or around the top of the wall to be filled with concrete and reinforced with steel rods. This is called a *bond beam,* Figure 9-14. To accomplish this, the inside webs of the blocks are cut out approximately ¾ of the way down. The cut block is then laid across a wooden form, if it is to serve as a lintel for an opening. Concrete and steel rods are installed in the center of the blocks. The bond beam can also be used in a structure to give added strength when necessary for special conditions. After the concrete or grout has cured, the form can be removed.

**Fig. 9-13 Mason cutting a rip block on the saw (wet saw with a diamond blade).
Notice the proper clothing and eye protection.**

Fig. 9-14 Bond-beam block cut out of a full block. Notice that the lower part of the block is left in for the placement of the steel rods and concrete.

CUTTING ANGLES

It is possible to cut masonry units on angles on the masonry saw. This is done by either turning the block to match the line of the blade on the table, or by using the miter attachment which is graduated in degrees. Two angle cuts are shown in Figure 9-15.

The cuts discussed here are only a few of the many that are needed on a masonry job. Although concrete blocks are shown in the illustrations, any masonry unit can be cut on the saw, if the proper blade is used.

> **Caution:** Good safety practices should be followed at all times regardless of the type of materials and the nature of the cut.

CARE OF THE SAW

A saw operates more efficiently and for a much longer period of time if given good care and maintenance.

Fig. 9-15 Concrete blocks cut on an angle or miter.

This list of good work practices should be closely followed at all times:

- Use the recommended blade for the materials being cut.
- Never operate the diamond blade dry, even for a short period of time. This causes permanent damage to the blade.
- Make sure that the electric pump is operating and a good flow of water is on the blade at all times.
- Shut saw off if it makes any unusual noise.
- Keep the water tray full of water.
- Change the water as often as needed to keep it clean and free from sludge buildup.
- Check the tension and wear of the belts and adjust or replace as needed.
- When there is an electrical problem, have an electrician fix it.
- Any operating problems that affect the use of the saw should be reported to the foreman.
- At the end of the workday, drain and flush the saw and pump with clean water.
- Roll up the saw cords, and lock the blade and cord in the toolshed at quitting time.

USING THE SAW TO THE BEST ADVANTAGE

The saw can be used to its best advantage by following these recommendations:

- Set up the saw as close to the work area as possible, but not so close that it interferes with the work of the mason. This eliminates unnecessary waiting for cuts.
- On a large job, have one person place the orders in advance for cuts with the saw operator. Having the same person placing the orders will make the work more accurate.
- When using the wet saw, let the cuts dry out as much as possible before laying them in the wall. This is done by laying out the walls in advance and determining all of the cuts needed. The saw

operator can then make the cuts and stack them near the wall to dry.

- If cutting with the wet saw, mark all cuts for electrical boxes or fixtures with a crayon or waterproof marker.
- Replace a dull saw blade with a sharp blade.
- Follow good safety practices and report any accidents to the foreman immediately.
- If the saw stops running while cutting from overheating, it may have thrown the overload switch in the motor. The recommended safety practice to follow is to wait five minutes for the motor to cool before pushing the reset button on the back of the motor, Figure 9-16. If the saw does not restart, never immediately reset it a second time. The saw may still be overheated or shorted out. Starting the saw quickly again can cause an accident or damage the motor. An electrician should be asked to check the saw if any problems persist.

Fig. 9-16 Safety reset button on rear of saw.

ACHIEVEMENT REVIEW

A. Select the best answer from the choices offered to complete each statement. List your choice by letter identification.

1. When wet cutting, the best type of blade to use is
   a. silicon carbide.
   b. fiberglass.
   c. diamond.
   d. carbon.

2. The most popular all-around size blade to use on a masonry saw is
   a. 10″
   b. 12″
   c. 14″
   d. 20″

3. Diamond blades that are used on the saw must be lubricated with
   a. oil.
   b. graphite.
   c. alcohol.
   d. water.

4. The main reason masons do not like a dry blade is that it
   a. cuts slower.
   b. has dust problems.
   c. will break.
   d. is high in cost.

5. The main disadvantage of a wet saw is that
   a. the blade cuts slower.
   b. the blade is high in cost.
   c. the unit being cut becomes wet.
   d. there is a great danger when using a diamond blade.

6. To prevent the operator from getting an electrical shock, the saw should
   a. be grounded.
   b. never be filled with water.
   c. be placed on a wood platform.
   d. only be operated on 115/230 volts.

7. A block that has the end of one cell cut out is called a
   a. halfback.
   b. quarterback.
   c. fullback.
   d. bond beam.

8. Concrete blocks that are cut lengthwise and used when a full-size block is not needed are called
   a. angle cuts.
   b. bond beams.
   c. lintel blocks.
   d. rips.

9. When cutting a concrete block, if the block is pinched or bound in the saw blade, the best practice to follow is to
   a. call the foreman.
   b. turn the saw off.
   c. release your hand from the block and let it go.
   d. try to free the block from the saw blade as quickly as possible.

10. A block that has the inside web cut out to receive concrete and steel rods is called a
    a. rip.
    b. lintel.
    c. half high.
    d. bond beam.

B. Identify the various parts of the masonry saw shown in Figure 9-17.

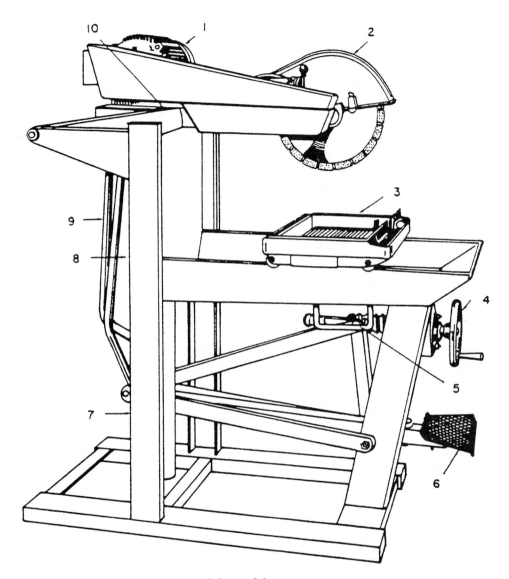

Fig. 9-17 Parts of the masonry saw.

PROJECT 4: CUTTING CONCRETE BLOCK WITH THE MASONRY SAW

OBJECTIVE

- The student will perform some of the more common cuts of concrete blocks required on the job using the masonry saw. The safety practices outlined in the unit must be followed. A wet saw equipped with a diamond blade should be used. If a wet saw is not available, a dry saw can be used.

EQUIPMENT, TOOLS, AND SUPPLIES

1 masonry saw equipped with a diamond or abrasive blade

Rubberized clothing and gloves

1 brick hammer

Safety goggles or glasses

Standard electrical receptacle box

Brush, to keep the saw table clean

Mason's rule

Yellow crayon or soapstone, to mark the masonry units

Small square

Supply of water close to the saw

Supply of 6″ or 8″ concrete blocks (to cut) and a few bricks

SUGGESTIONS

- In addition to having an old brush and a supply of water near the saw, also have handy a garden hose with a nozzle to flush the saw tray.
- Long-armed gloves are better than short gloves.
- Make sure that the saw frame is grounded to prevent electrical shock.
- Secure a wooden pallet or boards to stand on.
- Locate the saw close to the stockpile of materials.
- Keep a clean cloth handy to wipe the safety goggles off as needed.
- Do not force the saw when it is cutting; let it cut freely.
- Drain the saw tray when it is very dirty and when sludge has built up. Fill the tray with clean water.

PROCEDURE

1. Fill the saw tray with water.
2. Install a blade on the saw, making sure that it is tightened securely.
3. Check the conveyor cart to make sure it rolls smoothly back and forth and is free of chips.
4. Set a 6-inch block on the saw table.
5. Mark the webs with the soapstone or crayon to make a quarterback.
6. While rolling the saw table (tray) back and forth and pushing down with a foot on the foot pedal, cut the cell out.
7. Chip out the corners of the cut with the mason's hammer if the cut must be neat.
8. Next, cut a halfback with the masonry saw in the same manner.
9. Then cut a fullback with the saw.
10. To become more familiar with the masonry saw, cut a rip block, slab, bond beam block, and a few bricks.

11. Holding the electrical box even with the bottom and in the center of the block, mark the outline of the box on the block with the soapstone or crayon.

12. Cut out the box opening with the saw by making a series of vertical cuts where the box is marked on the block.

13. With the hammer, gently chip out the block so the box fits in the cut section and is flush with the face of the block.

14. Wear safety glasses at all times when cutting.

15. Have the instructor evaluate the work at the completion of the cuts.

16. Wash out and drain the saw when finished.

SUMMARY, SECTION 2

- Many hours of practice are required to perfect trowel skills. This is important as a mason's productivity is directly linked to good trowel skills.

- The beginning masonry apprentice should become familiar with the basic set of tools and acquire brand names.

- The trainee should be able to use either of the two methods described in Unit 7 on picking up mortar from the board—clock and cupping.

- It is essential to learn to use both hands at once when picking up and laying brick. This increases the work output of the mason, can be accomplished by continuous practice, and is necessary to earn a profit on the job.

- Always observe good safety practices when using tools. Masons have a responsibility to protect their fellow workers by not being careless with tools or taking dangerous shortcuts.

- There are various types of manual and power-driven machinery used in masonry work. The mason must become familiar with each tool and know related safety practices. Power-driven machinery has been developed to help masons in the performance of their work, not replace them.

- The use of masonry saws allows the mason to accurately cut materials that otherwise may not be able to be cut with hammers or chisels. It also speeds up the work, as cuts for a wall can be made ahead, increasing productivity.

- The operator of a masonry saw has to be especially careful because of the use of water for wet sawing and the possibility of electrical shock. The saw must be grounded, and the operator should be standing on a dry surface if sawing wet. A piece of wood works well for this purpose. The rules for operating a masonry saw presented in the unit should be studied before actually performing any sawing.

SUMMARY ACHIEVEMENT REVIEW, SECTION 2

Complete each of the following statements referring to material found in Section 2.

1. The tool used by the bricklayer more than any other is the _____.

2. Attempting to spread too much mortar at one time is a common mistake of the trainee. The result of this practice is _____.

3. Furrowing a mortar bed joint too deeply and causing bare spots in the bed joints is not acceptable because it may cause _____.

4. Stacking dry brick on the finished back section of a wall before laying it in the front section to form the wall is called _____.

5. When the brick is laid to the line, the best practice is to curl the thumb over the front top edge of the brick. This is done to prevent _____.

6. When tapping down brick with the trowel blade in the leveling process, the trainee should be especially careful not to chip the face of the brick or splatter mortar on the front face of the wall. The best position to hold the trowel to accomplish this is _____.

7. After the course of brick has been plumbed, the plumb rule level is used as a straight-edge to line up the brick from one end to the other of the course. The most effective location to place the level in doing this is _____.

8. Gasoline engine mortar mixers are more popular on a job than electric ones because _____.

9. When using a masonry saw, masons should always protect their eyes by wearing some type of _____.

10. When a course of concrete blocks to be laid across an opening is filled with concrete and steel rods, it known as a _____.

11. A concrete block that has one end cell cut out so that it will fit around a pipe in the wall is called a _____.

12. Concrete block can be cut to various angles by using the _____ attachment on the table or saw.

SECTION THREE
MORTAR

UNIT 10
The Development of Mortar

—————————————— OBJECTIVES ——————————————

After studying this unit, the student will be able to

■ explain the development of mortar from early times to the present.

■ list the ingredients of mortar.

■ describe the relation of lime to mortar.

A masonry wall is no better than the mortar that bonds its units together. Mortar must be capable of keeping the wall intact and it must create a weather-resistant barrier. Units 10 and 11 deal with the development of mortar and with different types of mortar. Masons should know how to recognize quality mortar and how to mix it.

Depending on the ingredients used, mortars have different compressive strengths, flexibility, and shrinkage rates. Therefore, masons should try to match the correct mortar to the needs of a particular job. The proper balance between the mortar and the structural requirements of a job is often the difference between a profitable job and an unsuccessful one.

HISTORY OF MORTAR

Clay was the first material to be used for mortar. It has been used through history in masonry walls of unburned brick, but the lack of a hard binding agent makes clay impractical in humid climates. In 2690 B.C. the Great

Pyramid of Giza was built in Egypt. The huge blocks of this structure were cemented together with mortar made from burned gypsum and sand. Many years later, the Greeks and Romans developed mortar from volcanic waste and sand. Structures built with this mortar still stand.

Mortar made from lime-sand was commonly used until the late nineteenth century. However, in 1824, portland cement was developed, an occurrence that marked the beginning of modern-day cement.

Portland cement was a much stronger material than had been used before, whether it was applied alone or combined with lime. Mortar used today is a combination of portland cement, hydrated lime, and masonry sand. More recently, masonry cements have been developed which require that only sand and water be added for the formation of mortar.

The formula or percentage of ingredients are usually not printed on bags of mortar. However, the contents of the bag of mortar must pass specifications for a mortar of medium strength. This mortar may be sold under various brand names, but portland cement is always specified by type.

Besides binding the masonry materials into a permanent structure, mortar seals the joints against penetration of air and moisture. Mortar acts as a bond for various parts of the structure such as reinforcement rods, anchor bolts, and metal ties so they may become an integral part of the wall. There are cases of leaks and cracks in masonry walls which can be attributed directly to the use of defective mortar.

INGREDIENTS OF MORTAR

To better understand how mortar reacts to different temperatures, stress, and prolonged use, the main ingredients of mortar must be examined. These main ingredients are portland cement, hydrated lime, sand, and water.

Portland Cement

Portland cement was discovered in 1824 by Joseph Aspdin, an English stonemason. He was attempting to produce mortar which would harden when water was added. The name *portland* was given to the new cement because Aspdin thought it resembled natural stone which was quarried on the Isle of Portland.

By the middle of the nineteenth century, portland cement was in wide use throughout England. It was manufactured in the United States from 1871 on and gradually replaced natural and lime-sand mortars. It became a major ingredient in mortar after 1880.

There are many lengthy procedures in the production of portland cement. The following are the most important operations:

- The quarrying of raw materials
- Raw grinding and mixing
- Initial blending of the raw materials
- Fine grinding in preparation for the kiln
- Burning in the kiln for the formation of clinkers
- Grinding of the clinkers
- Bagging and shipping

The end product is a fine, grayish powder made up of certain chemical compounds. The most common compounds include carefully blended limestone, clay, or shale. Although about 99% of all cement used in construction contains portland cement, it is always mixed with other materials. It is a common mistake for people to speak of a *cement sidewalk* or *cement highway*. There are always other ingredients present in the mixture.

Portland cement combined with water forms a paste. As the paste hardens, it binds the various materials in the mixture. The paste hardens more and more as it ages and eventually can become as hard as the rock or other aggregates mixed with it. This is one of the outstanding features of portland cement. However, portland cement must be kept dry in storage. If it becomes wet, lumps form and the cement cannot be used.

Some designers falsely assume that what is good practice for concrete is also acceptable for mortar. Actually, mortar differs from concrete in consistency, methods of application, and use. Mortar binds masonry units into a single mass; concrete is usually a structural element in itself. Most of the important physical properties of portland cement concrete concern its compressive strength. For these reasons, the requirements of masonry mortar are vastly different from those of portland cement concrete.

Note: All types of mortar have a percentage of portland cement in the mix.

Types of Portland Cement

The specifications of the ASTM list five types of portland cement which are used in masonry and the physical properties and chemical composition of each type. Any one of these types can be used for masonry mortar, but Types I and II are generally used in mortar. The other types are usually used for mixing concrete.

Type I. This is a general-purpose cement and is the one masons most often use. It may be used in pavements, sidewalks, reinforced concrete bridge culverts, and masonry mortar.

Type II (Modified Portland Cement). This cement hydrates at a lower heat than Type I and generates heat at a lower rate. It does, however, have better resistance to sulfate than Type I. It is usually specified for use in such places as large piers, heavy abutments, and heavy retaining walls.

Type III (High-early-strength Portland Cement). Although this cement requires as long to set as Type I, it achieves its full strength much sooner. Generally, when high strength is required in 1 to 3 days, this cement is recommended. In cold weather when protection from freezing weather is important, Type III often is specified.

Type IV (Low-heat Portland Cement). This is a special cement for use where the amount and rate of heat generated must be kept to a minimum. It is critical to hold the temperature down to ensure that the concrete cures properly. Since the concrete does cure slowly, strength also develops at a slower rate. Too much heat in the hardening process causes a defective or weak concrete. Low-heat portland cement is used in areas where there are huge masses of concrete, such as dams or large bridges.

Type V (Sulfate-resistant Portland Cement). This is also a special portland cement intended for use only in construction which is exposed to severe sulfate actions. It also gains strength at a slower rate than normal portland cement.

There are special air-entraining cements which contain small amounts of a chemical which enables them to hold air bubbles in the mix. The bubbles improve the workability and increase the resistance of the cement to freezing and thawing. These cements are used on jobs with special conditions. Air-entraining portland cements, covered by ASTM number C150, include Types IA, IIA, and IIIA.

Lime

The production of quicklime is the first step in the manufacture of lime. *Quicklime* is formed by calcining (or burning) pieces of limestone under controlled heat. The quicklime is then crushed and *slaked,* or mixed with water, in a hydrator where hydrated lime is produced. Figure 10-1 shows a worker bagging hydrated lime.

Prior to the twentieth century, quicklime was always used in the production of mortar. Today, however, hydrated lime is usually used in the production of mortar since it can be added directly to the mixer in dry form.

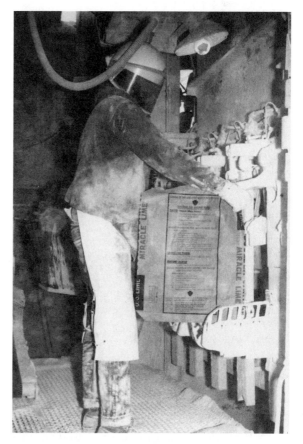

Fig. 10-1 Bagging hydrated lime in the plant.

Hydrated lime is essential to an all-purpose mortar because of the particular effects it has on the mortar.

The following are the effects hydrated lime has on mortar.

Bond Strength. Bond strength is probably the most important property of mortar. The addition of lime to mortar helps it fill voids and adhere to the unit. Figure 10-2 shows an example of good bond strength. Other factors in bond strength are water used in mixing (maximum should be used with retempering as needed), air content, and absorption and degree of roughness of the unit.

Workability. Lime helps mortar to be very workable, enabling the mason to fill joints completely without

Fig. 10-2 Good bond strength. Mortar adheres tightly to the brick even after the brick is picked up.

great effort. The result of this is better workmanship and more economical construction. Mortar with the correct lime content spreads easily with the trowel. Cement which is not mixed with lime is more difficult to spread. Sand graduation and air and water content also greatly aid workability.

Water Retention. Lime-based mortar is high in *water retention.* This means that the mortar resists drying effects when combined with the masonry units to build a wall. Mortar with a low lime content soon loses its moisture and sets prematurely when spread on the wall. When the mason does not have enough time to lay the material on the mortar bed before it dries out, a poor bond results, causing a weak structure. High water retention also minimizes the need for constantly retempering mortar.

Tensile Strength. Good mortar possesses adequate compressive and tensile strength with a substantial

safety factor. Mortar with an average strength tests 750 psi when cured for 28 days, which is the time factor used by engineers when checking mortar strengths. Mortar which tests stronger than 750 psi becomes progressively more rigid and brittle. Prepackaged masonry cement which requires that only sand and water be added to it should test within this range. Mortar should set with reasonable speed to enable construction to move forward without great delays. The addition of portland cement to a lime-based mortar helps to increase compressive and tensile strength.

Flexibility. For mortar to withstand stress by strong winds, lateral pressure, and hard jolts, it must have enough flexibility to prevent cracking. A tall masonry structure, such as a high chimney, may sway as much as 12″. Lime gives mortar plasticity, while portland cement adds strength.

Minimal Change in Volume. Very hard, high-strength mortar tends to shrink after hardening and can produce cracks between the mortar and units. This could result in a loss of bond strength. Of all the materials present in cement, lime undergoes the least change in volume.

Autogenous Healing. The ability of mortar to reknit or reseal itself if hairline cracks occur is called *autogenous healing.* Rainwater and atmospheric carbon dioxide react with the mortar to provide this feature. The hydrated lime dissolves and is then recarbonated by the carbon dioxide, which reseals hairline cracks in the mortar joint. A high-lime mortar shows little cracking, but it also has a lower compressive strength. A high-lime mortar is good when a tight sealing mortar joint, rather than a high compression mortar joint, is desired.

Resistance to Weather. Mortar should be able to resist strong winds, freezing temperatures, and alternate wet and dry weather. Wetting and drying cycles are actually beneficial to lime-based mortar and increase the overall strength of the mortar as it ages. The tightly knit mortar joints caused by the addition of lime to the mix help prevent water from entering the wall.

Sand

Sand, an aggregate of mortar, is the result of the deterioration of rock or stone. Good-quality sand is available in two types, *natural* and *manufactured,* and may be purchased almost anywhere.

Sand is sold by weight and delivered to the job by trucks. On an average job, ½ to 1 ton of sand will be lost and cannot be used. This should be considered when sand is purchased. Sand is one of the more inexpensive building materials and may be used on many different projects around the construction site.

Types of Sand. Manufactured sand is obtained by crushing stone, gravel, or air-cooled, blast furnace slag. Manufactured sand, with its sharp, angular grain shape, produces mortar with different properties than natural sand. Natural sand is used in most mortars. Manufactured sand is reserved for special purposes in most of the United States.

Natural sand is rounder and smoother than manufactured sand. It is usually found on lake bottoms or around riverbeds, pits, and riverbanks. Seashore sand is very high in salt content and is not recommended for use in masonry mortar. Most of the sand used in masonry work in the eastern United States is found near riverbanks or riverbeds and is mined by sand and gravel companies.

Characteristics of Sand. Sand acts as a filler in mortar which contributes to the strength of the mix. It decreases the shrinkage of mortar which occurs in setting and drying, therefore minimizing cracking. Water and cementitious materials form a paste which fills the voids between the sand particles and lubricates them to form a workable mix. The volume of voids or air spaces is determined by the range of the particle size and grade of sand being used.

It is important to use a good grade of sand. Over a period of time, it will be consistently more workable and form more satisfactory finished mortar joints. Grades of sand are determined by a standard screen sieve test. The Brick Industry Association has found that well-graded sand helps the workability of mortar.

The Importance of Choosing Good-Quality Sand. When sand is washed and treated, it is usually of good quality. The mason should never attempt to cut costs by using sand which has not been treated and washed.

It is important to understand why foreign substances are damaging to mortar. Silt causes mortar to stick to the trowel. This alone can cause a problem, as the mason's productivity is cut considerably. Silt also weakens mortar since it prevents the cementitious material from bonding to the sand particles. When tooling the mortar joints, silt causes mud pits and small holes on the surface of the joints. The mortar will possess a brownish, unnatural color. The color will vary from batch to batch, since the amount of impurities will not be the same in every batch mixed. Good-quality sand is worth the difference in price since it increases productivity and provides better-quality mortar.

The Siltation Test. If impurities (silt, clay, or organic matter) are suspected in the sand, a simple siltation test may be performed to determine if the sand should be used for masonry mortar.

Note: Many of these undesirable materials may be generally classified as loam.

1. To perform the siltation test, fill a glass jar (preferably quart-sized) half full of sand. Add about 3″ of water. Shake the jar vigorously and let it set overnight.
2. The sand will settle to the bottom of the jar and the foreign matter will rise to the top of the sand.
3. If the accumulation of silt or organic matter is more than ⅛″, the sand should not be used for masonry mortar. Figure 10-3 shows one example of a siltation test.

Water

Water used in mortar should be clean and as free as possible from alkalis, salts, acids, and organic matter. Water for mixing mortar should never be taken from a mud hole on the job. This is a very common practice on many construction jobs. Muddy water will cause the same reaction in mortar as sand with a high silt content. The amount of chemical deposits in the water should be determined by laboratory analysis if a problem is suspected. As a rule, water which is fit to drink

Fig. 10-3 The siltation test detects the presence of silt or loam. Notice the silt deposit on the top of the sand. This is in excess of the allowable ⅛″ limit.

is acceptable for use in mortar. Generally, purifying chemicals found in city water supplies have no adverse effect on mortar.

SUMMARY

The primary function of mortar is to bond masonry units into one integrated mass. The main components of mortars used today are portland cement, hydrated lime, sand, and water. No one combination of these ingredients yields the perfect mortar, however, since each particular job requires different properties in the mortar. Masons should be able to recognize good-quality mortar and mix ingredients in the proper proportions to obtain the best mortar for the particular job.

ACHIEVEMENT REVIEW

Select the best answer from the choices offered to complete each statement. List your choice by letter identification.

1. The early Greeks and Romans used a mortar made from
   a. lime and clay.
   b. volcanic waste and sand.
   c. portland cement and sand.

2. The addition of portland cement to mortar greatly increased mortar's
   a. strength.   c. workability.
   b. plasticity.   d. moisture resistance.

3. Quicklime may be defined as
   a. lime that has undergone a chemical change caused by treating it with water.
   b. crushed limestone.
   c. the fine, white powder resulting from burning in a kiln.

4. After lime has been slaked with water, it is called
   a. quicklime.
   b. hydrated lime.
   c. ground, burnt lime.

5. The term *bond strength* refers to
   a. compressive weight that the bond of mortar will tolerate before failing.
   b. tensile stress the bond of mortar will tolerate before failing.
   c. how well the mortar adheres to the masonry unit.
   d. the testing strength of the masonry unit.

6. Very high-strength types of mortar have one drawback. After hardening, they have a tendency to
   a. expand.   c. absorb moisture.
   b. shrink.   d. form pits.

7. One of the biggest advantages a lime-based mortar has is its autogenous healing. Briefly, that means it has
   a. the ability to resist excessive weight imposed on the mortar joint.
   b. the ability to withstand acids and alkalis from the atmosphere.
   c. the ability to reseal cracks in the joint by the process of recarbonation.

8. A good test used to check sand for foreign matter is the
   a. sieve test.
   b. siltation test.
   c. grading test.
   d. meter test.

9. Dirt in sand tends to make mortar difficult to handle with a trowel. The phrase best describing the condition of the mortar is
   a. sticky and gummy.
   b. gritty with a tendency to fall off the trowel.
   c. very liquid in form.
   d. rapidly setting and requiring excessive tempering.

UNIT 11
Types of Mortar and Their Characteristics

—————————————— OBJECTIVES ——————————————

After studying this unit, the student will be able to

■ identify the various types of mortar used in masonry work.

■ describe admixtures and how they are used in mortar.

■ explain the importance of the proper water content in mortar.

■ describe the causes of efflorescence and how to prevent its development in masonry work.

INTRODUCTION

The proper selection of mortar for masonry is very important. Choosing the right mortar type can lead to a durable masonry wall. Improper selection of mortar types for a particular masonry job can result in a leaky wall or deteriorating mortar. Mortar is the bonding agent that transforms the masonry unit and the mortar into a lasting masonry wall. It has to be durable, capable of keeping the masonry intact, and it must resist moisture penetration. Mortar also has to have plastic properties so that it is both economical and easy to use by a mason.

One important property of mortar that is overemphasized is its compressive strength. *Stronger is not necessarily better when specifying mortar types.* Many selections based on compressive strength alone ignore what is required for a particular job requirement. In general, it is the bond between the brick and the mortar that is the most important property. Bond strength refers to the force or stress necessary to separate the brick from the mortar joints. Brick with a higher rate of absorption of the mixing water in mortar tends to dry out the mortar too quickly before the mortar joints have had a chance to set properly. That is why it is necessary to dampen a highly absorbent brick with water in warm weather to achieve the best results for adhesion of the mortar to the brick. The performance of any

brick/mortar will be greatly influenced by the skill of the mason. The best types of mortar selected for a particular job requirement will not achieve its desired bond potential if the mason does not press the brick into a sufficient bed of fresh mortar mixed with the correct water content. Moving the brick after its initial set has taken place will damage the bond and therefore it will never achieve a proper bond.

CLASSIFICATIONS OF MORTARS

Two main classifications of mortars students should become familiar with are portland cement–hydrated lime mortars and masonry cement mortars. Both the ASTM and the BIA have established specifications these mortars must meet.

Portland cement–hydrated lime mortars are a combination of portland cement, hydrated lime, sand, and water, Figure 11-1. Depending on the proportions of the materials in the mix, the mortars will have different strengths and properties.

Masonry cement mortars are popular in masonry construction today because they come prepackaged. The mason needs to add only sand and water on the job, Figure 11-2. Correct proportions of sand and water are essential if masonry cement mortar is to meet standard specification ASTM Designation C270.

120

Fig. 11-1 Hydrated lime and portland cement combined with sand and water make mortar.

Fig. 11-2 A bag of masonry cement, mixed with sand and water, makes mortar.

Masonry cement mortar has certain additives to provide workability, flexibility, and water-retention properties. One major complaint of architects and engineers is that the proportions of these additives are not printed on the bag. Portland cement-lime mortars can be mixed in exact proportions to meet specifications and, therefore, be the correct mortar for a particular job. One mortar is not necessarily better than the other. However, the mason should be aware of the two main classifications of mortars used in masonry construction.

PORTLAND CEMENT-HYDRATED LIME MORTARS

The Brick Industry Association, the leading authority on brick masonry in the United States, has developed the following specifications for mortars used in the construction of brick masonry work. Four types of portland cement-lime mortars are covered under BIA Designation M1-72. The following is a description of the four types and their uses.

RECOMMENDED USES OF MORTAR

Type N

Type N is the mortar most often used. Type N is a medium-strength mortar suitable for general use in exposed masonry above grade and where high compressive or lateral masonry strengths are required. It is specifically used for the following:

- *Parapet walls* (that portion of a wall which extends above the roof line, usually used as a fire wall)
- Chimneys
- Exterior walls which are exposed to severe weather

Type N mortar has a compressive strength of at least 750 psi after being cured for 28 days, Figure 11-3.

When mixed, prebagged cement mortars sold under various brand names should conform to a mortar, like Type N, of at least medium strength.

Type M

Type M mortar has a higher compressive strength (at least 2500 psi in 28 days) and somewhat greater

Mortar Type	Compressive Strength*				Minimum Water Retention*** %	Maximum Air Content %	Efflorescence
	Minimum 7 days**		Minimum 28 days				
	psi	(kgf/cm^2)	psi	(kgf/cm^2)			
M	1600	(112)	2500	(175)			
S	1100	(77)	1800	(126)			
N	450	(32)	750	(53)	70	12	None
O	200	(14)	350	(25)			

*Average of three 2-9n. cubes.

**If the mortar fails to meet the 7-day compressive strength requirement, but meets the 28-day compressive strength requirement, it shall be acceptable.

***Flow after suction, percent of original flow.

Fig. 11-3 Physical requirements of laboratory mixed mortar.

durability than some of the other types. It is especially recommended for masonry which is below grade and in contact with the earth, such as foundations, retaining walls, walks, sewers, and manholes. It will also withstand severe frost action and high-lateral loads imposed by pressure from the earth.

Type S

Type S mortar is recommended for use in reinforced masonry and for standard masonry where maximum flexural strength is required. It is also used when mortar is the sole bonding agent between facing and backing units. Type S mortar has a fairly high compressive strength of at least 1800 psi in 28 days.

Type O

Type O mortar is a low-strength mortar suitable for general interior use in walls that do not carry a great load. It may be used for a load-bearing wall of solid masonry. However, it may only be used in those walls in which the axial compressive stresses developed do not exceed 100 psi, and which will not be subjected to weathering or to freezing temperatures. The compressive strength of Type O mortar should be at least 350 psi.

PROPORTIONS

Each type of portland cement-lime mortar should be mixed in proportions shown in Figure 11-4.

A recent improvement in some areas of the country in portland-lime cement is the introduction to the market of portland-lime cement mortar blended together in one bag and needing only sand and water to be added to form mortar. They are available in various classified types of mortar, Figure 11-5.

ADMIXTURES IN MORTAR

Mortar *admixtures* are materials which are added to mortar mix, usually in small amounts, to achieve a particular result. The results may include the following:

- Increased workability
- Added color
- A stronger bond between the mortar and masonry units
- An acceleration in setting time

Little information has been published regarding the effect of admixtures on mortar bond or strength. However, experience on various jobs has indicated undesirable results may occur in some instances. Air entrainment, for example, has a definite detrimental effect

Mortar Type	Parts by Volume of Portland Cement	Parts by Volume of Hydrated Lime	Sand, Measured in a Damp, Loose Condition
M	1	1/4	Not less than 2 1/4 and not
S	1	1/2	more than 3 times the sum
N	1	1	of the volumes of cement and
O	1	2	lime used.

Fig. 11-4 Mortar proportions by volume.

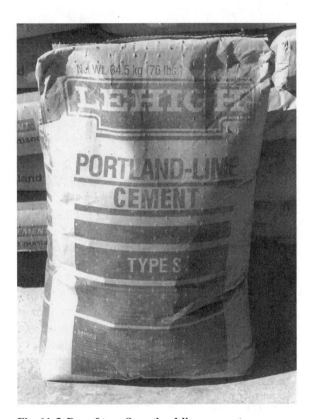

Fig. 11-5 Bag of type S portland-lime cement.

on the bond between mortar and the unit. Admixtures should never be used unless they are definitely specified in the contract and the manufacturer's mixing directions are followed exactly.

Color

Colored mortar is produced by adding colored aggregates or mortar pigments to the mixture. The pigments have a fine consistency to assure thorough mixing and even color in the mortar. Only the minimum amount of pigment should be used, as the strength of the mix could be affected. Color pigments should be of a metallic oxide composition in amounts not exceeding 15% of the portland cement content. Carbon black may be used as a coloring agent if it does not exceed 3% of the cement weight of the mix. Coloring agents that have not been tested commercially or by previous experience should not be used.

Precolored masonry cement which will blend with certain colors of brick or help create unusual effects in masonry walls may be purchased, Figure 11-6.

Air-Entraining Agents

Air-entraining agents are used in masonry cement mortar to increase its resistance to freezing and thawing, thereby increasing the life and durability of the mortar. The air-entraining admixture traps microscopic air bubbles in the mix. These air bubbles allow the mortar to contract and expand. When freezing and thawing occur, the mortar is less likely to crack and break. Mortar also has a tendency to hold water longer when an air-entraining additive is used, thereby increasing workability of the mix.

Mortar with air content over 12% by volume will weaken the bond strength of the mortar mix, according to tests by the BIA. Manufacturers must, therefore,

Fig. 11-6 Masonry cement can be obtained in several colors. Buff is very popular for colonial brick.

carefully control the amount of air-entraining agents in a mix.

Waterproofing Agents

Various types of waterproofing additives are added to mortar mixes to prevent moisture from penetrating the masonry wall. Bags of masonry cement or portland cement that contain waterproofing agents are marked to show this. Such cements are also recommended as a waterproofing coating for basement walls.

EFFLORESCENCE

Origins

Efflorescence is a major problem confronting architects, engineers, and masonry contractors. Efflorescence is a deposit of water-soluble salts upon the surface of a masonry wall, usually white in color. Efflorescence tends to spoil the appearance of the finished wall, Figure 11-7. It occurs many times just after the structure is completed, when architect and owner are most concerned with the finished appearance of the building.

Efflorescence appears as a white stain on brick walls. It is believed that the calcium chloride used to accelerate the setting time of the mortar is responsible for this condition. If the weight of the cement in the mixture does not contain more than 2% calcium chloride,

Fig. 11-7 Efflorescence on a brick wall.

efflorescence should not occur. Excessive amounts of calcium chloride may be responsible for efflorescence.

Efflorescence occurs when certain conditions are present. The wall must contain soluble salts and moisture must be present in the wall for sufficient time to take the salts into solution and carry them to the surface of the wall. If either of these conditions can be eliminated, efflorescence will not occur.

In most cases, the salts originate in the wall interior in either the masonry unit or the mortar itself, Figure 11-8. The entrance of water or moisture in the wall causes a reaction with the wall resulting in efflorescence. Sometimes chemicals in the atmosphere and the masonry combine and cause this action. Another source of salt is due to the masonry coming in contact with ground moisture, such as in basement walls or retaining walls. If the walls are not protected by some type of moisture barrier, the salts may be carried up through the first several courses of masonry above the ground level.

Principal Efflorescing Salt		Most Probable Source
Calcium sulfate	$CaSO_4 \bullet 2H_2O$	Brick
Sodium sulfate	$Na_2SO_4 \bullet 10H_2O$	Cement-brick reactions
Potassium sulfate	K_2SO_4	Cement-brick reactions
Calcium carbonate	$CaCO_3$	Mortar or concrete backing
Sodium carbonate	Na_2CO_3	Mortar
Potassium carbonate	K_2CO_3	Mortar
Potassium chloride	KCl	Acid cleaning
Sodium chloride	$NaCl$	Sea water
Vanadyl sulfate	$VOSO_4$	Brick
Vanadyl chloride	$VOCl_2$	Acid cleaning
Manganese oxide	Mn_3O_4	Brick
Iron oxide	Fe_2O_3 or $Fe(OH)_3$	Iron in contact or brick with black core
Calcium hydroxide	$Ca(OH)_2$	Cement

Fig. 11-8 Common sources of efflorescence.

Preventing Efflorescence

The following are several ways to prevent efflorescence:

- Reduce salt content by using washed sand.
- Use clean mixing water in the mortar.
- Keep materials on the job covered and stored off the ground at all times.
- After the wall is built, cover it with a canvas cloth or some other suitable waterproof covering.
- Overall good-quality workmanship and attaining a good adhesive bond between the unit and the mortar is the best way to eliminate the problem of efflorescence.

The appearance of efflorescence is dependent upon excessive moisture in the masonry work.

Removing Efflorescence

Efflorescence can be removed by washing the wall with water. If that fails, use a diluted solution of 1 part muriatic acid to 9 parts water and rinse thoroughly with clear water.

> **Caution:** When mixing the solution of muriatic acid into the water, make sure safety glasses or goggles are worn. Otherwise, eye injury could result.

This is only a temporary cure, however, and the stain may reappear after a period of time. Moisture must be kept from entering the wall.

WATER CONTENT OF MORTAR

Regardless of what type of mortar is used on a structure, one of the most important considerations is the water content.

Water content is possibly the most misunderstood aspect of masonry mortar. This is probably due to the confusion between mortar and concrete requirements. Many architects and builders assume that mortar requirements are the same as those for concrete. Therefore, they may make inaccurate specifications for mortar. The makeup of mortar and concrete is very different, especially in the water-cement ratio. Many specifications are incorrect. They require mortar to be mixed with the minimum amount of water to produce workable material and prohibit retempering of mortar. These specifications result in mortar which has maximum compressive strength but which has less than maximum tensile bond strength. A strong bond between the unit and the mortar is not achieved. Mortars should always be mixed with the maximum amount of water to provide maximum tensile bond strength of the mortar.

Mortar used to lay brick is somewhat thinner than mortar used to lay a concrete block, Figure 11-9. This is because the concrete block is heavier and has a tendency to sink into the mortar. There is also a difference in the water content of the mortar used with soft, sand brick and that for hard, dense brick. This is because the absorption of the two units will differ and the setting time will, therefore, differ.

Retempering mortar (adding water and remixing) is acceptable, but must be done correctly. Only that water which has been lost from evaporation should be added.

Fig. 11-9 The mortar shown here has been mixed with the proper water content. Note how it holds its body when pulled with a trowel.

In the past, masons commonly discarded mortar if it had become stiff from dryness. Modern-day specifications allow the mortar to be retempered as often as necessary, as long as it is used within 2 hours of the initial mixing. Pace the amount of mortar being mixed on the job with the amount of mortar being used. When tempering mortar, always wet the edges of the mortar pan or board with water. This helps prevent the mortar from drying out too quickly. Wetting the mortar board or pan before filling it with mortar also helps to reduce evaporation of water.

Mortar must be workable to permit the mason to be productive. It must also have good strength and durability. Mixing the correct amount of water initially is the best way to achieve mortar with the proper consistency. Since masons spend approximately 50% of their time working with mortar, it is important that they understand the role of admixtures and water in mortar.

REPOINTING MORTAR JOINTS IN HISTORICAL BUILDINGS

Mortar joints can be seriously eroded over a period of years due to the changing seasons and wetting and freezing cycles. This is especially true in the older historical buildings where the brick and the mortar joints are softer. The mortar joints can be renewed and replaced by repointing, which is the process of removing the damaged mortar and replacing it with fresh mortar. It is not a job for an amateur, and does require a high degree of skill and knowledge about matching the original mortar.

Proper Time to Repoint

Repointing should be started when the existing mortar joints are eroded or exposed to a depth of $1/2''$ or more, or when the mortar has shrunk away from the majority of the edges of the brick. Many of the older types of lime-based mortar, while soft, are still structurally sound, as this was the nature of lime-based mortars. It is not generally necessary to completely repoint all of the mortar joints in an old building, but rather to repoint only the bad places, taking care to match the original color, composition, and psi strength of the original mortar. The goal is to restore the brick work to its original appearance. The texture of the aggregate is also important in achieving that original look.

Ingredients of Historical Mortars

All repointing mortars were made of water, some type of aggregate (sand, crushed oyster shell, ground flint, shale, etc.), and binders materials. Admixtures found in today's masonry cement mortar are rarely used, with the exception of coloring materials. In early days, some clay was used that did have a certain amount of natural cement and lime in it. However, as time progressed lime became the most important binding agent. After portland cement was discovered in the late 1870s it was discovered that a small amount of portland cement added to the mix created a much longer-lasting mortar.

There is a common misconception that the harder the mortar, the better it is. This is totally wrong. Historical mortar should match approximately the compressive psi of the masonry unit or brick that it is being repointed around. Matching the aggregates is always one of the toughest jobs in repointing. This is done by inspecting a sample of the mortar joint in the wall and working on a sample section of the wall until it meets with satisfaction.

Color Pigments

Prepackaged color pigments are available from many building supply dealers that can be added following the manufacturer's directions for the mortar to obtain the proper color match. Pages 128 and 129 show a specifier's data sheet for concrete and mortar colors. Any pigment selected should not exceed a maximum of 10% of the volume or 6% of the weight of the binder. The color of the sand being used has a lot to do with the eventual color of the mortar joint. As a rule a light brown or buff sand matches best for repointing and it should be a washed sand which is available at the masonry building supplier on request. A simple test to make sure that it is clean is to moisten a handful of sand, squeeze it into a ball. Upon releasing it the sand should basically crumble away and not remain in a ball. If it does not, then loam or mud may be present and this sand should not be used, as it will affect the ultimate strength of the mortar. My recommendation for mixing a typical historical repointing mortar is to use 1 part of portland cement to 2 parts of mason's hydrated lime to 8 or 9 parts of sand or aggregate. This will result in a mortar with a strength of approximately 350 psi, which is considered soft. The addition of the extra lime to the mix causes the mortar to adhere to the edges of the brick and also provides for the maximum expansion and contraction of the joint without cracking. Any hairline cracks will usually reseal themselves when rainwater comes in contact with the joints. This process is known as *autogenous healing* and was mentioned in Unit 10 in the discussion of lime. Figure 11-10 shows a sample of portland cement, hydrated lime, and sand used in a typical repointing mix.

Repointing Process

The first step in repointing is to remove any of the old deteriorated mortar joints using a special plugging or joint chisel, being careful not to chip the edges any more than possible. It is recommended to cut the joints out to a depth of $\frac{3}{4}''$ to $1''$ for the best results. There are mechanical grinders made that will do this on a large scale, but they often damage the surrounding edges of the bricks. Now there is also on the market what is known as a pin grinder, which has a narrow $\frac{1}{4}''$ or $\frac{3}{8}''$

Fig. 11-10 From left to right, a sample of portland cement, mason's hydrated lime, and sand used to make repointing mortar.

pin coated with a diamond type of abrasive that works like a drill bit. As it rotates, it grinds the mortar out with a minimum of damage to the edges. However, this is a labor-intensive process and is expensive. Using a plugging chisel is still the best method. Figure 11-11 shows a mason cutting out mortar joints over a window with a hammer and plugging chisel.

After the joints have been cut out to the required depth, they should be brushed clean of all particles or old mortar and then wetted with a brush and water before repointing. The wetting of the joints is done to slow down the absorption of the moisture from the mortar mix as it cures, because this could cause premature

Fig. 11-11 Mason cutting out mortar joints over a window with a hammer and plugging chisel.

Davis Colors
True Tone™ Sweet 16™
Concrete and Mortar Colors

| SPECIFIER'S DATA SHEET | SECTION 04100 |

PRODUCT DESCRIPTION

TRUE TONE SWEET 16 concrete and mortar colors are prepackaged dry pigments used for integral coloring of concrete, plaster, stucco and masonry mortar. The pigments are finely milled and specially blended to ensure maximum color tinting with complete, rapid and uniform dispersion throughout the mix. True Tone Sweet 16 colors require no measuring or weighing, are easily mixed, and are environmentally safe.

Basic Uses: True Tone Sweet 16 colors are added to masonry mortar to provide lasting color with a natural finish. When used within recommended mix rates, True Tone colors will not decrease mortar strength or bond. They are also well suited for coloring precast or ready-mixed concrete, plaster, stucco, concrete block, concrete brick, terrazzo and similar materials.

Application Limitations: Mixing rates greater than 10% by weight of color to the total weight of masonry cement or portland cement and lime will not enhance the color, and may weaken the mortar mix. The addition of color does not change the concrete's chemical composition, nor does it cause or prevent efflorescence.

Composition and Materials: True Tone Sweet 16 colors are made from iron oxides of recycled metal or naturally occurring iron oxide ores, finely milled and blended with mineral conditioners. They contain

no filler or artificial adulterants, and are uniform from bag to bag and shipment to shipment. The pigments have been micro-pulverized to achieve optimum integration of particle sizes

True Tone Sweet 16 Sample Kit and Dose-Sized Bag

to produce both color intensity and color longevity. Because of their extremely small particle size, they have the highest possible tinting strength. The pigment particles physically bind to grains of cementitious material, changing the overall color of the mixture. True Tone Sweet 16 colors are lime-proof, inorganic, light-fast, weather-resistant, inert to atmospheric conditions, stable, and free of water soluble content.

Packaging: True Tone colors are weighed and packaged at the factory

to eliminate guesswork and color variation caused by inaccurate job site weighing. True Tone Sweet 16 colors are pre-weighed in convenient 1½ lb. "dose-sized" bags and shrink-wrapped into bundles of 16 bags. Boxes have been eliminated to reduce job site waste.

Colors: A sample card is available which shows 64 of the colors that can be made with True Tone Sweet 16 pigments. Each column of samples shows a different mix rate for each pigment. The samples show mortar made with 1½, 3 or 6 lbs. (1, 2 or 4 dose-sized bags) added for each bag of gray masonry cement. The lightest color samples are made with white masonry cement and white sand.

TECHNICAL DATA

Applicable Standards: True Tone Sweet 16 colors meet or exceed requirements of ASTM C979, *Standard Specification for Pigments for Integrally Colored Concrete*. Color pigments are accepted under ASTM C270, *Standard Specification for Mortar for Unit Masonry, Types M, S, N, O and K*.

Applicable Codes: Job site use of mineral oxide pigments is permitted by ACI 5301/ASCE 6/TMS 602, *Building Code Requirements and Specifications for Masonry Structures*, Article 2.2.2.2. Maximum percentage of pigment is limited to 10% by weight of cement in

					MORTAR MIX DESIGN			
MORTAR TYPE	PORTLAND CEMENT 94# BAG	HYDRATED LIME 50# BAG	MASONRY CEMENT 78# BAG	SAND (CU FT)	NUMBER OF TRUE TONE BAGS (1½ POUNDS EACH)			
					LIGHTEST*	LIGHT	MEDIUM	DARK
N,S,M	—	—	1	3	½	1	2	4
N	1	1	—	6	1	2	4	8
S	2	1	—	9	1½	3	6	12
S	1	—	2	9	1½	3	6	12
M	2	½	—	6	1½	3	6	12
M	1	—	1	6	1	2	4	8
O	1	2	—	9	1½	3	6	12

* White Cement

TRUE TONE SWEET 16	CONCRETE & MORTAR COLORS

either portland cement-lime mortar or masonry cement mortar.

INSTALLATION

Workmanship: Follow recommended masonry construction practices.

Test Panel: The shade or tint of the dry color pigment may not appear to match the final mortar color. The pigment must be completely mixed with cementitious materials, water and aggregate and then cured to fully develop the true color. Sample panels of selected colors should be constructed with job site materials and procedures and then fully cured to establish acceptable standards. All materials for sample panel must be mixed in full batch quantities using a mechanical mixer. Do not mix by hand. Retain sample panel until masonry work is complete and accepted.

Materials: The color of the finished mortar joint will be influenced by the color of the cementitious mortar ingredients and aggregate. White portland cement or white masonry cement and white sand will enhance lighter colors.

Sand of the same type and amount should be used in every batch. Aggregates should conform to ASTM C144, *Specification for Aggregate for Masonry Mortar*.

Use the same type and brand of cement from the same mill throughout the entire project. Masonry cement should conform to ASTM C91, *Standard Specification for Masonry Cement*. Portland cement should conform to ASTM C150, *Standard Specification for Portland Cement*. Lime should conform to Type S, ASTM C207, *Standard Specification for Hydrated Lime for Masonry Purposes*.

Water should be clean and free of deleterious or harmful acids, alkalis and organic materials.

Admixtures may affect mortar color. The use of calcium chloride based accelerators and other admixtures containing chloride ions are not recommended with colored mortar.

Mixing: Uniform color requires consistent material proportions and mix times.

A mechanical mortar mixer must be used to obtain proper dispersion of pigments. Mix only full batches. Add only full bags of color, using required amount (see Mortar Mix Design Table). Once a job site sample panel is approved, use the same mix design, the same type and brand of sand, cement and lime for every mortar batch. Always add color by weight, never by volume. Mix the same amount of color, cement, lime, sand and water in every batch.

Start mixer and add 2/3 batch water, 1/3 batch sand, all the cement, all the color (by weight), then the remaining sand and continue mixing for one minute. Then slowly add remaining water to bring mortar to the proper consistency. Mix the full batch for at least five minutes.

Use all mortar within 2½ hours of original mixing (1½ hours if ambient temperature is above 80° F). Do not re-temper or add water to colored mortar, as this will cause colors to lighten.

Tooling: Do not tool joints too early or too wet. Tool all mortar joints at the same degree of hardness and same moisture content. Wetter mortar tools lighter, and drier mortar tools to a darker color.

Cleaning: Let mortar splatters harden for 7-14 days, then remove with chisel, trowel, or stiff brush and water before they bond rigidly to units. Prior to further cleaning, allow walls to continue curing for at least three weeks in summer, or four weeks in winter. Do not use muriatic (hydrochloric) acid to clean colored mortar joints.

If cleaning agents are used, pre-wet wall, test and check effects on a small inconspicuous area prior to proceeding. Begin cleaning at the top and work down. Thoroughly rinse wall afterwards with clean water. Follow cleaner manufacturer's instructions.

AVAILABILITY

True Tone Sweet 16 colors are sold by recognized building material

dealers, lumber yards, and concrete block and brick distributors throughout the United States.

WARRANTY

Information contained in this brochure is, to our best knowledge, true and accurate, but all recommendations or suggestions are made without guarantee. Since the conditions of your use of our products are beyond our control, the Davis Colors Company disclaims any liability incurred in connection with the use of our products and information contained herein. No person is authorized to make any statement or recommendation not contained herein, and any such statement or recommendation so made shall not bind the Company. Furthermore, nothing contained herein shall be construed as a recommendation to use any product in conflict with existing patents covering any material or its use, and no license is implied or in fact is granted herein under the claims of any patents.

MAINTENANCE

No maintenance is required.

TECHNICAL SERVICES

Complete technical information and literature is available from Davis Colors. For other colors, special matches, custom colors or application questions, contact our service lab.

Setting the Standard for Concrete Colors

East
12116 Conway Road
Beltsville, MD 20705
(800) 638-4444
(301) 210-3400 In Maryland
(301) 210-4967 Fax

West
3700 East Olympic Blvd.
Los Angeles, CA 90023-0100
(800) 356-4848
(213) 269-7311 In California
(213) 269-1053 Fax

cracking of the new joints. On large surfaces, a garden sprayer set on a fine mist works very well. Do not soak the wall; only moisten the area to be repointed. Re-dampen the joints as needed if they start to dry out again. Figure 11-12 demonstrates cleaning out the joints with a brush and dampening the joint area.

The repointing mortar should be mixed at a slightly drier consistency than that used when laying brick in a wall, and should stick to the trowel and slicker pointing tool without falling off. If not, add a little more lime. Using a flat blade slicker pointing tool mentioned in Unit 6 on tools, pack or press the fresh mortar back into the mortar joints until flush. Various sizes of slickers are available to match large or small joints. If the joint is especially deep, it may require several applications of repointing mortar to fill the joints out completely. When the joint surface reaches a thumbprint hardness stage, tool the joint with the desired finish. Generally a flat joint is most popular for historical joints but any can be used depending on what the original joint finish was.

Figure 11-13 shows a mason repointing the mortar in the joints with a slicker tool.

Complete the repointing process by brushing the area down with a medium soft brush. The mortar will cure and age with a little help from Mother Nature and

Fig. 11-13 Repointing mortar in the joints with a flat slicker tool.

some gentle rains. If the weather is dry, it will help the mortar joints cure better to gently spray the area with a fine mist from a garden sprayer or hose, being careful not to wet the area excessively. In cold or freezing weather, it is not recommended to wet the area unless heat is applied to the wall. Figure 11-14 shows a mason brushing the wall to complete the job.

If the wall is to be repainted, it should not happen until at least 2 weeks have passed, allowing the mortar a chance to cure properly.

Fig. 11-12 Cleaning out and dampening the mortar joints with water before repointing.

Fig. 11-14 Brushing the wall to complete the job.

ACHIEVEMENT REVIEW

The column on the left contains a statement associated with the composition and mixing of mortar. The column on the right lists terms. Select the correct term from the right-hand list and match it with the proper statement on the left.

1. Mortar used for most masonry work such as exterior walls, chimneys, parapet walls, and exposed walls above the finished grade; tests 750 psi

2. Mortar with a high compressive strength and greater than average durability; used for masonry below grade, sewers, manholes, and retaining walls; tests 2500 psi

3. Low-strength mortar used for interior walls; tests about 350 psi

4. High compressive strength mortar used to reinforce masonry walls and where high bond strength in the wall is required; tests 1800 psi

5. Materials added to mortar mixture to achieve particular results such as color or better workability

6. Chemical used to speed the setting time of mortar

7. Deposits of water-soluble salts on the surface of masonry work; whitish in color

8. Principal cause of efflorescence

9. Chemical used to remove efflorescence from masonry work

10. The process of adding water to mortar to make it more workable after it has become stiff from dryness.

11. Mortar that has been mixed with a minimum amount of water acquires a high compressive strength but is weak in this strength measurement

a. Type S mortar
b. Calcium chloride
c. Tensile strength
d. Type N mortar
e. Moisture
f. Type O mortar
g. Tempering
h. Muriatic acid
i. Type M mortar
j. Admixtures
k. Efflorescence
l. Bond

UNIT 12
Mixing Mortar

OBJECTIVES

After studying this unit, the student will be able to

■ proportion mortar ingredients for specific mixes.

■ mix mortar manually with hand tools and equipment.

■ mix mortar with a power mixer.

■ follow correct safety practices when mixing mortar.

Student masons or apprentices must be able to mix mortar properly for different job requirements. As previously mentioned, masons usually do not mix their own mortar or mud, as it is known in the trade. However, on small jobs or repair jobs, it is often economically practical for masons to mix their own mortar. Small jobs or repairs include repairing a window sill, pointing up cracks in a wall, or replacing the last course of brick on a retaining wall.

When only a small amount of mortar is needed, it can be mixed in a contractor's wheelbarrow or a small mortar box. For larger jobs, a mechanical mixer is used. Setting up the job with the necessary materials beforehand offers a savings when the work is started. This unit presents procedures for mixing mortar both manually with a shovel and hoe and mechanically using a power mixer.

> **Caution:** Persons mixing mortar should wear safety glasses to protect against splashing mortar. When mixing in a poorly ventilated area, wear a dust mask.

ASSEMBLING AND STORING MIXING MATERIALS

The materials needed for mixing mortar should be placed close to the area where the work will be done. However, the materials should not be so close as to in-

terfere with the laying of the masonry walls. Keeping materials close to the working area speeds up the work process.

All of the mixing ingredients should be covered with a plastic or canvas tarpaulin to protect them from the elements, Figure 12-1. It is not necessary to protect the sand from moisture in warm weather. During winter, however, masons must keep the sand pile covered to prevent freezing. If sand freezes, it must be

Fig. 12-1 A roofed frame protects the materials and worker. Such protection is especially valuable in cold weather.

132

completely thawed before it is usable. Cement and lime must be kept dry. They may be stored in a shed or placed on wooden pallets or boards off the ground to ensure complete protection from moisture.

Place mixing materials close together and have a source of water available for mixing materials and cleaning mixing equipment. A 50 gallon (gal) steel drum filled with clean water is placed near the mixing area on many jobsites. A length of rubber hose is required to fill the water barrel. The hose is also useful in cleaning mortar from the mixing equipment.

STANDARD PROPORTIONS FOR MORTAR MIXTURES

Proportions of ingredients in mortar are based on volume measurements. Mortar materials are manufactured, bagged, and delivered to the jobsite accordingly. Portland cement is available in 94 lb bags containing 1 cu ft. Mason's hydrated lime comes in bags which contain 50 lb. Sand is sold by weight but is measured by the cubic foot (cu ft) or yard (yd).

In actual practice, portland cement-lime mortar is mixed on the job by combining a certain number of parts of sand, lime, and portland cement. The ingredients may be measured by the shovel and then combined in the mixing box or mechanical mixer. The cement and lime may be measured by the bag.

Specifications of some jobs require that the mortar ingredients be measured very carefully by using a material batcher. This is done by counting the number of shovels of sand needed to fill a box 1′ long × 1′ wide × 1′ deep (inside measurements), Figure 12-2. After the number of shovels of sand for 1 cu ft has been determined by filling the batcher, a matching amount is placed directly in the mixing box or mixer for each cubic foot required. The number of shovels required to fill the batcher should be rechecked each day to allow for any moisture or other factors which may cause the measurement to vary.

Masonry cement mortars require that only sand and water be added to the mix. Usually 1 shovel of masonry cement is added to 3 shovels of sand, or 18 shovels of sand to one 70 lb bag of masonry cement. This combination should be equal to a Type N mortar by ASTM

Fig. 12-2 Cubic foot box used to measure dry mortar ingredients.

standards and is a standard proportion for masonry cement mortar throughout the United States.

Different proportions and combinations of materials for mortar can be reviewed by referring to Unit 11, Figure 11-5.

MIXING MORTAR MANUALLY

On small jobs, mortar may be mixed manually rather than with a mechanical mixer. To make mixing mortar with hand tools and equipment as easy as possible, specific steps and procedures should be followed.

The necessary materials must first be assembled. The ingredients for this mix in this example are portland cement (Type 1), lime, sand, and water. The proportions of the mix are expressed as *1:1:6 (Type N)*, or 1 part portland cement, 1 part lime, and 6 parts sand. The mixing process may be followed in Steps 1 through 9, Figure 12-3.

1. Before mixing materials, the mortar box should be blocked up and leveled. There is no definite height at which the mortar box should be raised; it is determined by the existing ground level. Leveling prevents the water used in the mix from running to one end of the mortar box. Leveling also positions the mortar box more conveniently for the mason to use the hoe.

STEP 1

STEP 5

STEP 2

STEP 6

STEP 3

STEP 7

STEP 4

STEP 8

Fig. 12-3 Steps in making mortar.

STEP 9

Fig. 12-3 Steps in making mortar (continued).

2. Put one-half the desired amount of sand needed for the batch in the mortar box. Measure all proportions for this mix with a standard dirt shovel. (This is standard practice on the job unless otherwise specified.) Spread the sand evenly over the bottom of the mortar box.

3. Use a shovel to spread the desired amount of portland cement and lime over the sand. If a large batch of mortar is being mixed, the portland cement and lime may be added by the bag rather than by the shovel. Add the remaining half of the sand to the mortar box over the portland cement and lime. Adding the sand in halves helps to distribute materials evenly throughout the mix.

4. Blend the dry materials together with the shovel or hoe and push them to one end of the mortar box.

5. Add about $2\frac{1}{2}$ gal of clean water to the other end of the mortar box from a 5-gal bucket. Do not use a hose to add water since the water cannot be measured accurately. The total amount of water for a mix is determined by the moisture content of the sand.

6. Combine the dry mix and the water with a chopping action. Pull and push the mortar back and forth through the mix with the hoe at a 45° angle until all of the mortar is well mixed. Add water as needed to complete the mixing process. Tilting the hoe helps eliminate strain on the arms and wrists.

Caution: Wear safety glasses or some other eye protection when mixing mortar. If mortar splashes in your eyes, do not rub them, as the sand may scratch your eyes. Wash your eyes immediately with clean water. If you cannot remove all of the mortar by rinsing with water, see a doctor as soon as possible for further treatment.

7. After the mortar has been mixed, pull all of the mortar to one end of the box with the hoe. This will help prevent the mortar from drying out before it is removed from the box and placed in the wheelbarrow. The mortar can be checked at this time by a simple test to see if it has the proper workability and adhesiveness. Pick up a small amount of mortar with the trowel and set it firmly on the trowel by flexing the wrist downward with a sharp, jarring movement. Turn the trowel upside down. If the mortar is of a proper consistency, it will adhere to the blade of the trowel.

8. Wet the wheelbarrow with a small amount of water before shoveling the mortar into it. Do this so that the mortar does not stick to the sides of the wheelbarrow. Move the mortar to the wheelbarrow with a flat shovel. When only a small amount of mortar is needed for a job, mix it in the wheelbarrow instead of the mortar box.

9. The final step in mixing mortar is to clean the mixing equipment with water. Do this as soon as possible after use. Mortar setting and drying on mixing tools can cause a buildup of mortar and permanent damage to the mixing equipment. A hose with a spray nozzle does the best job of cleaning because the water is under pressure. To scrub a stubborn spot or particles of mortar from the tools, use a stiff brush.

MIXING MORTAR WITH THE POWER MIXER

When using a mechanical mixer (gasoline-powered or electrically driven), place all materials near the mixing area, as in the manual operation. The mechanical mixer should be in the center of the assembled materials. On many jobs, to avoid excessive wear on the tires, prop up the mixer with blocks. Secure the mixer carefully, since on a large job, it may stay in one location for over a year.

When mixing mortar with a power-driven mixer, the following steps should be followed:

1. Add a small amount of water to prevent the materials from building up and sticking to the sides of the mixing drum, Figure 12-4. This also helps to prevent the mortar from caking on the machine paddles as they whip through the mix.

Caution: Wear safety glasses or goggles to protect the eyes.

2. Add one-half to one-third of the sand needed for the total mix. Add sand to the mixer while the paddles are turning to prevent excessive strain on the motor, Figure 12-5.

Caution: Be extremely careful not to place the shovel inside the mouth of the mixer where it may become caught in the turning paddles. You could be severely injured by the uncontrolled shovel handle. Place a safety grate in the mixer when adding materials to help prevent accidents.

3. Add the necessary amount of portland cement, lime, or masonry cement to the mixing drum, Figure 12-6. Usually this is done by the bag. Lay the bag on the safety grate and cut it with a penknife, an old trowel, or by pulling the string on the corner of the bag to release the material. Some mixers have a sharp piece of metal on the top of the grate to puncture the bag.

Fig. 12-4 Adding water to the mix to prevent mortar from sticking to the drum.

Fig. 12-5 Adding the sand to the mixer. Sand is thrown on the grate, never into the mouth of the mixer.

Fig. 12-6 Adding the masonry cement to the mixer. One bag of cement makes one batch.

ure 12-7. The turning of the blades clears most of the mortar from the mixer.

Caution: Do not scrape remaining mortar from the mixer drum until you take the blades out of gear and shut off the mixer.

6. Start the mixer again. Put enough water in the mixing drum to clean the inside while the paddles are turning. When no more mixing is to be done, scrub the mixer with a stiff-bristle brush and flush with enough water to remove all of the mortar, Figure 12-8.

Caution: Never attempt to scrub the mixer with the brush when the mixer is running.

Power mixers are necessary on larger jobs because they complete the job with relative speed and save labor since they require only one operator. However, the use of good safety practices cannot be overemphasized

Caution: Hands or fingers should never be put inside the mixer while it is running. If part of the torn bag falls into the mixer, do not try to remove it while the mixer is running.

4. Add the rest of the sand to the mixer. Add additional water at this time as needed to bring the mortar to the desired consistency. Allow the mixer to blend all materials thoroughly for 3 to 5 minutes. Prolonged mixing of mortar should be avoided. Excessive amounts of air will become trapped in the mix and the mortar will be spongy. Mortar of this consistency will not have the proper body and may cause problems in supporting the masonry unit.

5. At the completion of the mixing process, grasp the mixing drum by the dumping handle while the mixer is still running and dump the mortar into the wheelbarrow or mortar box, Fig-

Fig. 12-7 Dumping the mortar from the mixer into the wheelbarrow.

Fig. 12-8 Rinsing the mixer with water.

when operating a power mixer. Power mixers also ensure a uniform mixture of materials. Usually, if the same person mixes a batch of mortar, the proportions and consistency should remain relatively the same.

Power mixers can be obtained in single-batch or double-batch sizes. A *batch* is defined as 1 complete mix of mortar, usually consisting of 1 bag of portland cement or masonry cement and other necessary materials.

PROBLEMS ENCOUNTERED IN MIXING MORTAR

Three problems masons face when mixing mortar are proportioning materials improperly, using poor-quality materials, and working in cold weather.

Proportioning Materials Improperly

If an excessive amount of water is added to a mortar mixture (known as *drowning* the mortar), the proper water-cement ratio is destroyed. To correct this prob-

lem, cementing materials, lime, and sand should be added in the correct proportions until the desired consistency is reached.

Producing mortar of the proper consistency for specific job conditions comes only with experience. The job will be easier for the trainee if water is added very slowly to the mix. It is considerably more trouble to add all the necessary ingredients again in their proper proportions. The same problem results when too much water is used to temper mortar.

Adding too much sand is one of the more common problems encountered when mixing mortar. This is caused primarily by poor measurement of materials. If a shovel is used for proportioning, the same type of shovel should be used all of the time with the same amount of material placed on the shovel each time. Usually, a standard dirt shovel is used. The amount of sand that a shovel holds will vary slightly depending upon the moisture content of the sand. The batcher measuring device is the most accurate.

Masons should never add sand to stiffen a mortar mixture without adding the proper proportions of other ingredients at the same time. This is a common problem in the trade, but should be avoided. If only sand is added, the compressive and tensile strength of the mixture changes drastically. The result is inferior mortar. It also causes variations of color that are noticeable in the finished mortar joint. Over-sanded mortar is harsh, difficult to use, and forms a very weak bond. All of these factors result in a poor job.

Mortar that contains a high percentage of cementitious materials, known as *fat mortar,* is sticky and hard to dislodge from the trowel. Mortar that is lacking in cementitious materials is called *lean mortar.* Avoid both of these conditions.

Using Poor-Quality Materials

Cement that is old or that contains hard knots or lumps should not be used. The knots or lumps are caused by moisture penetrating the material and hydration taking place. The mason must remove all the knots and hard lumps from the mortar before the masonry unit can be laid in the wall. This is a tedious practice that wastes labor. It is more economical to discard the hardened mortar. Proper storage prevents this condition.

Working in Cold Weather

Cold weather presents special problems when mixing mortar. The use of admixtures or antifreezes is discouraged since they sometimes decrease the strength and affect the color of the mortar. If used, calcium chloride sometimes causes corrosion of metal ties.

When the temperature falls below 40°F, the mortar materials may require heating. Sand must be heated slowly and evenly to prevent scorching, which turns sand slightly red. Heating is usually done by piling the sand over a metal tube or pipe and lighting a fire inside the tube.

Mixing water should also be heated in cold weather. The water should never be heated above 160°F, however, because of the danger of a flash set when portland cement is added to the mix. A *flash set* occurs when the portland cement sets prematurely due to excessive heat. After combining all of the ingredients to the mix, the temperature should be between 70°F and 105°F. If mortar temperatures exceed this limit, premature hardening may occur, which results in greatly reduced compressive and bond strength. If it is less than 70°F, mortar sets too slowly.

When a problem concerning mortar occurs on the job, analyze and correct it as soon as possible. Good-quality mortar requires strict adherence to specific proportioning of materials. This care results in strong, watertight masonry walls.

MORTAR FOR TRAINING PURPOSES

If mortar is being used for training in a shop, only hydrated lime (no cement) is added to the mix. Lime is available in 50-lb bags. For good-quality mortar for training purposes, mix 1 part lime to $2\frac{1}{2}$ to 3 parts sand, depending on the fineness of the sand.

Lime is used without cement in a training mortar because the mortar does not harden and may be reused.

Lime gives the mortar the necessary plasticity and workability. It may be necessary to add a small amount of lime to the rescreened mortar from time to time to increase the plasticity and workability. The amount to be added must be determined by the trainee according to the needs of the job.

Recently material other than lime has been found to be excellent for the mixing of practice-shop mortar. Many instructors in the eastern United States have changed to this in place of hydrated lime. It works as real mortar but is much smoother to use and has very little dust, which has been one of the major problems with the use of lime for practice mortars. Its brand name is CAROTEX, and it is a sedimentary aluminum silicate of very fine particle size (90% below .2 microns) called kaolin clay. It has been used for about 20 years in the ceramic tile trades. In addition, approximately one-half the quantity of CAROTEX (by volume) is used to achieve the same quality mortar mixture as obtained when using lime. As it is an inert substance, it is not harmful to the skin. The manufacturer stresses to masonry teachers or trainees that CAROTEX is not merely a convenient substitute for practice mortar, but also a fully acceptable masonry product used with or in lieu of hydrated lime. If available in your area, it is highly recommended in place of lime. It is available from Southeastern Clay Company, Aiken, South Carolina 29801.

When mixing mortar, the student mason should observe certain safety precautions.

- Always protect eyes and hands from injury.
- Wear safety glasses or goggles.
- Do not place hands in a power mixer when the machine is operating.
- Report all accidents at once to the class instructor or job supervisor.

ACHIEVEMENT REVIEW

A. There are specific procedures concerning mortar mixing which the trainee should have mastered by the time he or she has reached the conclusion of this unit. Mixing good-quality mortar requires a knowledge of the material and the proper technique to form the finished mix. Review the following procedures until each one is learned.

Mixing by Hand

- Level the mortar box.
- Measure ingredients for the desired mix.
- Dry mix the ingredients.
- Add water.
- Mix the mortar with the hoe.
- Clean mixing tools.

Mixing Mortar by Machine

- Add water before combining ingredients.
- Proportion ingredients in the mixer.
- Add additional water.
- Mix 3 to 5 minutes.
- Unload mixer.
- Clean mixer.

B. Select the best answer from the choices offered to complete each of the following statements. List your choice by letter identification.

1. The trade name for mortar is
   a. cement.
   b. concrete.
   c. mud.
   d. gypsum.

2. Proportions of masonry mortar are based on
   a. volume.
   b. square feet.
   c. weight.
   d. liquid measurement.

3. Usually, the type of portland cement used to mix portland cement mortar is
   a. Type 1.
   b. Type 2.
   c. Type 3.
   d. Type 4.

4. The total amount of water needed for a mixture of mortar is determined by
   a. the amount of dry material being mixed.
   b. the thickness of the mortar joint.
   c. the moisture content of the sand.
   d. the type of mortar being mixed.

5. After the mortar has been mixed, it is tested by filling a trowel with mortar and turning it upside down to see if the mortar remains on the trowel. The property of the mortar being tested is
   a. water retention.
   b. plasticity.
   c. strength.
   d. adhesive quality.

6. When mixing mortar with the mechanical mixer, water is always added to the mixer first. The main reason for this is
   a. to help the mortar mix more thoroughly.
   b. to speed the mixing process.
   c. to prevent the mortar from sticking to and building up on the mixer.
   d. to increase the workability of the mix.

7. If mortar is mixed for an excessive amount of time in a mechanical mixer, it becomes very spongy. The correct mixing time is
   a. 8 minutes.
   b. 3 to 5 minutes.
   c. 2 minutes.
   d. 10 minutes.

8. If the mortar has been mixed with too much water, the mason
   a. adds more sand.
   b. adds more cement.
   c. adds more lime.
   d. adds in proportion all of the materials that were originally in the mix.

9. Mortar that contains a high percentage of cement is sticky and difficult to dislodge from the trowel. The trade term for this type of mortar is
   a. lean mortar.
   b. fat mortar.
   c. sharp mortar.
   d. mud.

10. In cold weather, masonry materials must be heated so that the proper curing of the mortar may take place. Heating the materials is recommended after the temperature drops below
    a. 32°F.
    b. 40°F.
    c. 45°F.
    d. 50°F.

SUMMARY, SECTION 3

- Portland cement added strength to the traditional lime-based mortar and changed the entire building industry.
- Lime-based mortar has many advantages over other types, including high water retention, workability, plasticity, greater bond strength, elasticity, flexibility, economy, high resistance to moisture from weathering, and the ability to reseal cracks in joints by autogenous healing.
- Sand for masonry mortar should be clean and free from organic matter.
- Water used to mix mortar should be clean and free from alkalis, salts, acids, and organic matter.
- When selecting a mortar, consider the needs of the project.
- The different types of mortars include Types M, N, S, and O. The different types are identified by the proportions of sand, lime, and portland cement in the final mix.
- Standard masonry cement is obtained prepackaged and requires only the addition of sand and water. It is very popular with masons, as there is less material to combine when preparing mortar on the job.
- Admixtures are available for use in mortar. Their benefits vary, but there is the possibility of the admixture affecting the strength of the mortar. Admixtures should be used only when there is a definite need. If they are used, the manufacturer's instructions must be followed to the letter.
- Efflorescence is one of the major problems encountered in the masonry trade. It is caused mainly by moisture coming in contact with soluble salts either in the mortar or the masonry unit. It can be prevented somewhat by keeping the masonry units dry.
- Always mix mortar with the maximum amount of water to maintain proper workability.
- Place all necessary mortar mixing materials as close together as possible without hindering the work.
- When mixing mortar manually or by machine, proportion the ingredients as accurately as possible.

- Always observe good safety practices.
- Do not use old or defective cement.
- Mortar materials can be heated in cold weather but strict adherence to designated temperatures must be observed or the strength of the mortar will be affected.

SUMMARY ACHIEVEMENT REVIEW, SECTION 3

Complete each of the following statements that refer to material found in Section 3.

1. The ingredients of mortar include _____.
2. Quicklime is changed to hydrated lime by a chemical reaction caused by the addition of _____.
3. The ability of mortar to hold moisture when applied to the masonry unit is called _____.
4. The use of portland cement in mortar gives the mortar _____.
5. Sand is defined as the deterioration of _____.
6. The siltation test is designed to detect the presence of _____ in sand.
7. The total number of portland cement-lime mortar types discussed is _____.
8. So that mortar may better resist freezing and thawing, _____ are added.
9. Efflorescence can be removed temporarily from masonry work by using either plain water or muriatic acid and water. The cause of efflorescence is _____.
10. Retempering of mortar should only be done to replace water lost through _____.
11. When using a material batcher for measuring mortar ingredients, the unit of measurement is _____.
12. In cases in which mortar is to be mixed by hand, it is mixed with a (an) _____.
13. Water is added to the mix slowly to prevent _____.
14. Before scraping excess mortar from a mechanical mixer, the mason should be sure that the mixer is _____.
15. Portland cement that sets prematurely due to excessive heat is called a (an) _____.
16. If only one type of portland cement-lime mortar could be chosen for a variety of conditions, the best choice would be _____.

SECTION FOUR
ESSENTIALS OF BONDING

UNIT 13
Introduction to Bonding

OBJECTIVES

After studying this unit, the student will be able to

- discuss the different meanings of *bonding*.
- describe various types of bonds.
- describe various brick positions and their application.

BOND

The term *bond* has several different meanings to the masonry student or apprentice. The following three types of bonds are important for masons to know:

- The adhesion of the masonry unit with the mortar joint is known as a *mortar bond*.
- The interlocking of masonry units to each other to distribute the weight of the wall is called a *structural bond*.
- The arrangement of masonry units to form a pattern, design, or texture is called a *pattern bond*.

All bonds should be structurally strong, provide a good adhesive bond between the mortar and the masonry unit, and maintain the pattern or design specified. These qualities are present in varying degrees, depending on the type of bond used.

Once the bond type has been established and construction of the wall has begun, the bond must be carried true for the entire length and height of the wall. *True,* in this sense, means that in every other course the unit should be in line vertically with the unit underneath

it or so that the pattern or design is not changed. All masons share the responsibility of building a true, strong bond if the finished job is to be an acceptable one.

MORTAR BOND

Mortar is the first material the mason uses when laying any masonry unit. Therefore, establishing a good bond between the unit and the mortar should be a prime consideration. The adhesion of mortar to the masonry unit, wall ties, or other means of reinforcement is known as the *mortar bond*.

The stronger the mortar bond is, the less chance there is of the wall leaking due to space forming between the mortar joints and masonry units and shrinkage of the joints, Figure 13-1. When reinforcing is used to strengthen the wall, a good bond must be formed between the steel and the mortar. Lime-based portland cement mortar provides a very high bond strength. Mortar must not be allowed to dry too quickly or a very poor bonding between the masonry units results and the strength of the wall is affected. Spreading mortar too far ahead of laying the masonry units results in a poor

Fig. 13-1 Bricks from an old building demonstrate that lime-based mortar adheres much better to brick (mason's left hand) than mortar with a high cement content (mason's right hand). Notice that the brick in the right hand has separated completely from the mortar.

Fig. 13-2 A 4″ brick wall laid one-half lap stretcher bond. This bond provides excellent distribution of weight.

mortar bond. Spreading mortar so that it does not dry too quickly is a skill that every mason must master.

Once the masonry units have been laid and the initial set has taken place, the units should not be moved or shifted in the wall. If the unit is moved after the mortar has partially set, the bond will be seriously weakened or broken. In these cases, the broken bond between the mortar and masonry unit never reunites completely. The high water retention of lime-based mortar allows more time to lay the unit before the mortar dries on the wall.

THE STRUCTURAL BOND

Structural bonding is the method by which individual units are interlocked or tied together when laid so that the entire mass acts as a single structural unit. Structural bonding of brick and concrete block may be accomplished in three different ways.

Overlapping of Units

Masonry units may be lapped over one another to provide structural strength. Lapping one unit half-way over the next unit provides the best distribution of weight, Figure 13-2.

Bonding Ties

Metal ties embedded in the mortar or steel wire reinforcements embedded in the horizontal mortar bed joint are used to strengthen and bond walls together. Modern building codes permit the use of rigid steel bonding ties, Figure 13-3. At least 1 metal tie should be used for every $4\frac{1}{2}$ square feet (sq ft) of wall surface. Ties in alternate courses should be staggered so that no 2 ties form a continuous vertical line. The maximum distance between ties in a horizontal position is 36″. In a vertical position, ties should be installed every 16″ for best results. Since there is a high degree of moisture present in all masonry walls due to condensation, ties should be coated to prevent deterioration. The most popular method is to galvanize the ties.

Structural bonding with wall ties is also used when building *cavity walls* (walls which have a cavity or air space of a minimum 2″ between them). This type of tie may have a *drip crimp* in the center. This drip crimp allows the moisture to drip from the center of the tie into the cavity rather than travel across the cavity to the other wall, Figure 13-4.

Structural bonding can also be accomplished by use of a header brick, considered the strongest tie in masonry. A *header brick* is a brick laid in such a position that it crosses over and rests on both sides of an 8″ wall. Its name is derived from the fact that the head or end of

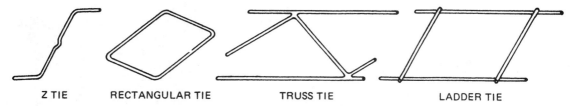

Z TIE   RECTANGULAR TIE   TRUSS TIE   LADDER TIE

Fig. 13-3 Various ties used to bond masonry walls together.

the brick is exposed on the face of the wall. Header ties are used in the construction of a solid masonry wall or cavity wall, Figure 13-5.

Reinforcement Rods

The third method of structural bonding is accomplished by building a double wall and leaving a space in the center, usually 2″ to 4″ wide. Steel reinforcement rods are installed in the air space and mortar or concrete grout is mixed and poured in the space to fill all voids

or holes, Figure 13-6. This type of mortar or concrete is known as *grout*. Although the grout should be thin enough to be poured, it must also be strong. There are specific proportions for the mixture of grout. This type of masonry work is known as *reinforced masonry*. This is a good method to use in structures which will have a great lateral load on the wall.

The method of structural bonding which is used depends on the requirements of the building, the wall type, and other factors. The metal tie method is highly recommended for exterior walls. Advantages include greater resistance to rain penetration and ease of construction. Metal ties also allow slight differential movement of the facing and backing walls with a minimum

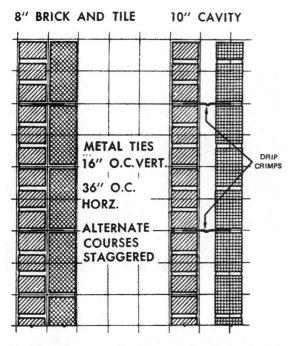

Fig. 13-4 Masonry wall with metal ties. Notice the drip crimps which allow moisture to drip into the cavities.

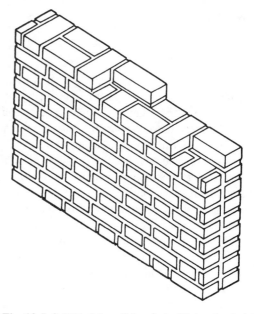

Fig. 13-5 Solid brick wall bonded with header bricks.

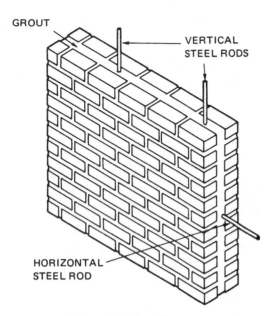

Fig. 13-6 Reinforced brick wall.

Fig. 13-7 Recessing and projecting of brickwork can be used to add variety and interest to structures.

of cracking. On walls with headers, moisture may collect on the inside of the wall where the header is installed. Any shifting of the wall also may cause cracks, as the brick header does not have the flexibility of the metal tie.

THE PATTERN BOND

Pattern bonding is defined as the arrangement of masonry units in a wall to form a pattern or recurring design. *Pattern* can also refer to a change in color and texture of the units. It may be possible to secure many different patterns with the same structural bond by mixing various colors and textures. For example, placing dark brick headers in specified positions throughout a wall has an interesting effect.

Recessing bricks or concrete block from the face of the wall and then returning the next course to the original wall line causes shadows and adds depth to the wall by suggesting a three-dimensional appearance. This method can be used to add interest and character to a structure, Figure 13-7.

Joints can be finished in a variety of ways with the use of striking tools. The raked joint adds depth to any wall. The V-joint highlights a rough-textured brick. The grapevine joint creates a pleasing effect with an irregular line running through the bed and head joints, Figure 13-8. The effect is especially appealing when used with pink colonial sand bricks.

Fig. 13-8 This section of brick from a building on the campus of the University of Virginia shows the colonial grapevine joint. The building, dating from the late eighteenth century, was constructed with a lime-based mortar, since portland cement had not yet been developed. The structure is still in good condition.

There are five basic structural bonds that form pattern arrangements in masonry work. Included are the *running bond,* more commonly called the *all stretcher bond, common* or *American bond, Flemish bond, English bond,* and *stack bond.* These bonds will be explained further in Unit 14. There are, of course, more than five bonds that can be used, but the majority of all masonry work specifies one of these five. Through the use of bonds, mixing of color and texture, recession and projection of the masonry units, and special treatment of mortar joints, an almost unlimited number of patterns can be developed.

Brick Positions

To create the various bonds and patterns found in masonry work, the masonry unit must be laid in a particular position. These positions of bricks have been given trade names over the years by the masonry industry. Architects specify the arrangements of patterns and bonds by indicating the position of the bricks or blocks when laid in the wall.

There are six different positions in which a brick or masonry unit may be laid, Figure 13-9.

The *stretcher* is laid in a horizontal position with the longest, narrowest side exposed at the front of the wall. This is the most frequent position in which bricks are laid.

The *header* is laid with the 4″ face of the brick exposed at the front of the wall. The widest part of the header is laid down in the mortar bed joint. Headers are used primarily for tying two separate masonry walls together. They can also be used for capping walls, contributing to different pattern arrangements, and on flat windowsills.

The *soldier* is laid in a vertical position on its narrowest and shortest side, with the longest side of the brick exposed on the front side of the wall. Soldiers are most frequently used over doors, windows, or openings to simulate an arch. Do not, however, confuse a soldier course with an arch, as they are entirely different. The soldier is also used in different types of pattern arrangements.

The *shiner* is laid with the widest side of the brick in a horizontal position exposed on the front of the finished wall. It is also used for leveling when a material 4″ high is needed, or for decorative purposes when constructing certain pattern bonds. The most common use of a shiner is in flat paving work, such as brick walks.

The *rowlock* is laid in a vertical position with the end of the brick facing the front of the finished wall. This is different from the header in that it is laid on its narrowest edge. The most common use of the rowlock is brick windowsills. Rowlocks are also used to cap walls, in pattern arrangements when building complicated bonds, and in the ornamental cornices of brick buildings.

The *sailor* is laid with the widest part of the brick facing the front of the finished wall in a vertical position. Sailors have very limited use, but are found in different pattern bonds. The most common use of the sailor is in brick walks and to form bond patterns in a wall.

It has been said that much of the old charm and appeal of masonry work has been lost since prefabricated materials have become popular. However, through the use of various bonds and patterns, the mason can exercise creativity and add interest to structures. However, masons should remember that as the design becomes more complicated, the cost involved in creating the structure rises.

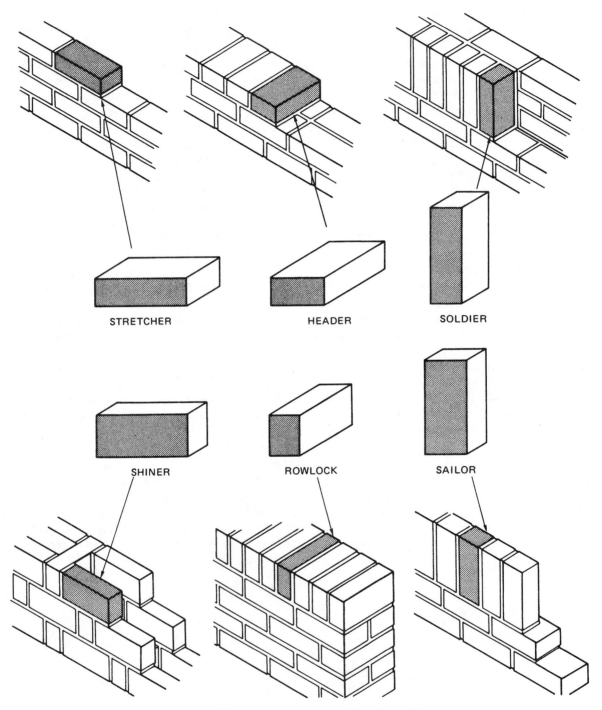

STRETCHER   HEADER   SOLDIER

SHINER   ROWLOCK   SAILOR

Fig. 13-9 Brick positions as they appear in a wall.

ACHIEVEMENT REVIEW

A. Select the best answer from the choices offered to complete the statement or answer the question. List your choice by letter identification.

1. Mortar bond is related to which of the following terms?
   a. Pattern
   b. Adhesiveness
   c. Interlocking units
   d. Tying

2. The term *bond* means several different things to the mason. Which of the following is the most important type of bonding that is related to distribution of weight?
   a. Mortar bond
   b. Structural bond
   c. Pattern bond
   d. Finish and texture bonding

3. Which of the following describes the term *breaking the set?*
   a. Cutting the masonry unit
   b. Laying out the bond without mortar
   c. Moving the masonry unit after the mortar begins to harden
   d. Arrangement of the masonry unit to form a pattern

4. Masonry units may be overlapped when building a wall. The strongest overlap is the
   a. one-quarter lap.
   b. three-quarter lap.
   c. half lap.

5. The maximum distance between horizontal wall ties in a masonry wall should be
   a. 36″.
   b. 16″.
   c. 24″.
   d. 48″.

6. The maximum distance between vertical wall ties in a masonry wall should be
   a. 24″.
   b. 36″.
   c. 48″.
   d. 16″.

7. Structural bonding can be accomplished by tying two walls together with a brick. When a brick is laid in this position it is called a
   a. stretcher.
   b. snap header.
   c. soldier.
   d. header.

8. Mortar proportioned so that it is thin enough to pour into a cavity wall and fill all voids is called
   a. rich mortar.
   b. grout.
   c. portland cement.
   d. lean mortar.

9. A wall can be given the appearance of pronounced shadows and a sense of depth by
   a. staggering the brick alignment in the bond.
   b. rough-textured brick.
   c. recession and projection of brick.
   d. use of a darker mortar in the joints.

10. It is possible to design and construct many different patterns while using the same structural bond. This is done by
    a. mixing the colors and textures of the bricks.
    b. cutting the bricks.
    c. tooling the joints differently.
    d. changing the color of the mortar joint.

B. Study the different brick positions shown in Figure 13-9. Make a drawing showing each of the brick positions as they may be used in a brick wall.

Example (Figure 13-10):

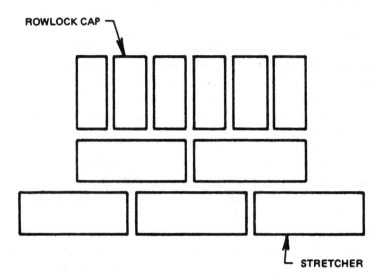

Fig. 13-10 Illustration for achievement review, part B.

PROJECT 5: BRICK POSITION WALL

OBJECTIVE

- The masonry student will dry bond and build an 8″ brick wall 16 courses high, including the rowlock cap course, utilizing the various brick positions shown in Figure 13-11.

EQUIPMENT, TOOLS, AND SUPPLIES

Mortar pan or board	Brick hammer
Mortar	Convex sled runner jointer
Standard modular face bricks	4- and 2-ft level
Chalk box	Modular rule
Tempering water	Brush
Mason's trowel	Line and line blocks
Brick set chisel	Slicker jointer
Square	Pencil

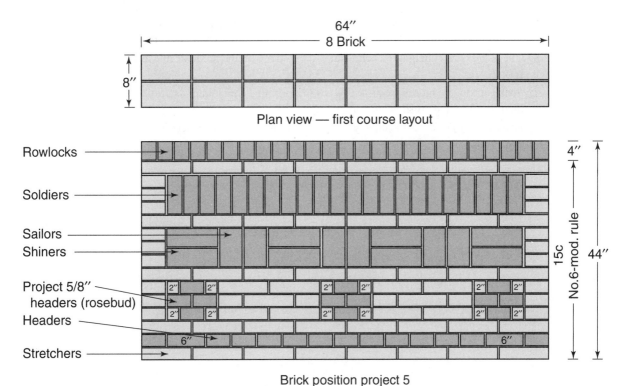

Brick position project 5

Fig. 13-11 Brick position Project 5.

SUGGESTIONS

- Space materials from wall line.
- Keep all tools not being used away from immediate work area to avoid accidents.
- Have a concrete block handy to set levels in when not in use.
- Study the plan carefully before laying out project.
- Return all excess mortar as you work to mortar pan or board and re-temper mortar as needed.
- Observe proper safety practices at all times.

PROCEDURE

1. Strike a chalk line on the base for the project.
2. Lay out the first course dry using the end of the index finger as a spacer for the head joints.
3. Spot a brick (lay) in mortar to the proper height on each end of the project. Level, plumb, and square the end brick.
4. Attach a line block and line to the end brick and lay the first course out to the line.

5. Continue laying each succeeding course of brick according to the plan. Do not forget to project or set out the header rosebud pattern brick 5/8″ as shown on the plan.

6. Check the correct height of each course to number 6 on the modular rule as the project is built.

7. Strike or tool the mortar joints with a convex or round sled jointer as needed and brush the project with a brush when completed.

8. Recheck the project for accuracy according to the plan. The project is now ready for inspection.

UNIT 14
Traditional Structural and Pattern Bonds

OBJECTIVES

After studying this unit, the student will be able to

■ describe the five basic bonds and how to start each.

■ lay out and construct a brick panel which includes the five basic bonds.

■ dry bond masonry projects in preparation for construction.

As mentioned in Unit 13, there are five basic bonds with which the mason should be familiar. They are the *running* (or *all stretcher*) *bond, common* (or *American*) *bond, Flemish bond, English bond,* and the *stack bond.* Although the mason may not build all of these bonds frequently, a thorough knowledge of the construction of each is necessary, since every mason will lay out and construct each bond at some time. Remember that the arrangement of texture and color can accentuate all bonds and pattern designs.

The most important step in building any bond is to start the course correctly. Starting a course involves using different pieces or cuts of brick on the corner jamb of the structure. The mason should be aware of the requirements of brick pieces for each bond.

The bond must be built true to the pattern or layout course. Breaking up an established pattern in a decorative bond destroys the beauty and meaning of the bond. Architects and builders are also very conscious of good bond pattern development, since even people unfamiliar with masonry construction can detect broken bond patterns when the job is completed.

THE RUNNING OR ALL STRETCHER BOND

The simplest of all bonds to build is the *running* or *all stretcher bond.* This bond consists entirely of stretcher bricks (full bricks) with the exception of pieces that would occur at windows, doors, or other openings.

Since there are no headers, metal ties are used when the wall must be tied to another masonry backing or framework such as the sheathing of a frame house. The running or all stretcher bond is simple, fast, and economical to construct. It is, therefore, a favorite of builders for masonry construction.

The majority of brick masonry houses constructed today are brick veneers, Figure 14-1. Brick veneer walls are usually 4″ thick and built against a frame or a comparable substitute. The running bond is a natural choice for veneer work, since the standard brick is 4″ thick.

The running bond is constructed with either the *half lap* (half bond) or *one-third lap* (one-third bond). The half lap is used when laying standard 8″ bricks, since the return brick on the corner measures 4″ and the standard brick will, therefore, bond evenly over the next brick, Figure 14-2.

The one-third lap is generally used when laying a 12″ long Norman brick, Figure 14-3. It forms a weaker bond than the half lap because the weight is not distributed as evenly and the half lap covers more area. The one-third lap is, however, a very sound lap which forms a good bond.

THE COMMON BOND

The *common* or *American bond* is a variation of the running bond, with a course of headers at regular intervals tying the units together. The headers may be installed

153

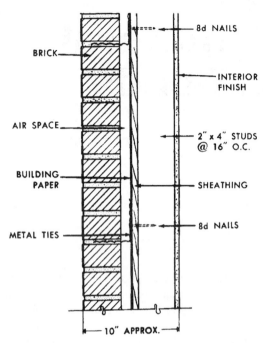

Fig. 14-1 Brick veneer wall section. Note the wall ties in the framework.

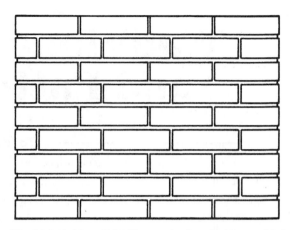

Fig. 14-3 Brick wall laid in running bond with one-third lap.

on the fifth, sixth, or seventh course, depending on the size of the masonry or backup units, Figure 14-4. As a rule, the common bond is only used when a solid wall is being built. The common bond was the most often used bond until the demand for veneer masonry became so great.

To start the corner for a header course, lay a 6″ piece of brick (often called a *three-quarter*) on the corner. Two three-quarters are required, as one must be laid on each side of the corner. The header brick is laid against the three-quarter and carried throughout the course. This is known as a *full header course,* Figure 14-4. The three-quarter on the corner allows the header to center over the head joint in the stretcher course below. Every other header laid will then be centered over the head joint of the stretcher beneath it.

If a variation of the header is desired for a pattern or design, a *Flemish header* may be used. The Flemish header is also started from the corner with a three-

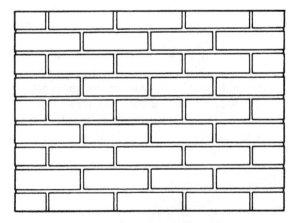

Fig. 14-2 Running bond (all stretcher) wall laid with a half lap over brick underneath.

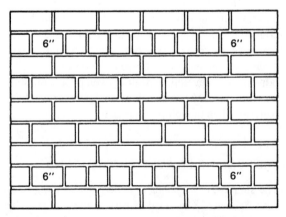

Fig. 14-4 Common bond. Note the full header courses every sixth course.

quarter piece. Instead of using headers throughout the course, however, a stretcher is alternated between every header. The Flemish header does not form as secure a bond as the full header since only every other brick is tying the wall together, Figures 14-5 and 14-6.

When the mason reaches the height of the wall at which the header is to be installed, the wall is said to be *header high.* When using the common or American bond on a wall, it is important that the front of the wall

Fig. 14-5 Common bond. Flemish headers are on the sixth course.

Fig. 14-6 Brick wall laid in common bond with a Flemish bond header every seventh course.

and the back portion of the wall be kept level at all times so that the header course is level across the two walls.

THE FLEMISH BOND

The *Flemish* bond is one of the truly beautiful pattern bonds. Each course of bricks consists of alternate stretchers and headers, with each header centered over the stretcher that is directly beneath it. The headers, located on every other course, are in an even vertical line. This presents a very pleasing appearance. Since the wall is tied on each course, due to the fact that each header is tied into the backing course, a wall with a Flemish bond will always be built of solid brick. The only exception to this rule occurs when a 4″ wall is built with a Flemish bond and *dummy headers* (or *snap headers*) are cut and used on the wall.

Since the wall with the Flemish bond must be built of solid brick, it is a more costly type of wall to build. Operators of many buildings such as banks, libraries, schools, and museums prefer the architectural beauty that the Flemish bond provides and are willing to pay the extra cost. The Flemish bond was very popular in colonial America when the cost of labor and materials was not a factor as it is today.

There are two different methods of starting a Flemish bond from the corner, known as the Dutch corner and the English corner, Figure 14-7. In the Dutch corner, a 6″ piece of brick (three-quarter) is used to start

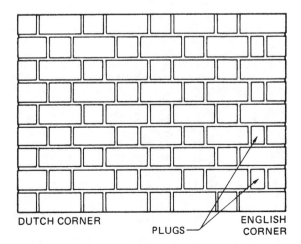

Fig. 14-7 Flemish bond with Dutch and English corners.

the corner or is laid against the head of the return brick on the corner, depending on the structure. The header is then laid against the three-quarter in the same way as the full-header bond is laid. The English corner uses a 2″ piece (also known as a *plug*) against the corner return brick, Figure 14-8. Very small pieces of brick are normally not used in masonry. It is very seldom that a piece smaller than a half brick is specified by an architect unless a true colonial job is to be built and then a 2″ piece is used.

The Dutch corner is considered a more modern design than the English corner. The Dutch method is more commonly used since it is easier to cut and is more firm in the mortar joint, thereby being more economical to build. Both methods serve the same purpose of breaking the bond of the brick that is located underneath it.

ENGLISH BOND

The *English* bond is composed of alternate courses of headers and stretchers, Figure 14-9. Do not confuse this bond with the Flemish bond, which has a header and stretcher alternating on the same course. In the English bond, the headers are centered on the stretchers and the joints between the stretchers in all courses are in line vertically. This bond is usually used when a

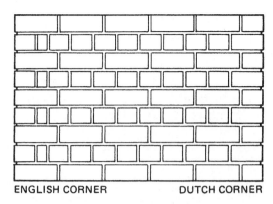

ENGLISH CORNER   DUTCH CORNER

Fig. 14-9 English bond with English corner (1″ piece) and Dutch corner (6″ piece).

structure with an 8″ solid brick wall is being built. If a 4″ wall is specified, snap headers must be cut and used in place of the header bricks.

The English bond is seldom used in masonry today due to the expense of constructing a solid brick wall and the high cost of laying an intricate pattern. It was a favorite in colonial period masonry and is still preferred when colonial reproduction is required, Figure 14-10.

Fig. 14-8 Corner of a brick building built with Flemish bond. Note the 2″ starter pieces near the corner.

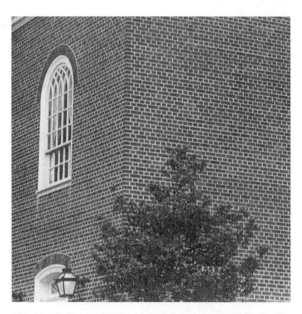

Fig. 14-10 Example of an English bond wall. Note the diamond pattern effect of this bond.

The mason must be especially precise in keeping the head joints in true plumb alignment. If the bond staggers even slightly, the effect of the pattern is seriously affected. The cross joints between headers must be kept full of mortar to prevent any water from leaking through the wall. Any bond that has many headers is more subject to moisture leakage than a running bond, since moisture which collects in the center of the wall leaks out at the header level.

The English bond can be started from the corner or jamb with a 2″ piece of brick (English corner) or a three-quarter (Dutch corner), just as a Flemish bond is started.

THE STACK BOND

The *stack* bond, Figure 14-11, is used basically for decorative purposes. In this bond, there is no overlapping of masonry units when they are laid in the wall. This pattern must be bonded or tied with ladder or truss-type steel reinforcing wire, whether the wall is *load bearing* (a wall which supports part of the weight of a building, such as steel beams, concrete floors, or joists) or *non-load bearing* (a wall that supports only its own weight).

Masonry units in stack bond must be of the same length and height since the vertical alignment of the head joint must be perfectly straight. A truer size of face brick is usually selected when building a project

in the stack bond to help assure that the joints will be plumb. The mortar head joint will compensate for only a minor difference in brick size before it becomes noticeable in the wall. The stack bond takes more time for the mason to construct than a conventional running bond because of these factors.

If a stack bond is specified, it is for design purposes. Structurally, the stack bond is the poorest of all of the bonds discussed since there is no overlapping of the masonry units. A slight movement of the units caused by expansion or contraction of the wall could cause cracks to appear in the vertical head joints.

However, a wall in the stack bond is a safe and acceptable method to use when constructing a masonry wall. If specifications are followed and the units are tied together with ladder or truss-type wall ties, the wall will be structurally sound. Many architects design and recommend the use of stack bonded walls in modern buildings.

GARDEN WALL BOND

An adaptation of the Flemish bond, the *garden wall bond,* is often used in walls to enclose or screen a garden, courtyard, or estate. There are two versions of the garden wall bond. A bond with 2 stretchers and a header alternating on the same course is known as a *double stretcher garden wall bond,* Figure 14-12.

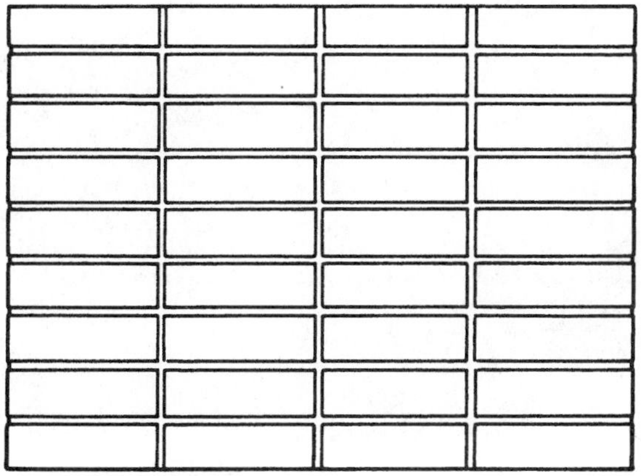

Fig. 14-11 Stack bond. Notice the plumb vertical head joints.

Fig. 14-12 Double stretcher garden wall bond with units in diagonal lines.

A garden wall bond with 3 stretchers and a header alternating on each course is known as a *three stretcher garden wall bond,* Figure 14-13. The diamond effect created appears larger in the three stretcher garden wall bond.

Combinations of dark headers and stretchers emphasize the pleasing diamond pattern, making the garden wall bond popular for decorative work. Projecting or recessed bricks also achieve special effects in the pattern. However, each brick must be properly placed or the pattern will be ruined. Masons should follow an elevation view of the project to ensure that the pattern is correct.

Fig. 14-13 Three stretcher garden wall bond with brick in dovetail fashion.

Fig. 14-14 A combinations of intricate bonds, patterns and colors can be used in the same building to create very beautiful unusual brick patterns such as shown in the old First Union Bank in Key West, Florida.

DRY BONDING THE MASONRY UNIT

Bricks

Dry bonding is another important type of bonding that the mason must be able to perform. This involves laying out the masonry unit dry (without using mortar) to determine if a unit will fit in a given area without cutting small bricks. Dry bonding by the proper method is the sign of a true craftsperson.

It is always good practice to lay out the bond dry before spreading mortar and actually laying bricks. Since bricks are products which have been burned under intense heat, the sizes are going to vary somewhat even though they are initially constructed in the standard size. Architectural plans for wall length cannot be changed to accommodate brick length.

How to Dry Bond

To dry bond, select 1 brick of each size that will be used in the finished wall. Three different sizes of bricks may be necessary to lay out the course. Starting at the beginning point, usually the corner, lay a brick even with the edge. As this is being done, insert your index finger sideways between the head of the bricks as a gauge for the head joint, Figure 14-15. Lay the first course of bricks across the project to the opposite corner. Using the index finger as a gauge, the head joint will usually measure $\frac{3}{8}''$, which is the standard preferred head joint. It may be necessary to open the head joints further to work with the bond or tighten the joints to prevent cutting the bricks. If the head joints are too big to open up, reverse the corner and tighten the joints to make up for the difference.

After the bricks have been laid out dry across the entire course and adjusted to accommodate all full-sized bricks, the mortar is spread and the bricks are laid permanently into position. Do not remove all the bricks from the dry bonding position before spreading the mortar. If all the bricks are removed at once, the mason cannot possibly remember all the proper positions and the dry-bonding technique would be useless.

Spread mortar for 2 or 3 bricks at one time and lay them in their proper position without disturbing the others. Continue to pick up and lay bricks across the course until all have been laid. If all the bricks must be removed

Fig. 14-15 When dry bonding brick, use the index finger as a gauge for the head joint. The joint should be about $\frac{3}{8}''$.

across the entire course before spreading the mortar, the end of every third brick could be marked with a pencil or crayon on the base where it is to be laid.

At no time should a piece of brick smaller than $6''$ be laid in the wall unless, of course, the bond specifies its use. If it is not possible for the wall to be entirely of full-sized bricks, cut as many three-quarters as are needed to make up the difference.

For example, suppose that a mason dry bonds a brick wall and finds that full bricks will construct it. The corners are laid up and the course filled in with bricks. However, upon reaching the center of the wall, the mason discovers that either due to a mistake in dry bonding or variations in brick sizes, a *bat* (half brick) is in the middle of the wall, Figure 14-16A. Since tearing down both corners would be expensive and impractical, the mason should cut 3 bricks of equal size to replace the half brick, Figure 14-16B. If the bricks are cut neatly, it is difficult to distinguish them from full bricks. A wall of full bricks is preferable, but faced with this field problem, a mason must deal with it in the most practical way.

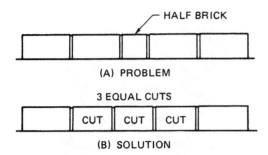

(A) PROBLEM

3 EQUAL CUTS

(B) SOLUTION

Fig. 14-16 Proper method of replacing a bat (half brick) in the wall.

ACHIEVEMENT REVIEW

The column on the left contains a statement associated with bonding. The column on the right lists terms. Select the correct term from the right-hand list and match it with the proper statement on the left.

1. Number of basic bonds the mason should know and be able to build

2. Bond built completely of whole bricks or stretchers

3. Double wall with air space in the center

4. Reason that the half lap of one brick over another is considered the strongest type of bond

5. Variation of the running bond with a course of headers being laid at the fifth, sixth, or seventh course

6. 6″ piece of brick

7. Header course with alternating stretchers on the same course

8. Header course that is composed completely of headers

9. Height of the wall where the header is to be installed

10. Bond consisting of alternating headers and stretchers on every course

11. Dummy headers that have been cut for use in a 4″ wall

12. Method of starting a corner by using a 2″ piece of brick against the return brick

13. Method of starting a corner by using a 6″ piece of brick

a. Snap headers
b. Full header
c. Three-quarter
d. Dry bonding
e. Bat
f. Cross joint
g. English corner
h. Header high
i. Flemish header
j. Cavity wall
k. Common or American bond
l. Garden wall bond
m. Stack bond
n. Dutch corner
o. Best distribution of weight
p. Five
q. Flemish bond
r. Plug
s. English bond
t. Running bond
u. Nine
v. Economy

14. Bond constructed of 3 stretchers alternating with a header on the same course

15. Bond constructed of alternate courses of headers and stretchers

16. Bond in which there are no overlapping bricks and all of the head joints are in true plumb alignment

17. Procedure of laying out bricks to check the bond without using mortar

18. Common term for a half brick

19. Mortar joint between headers

20. Common term for a 2″ piece of brick

BOND DRAWING ASSIGNMENT

Use a number 2 pencil, ruler, and one-fourth square graph paper for the assignment.

A. Draw a brick wall section that is seven bricks long and 18 courses high. Let every four squares represent one brick in length and each square in height represent a course of brick. Do not try to draw in mortar joints in thickness; use only a single line for head and bed joints. Start the drawing by using all full bricks on the first course and alternating in a half bond for the rest of the drawing. Shade in with pencil all half bricks used. Your drawing can be improved by using a red or pink colored pencil to shade bricks and a yellow pencil to highlight the mortar joints. Label the drawing RUNNING BOND when completed.

B. Draw a brick common bond wall section that is six bricks long and 13 courses high. Start the wall with six bricks long on the first course and a full header course on the second course. Refer to the common bond wall illustration in the unit for other information. Continue drawing the wall until the 13 courses are completed. Make sure that you mark all 6″ pieces on the header courses. As in the Running Bond Assignment, do not attempt to show the mortar joints in thickness; use only a single line for all head and bed joints. Shade in all of the headers lightly with the pencil when completed. Label this drawing COMMON BOND WALL WITH A FULL HEADER.

C. Draw a Flemish bond wall section that is six bricks in length and 15 courses in height, using the illustration of a Flemish bond that is shown in the unit. Make sure that the left side of your drawing is labeled and uses a Dutch corner (6″ piece) and the right-hand side uses an English corner (2″ piece, or plug as it is known in the trade). Shade in all of the headers lightly with a pencil so they will stand out and be sure to mark the 6″ pieces by marking 6″ on them and the 2″ pieces by marking 2″ on them. The drawing can be highlighted by shading the headers in lightly with a light pink or red pencil. Label this drawing FLEMISH BOND.

PROJECT 6: DRY BONDING AND CONSTRUCTING A 4″ WALL

OBJECTIVE

- The masonry student will dry bond and build a 4″ brick wall 10 courses high and 4′ long in the running bond, Figure 14-17.

EQUIPMENT, TOOLS, AND SUPPLIES

Mortar pan or board   Brick hammer

Mortar   Convex striker

Face bricks   Plumb rule

Chalk box   Modular rule

Tempering water   Brush

Mason's trowel

SUGGESTIONS

- Space materials being used approximately 2′ from wall.
- Keep all tools not being used away from the immediate work area to avoid accidents.
- Have a concrete block handy in which to set plumb rule when not in use.
- Be careful not to splash mortar over the wall.

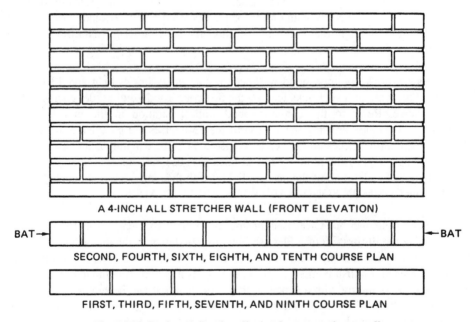

A 4-INCH ALL STRETCHER WALL (FRONT ELEVATION)

BAT→ ←BAT

SECOND, FOURTH, SIXTH, EIGHTH, AND TENTH COURSE PLAN

FIRST, THIRD, FIFTH, SEVENTH, AND NINTH COURSE PLAN

Fig. 14-17 Project 6: Dry bonding and constructing a wall.

- Use as little pressure as possible when leveling and plumbing. The higher the wall is built, the more delicate the wall is until the mortar sets.
- Return all mortar that falls on the base to the pan and temper the mortar as necessary. Keep the work area neat and clean.
- Observe proper safety practices at all times.
- If any mortar splashes in the eyes, wash out immediately with water. Avoid rubbing the eyes.

PROCEDURE

1. Strike a chalk line a little longer than 4′ on the base for a reference point.
2. Lay out 6 bricks dry using the index finger as a guide for the head joints.
3. Bed up a brick on each end of the wall, level, and plumb.
4. Lay bricks between the two end bricks. Level and plumb. Be sure to straighten the top edge of the course with the level.
5. Do not be concerned if the wall is slightly over 4′ in length. Be certain, however, that all of the head joints are ⅜″ in width and well filled with mortar.
6. Lay a bat on each end of the wall on the second course, as shown on the plan. Run the second course. Follow the same procedure of leveling and plumbing.
7. Check the height of the wall with the modular rule. Use the number 6 on the rule as your checking point.
8. Build the wall following the described procedures until it is 10 courses high.
9. Strike the mortar joints as needed with a convex jointer and brush the wall.
10. Double-check the wall to be certain that it is built according to directions and is ready for inspection.

PROJECT 7: BUILDING AN 8″ WALL IN COMMON BOND

OBJECTIVE

- The masonry student will lay out and build an 8″ wall in the common bond and lay a header course to tie the wall together, Figure 14-18.

EQUIPMENT, TOOLS, AND SUPPLIES

Mortar pan or board

Mortar

Face bricks

Mortar hoe to temper mortar

Tempering water

Brick tongs

Chalk box

Mason's hand tools (trowel, convex or round striker, plumb rule, brick hammer, modular rule, square, and brush)

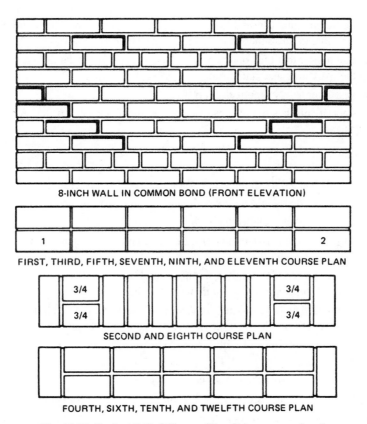

8-INCH WALL IN COMMON BOND (FRONT ELEVATION)

FIRST, THIRD, FIFTH, SEVENTH, NINTH, AND ELEVENTH COURSE PLAN

SECOND AND EIGHTH COURSE PLAN

FOURTH, SIXTH, TENTH, AND TWELFTH COURSE PLAN

Fig. 14-18 Project 7: Building an 8" wall in common bond.

SUGGESTIONS

- Cut the header starter pieces 6″ long.
- Be sure that the header is laid level and straight across both walls.
- Use solid joints, particularly on header courses.
- Check mortar joints constantly and strike when the thumbprint is firm.
- Prevent excess mortar from dropping between the walls.
- Build to the brick header a height on one side of the wall, then build this height on the other side to keep the wall in alignment.
- Keep the work area clean and free from hazards.

PROCEDURE

1. Assemble materials in the work area.
2. Set the mortar pan back from the wall approximately 2′ to provide sufficient working space.
3. Strike a chalk line a little longer than 4′ on the base as a reference point.

4. Dry bond the project.

5. Spot (lay) a brick in mortar on each end of the project level and plumb. Square jambs (sides).

6. Lay the course level and plumb with the spotted bricks.

7. Back the wall with another course of bricks, being certain that the wall is the same width as the header bricks. This measurement will be approximately 8″ in width but must be the exact size of the header bricks being used. Use a header brick laid dry to gauge the width.

8. Lay the second course of bricks as shown on the plan. Be sure to follow the course layout for the header course. Follow the same procedure of leveling and plumbing.

9. Continue building the wall until completed as the plan indicates.

10. Check the height of the wall with the number 6 on a modular rule as needed while constructing the wall.

11. Strike the work as needed with a convex or round jointer and brush the wall upon completion.

12. Recheck the work for accuracy according to plan. The work is now ready for inspection.

UNIT 15
Bonding Concrete Block and Rules for Bonding

OBJECTIVES

After studying this unit, the student will be able to

- lay out a block wall in the running bond.
- lay out a block wall in the stack bond.
- apply rules for bonding brick and block.

Concrete masonry units, like bricks, can also be laid in different bonds. As a rule, they are laid in either the running bond or the stack bond. Various effects can be achieved depending on the bond, the finish of the mortar joint, and the choice of the many types of decorative blocks now available. You may wish to review the types of blocks described in Unit 4.

RUNNING BOND

The procedure for establishing a running bond in block is the same as that described for brickwork. That is, lapping half a unit over the unit underneath. As a rule, this allows the best distribution of weight and forms the strongest wall.

Laying Out the Corner in Running Bond

The size of the block used to start the corner depends on the size of the block being laid. *Size of the block* refers to its width, as all standard blocks are the same length: $15\frac{5}{8}''$, with $\frac{3}{8}''$ for the mortar joint for a total of $16''$.

Eight-inch concrete blocks are the only blocks that lap perfectly over each other without the necessity of cutting the corner block. This is because the corner end of the $8''$ block is exactly half the length of the $16''$ stretcher block. Each course laid will reverse and alternate as the wall is built, Figure 15-1.

When the concrete block is wider than $8''$, special L-shaped corner blocks are available. The L-shaped corner block enables the adjacent block to make the correct lap of $8''$ over the block beneath. Figure 15-2 shows how typical L-shaped corner blocks appear when they are laid in position.

L-shaped corner blocks are usually available only in $10''$ and $12''$ sizes. All other size block walls must be started with a cut block on the corner to produce the $8''$ lap bond.

Figure 15-3 illustrates three different block sizes and the proper corner block lengths for each size. The cut piece is different for each size block shown. To determine what length to cut a block corner, add the width of the block being laid to the standard figure of $8''$. For example, a mason working with $4''$ block cuts a block corner $12''$ long—the standard lap of $8''$ plus the width of the $4''$ block makes a total of $12''$, Figure 15-4.

STACK BOND

A concrete block wall built in the stack bond is constructed the same as a brick wall. Full stretchers can be laid or, for decorative purposes, all half bricks may be used. Concrete blocks are very accurately sized and it is much easier to maintain them in true vertical alignment than brick units. On many jobs, the blocks are cut with the masonry saw. After the first course is laid out, cuts for the entire wall can be made since the size of the cuts from the first course to the top course never changes. This speeds up the job by allowing the cuts

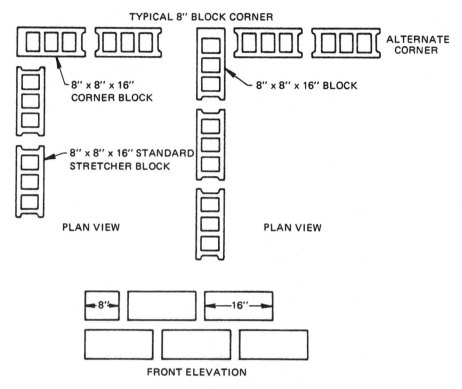

Fig. 15-1 Bonding an 8″ corner block. The plan views show alternating courses; front elevation shows view facing the wall.

to be made beforehand and the blocks to be set to dry in the work area.

PLANNING THE BOND

Block walls and bonds must be carefully planned for economy and efficiency. Because of the time factor and the cost of labor involved, the cutting of block should be kept to a minimum and all openings and heights should be planned specifically for the size of the unit being laid. Concrete block buildings should be laid out by the architect on an 8″ grid since either a half or whole block can be used around windows, doors, and other openings. The fewer pieces there are in the wall, the more efficient the project is.

A mason (or on larger jobs, a group of masons) is selected to work ahead of the rest of the crew and lay out major features of the structure. The layout mason determines if any dimensions can be changed so that the masonry can be built with as few cut units as possible. The bond is also laid out for the crew in advance, Figure 15-5. Since they do not have to figure out the bond pattern, the crew's efficiency is greatly improved.

Masons can use several tools to mark off concrete block bond on the base before laying any of the blocks in mortar:

- Mason's folding rule marks should be placed every 16″ on the layout line. These are known as *checkpoints* in the trade.

- The steel tape is used to mark off the bond over long distances.

- Use the 48″ plumb rule as a gauge by laying it down on the layout line and marking the end of it. Then shift the plumb rule one length and mark again. Three concrete blocks equal 48″, the length of the plumb rule.

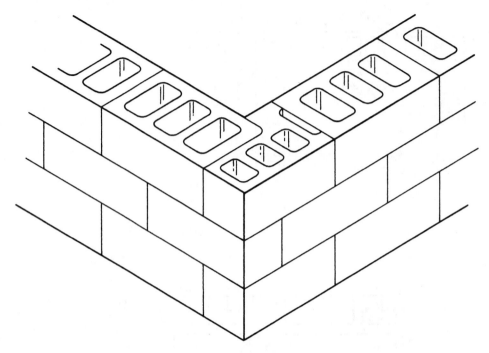

Fig. 15-2 L-shaped, 12″ corner block laid in position.

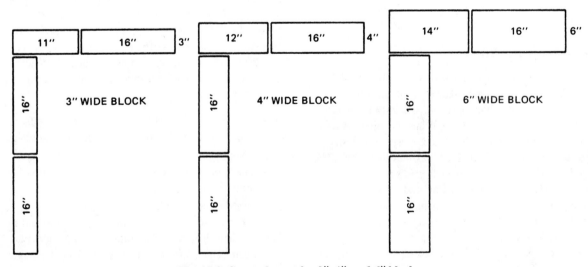

Fig. 15-3 Corner layout for 3″, 4″, and 6″ block.

Fig. 15-4 Student mason laying out 4″ block corner with a 12″ cut block.

Fig. 15-5 First course of block laid out and wall stocked with materials ready for masons.

To begin the project, lay the first course and establish the bond for the wall. Study the floor plans and determine locations of windows, doors, and other openings. With a steel tape or rule, mark these openings on the first course with a crayon or pencil. Windows and doors may often be shifted so that fewer blocks require cutting. If this may be done without changing any critical dimensions, a considerable lowering of costs is possible. However, the supervisor should always be consulted before any changes are made.

The height of concrete block walls should be laid out in the same way, Figure 15-6. Since most windows, doors, and openings are estimated by using the modular grid, the space between the top of openings and the floor level should be made up of full blocks. To accomplish this, it may be necessary to start with less than a full-sized block on the first course. (Blocks

less than 8″ in height are referred to as *rip blocks*. The 4″ block is the most commonly used.) It is impossible, in some cases, to change wall dimensions to accommodate only whole block. If this is the case, a cut block must be used. However, masonry work should never be started without careful consideration and planning. Without careful planning, economy and appearance are likely to be sacrificed.

BONDING BRICKS AND CONCRETE BLOCK TO FORM A WALL

Another form of bonding involves the combination of two different masonry materials to form a single structure. This is known as *composite masonry*.

Many different masonry materials may be bonded together, the most common being bricks and concrete block. There are two basic methods of bonding bricks and concrete block together. One method has a specified number of courses of bricks laid to a given height. The bricks are then backed with concrete block which are the same height as the brickwork. Each single vertical section of masonry 1 unit in thickness is called a *wythe* or *withe*. A brick header may be utilized to tie 2 wythes together into 1 single wall, Figure 15-7.

The second method involves construction of the brick wall backed by concrete blocks and tied together with metal ties or reinforcement wire, Figure 15-8.

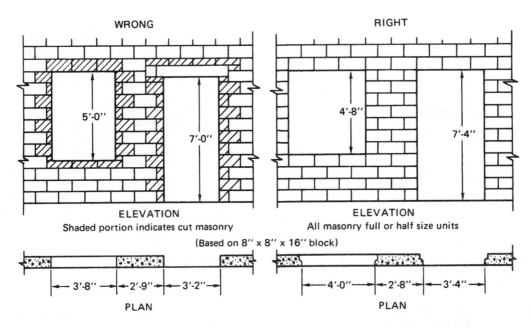

Fig. 15-6 Planning the wall using concrete block. Notice that the illustration showing the correct procedure calls for only full-size block in the space between the top of the openings and the ceiling.

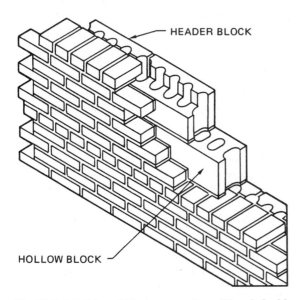

Fig. 15-7 A brick and block composite wall bonded with a brick header.

Fig. 15-8 Brick and block composite wall tied together with metal wire joint reinforcement.

RULES FOR BONDING

When bonding masonry materials, masons should follow certain rules. Following these rules will ensure that the structures built are strong, well designed, and economically constructed with no more pieces than necessary. The following three particular rules will help the mason achieve high-quality bonds:

- Keep the bond plumb.
- Keep walls level with backing materials.
- Maintain the set of the mortar bond.

Keeping the Bond Plumb

In high-quality masonry work, every head joint must be kept exactly plumb with those on alternating courses below. This is known as creating a *plumb bond*. The distribution of weight is more uniform in a plumb bond, resulting in a stronger wall. In decorative bond patterns, it is particularly important that the pattern be laid accurately on each course or the effect of the pattern will be lost.

When the architect has specified a very exact plumb bond, the mason sometimes plumbs the head joints with the plumb rule at given points on specific courses. This will assure that the joints are kept in true vertical alignment. It is the responsibility of the mason who is laying out a bond to maintain the bond until the wall is completed.

When masonry units are laid or stacked directly over one another with the head joints in plumb vertical position, they are laid, in the trade term, *jack-over-jack*. Only in the stack bond is this practice acceptable.

Keeping Walls Level

Backing courses should always be level with the face of the wall. This allows the proper placing of headers or wall ties to bond 2 wythes together to form a solid structural unit. The backing wall should be kept level as each course is laid, rather than being constructed all at once with 1 mortar joint. This is a common practice in the trade, but can cause problems in maintaining a level wall.

Maintaining the Set of the Mortar Bond

A masonry unit should not be moved or shifted once it is laid in mortar and takes the initial set. (*Set* means the mortar adheres to the unit.) If the unit is disturbed, the bond between mortar and unit is said to be *broken* or *lost*. The lack of a seal may result in a leaky mortar joint. If a unit must be moved after taking the set, it must be removed from the wall, fresh mortar spread, and the unit relaid. Mortar spread too far ahead will dry too quickly and lose its strength, resulting in a poor bond, Figure 15-9.

These rules of bonding apply to all masonry units. If the rules are followed strictly, the final result will be strong, waterproof masonry construction.

Fig. 15-9 Notice how mortar does not stick properly to brick after initial set has taken place. This brick must be relaid in fresh mortar to obtain a good mortar bond.

ACHIEVEMENT REVIEW

Select the best answer from the choices offered to complete each statement. List your choice by letter identification.

1. The only concrete block that can be laid one over the other without cutting the corner block is the
   a. 12″ block.
   b. 8″ block.
   c. 6″ block.
   d. 4″ block.

2. The correct starting piece for a 6″ concrete block wall is the
   a. L-shaped corner block.
   b. 14″ piece.
   c. 12″ piece.
   d. 11″ piece.

3. Concrete block should be laid out by the architect on a grid that would be suitable for structures with windows and doors. The correct grid for concrete block is the
   a. 6″ grid.
   b. 4″ grid.
   c. 8″ grid.
   d. 12″ grid.

4. Before changing the locations of openings of a building, such as windows and doors, the mason should always consult the
   a. architect.
   b. supervisor.
   c. fellow mason.
   d. contractor.

5. Many times it is necessary to use a rip block when starting a concrete structure to attain the correct height. The standard height for a rip block is
   a. 4″.
   b. 6″.
   c. 8″.
   d. 2″.

6. The term *plumb bond* indicates that
   a. all bricks are laid perfectly horizontally.
   b. wall ties are used in the wall.
   c. vertical joints are aligned in every other course.

7. A term used frequently to refer to a wall of a single thickness is
   a. stack.
   b. column.
   c. wythe.
   d. unit.

8. If the set of the mortar has been broken, the best procedure is to
   a. relay the units with fresh mortar.
   b. wet the bricks.
   c. tap the units until contact is re-established.

MATH CHECKPOINT

1. A job requires 600 concrete blocks for the north wall, 800 for the east wall, and 900 for the west wall. How many blocks are needed for the job?

2. A foundation has six corners counting the angle returns. Each corner requires one L-shaped corner block for each course. The total number of courses high is 12. How many corner blocks will be needed to do the job?

3. A mason lays 188 blocks the first day, 210 the second day, 300 the third day, 275 the fourth day, and 320 the fifth day. What is the total number of blocks laid during the week?

4. A mason lays out a foundation that measures 64′ long by 34′ wide. How many linear or running feet are there when the whole foundation is measured?

5. A mason lays 30 blocks per hour. How many blocks will be laid in 24 hours of work?

6. A concrete block building that measures 20 ft. × 40 ft. is being laid out. How many blocks will it take to lay one course completely around the building? Do not allow any openings!

7. If the completed walls are 12 courses high, how many blocks will it take to build it up to the top? No openings are being allowed.

8. Assuming that 1 bag of masonry cement mortar will lay 30 blocks, how many bags of masonry cement will be needed? Round off to the nearest bag over what you need.

9. If one ton of sand is required to every 8 bags of masonry cement needed, how much sand will be needed? Round off to the nearest one-half ton over your figure to allow for waste.

10. If 8″ × 8″ × 10″ concrete blocks are required and each one costs $.98 delivered, what will be the total cost of the blocks?

SUMMARY, SECTION 4

- The term *bond* has three meanings to the mason. The mortar bond is the adhesion of mortar to the masonry unit. The structural bond is the interlocking of masonry units to each other to distribute the weight of the wall. (The half lap is an extremely strong structural bond.) The pattern bond is the arrangement of masonry units to form a pattern or design.

- Patterns can also be achieved by projecting and recessing units on the face of the wall. Tooling the mortar joint in various ways also contributes to the design and beauty of the structure.

- The different meanings of *bond* apply to all types of bond in varying degrees.

- There are several different brick positions which architects specify by the trade term.

- The most commonly used bond for brick or block work is the running bond. This is due to the simplicity of the running bond, which requires no cutting or designing of complicated patterns. Of all the bonds discussed, the running bond is the most economical for the contractor to construct.

- Masonry work should be planned and dry bonded before bricks or block are laid.

- Eight-inch concrete block is the only size block that does not require cutting of the corner block.

- Concrete block is commonly used for walls built in the stack bond because the units do not differ in size.

- Jobs should be well planned so that the cutting of blocks is kept to a minimum.

- The mason's main responsibility concerning bonding is to maintain the bond as it was specified and laid out, and to be certain that it remains plumb throughout the job. Changes in the bond should not be made without approval of the supervisor.

SUMMARY ACHIEVEMENT REVIEW, SECTION 4

Complete each of the following statements referring to material found in Section 4.

1. The adhesion of mortar to the masonry unit is called _____.

2. A type of bonding in which the units are interlocked over one another and which is designed primarily for strength is called _____.

3. The arrangement of masonry units to form a particular design is called _____.

4. A standard brick laid in a horizontal position in a wall is known as a (an) _____.

5. The most common brick position in windowsills is the _____.

6. The most commonly used bond in brick veneer work is the _____.

7. The bond which is a variation of the running bond with headers installed on the fifth, sixth, or seventh course to provide strength and design is the _____.

8. Laying one masonry unit halfway over the next is called the _____.

9. The best method of starting a header course from the corner of a structure is to cut and install a piece measuring _____.

10. One of the most beautiful bonds in brickwork calls for alternating headers and stretchers on the same course. When this occurs in every course of a wall, the bond formed is the _____.

11. The Flemish bond may be started from the corner in one of two different ways. If a 2″ piece is used, it is called a (an) _____ corner; and, if a 6″ piece is used, it is called a (an) _____ corner.

12. The Flemish bond can be adapted to present various appearances. If 3 stretchers are alternated with a header, it is known as a (an) _____.

13. One of the least used bonds in brickwork consists of alternate courses of headers and stretchers. This bond is called the _____.

14. A bond that is specified by many architects for decorative purposes and in which there is no overlapping of masonry units is called a (an) _____.

15. The method of laying masonry units dry (without mortar) to check the bond is called _____.

16. Concrete blocks are usually laid in the running bond because _____.

17. Concrete blocks measuring 10″ and 12″ in width are usually used for basement walls. To avoid cutting such large blocks to start the corner, special blocks are used. These blocks are known as _____.

18. When building a brick wall which is backed by another masonry wythe, the most important consideration is to _____.

19. A mason's plumb rule can be used to space concrete blocks on the layout course. The length of the plumb rule equals the length of _____ blocks.

20. When laying out a concrete block corner built of 6″ blocks, it is necessary to start the corner off with a piece measuring _____.

SECTION FIVE
LAYING BRICK AND
CONCRETE BLOCK

UNIT 16
Laying Brick to the Line

OBJECTIVES

After studying this unit, the student will be able to

- attach a line block, nail, and line pin to a wall.
- set a trig.
- lay bricks to the line while spacing them the correct distance from the line.

The first tasks of a beginning mason usually include spreading mortar and striking joints. These practices acquaint apprentices with the properties of mortar and proper tooling of joints. Beginning masons may also help stock materials near the work area and cut materials with the saw. In this way, they learn material requirements for various jobs and the ways in which the materials are handled. The various sizes of bricks and block are quickly learned.

After learning the particulars of these jobs, apprentices then learn to lay bricks to the line. Masonry contractors always require that apprentices learn the skill of laying to the line before learning to use the plumb rule. The line, which acts as a guide for the wall, is used as the best means by which apprentices can learn to lay bricks. After learning how to lay bricks, apprentices proceed to the more difficult task of building the corner. By progressing from the simpler jobs to the more

difficult, apprentices learn as they work and master each task independently of others.

LAYING BRICKS

The Corner Pole

There can be difficulty for apprentices learning to lay bricks in the shop because there are not enough experienced masons to build corners so that trainees can finish the wall. To overcome this problem, a line can be attached to a corner pole so that the wall can be constructed without building corners. A *corner pole* (or *dead man*) is any type of post that is propped and braced into a plumb position so that a line can be fastened to it. A manufactured corner pole can be bought or a wooden pole can be constructed in the shop from a $2'' \times 4''$. Angle-iron frames or stacks of block can also be used.

On brick veneer homes, the manufactured corner pole can be braced against the frame of the house and the base masonry, Figure 16-1. This completely eliminates the need for a corner to be built before the bricks are laid. This is a very popular practice in the building trade, as labor costs are considerably reduced.

Use of the Line

When building walls longer than 4′, a line should be used as a guide for laying the units, since the plumb rule is only 4′ in length and, therefore, cannot accommodate a longer wall. The line serves to keep the wall level and plumb and greatly increases the speed by which the bricks may be laid. Most masonry work is laid to the line.

Note: In the practice situations that follow this unit, the plumb rule will be used only to spot check walls to be sure they are level and plumb and for work around window jambs and doorways.

Preparing the Work Area and Beginning the Job

Sweep or brush the area where the wall is to be built. It is important in the beginning to have a clean surface

Fig. 16-1 Masons laying brick with a corner pole as a guide. Notice the line attached to the pole at the top of the bricks. The wall is being laid in the running bond.

to achieve a good mortar bond with the base. Set the mortar pans approximately 2′ away from the wall line to allow for sufficient working space.

> **Caution:** When stacking the bricks on each side of the mortar pan, be sure to alternate the bricks on every other course to prevent the pile from falling over and injuring yourself or a fellow worker.

Strike a chalk line on the base if necessary to establish the wall line for laying out the first course. At times, a nail may be found in the concrete base to mark the wall line. It is usually placed there by the building engineer or carpenter foreman. Dry bond the wall to establish the proper spacing of the bricks.

Lay 1 brick in the mortar bed at one end of the wall or corner, lining it up with the mark established by the dry bond. Level and plumb the brick with the wall line. Repeat the steps at the other end of the wall. Be sure that the brick is laid to the proper height by checking with a rule or a gauge rod. This process is known as *spotting the brick.*

If a corner pole is to be used, it may be placed in position at this time.

Attaching the Line with the Line Block

Line blocks have a slot cut in the center to allow the line to pass through, Figure 16-2. Fasten 1 block against the left corner of the wall either by passing the line through the slot and tying a nail or large knot on the end to secure the line, or by making several wraps around the block with the line.

The person on the left end of the line must hold the line block in line with the top of the course to be laid until the person on the right end of the line is ready to pull the line tight. Pull the line as tight as possible without breaking it, being certain that it is passing through the slot in the block. When the line is pulled tight, wrap the line about 3 or 4 turns around the line block, Figure 16-3. Hook the line block on the right corner, Figure 16-4. Make sure that the line is perfectly level with the height of the course being laid.

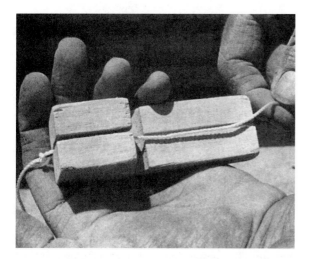

Fig. 16-2 Knot on end of line prevents it from slipping through line block.

Caution: Use extreme care when working with line blocks. If a line block is used on a long wall, there is a danger of it being jolted loose and causing injury. For this reason, use line blocks only on short walls or with a corner pile.

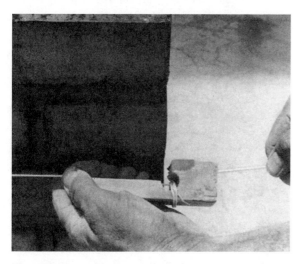

Fig. 16-3 Wrapping line around line block and pulling it tight to place on brick corner.

Fig. 16-4 Attaching a line to a brick corner with a line block.

When line blocks are attached to a corner pole and the line is tightened, be sure to check that the line is the correct height before laying any bricks. This is very important, since the height of the finished wall is directly affected by the height of the line.

Pulling and Attaching the Line with the Nail and Line Pin

Use caution when attaching the line to a corner with line pins or with a nail and line pin. Traditionally, a nail is used on the *peg end* of the line (end against which the line is pulled), and a steel line pin is used on the pulling and wrapping end. Many masons prefer to use line pins on both ends because the line pin, due to its shape, is less likely to pull out of the joint. If a nail is used on the peg end of the wall, make sure it is a large nail (at least size 10d Common). Not many things are more dangerous to masons than a flying nail.

As the mason faces the wall, the peg end of the line is always driven into the left-hand side of the wall. Most people are right-handed and it is easier for them to pull a line to the right. Left-handed persons must adjust to this traditional practice. Drive the nail or pin securely into the head joint, making sure that its top is level with the top of the course of brick to be laid, Figure 16-5. Always place the nail at a downward angle (45°) in the head joint and several bricks away from the corner. This prevents the nail from coming loose as the line is pulled.

Fig. 16-5 Nail (peg end) of line properly driven in mortar head joint, even with top edge of brick course.

Fig. 16-6 The mason pulls the line with the left hand and wraps the line around the pin with the right hand.

The person setting up the nail end of the line is responsible for making certain that the nail will remain secure in the mortar joint when the other person tightens the line. Testing the nail is done by giving the nail a few sharp tugs after driving it into the joint. This practice can reduce accidents and time lost on the job.

After the line has been tested, the person at the nail calls *tight away* to the person located by the pin. This indicates that it is safe to pull the line tight. If it is a very long wall, the masons working at the center of the wall may relay the call to tighten the line or give two or three short tugs on the line as a signal.

The mason on the pulling end of the line drives or pushes the line pin securely into the head joint, also even with the top of the course. Great care should be taken to see that the pin is secure and in the correct position before pulling the line, Figure 16-6. The line is then pulled tight and wrapped around the line pin.

> **Caution:** When tightening the line, the left hand, which pulls the line, should be angled away from the wall in case the line breaks under the strain.

When the line pin is moved up for another course, immediately fill the holes where the nail and pin were located with fresh mortar. Waiting until the project is finished to fill the holes usually results in an added ex-

pense. There is also a noticeable difference in color where the holes were pointed.

Setting the Trig Brick

To prevent the line from sagging or being blown out of alignment by the wind, it may be necessary to set a *trig brick,* Figure 16-7. A trig brick is also set to stabilize the line when there are many masons working on the

Fig. 16-7 Setting the trig brick. The mason is adjusting the brick to the line before laying brick on the wall.

wall. The first step in setting a *trig* (metal fastener or loop of line) is to be sure that the brick with the trig is set with the bond pattern of the wall. Be sure that the brick is level and plumb with the face of the wall. Check the trig brick for the proper height with a mason's rule or a course pole which is marked with the proper spaces for the masonry courses.

Sight down the wall every 3 or 4 courses to be sure that the trig brick is in line with the wall line. This sighting is usually done by the person located by the pin and another person who stands by the trig brick. One person holds the line in position with the trig brick, as the person sighting the line bends down and checks the top edge of the brick on one corner with the other corner to be sure they are in line. This method is usually accurate within $\frac{1}{8}''$. The trig brick may not always be exactly on number 6 of the modular rule, but should always be the same height as the corners. A straight wall can be maintained by these sighting techniques.

After the trig has been correctly set, place a dry brick on top of the trig to hold the line in position with the top outside edge of the trig brick, Figure 16-8.

The trig brick should always be set ahead of the line being raised for the next course. This is so that other masons working on the line do not have to wait for the line to be erected to proceed with their work. This can easily be done by setting the trig brick immediately after the line has been run. While one mason sets the trig brick, the other masons can be walling bricks for the next course and striking the mortar joints. This procedure calls for timing and teamwork and adds to the productivity of all masons working on the line.

Setting the trig brick (or *carrying the trig*) is a job which calls for a great amount of accuracy and responsibility. It must be laid perfectly plumb and level and kept in alignment with the wall. An incorrectly set trig brick will cause the wall to be incorrectly constructed even though the corners may be perfectly built.

Laying the Units

Spread mortar on the wall with the trowel. Try not to disturb the line. If the trowel is held on a slight angle as described in the unit on basic tools, it should not touch the line. Grasp the brick with the fingers and thumb. Position the fingers on the brick so that the line may pass between the fingertips and brick when the hand is released, Figure 16-9.

This process demands good coordination and timing, which is developed only with practice. If the brick is held too long, the fingers or thumb may push the line

Fig. 16-8 Trig brick in position with the line attached to the metal trig. Notice that the trig holds the line the correct distance ($\frac{1}{16}''$) away from the face of the brick.

Fig. 16-9 Position of the fingers when laying brick to the line. Notice that the mason's fingers grasp only the top edge of the brick.

out of alignment, preventing other masons from properly aligning bricks. Releasing the fingers or thumb from the brick too soon will result in the brick being laid unevenly on the mortar bed. If this happens, the brick will not be level and plumb and will require relaying in a fresh bed of mortar.

When bricks are protruding past the wall line, they are said to be *hard to the line*. When a brick is too far away from the line, it is said to be *slack to the line*. Either condition results in an unacceptable job. Bricks should be laid about $\frac{1}{16}''$ from the line, or as some persons in the trade say, one should be able to see a little daylight between the line and the brick. With practice, the mason will be able to leave the proper amount of space without using a rule. When laid correctly, the bottom edge of the brick should be in line with the top of the course beneath it, and the top edge even with the top of the line and $\frac{1}{16}''$ back from the line, Figure 16-10.

Laying bricks so that they touch the line (known as *crowding the line*) is one of the biggest problems the apprentice must overcome when learning to lay bricks. Developing this technique requires a great deal of practice.

Pressing the Brick into Place

It is sometimes permissible to tap the bricks into final position. However, to develop a good rate of speed in laying bricks, the trainee must perfect the technique of pressing the brick into place, Figure 16-11.

With the trowel, cut off the mortar that has squeezed out of the joints after pressing down and apply it to the head of the brick just laid. Time can be saved by applying the head joint in this manner, since the mason does not have to return to the mortar pan for each individual head joint. When applying the mortar for the head joint, hold the trowel blade at an angle so as not to move or cut the line. It is not possible to lay bricks to the line without disturbing the line to some degree, but with the correct movement, it can be held to a minimum.

Laying the Closure Brick

The last brick to be laid in a wall is the *closure brick,* usually located near the center of the wall. To eliminate any chance of leaking or moisture penetration, the mortar joints must be well filled. Apply mortar for the head joint (also known as *throwing* a joint) on each end of the two bricks adjoining the area where the closure brick will be placed. Then *butter* (apply mortar to) each end of the closure brick and lay it into position, Figure 16-12. When laying the brick, press it carefully so that the rest of the bricks are not disturbed. This practice is known as *double jointing*. Immediately fill any remaining holes to ensure a watertight joint.

Fig. 16-10 The correct spacing of brick is $\frac{1}{16}''$ from the line.

Fig. 16-11 Adjusting brick to the line by pressing down with the hand.

Fig. 16-12 Laying the closure brick in the wall. Notice that the mason has double jointed both the closure brick and the two bricks surrounding it. This action protects the wall against leaks.

POINTS TO REMEMBER

Most of the mason's time is spent laying bricks to the line. Since the procedure involves repetition, needless actions can result in a great deal of lost time. For efficiency on the job, remember the following points:

- A brick should always be picked up with the face out so that it is in the position in which it will be laid in the wall.
- A mason should be approximately 2′ from the stocked brick or within arm's reach. Unnecessary walking tires the mason and reduces efficiency.
- If the brick contains a depression or frog, it should be picked up with the frog facing down, since this is the way it will be laid in the wall. The mason should refrain from unnecessary turning of the bricks in the hand.
- Well-filled mortar bed and head joints should be formed. This greatly decreases the amount of time needed to strike joints later in the job. Solid joints also ensure stronger, waterproof walls.

ACHIEVEMENT REVIEW

Select the best answer from the choices offered to complete the statement. List your choice by letter identification.

1. A deadman is
   a. a method of bonding.
   b. a post to which the line is fastened.
   c. a poor safety practice.
   d. inefficiency on the job.

2. A line should always be used on a wall if it is longer than
   a. 6′.
   b. 2′.
   c. 4′.
   d. 8′.

3. The term *spotting a brick* is the
   a. addition of color to the brick.
   b. arrangement of bricks to form a pattern in the wall.
   c. laying of a brick in mortar to establish the wall line.
   d. spreading of mortar on the brick.

4. A nail no smaller than 10d Common should be used for attaching the line because
   a. it helps to prevent the line from becoming dislodged.
   b. it matches the size of the mortar joint.
   c. it does not leave an excessively big hole in the mortar joint.
   d. it is easier to tie the line to a nail of this size.

5. The main reason for tugging on the line before calling to the mason on the other end to tighten it is
   a. to check that the line is at the correct height.
   b. to be sure that the line is securely in position and will not pull loose.
   c. to take the initial stretch out of the line.

6. When setting a trig brick, the line should be sighted periodically from corner to corner to assure that
   a. the wall is plumb.
   b. the bond pattern is correct.
   c. the trig brick is in proper alignment with the wall.
   d. the trig brick is the same size as the other bricks.

7. The correct distance to position bricks back from the line is
   a. $1/8''$.   c. $1/16''$.
   b. $1/4''$.   d. $3/16''$.

8. The last brick to be laid in the center of the wall is called the
   a. filler brick.   c. key brick.
   b. closure brick.   d. header.

9. Double jointing the last brick laid in the wall is important because it
   a. forms a stronger wall.   c. forms a more waterproof joint.
   b. allows for expansion in the wall.   d. divides the joints more evenly.

MATH CHECKPOINT

1. A contractor used 50,000 bricks and ordered 2,600 more to complete the job. How many bricks does the job require?

2. If a mason can lay 500 bricks in an 8-hour day, how many would be laid per hour?

3. Four masons working together laying brick to a line laid 2,464 bricks in one day. What is the average number of brick each laid?

4. Five masons work a total of 820 hours on a job. How many hours did each work?

5. In estimating brick for a job, a contractor allowed 7,000 bricks. When the job was completed, he discovered 3% were left over. How many bricks were used in the actual job?

PROJECT 8A: CONSTRUCTION OF THE CORNER POLE

Contractors who employ apprentices prefer to have the beginner lay bricks and concrete block to the line before attempting the more difficult task of building a corner. There are two reasons for this. First, it is more profitable for the contractor, since it is accomplished more quickly. Secondly, easier skills are mastered before the more difficult ones.

In a regular training situation, it is usually difficult to construct enough corners so that each student can lay units to the line. To remedy this situation, corner poles can be constructed very inexpensively from framing lumber. The only necessary tools are basic carpentry hand and power tools. A table or radial arm saw is also needed to rip the lumber to the correct width (see Figure 16-13).

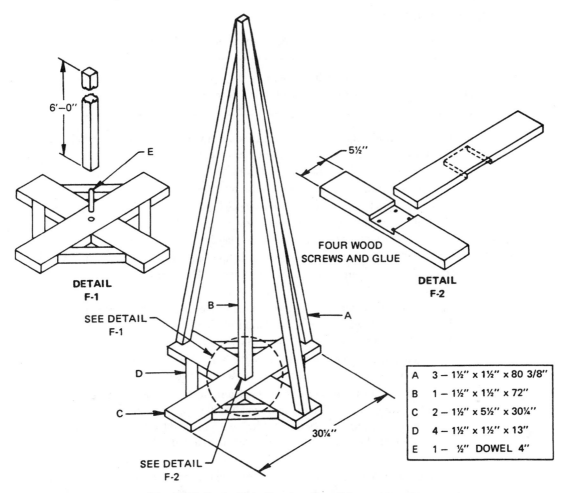

A	3 — 1½″ x 1½″ x 80 3/8″
B	1 — 1½″ x 1½″ x 72″
C	2 — 1½″ x 5½″ x 30¼″
D	4 — 1½″ x 1½″ x 13″
E	1 — ½″ DOWEL 4″

Fig. 16-13 Project 8A: Construction of the corner pole.

EQUIPMENT, TOOLS, AND SUPPLIES

The following are materials needed for one pole. (One wall will require one pair of poles.)

1. Select 3 straight, 8-foot $2'' \times 4''$s to serve as the upright poles and braces. These parts are indicated in the plan as A, B, and D.

2. One $2'' \times 6''$ at least 61″ in length. This board will be crosscut in half to form the base. (Allowance has been made for loss.) This part of the base is indicated as C in the plan.

3. A supply of 1¼″ #6 or #8 wood screws. (Nails may be used. Screws require more time to install but are more secure.)

4. One wood dowel ½″ in diameter. It is shown in the plan as E.

5. One can of wood glue which will withstand moisture, as the pole may be used in an outside area. Epoxy glue is effective.

6. One can of plastic clear varnish and brush for sealing the wood.

PROCEDURE: CONSTRUCTION

Step 1. Lay the $2'' \times 4''$s on their widest side on the table saw. Rip in half lengthwise.

> **Caution:** Wear eye protection when operating the saw to prevent an eye injury. Have another person help you take the lumber off the saw.

If a jointer is available, joint the edges. If one is not available, use a hand plane. This decreases the need for sanding. Cut the center pole and braces to the correct length shown on the plan (indicated as A and B).

Note: Outside braces are longer than the center pole because of the angle.

Step 2. Cut the $2'' \times 6''$ to lengths as shown on the drawing, indicated as C. Mark the center of each piece and cut a dado joint, allowing the two parts of the base to lap halfway over each other. (See detail F-2 on plan.)

Step 3. Cut 4 angled braces to secure the base, on an approximate 45° angle. Follow the specified length shown on the plan, indicated as D. Glue, screw, or nail into position.

Step 4. Drill a $\frac{1}{2}''$ hole $1''$ in depth through the center of the base. Insert a piece of dowel rod $4''$ long coated with glue (E), and tap firmly in place. This prevents the pole from shifting after the base and rod are fastened together.

Step 5. Set the center pole in glue firmly down on the base until contact is made. Be sure the center pole is plumb with the base, using a metal square.

Step 6. The angle of the outside braces (A) can be determined by cutting a 45° angle. To achieve the proper angle, hold the brace against the side of the center pole while another person holds the pole straight. Mark with a pencil the portion which projects past the end of the pole. Attach the side braces to the pole with glue, screws, or nails.

Step 7. Using number 6 on the mason's modular folding rule, mark the courses from the bottom of the base. Number the courses consecutively from the bottom up. The marks can be cut in with a hand saw and marked with a black felt pen for greater visibility.

Step 8. Finish the construction process by applying two coats of plastic base varnish. The pole will have more than enough height to erect training walls to scaffold height ($4'$ to $5'$ high). If concrete block is to be laid instead of bricks, simply raise the line 3 courses of bricks for every course of concrete block.

PROCEDURE: ERECTING THE POLE

Step 1. Strike a chalk line on the base where the wall is to be built. Determine if the base is level at each end where the center pole is to be placed. This can be done with a regular level and a straightedge or a builder's level.

Step 2. Assuming the base is level, start at the highest end and set up the corner pole, being sure that the pole is in line with the chalk line where the line will be attached. This is a very important step.

Step 3.   Go to the other end of the wall and block up the opposite corner pole with scraps of wood so that it is level with the pole first set.

Step 4.   Lay several concrete blocks on top of the base of the corner poles to prevent them from tipping over when the line is pulled tight. Attach the line blocks and line to the corner pole, being sure they are the same height, and pull tightly.

The corner pole is now ready for use. Dry bond the wall and proceed to lay out the first course. The wall can now be built to scaffold height by simply raising the line up for additional courses as needed. It is good practice to recheck the pole occasionally to be sure it remains plumb.

Note: Any difference in mortar joints which occurs when laying out the first course should be adjusted before beginning construction of the wall.

The construction of a corner pole allows the student to learn tool skills and the techniques involved in laying units to the line without the necessity of building a corner. After the necessary skills have been learned, the student can begin learning the basics of corner construction.

PROJECT 8B: USING THE CORNER POLE

OBJECTIVE

- The masonry student will be able to lay a 4″ brick wall in the running bond, using the corner pole as a guide. The corner pole can be used as a replacement for the corner on almost any job, Figures 16-14, 16-15, and 16-16.

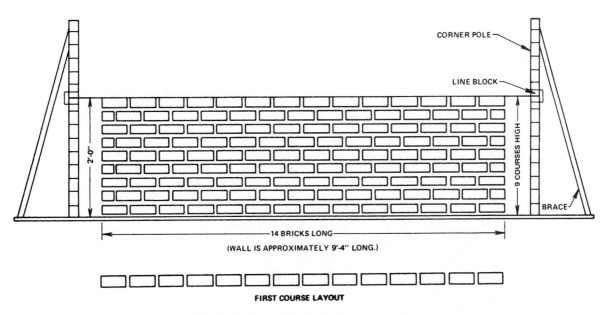

Fig. 16-14 Project 8B: Using the corner pole.

Fig. 16-15 Student masons can work together as a team utilizing the corner pole to build a brick wall. (Photo by author)

Fig. 16-16 Students laying brick to the line using corner pole as a guide.

EQUIPMENT, TOOLS, AND SUPPLIES

2 mortar pans or boards

Mixing tools

Mason's trowel

Brick hammer

Plumb rule

2 corner poles and braces

Chalk box

Ball of line

Convex sled runner

2′ square

Mason's modular rule

1 pair line blocks

Brush

Approximately 133 standard-sized bricks

$1\frac{1}{2}$ bags cement or lime*

27 shovels of sand

Supply of clean water

1, $8'' \times 8'' \times 16''$ block to use as holder for level

*Lime or Carotex may take the place of cement in a training situation, since it can be reused.

SUGGESTIONS

- Position materials and mortar pans approximately 2′ from the wall to allow for working room.
- Be sure the line is on the correct coursing mark on the corner pole.
- Spread the mortar as uniformly as possible.
- Use solid head joints on all work.
- Remove the excess mortar from the back of the wall at the completion of each course.
- Strike the joints when mortar is thumbprint hard.
- Keep the work area neat and free from debris.

PROCEDURE

1. Mix the mortar.
2. Stock materials in the work area and load the mortar pans.
3. Strike a chalk line approximately 11′ long on the concrete floor to act as the wall line.
4. Set up the corner poles at each end of the chalk line. Plumb and brace them into position with the wall line.

   Note: If corner poles are not available, a stack of block set up plumb or an angle-iron brace also work as guides.
5. Dry bond the first course which is 14 bricks in length.
6. Attach the line and blocks to the corner pole at the height of the first course.
7. Lay the first course in the mortar.
8. Move the line up 1 course.
9. Cut bats for the second course and finish laying the course as shown on the plan.
10. Build the wall to the required height as shown on the plan, plumbing *jambs* (ends of wall) on every course.
11. Strike the mortar joints with a convex sled runner jointer and brush.
12. Recheck the wall with a level (plumb rule) at the completion of the work.
13. Always follow good safety practices.

UNIT 17
Building the Brick Corner

——————————— OBJECTIVES ———————————

After studying this unit, the student will be able to

■ lay out the first course of a wall with the correct number of stretchers in preparation for building a corner.

■ construct a rack-back lead.

■ construct an outside and inside corner.

Building a corner is one of the more difficult tasks that the mason performs. If the corner is built incorrectly, the wall is also built incorrectly, since the corner serves as a guide for the wall. Building corners (also known as *leads*) on a job requires a great deal of responsibility and care on the part of the mason. Care must especially be taken in leveling and plumbing to assure that the corner is true.

The person building the corner must work far enough ahead of the other masons so that they are not detained from laying bricks to the line. However, rushing the work unnecessarily may result in a shoddy job. Each time a line is run and bricks are stacked for the next course, the mason building the corner should call for the line to be raised another course. It is the responsibility of the mason building the corner to set the proper working pace for all the masons on the line. This assures a steadily moving job and uniform productivity of all masons on the job.

When learning to build corners, the mason should be more concerned with technique and good workmanship than speed. The mason should be certain that a course is satisfactory before proceeding to the next course. Once a corner is out of alignment, it is very difficult to straighten it without breaking the set of the mortar.

THE RACK-BACK LEAD

Very often, it is necessary to build a lead between corners on a long wall. The simplest type of lead or guide for a wall is called a *rack-back lead*. This is usually the type of lead with which the mason first becomes familiar. This type of lead does not have any angles or *returns*. It is merely a number of brick courses laid to a given point by racking back a half brick on each course. The fundamentals of corner building, with the exception of turning the corner, apply to building a rack-back lead.

The first course of a rack-back lead rarely exceeds 6 bricks in length. This is because the plumb rule is 48″ in length, the same length as 6 bricks. After laying the first course, the mason must be sure the bricks are plumb and straight. Each succeeding course is racked back a half brick on each end of the lead until the last brick is laid. Excess mortar should be removed as each course is laid and returned to the mortar pan for reuse. Figures 17-1A through 17-1F show six major steps in the construction of a rack-back lead.

Fig. 17-1A Laying a course in a rack-back lead. Note how the end bricks are racked back a half lap.

Fig. 17-1B Checking height with modular rule. Note how the bricks are laid to number six on the rule.

When the lead is completed, check the alignment of the tail end of the lead with the plumb rule. This is done by holding the plumb rule at an angle on the edge of the corner of the bricks at the racking point. This is known as *tailing a lead.* If this is not done, protruding bricks may cause a bulge in the wall when the line is attached to the lead and the wall is built.

Fig. 17-1C Tooling the mortar joints with a convex sled runner.

Fig. 17-1D Brushing the lead.

Fig. 17-1E Tailing a rack-back lead. Notice that the plumb rule is in line with the corner of each top brick to assure proper alignment.

Fig. 17-1F Completed rack-back lead built on a wooden $2'' \times 4''$.

The mason should tool the joints on the completed lead with a striking tool and fill any holes. The mason should then brush the head with a soft bristle brush. Finally, the lead should be rechecked to be certain that it is still in proper alignment.

BUILDING AN OUTSIDE BRICK CORNER IN THE STRETCHER BOND

The number of courses high of a corner should be determined before laying out the first course. Estimating the number of courses is a simple procedure. The sum of the stretchers used on the first course corner layout equals the number of courses high that the completed corner is.

For example, assume that a corner is to be built 9 courses high. To build the corner, 5 bricks must be laid in one direction and 4 bricks in the other direction. A corner that is to be 11 courses high requires 5 bricks on one side and 6 bricks on the other. Corners are usually built the height of a scaffold (4′ to 5′). When that section of the wall is laid, the scaffolding is erected and the corner is built again. This procedure is repeated until the wall reaches the desired height. Extending the lead any further than necessary when building a corner is considered poor practice, since these bricks could be laid more efficiently with the aid of a line. Figures 17-2A through 17-2F show the steps in laying the first course.

Corners should always be ranged with a line. *Ranging* is the horizontal alignment of masonry units by use of a line tightly drawn between two points. It is done after the first course has been laid. To range corners, fasten one end of the line to the edge of one corner and the other end of the line to the outside edge of the other corner. Adjust any bricks in the line that are not in horizontal alignment until the corners line up perfectly with the range line. The line may then be stored until it is time to lay bricks to the line.

The mortar must be spread as uniformly as possible and should be the thickness of the specified mortar joint. If this is not done properly, the bricks must be removed and laid again. The ability to sense the proper joint thickness is acquired by the mason with practice, as every course is not measured. Many times, an experienced mason building a corner checks for height every third course, using a mason's rule or story pole. How-

Fig. 17-2A Laying out the corner with a steel square.

ever, if it is critical that the courses be perfect for height, the mason checks every course. The apprentice should check every course on every job until the technique is perfected.

The face of the bricks must be initially laid plumb with the face of the wall without too much shifting of the bricks on the mortar bed. This can be done by constantly keeping the eyes in line with the face of the wall and sighting down the corner. This is especially important on the outermost edge of the corner or wall, Figure

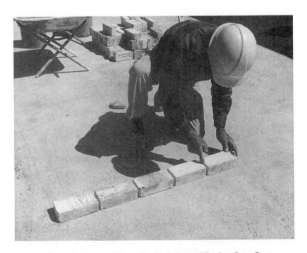

Fig. 17-2B Dry bonding the bricks with the forefinger.

Fig. 17-2C Dry bonding the proper number of bricks for a corner nine courses high. Notice that there are five bricks leading in one direction and four in the other direction.

Fig. 17-2E Reversing the procedure and leveling the opposite side. The mason has not yet disturbed any bricks between the leveling points.

17-3. If done correctly in the beginning, the amount of plumbing to be done later in the job can be decreased greatly and the work made easier. Sighting is a very important skill to the mason, especially when constructing corners. Sighting with efficiency and speed comes only with practice.

All excess mortar should be removed from the ends and inside the corner immediately after the course has

been leveled and plumbed. All measurements of height should be done on the edge of the return of the corner. The work is then leveled from that point. Strike the joint in the corner as soon as the mortar has set sufficiently, or is thumbprint hard, Figure 17-4.

To be sure that the corner is plumb, hold the plumb rule against the bottom brick and adjust the other bricks.

Brush the corner carefully after the striking has been completed. This must be done gently so that mortar is not brushed from the joints. Check the corner with

Fig. 17-2D Laying the first course of bricks in mortar. The mason is leveling the corner brick and the end brick before moving bricks in the center. This is the procedure to follow when working on a base that is not level.

Fig. 17-2F Checking the corner on a diagonal to be sure that the first course is level.

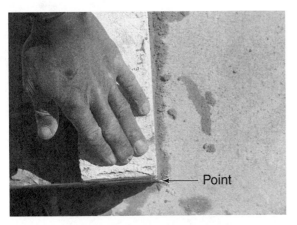

Point

Fig. 17-3 Sighting down the outermost point of the corner bricks to be certain they are plumb. The corner shown, five courses high, is in perfect alignment.

Fig. 17-5 Checking the corner with the plumb rule to be sure it is properly aligned. This is usually the last step in building a corner.

the modular rule for the specified height at the completion of the job.

Recheck the corner to be sure that it is plumb and level and is neatly struck. Always recheck the tail end of the lead at the completion of work, Figure 17-5.

With practice, the student will be able to lay the first brick on the corner to the prescribed height and proceed to lay out the first course level, plumb, and on the squared line.

Fig. 17-4 Striking the corner with a V-jointer. Mortar joints should be tooled before they are too stiff.

PARGING THE CORNER

Parging is the application of mortar to a masonry wall to prevent water penetration. The mortar, applied with a brick trowel, is known as *back parging*. Parging a brick wall should be done before a backing material is installed.

The mortar coat should be about $\frac{3}{8}''$ thick for best results. Mortar that is too thin cracks. Mortar that is too thick cracks and pulls loose from the surface of the wall. A good guide for parging is a piece of wood held on top of the last course and allowed to project over the back of the wall about $\frac{3}{8}''$, Figure 17-6.

TOOTHING THE CORNER

Toothing is the construction of a temporary end of a wall by allowing every other end of stretcher bricks to project halfway over the stretcher below. This is done by laying a bat in the mortar on the end of the course previously laid and laying the next brick directly over it, Figure 17-7. After the corner has set a reasonable length of time, the bats are removed very carefully so that the corner is not knocked out of alignment. The bats should not be allowed to set completely in the mortar.

Fig. 17-6 Parging the back of the corner. Notice the board at the top of the corner serving as a guide.

Fig. 17-7 Toothing of a brick corner. The supporting bats and excess mortar have been removed so that the toothing can receive the wall at a later time. Toothing is not a common practice but is necessary in some cases.

Toothing is used when a corner cannot be extended far enough so that it can be built to the specified height. Toothing may also be done when an opening is needed through which wheelbarrows may pass, or when windows or doors have not been delivered to a job on time.

Although toothing is necessary in specific instances, it is not generally recommended. If a joint is not filled with mortar and finished properly after toothing, there is a much higher chance of water leaking through the joint. In addition, toothing construction requires more time than other brickwork.

When filling around a brick toothing with mortar, the mason must make sure the brick that is laid under the overlap is well buttered with mortar. After the brick is laid, the mason points up the mortar joints around the toothing with a trowel and the slicker striking tool. Finally, the mason parges the back of the toothed area with mortar to seal the joints. If a brick wall leaks, it does so most often at a faulty toothing. Following these steps will ensure a leak-proof wall.

Architects and builders permit toothings infrequently, and they are built only under the strictest supervision. Since it is necessary to use toothings in masonry work, however, the apprentice should be familiar with them and be able to install them properly.

INSIDE CORNER

The construction of an inside angle corner is the same as that for an outside corner except that it is reversed. Inside corners are not built as frequently as outside corners, but their use is common in structures that have exposed brickwork in the interior. There may also be inside corners in places where two exterior walls join at an inside angle. If they seem more difficult to build, it is only because they are not built as often as outside corners.

The inside corner should not be attempted until after the construction of a rack-back lead and outside corner have been mastered.

Remember, all types of corners require quality workmanship. Since the corner is the guide for the wall, it must be built level, plumb, and to the specified height.

PROGRESS CHECK

At this point, masonry students should have acquired certain basic skills. They must practice these skills as every mason must have them. Students should demonstrate for their instructor each of the following skills. Any skills that are not mastered should be practiced until they can be demonstrated correctly.

- Dry bond a corner.
- Lay out a level course of bricks.
- Plumb a course of bricks.
- Spread mortar to uniform thickness.
- Use a square correctly to square a corner.
- Lay bricks to the line correctly.
- Attach a line pin, nail, and line block to masonry.
- Strike a joint correctly.
- Use a chalk line correctly to lay out a wall.
- Read the modular rule and build to a specified height.

ACHIEVEMENT REVIEW

Select the best answer from the choices offered to complete each statement. List your answer by letter identification.

1. Another name for a lead is a
   a. return.
   b. corner.
   c. bond.
   d. wall.

2. The simplest type of lead to build is the
   a. rack-back lead.
   b. outside corner.
   c. inside angle corner.
   d. Flemish bond lead.

3. The tail of the lead is located at the
   a. point where the lead returns on an angle.
   b. top of the last course laid.
   c. outermost corner of the racking point.
   d. first course of bricks laid on the lead.

4. The number of courses on a corner can be determined by
   a. adding the sum of the stretchers used in the first course layout.
   b. measuring the total length of the layout course.
   c. dividing each course into the total height.
   d. dividing the height by the total number of bricks.

5. A range line is used to
   a. lay the wall.
   b. align the corner when starting the first course.
   c. check the corner to be sure it is plumb.
   d. check the corner for the proper height.

6. Apprentices can learn the skill of plumbing a corner much faster if they master the knack of
   a. spreading mortar.
   b. leveling before the mortar dries.
   c. sighting each course laid.
   d. checking each course for height.

7. The principal reason for parging a corner is to
   a. add beauty to the corner.
   b. strengthen the corner.
   c. create a bond pattern.
   d. make the corner water resistant.
8. Toothing should not be done frequently but is sometimes necessary. The most common fault associated with toothing is
   a. change in the color of the wall.
   b. leakage of the joint where the toothing is.
   c. an unappealing design.
   d. change in the texture of the wall.

PROJECT 9: CONSTRUCTING A 4″ RACK-BACK LEAD IN THE RUNNING BOND

OBJECTIVE

- The student will be able to lay out and build a 4″ brick rack-back lead in the running bond, Figure 17-8. This requires the skills involved in leveling, plumbing, and building to a specified height. This is a typical lead that masons build between two outside corners on a long wall.

EQUIPMENT, TOOLS, AND SUPPLIES

Mortar pan or board   Plumb rule

Mixing tools   Modular rule

Mason's trowel   Chalk box

Brick hammer   Pencil

Convex sled runner (medium size)

1 batch of mortar (1 bag lime to 16 shovels sand)

24 standard bricks (3 of these are allowed for waste)

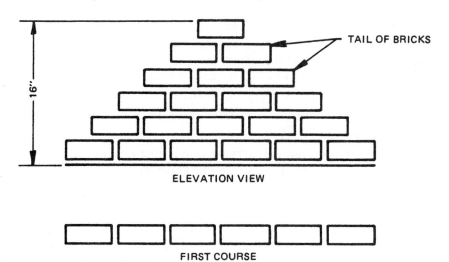

Fig. 17-8 Project 9: Constructing a 4″ rack-back lead in the running bond.

3, $4'' \times 8'' \times 16''$ concrete blocks on which to lay $2'' \times 4''$

1, $8'' \times 8'' \times 16''$ block in which to set the level

Supply of clean water

1, $2'' \times 4''$, $6'$ long board on which to build the project (project could also be built on a concrete floor)

SUGGESTIONS

- Apply a little more mortar than the amount required for a solid joint.
- Apply mortar evenly when forming the head joints.
- Sight each brick which is laid even with the edge of the $2'' \times 4''$ or layout line.
- Clean all excess mortar from the project as the work progresses and return it to the pan to be tempered.
- Plumb the lead gently to prevent dislodging the lead.
- When striking mortar joints, hold the hand on top of each course of bricks to hold the lead securely.
- Always use safe working practices.

PROCEDURE

1. Mix the mortar and place it in the mortar pan. The mortar pan should be located approximately $2'$ from the immediate work area.
2. Lay out the lead with a chalk line if it is being built on the floor. If using a $2'' \times 4''$, the edge of the wood serves as the wall line. Lay the $2'' \times 4''$ on 3, $4''$ blocks.
3. Spread enough mortar on the $2'' \times 4''$ or floor for 6 bricks.
4. Lay the first course as shown on the plan.
5. Check the height of the first brick laid, using the number 6 on the modular rule as a reference.
6. Level and plumb the course using the first brick as a guide.
7. Racking back one half brick on each end of the succeeding courses, continue laying the lead until the specified height of $16''$ (6 courses) is reached. Check each course with the rule. Level, plumb, and align the tail end of the lead.
8. Strike the joints with a jointer as needed. Brush the wall at the completion of the job.
9. Recheck the project with a plumb rule before it is inspected.

PROJECT 10: CONSTRUCTING AN OUTSIDE AND INSIDE BRICK CORNER FOR A 4″ WALL IN THE RUNNING BOND

OBJECTIVE

- The student will be able to lay out and build an outside and inside corner 9 courses high in the running bond, Figure 17-9. These are the two basic corners used in brickwork. Build the outside corner first and the inside corner second. Procedures for both corners are the same.

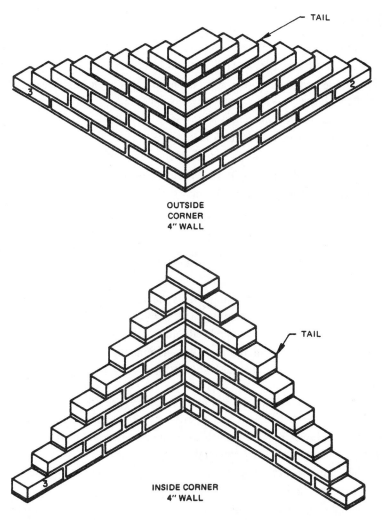

Fig. 17-9 Project 10: Constructing an outside and inside brick corner for a 4″ wall in the running bond.

EQUIPMENT, TOOLS, AND SUPPLIES

Mortar pan or board

Mixing tools

Mason's trowel

Brick hammer

Plumb rule

2′ square

Chalk box

Convex sled runner striker

Modular rule

Brush

Pencil

Approximately 45 standard-sized bricks

½ batch of mortar (½ bag lime to 8 shovels sand)

Clean water

1, $8'' \times 8'' \times 16''$ concrete block in which to set plumb rule

SUGGESTIONS

- Form solid head joints to prevent leakage.
- Select straight, square, unchipped bricks for the corner.
- When building the inside corner, form the head joints in the angle neatly since this is the center of focus. This also ensures against leaking.
- Recheck the corner for squareness after laying the first course of bricks.
- Sight down the corner as each brick is laid.
- Remove the excess mortar as soon as each course is laid.
- Keep all tools and debris away from the immediate work area.
- Observe good safety practices at all times.

PROCEDURE

1. Stock the bricks and place the mortar pan approximately 2′ from the work area.
2. Mix mortar and place into mortar pan.
3. Lay out the corner on the floor using a steel square and pencil. Extend the line a little longer than the actual corner measures with chalk.
4. Lay out the first course dry keeping head joints uniform ($\frac{3}{8}''$).
5. Lay brick #1 and brick #2 in mortar without moving the bricks in between. (See illustration.) Level and plumb brick #1 with brick #2 and straighten the edge with the plumb rule.
6. Lay brick #3. Level, plumb, and straighten the edge with brick #1. Lay the remaining bricks to complete the course. Level, plumb, and straighten the edge of each corner.
7. Lay the corner brick (#1) first in each succeeding course and work toward the end of the lead. Level, plumb, and straighten the edge of each course.
8. Check the outermost corner of every course laid with the number 6 on the modular rule.
9. Build the corner 9 courses high. It will measure 2′ in height.
10. At the completion, check the tail of the lead on each side with a plumb rule.
11. Strike the corner with a convex sled runner striker. Brush after striking.
12. Parge the back of the corner with mortar. Be sure that the parging is $\frac{3}{8}''$ thick. Remove excess mortar from the corner after parging.
13. Recheck the corner with a plumb rule before it is inspected.

PROJECT 11: BUILDING A FLEMISH BOND BRICK 12″ RETURN CORNER

OBJECTIVE

- The student will be able to lay out and build a 12″ Flemish bond brick corner 11 courses high. It is important that every header brick on every other course lines up vertically with the one below, maintaining the pattern of the bond, Figure 17-10.

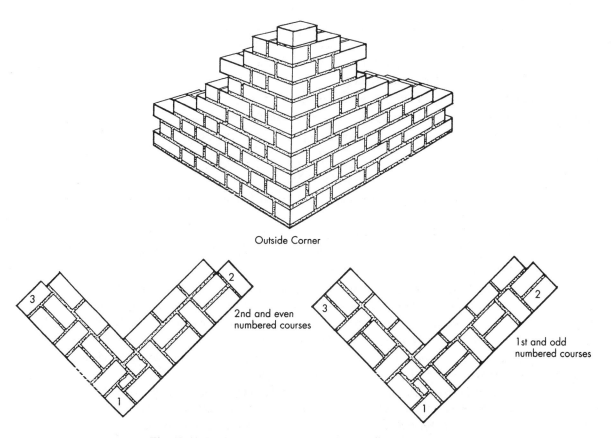

Outside Corner

2nd and even
numbered courses

1st and odd
numbered courses

Fig. 17-10 Project 11: Flemish bond brick 12″ return corner.

EQUIPMENT, TOOLS, AND SUPPLIES

Mortar pan or board

Mixing tools

Mason's trowel

Brick hammer

2′ square

Chalk box

Training mortar (Use hydrated lime or
Carotex if available in a 1 part Carotex
to 3 parts sand with water)

Grapevine jointer

Modular rule

Brush

Pencil

Brick set chisel

Approximately 150 standard size brick
1, 8″ × 8″ × 16″ concrete block in which
to set level when working

SUGGESTIONS

- Form solid full head joints to prevent leakage.
- Select straight, square unchipped brick for the corner.

- Recheck the first course with the square before proceeding to build the corner.
- Sight down the corner as each corner brick is laid.
- Make sure that the header bricks are laid level also on their longest side so they do not appear twisted on the face side.
- Wear safety eye protection when cutting any brick.
- Keep all tools and debris away from under your feet and observe safe work practices.
- Don't forget to cut a 6″ piece for the corner brick on every course.

PROCEDURES

1. Stock the brick and place the mortar pan or board approximately 2′ back from where the corner will be built.
2. Mix the mortar and place in pan or board.
3. Lay out the first course line using a steel square and a pencil. Extend the lines a little longer than the actual corner with the chalk box.
4. Cut the 6″ corner piece as shown on the plan and dry bond the first course out on the layout line using the forefinger as a gauge. The head joints should be approximately ⅜″.
5. Lay brick #1 and brick #2 in mortar without moving the brick in between. Level and plumb brick #1 with brick #2. Straightedge the brick between with the level.
6. Lay brick #3. Level, plumb, and straighten the edge with brick #1. Lay the remainder of the course. Level, plumb and straightedge the course.
7. Use the modular rule and scale number 6 for the coursing height. Always set the rule on the top outside edge of the first corner brick laid when checking the height.
8. Lay the corner brick #1 in succeeding courses and level out from that point.
9. Be sure to check the tail ends of the lead with the level to make sure they stay in alignment as they are built.
10. Build the corner to the number of courses shown on the plan.
11. Remove all excess mortar from the ends of the corner as it is built.
12. Tool the mortar joints with the grapevine tool when they are thumbprint hard.
13. Brush the corner to complete the job.
14. Recheck the corner with the level before it is inspected.

PROJECT 12: LAYING A BRICK CORNER AND BUILDING A WALL IN THE RUNNING BOND WITH A LINE

OBJECTIVE

- The student will be able to build brick leads and a 4″ wall in the running bond, maintaining a uniform thickness of all head and bed joints, Figure 17-11. *Note:* This project, which combines the construction of corners and laying bricks to the line, is typical for a home or small building.

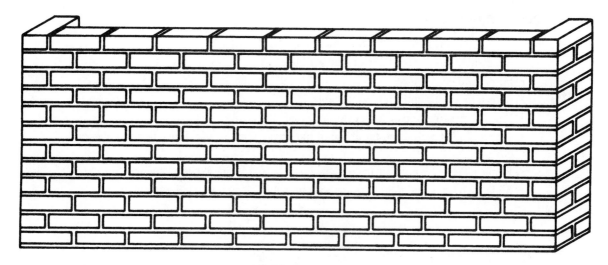

ELEVATION

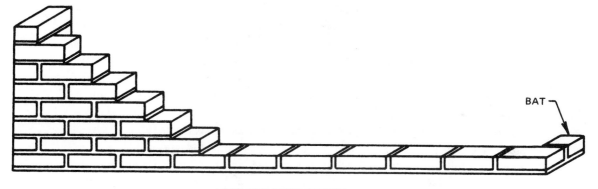

BAT

LEAD AND FIRST COURSE

Fig. 17-11 Project 12: Laying a brick corner and building a wall in the running bond with a line.

EQUIPMENT, TOOLS, AND SUPPLIES

2 mortar pans or boards

Mixing tools

Mason's trowel

Brick hammer

Plumb rule

2′ square

Ball of nylon line

Line pin and nail

Pencil

V-joint sled runner

Brush

Chalk box

Approximately 145 standard-sized bricks (waste included)

$1\frac{1}{2}$ bags of lime; 27 shovels of sand

Supply of clean water

1, $8'' \times 8'' \times 16''$ concrete block in which to set level

SUGGESTIONS

- Be sure that the base for the construction is clean.
- Keep all materials about $2'$ from the wall line.
- Form solid head joints.
- Spread mortar as uniformly as possible.
- Practice pressing bricks into position. Avoid unnecessary tapping.
- Do not spread mortar too far ahead of the brick laying or it will dry before it can be covered.
- Remove the excess mortar from each course before laying the next course.
- Sight each brick laid to be certain the bond pattern is in correct alignment.

PROCEDURE

1. Assemble the bricks in the work area.
2. Mix the mortar and load the mortar pans.
3. Snap a chalk line longer than $8'$ on the floor.
4. Mark a line for an $8''$ jamb at one end of chalk line with a $2'$ square.
5. Dry bond 10 bricks from the point which was squared off to the other end of the line. (The line will be slightly longer than necessary.) This is used as a reference point when laying out the bricks. Use the forefinger for correct spacing.
6. Place a mark at the end of the tenth brick. Use a square to square off this point for the jamb on the other side of the wall.
7. Use a brick of average length as a gauge when laying out the jamb of the lead at each end.
8. Bed the corner bricks at each end of the wall to the correct height (number 6 on the modular rule). Level the two bricks with each other by setting another brick temporarily in the middle as a check point. (A long, straight $2'' \times 4''$ with a plumb rule laid on top could also be used as a leveling board.) Be sure that the bricks are plumb, aligned, and ranged with the layout line.
9. Attach the line by pushing a nail to which the line is attached under the end of the brick on the left end of the wall. Pull the line up over the top of the brick and push the line to the face of the brick. Lay a couple of bricks on top to hold the line in place. Push the line pin under the end of the brick at the right end of the wall. Pull the line tight and wrap it around the pin. Lay a couple of bricks on top of the line to prevent it from dislodging.
10. Pick up the dry-bonded bricks as needed and place them in mortar to lay the first course.
11. Cut bats for jambs with the hammer and lay one at each end. Level, plumb, and square the bats with the wall line.
12. Build a lead projecting $3\frac{1}{2}$ bricks at the end of each wall. Check the height of each course with the number 6 on the modular rule. Cut bats as they are needed for jambs. Check jambs periodically to be sure they remain square, plumb, and level.
13. Fill in the space between leads, using the line as a guide.

14. Resume building leads up to the specified 12 courses in height and fill in the wall to the line. Check the work for the correct height (number 6 on the modular rule).
15. Strike the wall with a V-joint sled runner striking tool. Brush the project at completion.
16. Parge the back of the wall and the inside of the jambs with mortar.
17. Recheck the project with a plumb rule before inspection.

UNIT 18
Estimating Brick Masonry by Rule of Thumb

OBJECTIVES

After studying this unit, the student will be able to

■ determine the number of square feet in a brick wall by the wall area method.

■ estimate brick, masonry cement, and sand for small-sized and average-sized jobs.

Estimating by rule of thumb is not intended to be a mathematically perfect method of estimation. It is workable, however, when estimating materials for the construction of small-sized and average-sized jobs. Some examples of these small-sized and average-sized jobs are a home, garage, retaining wall, or chimney. Apprentice masons are not expected to estimate materials for large or complicated jobs. However, they should be able to figure the necessary materials for daily work projects.

THE IMPORTANCE OF ESTIMATING

The rule of thumb method of estimation was developed through years of practical experience on the job. It is designed to allow for some waste on the job. Most small contractors estimate their materials by rule of thumb.

The apprentice should be able to apply basic arithmetic, including addition, subtraction, multiplication, and division, to solve typical estimating problems. The majority of mistakes made in estimating occur in the process of simple arithmetic. The mason should always recheck work before accepting the results as final.

It is important that materials not be grossly overestimated or underestimated. When materials are overestimated, the materials that remain must be moved to another job. This results in wasted labor, time, and money. It may be some time before the materials are used again, especially if they are a special type. Underestimating, on the other hand, is also wasteful of time and energy.

THE WALL AREA METHOD

Most masonry materials used today are based on the modular system. This greatly simplifies estimating for the mason.

For great simplicity and accuracy, the most widely used method of estimation is the *wall area* or *square foot* method. It consists of multiplying the length and height of the wall minus openings.

Rule of thumb estimating can best be explained by using a typical problem as an example. In the discussion of estimating techniques, which follows in this unit, a specific problem will be used as an example. Assume that the required joint size is $\frac{3}{8}''$.

ESTIMATING SQUARE FEET

A $4''$ thick wall is to be built $20' \times 8'$. It is built in a running bond, so no extra amounts must be considered for header courses. Estimate the total number of square feet as follows:

$$20' \text{ (length of wall)}$$
$$\times 8' \text{ (height of wall)}$$
$$160 \text{ total square feet (sq ft)}$$

ESTIMATING BRICK

After finding the total square feet in the wall, the mason may then estimate the number of standard-sized bricks required. There are 6.75 standard bricks to each square foot of wall area. When estimating by the rule of thumb, round off this figure to 7. This allows for waste and broken bricks.

To estimate the number of bricks, multiply 7 by the total wall area.

$$160 \text{ sq ft of wall}$$
$$\underline{\times 7} \text{ bricks per sq ft}$$
$$1120 \text{ total amount of bricks needed}$$

ESTIMATING MASONRY CEMENT

Masonry cement can be estimated in two different ways. The first way is by using a base figure of 8 bags per 1000 bricks. The second way is by calculating the number of cubic feet per 100 sq ft of wall area. If the sand is coarse, it may be necessary to increase the amount of masonry cement to improve the workability. Even though the mortar has great strength, if it does not handle well on the trowel, the cost for the job greatly increases since the mason's productivity is cut. This depends on the particular situation.

When using portland cement and lime-based mortars, a table must be consulted to arrive at the correct proportions. A table must be consulted because the percentage of portland cement and lime differs according to the type of mortar specified. For estimating purposes, only masonry cement will be dealt with in this unit, since it is the most commonly used throughout the masonry business.

To estimate masonry cement for a wall measuring $20' \times 8'$, figure 8 bags of masonry cement to every 1000 bricks. Waste is included. By dividing 1000 by 8, it is found that 1 bag of masonry cement is sufficient for 125 bricks. To find the number of bags needed for the wall in question, divide as follows:

$$
\begin{array}{r}
8.96 \text{ bags of masonry cement} \\
125 \overline{)\ 1120.00} \\
\underline{1000\ \ } \\
120\ 0 \\
\underline{112\ 5} \\
7\ 50 \\
\underline{7\ 50}
\end{array}
$$

Round off the answer to 9 bags.

ESTIMATING SAND

Sand, one of the most inexpensive materials used in masonry work, is sold by weight. Because it is inexpensive, it is not important to estimate to the exact pound for a job. As a rule, masons order sand by the ton unless the job is very small, and then in hundreds of pounds. The base figure for estimating sand is 1 ton per 1000 bricks, allowing for a normal amount of waste.

There are several factors to consider when estimating sand. If more than 3 tons of sand are needed, $\frac{1}{2}$ ton should be allowed for waste. First to consider is the loss of sand when it is piled on the ground. (This is the usual method of storing sand.) Since it is on the ground, some of the sand will mix with dirt and foreign particles.

Also, if the sand is located in a populated area with small children and is unprotected, the mason must assume that a certain amount will be lost. The estimated loss depends on the project.

The wall in the original problem, which required 1120 bricks, would require $1\frac{1}{2}$ tons of sand. This figure allows a sufficient amount for waste. This is as close as the mason would estimate on this particular job.

ESTIMATING MATERIAL COST ON THE JOB

The materials needed for a $20' \times 8' \times 4'$ brick wall would be listed:

$$1120 \text{ bricks}$$
$$9 \text{ bags masonry cement}$$
$$1\frac{1}{2} \text{ tons sand}$$

To estimate the total price of the material, multiply each by the price per unit. Assume that the brick costs $.28 each, the masonry cement costs $4.75 per bag, and the sand is $20.00 per ton.

$$
\begin{array}{r}
1120 \text{ brick} \\
\underline{\times 0.28} \text{ cost per brick} \\
89.60 \\
224\ 0 \\
\underline{0000\ \ } \\
\$313.60 \text{ total cost of brick}
\end{array}
$$

$$
\begin{array}{r}
\$\ 4.75 \text{ cost per bag of cement} \\
\underline{9} \text{ number of bags} \\
\$42.75 \text{ total cost of cement}
\end{array}
$$

$20.00 per ton of sand
  1.5   tons sand needed
10.000
20 00
$30.00 total cost of sand

Add all costs together for total cost

$313.60 brick
 42.75   m. cement
  30.00 sand
$386.35 total material costs

ESTIMATING LABOR COSTS FOR BRICK MASONRY WORK

After materials have been estimated, labor costs must be added before a final price for the job can be determined. Many factors affect costs for labor, including weather conditions, accessibility of materials, and other hidden costs. The contractor must expect some hidden costs for such things as equipment repair, insurance, and depreciation of equipment. In addition, the contractor must estimate how much work the masons and laborers will produce. Finally, the contractor must realize that costs will vary depending on geographical location and the availability of skilled masons.

Labor Wage Rates

Factors like unionization, geographical location, and the amount of work available determine wage rates. Rates vary and change constantly in response to these and other factors.

In estimating labor costs, it also helps the contractor to know how many bricks a mason can lay per day. This amount depends on the type of bond being used. However, according to the International Union of Bricklayers and Allied Craftsmen, a national average for masons using running bond is about 675 bricks per 8-hour day.

In addition, the contractor must estimate the number of laborers needed. Generally, this figure is one laborer for every four masons, but it may have to be adjusted for certain job conditions.

After determining costs for labor and materials, the contractor adds them together to reach a final bid or price for the job. As you can see, estimating can be a complicated process, and estimating large jobs should be left to a professional estimator or contractor, who often use computers to help cost out jobs, Figure 18-1. However, whether the job is large or small, careful attention to mathematics will eliminate many estimating mistakes that often appear.

Using Estimating Tables for Larger Amounts or Big Jobs

According to the Brick Industry Association (formerly The Brick Institute of America), except for the non-modular "Standard Brick" (3¾ by 2¼ by 8 in.) and some oversize brick (3¾ by 2¾ by 8 in.), virtually 100 percent of the brick produced in the United States is sized to fit the modular system. Even the "Standard Brick" is also available in a modular size (nominal dimensions: 4 by 2⅔ by 8 in.). Since there is still a considerable production of non-modular brick as well as modular brick, both types are shown in the tables in this unit, which are courtesy of The Brick Industry of America, the most respected technical brick industry organization in the United States.

Fig. 18-1 Professional estimators taking off a job from a set of plans.

Estimating Procedure Using Tables

As mentioned previously in the beginning of this unit, because of its simplicity and accuracy, the most widely used estimating procedure is the "wall-area" method. It consists of multiplying known quantities required per square foot by the net area (gross areas less areas of openings), regardless of whether modular or non-modular brick are being estimated.

Estimating materials quantities is greatly simplified under the modular system. For a given nominal size, the number of brick per square foot of wall area will be the same regardless of mortar joint thickness, assuming of course that the brick are to be laid with the thickness of joint for which they are designed. *There are only three standard modular mortar joint thicknesses: $1/4$ in., $3/8$ in., and $1/2$ in.* In contrast, the number of non-modular standard brick required per standard foot of wall will vary with the thickness of the mortar joint.

When estimating, always determine the net quantities of all materials before adding any allowance for waste. Allowance for waste and breakage varies, but, as a general rule, at least 5 percent should be added to the net brick quantities and 10 to 25 percent to the mortar quantities. Certain job conditions may dictate increasing the allowance factor.

Table 1, Figure 18-2, gives net quantities of brick and mortar required to construct walls one wythe in thickness with various modular brick sizes and the two most common joint thicknesses ($3/8$ in. and $1/2$ in.). Mortar quantities are for full bed and head joints.

Table 2, Figure 18-3, also provides similar information for walls constructed of only non-modular brick.

The brick and mortar quantities in Tables 1 and 2 are for running or stack bond, which contains no headers. For bonds requiring full headers, the correction factors given in Table 3, Figure 18-4, must be applied. Also, when estimating quantities for multiwythe walls, the mortar quantities for interior and/or longitudinal collar joints given in Table 4, Figure 18-5, must be added.

Nominal Size of Brick in.			Number of Brick per 100 sq ft	Cubic feet of Mortar			
				Per 100 Sq Ft		Per 1000 Brick	
t	h	l		3/8-in. Joints	1/2-in. Joints	3/8-in. Joints	1/2-in. Joints
4 × 2 2/3 ×	8		675	5.5	7.0	8.1	10.3
4 × 3 1/5 ×	8		563	4.8	6.1	8.6	10.9
4 × 4	8		450	4.2	5.3	9.2	11.7
4 × 5 1/3 ×	8		338	3.5	4.4	10.2	12.9
4 × 2	12		600	6.5	8.2	10.8	13.7
4 × 2 2/3 ×	12		450	5.1	6.5	11.3	14.4
4 × 3 1/5 ×	12		375	4.4	5.6	11.7	14.9
4 × 4	12		300	3.7	4.8	12.3	15.7
4 × 5 1/3 ×	12		225	3.0	3.9	13.4	17.1
6 × 2 2/3 ×	12		450	7.9	10.2	17.5	22.6
6 × 3 1/5 ×	12		375	6.8	8.8	18.1	23.4
6 × 4	12		300	5.6	7.4	19.1	24.7

Fig. 18-2 Table 1: Modular brick and mortar required for single wythe walls in running bond. (No allowances for breakage or waste.) (Courtesy Brick Industry Association)

Size of Brick in.			With 3/8 in. Joints			With 1/2 in. Joints		
			Number of Brick per 100 Sq Ft	Cubic Feet of Mortar per 100 Sq Ft	Cubic Feet of Mortar per 1000 Brick	Number of Brick per 100 Sq Ft	Cubic Feet of Mortar per 100 Sq Ft	Cubic Feet of Mortar per 1000 Brick
t	h	l						
2 3/4 × 2 3/4 × 9 3/4			455	3.2	7.1	432	4.5	10.4
2 5/8 × 2 3/4 × 8 3/4			504	3.4	6.8	470	4.1	8.7
3 3/4 × 2 1/4 × 8			655	5.8	8.8	616	7.2	11.7
3 3/4 × 2 3/4 × 8			551	5.0	9.1	522	6.4	12.2

Fig. 18-3 Table 2: Non-modular brick and mortar required for single wythe walls in running bond. (No allowances for breakage or waste.) (Courtesy Brick Industry Association)

Bond	Correction Factor
Full headers every 5th course only	1/5
Full headers every 6th course only	1/6
Full headers every 7th course only	1/7
English bond (full headers every 2nd course)	1/2
Flemish bond (alternate full headers and stretchers every course)	1/3
Flemish headers every 6th course	1/18
Flemish cross bond (Flemish headers every 2nd course)	1/6
Double-stretcher, garden wall bond	1/5
Triple-stretcher, garden wall bond	1/7

Fig. 18-4 Table 3: Bond correction factors for walls of Tables 1 and 2. (Add to facing and deduct from backing.) (Courtesy Brick Industry Association)

(*Note*: Correction factors are applicable only to those brick which have lengths of twice their bed depths.)

Table 5, Figure 18-6, contains the quantities of portland cement, hydrated lime, and sand required for 1 cubic foot of four types of mortar. Although ASTM Standard Specifications for Mortar for Unit Masonry (ASTM Designation C270) permits a range of proportions for each mortar type, the quantities in Table 5 have been based on a single set of proportions for each of these types. For convenience in estimating, quantities based on both weight and volume are included in the table. Mortar is generally proportioned by volume on the job, although proportioning by weight is the most accurate method.

Brick Masonry Estimator Slide Rule

The Brick Industry Association has a very handy Brick Masonry Estimator Slide Rule available for estimating the amount of brick and mortar needed for modular and non-modular brick, Figure 18-7. I bought mine from the Bon Tool Company through their catalog at a very small price. You can contact them at their phone order hotline, 1-800-444-7060, or at their Website, www.bontool.com.

Cubic Feet of Mortar Per 100 Sq Ft of Wall

1/4-in. Joint 2.08	3/8-in. Joint 3.13	1/2-in. Joint 4.17

Note: Cubic feet per 1000 units = $\dfrac{10 \times \text{cubic feet per 100 sq ft of Wall}}{\text{number of units per square foot of Wall}}$

Fig. 18-5 Table 4: Cubic feet of mortar for collar joints. (Courtesy Brick Industry Association)

Material	Quantities by Volume				Quantities by Weight			
	Mortar Type and Proportions by Volume				Mortar Type and Proportions by Volume			
	M 1:1/4:3	S 1:1/2:4-1/2	N 1:1:6	O 1:2:9	M 1:1/4:3	S 1:1/2:4-1/2	N 1:1:6	O 1:2:9
Cement:	0.333	0.222	0.167	0.111	31.33	20.89	15.67	10.44
Lime	0.083	0.111	0.167	0.222	3.33	4.44	6.67	8.89
Sand	1.000	1.000	1.000	1.000	80.00	80.00	80.00	80.00

Fig. 18-6 Table 5: Material quantities per cubic foot of mortar. (Courtesy Brick Industry Association)

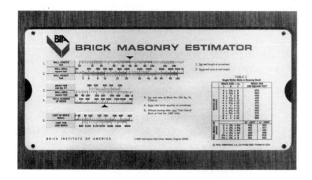

Fig. 18-7 Brick masonry estimator slide rule. (Courtesy Brick Industry Association)

ACHIEVEMENT REVIEW

Solve each of the following problems concerning the estimating of materials and cost and determining total area.

1. a. A brick wall 40′ long × 8′ high is to be built in the running bond. What is the total area in square feet?
   b. Determine the number of bricks needed to build the wall.
   c. Find the number of bags of masonry cement needed for the wall.
   d. Determine the amount of sand needed for the wall.

2. An 8″ brick retaining wall is being built. Figure the amount of materials needed to build the wall if it measures 100′ long × 4′ high.

3. A 4″ brick veneer wall is to be built across the front of a store. The wall is 50′ long × 5′ high. Estimate the total amount of materials needed for the total cost of the wall if the bricks cost $.28 each, masonry cement costs $4.75 per bag, and sand costs $20.00 per ton.

4. An 8″ brick garden shed 10′ wide, 10′ long, and 8′ high is to be built. There is one door which measures 3′ × 7′. Estimate the amount and cost of materials for the job, deducting the door area from the total wall area. Use the unit prices given in question 3.

5. A job calls for 7500 bricks. How much sand and masonry cement is needed for the job?

6. How many bricks could be laid with 25 bags of masonry cement?

7. If a mason lays 675 bricks a day, how many bricks would the mason lay in a 20-day period?

8. A large brick masonry job employs 44 masons. If one laborer is required for every four masons, how many laborers are needed on this job?

9. A mason orders 17,000 bricks for a house job. If the bill is paid in 7 days, a 9% discount is deducted. What is the total amount of the discount? Use a brick price of $.28 each.

10. Five bricklayers working together laid 675 bricks each in one 8-hour working day. What is the average of each for 40 hours?

11. Using the answer from problem 10, what is the total number of brick laid by all five bricklayers in a 40-hour work week?

12. Based on the fact that one bag of masonry cement mortar is needed to lay 125 bricks, how many bags will it take to lay 17,000 bricks? Round off your answer to the nearest bag over what is required.

13. Using the figure of 1 ton of sand for every 1000 bricks, how much sand will be needed to lay 17,000 bricks?

14. A bricklayer pays $1,680.00 for brick, $66.50 for masonry cement, and $120.00 for sand for a small job. What is the total cost of materials?

15. Using the answer from problem 14, add 5% sales tax. What is the total final cost of the materials?

UNIT 19
Laying Concrete Block to the Line

OBJECTIVES

After studying this unit, the student will be able to

- make bed joints and head joints for concrete block.
- lay concrete block to the line.
- cut concrete block with a brick set and hammer.
- describe and lay insulated concrete block.

The process involved in laying concrete block to the line is very similar to that of laying bricks to the line. There are some important differences, however, due to the fact that concrete block is much larger and heavier.

The placement of materials in the work area is of prime importance. Concrete block must be stacked on the pile with the bottom side down. The top of the block has a larger shell and web, which makes it easy to distinguish. The units must be handled carefully, as they easily chip and crack.

Concrete block should be kept dry at all times. When highly absorbent bricks become very dry, they are wet with water. This is not the case with concrete block. Moisture causes block to expand in size. If they are built into the wall when they are wet, they will later dry and shrink, causing cracks between the mortar and the block. Plastic covering is an inexpensive way to protect block. The block should be stored on pallets off the ground to prevent dampness from being absorbed.

SPREADING MORTAR

As a rule, mortar is bedded only on the outside edges of concrete block. Very seldom is a solid bed joint used. Bedding on the outside webs of block is known as *face shell bedding.* It is swiped on with the edge of the trowel, Figure 19-1.

All joints and holes should be filled when the block is laid so that all of the mortar adheres and dries at the same rate of speed. The mortar must be of the proper consistency. Mortar that is too runny or soft causes the block to sink below the line, which requires relaying of the block. Mortar that is too stiff also presents a problem, since the block requires a great amount of pressure to lay it in the mortar. This could cause damage to the block. Blocks should never be moved after they

Fig. 19-1 Swiping mortar bed joint on block with blade of trowel.

211

have been laid. Movement after the initial set may cause leaky walls and loss of bond strength.

The mortar on bed joints should never be dusted with dry cement to stiffen the mortar. The cement dust prevents the formation of a strong bond between the mortar and the block. It is also a time-consuming and expensive method of stiffening mortar. Nothing but mortar should be placed in bed joints, except wire reinforcement if specified.

A well-built block wall can result with good planning, proper care and spacing of materials, and accurate use of tools.

LAYING THE FIRST COURSE

The apprentice should learn to lay block to the line before attempting to build a corner. As is the case in brickwork, block is laid to the line, $1/16''$ away. The line is attached either to a corner pole or to a prebuilt concrete block corner.

Block should always be laid in the wall with the wider web of the block facing up, Figure 19-2. Block laid with the wider web down will not be any less strong. However, it is easier to apply mortar on the

wider web and less mortar is wasted if it drops into the cell of the block.

The first course should be laid with great care, since a level and plumb wall depends to a great extent on the first course. Sweep or brush the base before spreading the mortar. Spread a solid bed of mortar on the base. Do not furrow the mortar on the first course. A solid joint ensures a watertight wall and proper bonding with the base.

APPLYING THE HEAD JOINT

Full head joints should be formed on both *ears* (end edges) of the block to be laid. With the trowel, pick up enough mortar from the mortar board to form the head joints. Do not fill the entire trowel with mortar, since this much mortar is not needed for head joints. Stand the block on its end in a vertical position and apply head joints on both ears of the block with a downward swiping motion of the trowel. The mortar should then be pressed down on the inside of the ears of the block so that it will not fall off when lifted up and placed in the wall, Figure 19-3. If a head joint becomes dislodged from the block as it is picked up, reapply fresh mortar before laying.

Fig. 19-2 The two major types of concrete block, the two-celled (left) and the three-celled (right). Both are 16″ in length. The two-celled block is shown with the thicker web facing up, the position in which it is laid in the wall. The three-celled block is shown with the bottom side facing up. Notice the difference in web thickness of the two blocks.

Fig. 19-3 Forming a head joint on a concrete block. The mortar is applied to the ears, or ends, of the block.

Fig. 19-4 Lifting block and keeping trowel in the hand at the same time.

Lift the block firmly by grabbing the web at each end of the block and lay it on the mortar bed joint. Do not move the block with jerking motions. The mortar should stay intact. The trowel should remain in the mason's hand when laying block to save time, Figure 19-4. The hand should curl over the handle of the trowel and grasp the web of the block at the same time. This may be difficult in the beginning but will come with practice.

POSITIONING BLOCK

Practicing the techniques involved in laying block will determine the easiest methods for the individual mason. By tipping the block a little toward the body and looking down the face side of the block, the mason can position the block in relation to the top edge of the block in the course below. The block should then be rolled back slightly so that the top of the block is in correct alignment with the line. At the same time, the block is being pressed back toward the last block laid so that the proper amount of mortar oozes from the head joint.

Concrete block should be laid so that the top of the outside edge is level with the top of the line and located $\frac{1}{16}''$ back from the line, Figures 19-5A and 19-5B. The bottom outside edge of the block is then in line with the top edge of the course below.

The method in which the block is set on the bed joint is very important. The block should be laid gently in the mortar so that it does not sink too far. The block should not be released too quickly, or it will have to be relaid. By slightly delaying the release of the hand from the block, the block absorbs the moisture from

Fig. 19-5A Laying the block gently in position so that it does not sink below the line. When properly positioned, it is even with top of the line and approximately $\frac{1}{16}''$ back from the line.

Fig. 19-5B Block laid correctly to line, even with top of line and $\frac{1}{16}''$ back from line.

the mortar and takes an initial set. After the block is set, mortar should ooze out of the head and bed joints. If it does not, there was not enough mortar used in the joints. This presents problems in tooling joints and may also prevent a watertight joint from being formed.

To ensure a good bond, do not spread mortar too far ahead of the actual laying of the block, since the mortar stiffens and loses its plasticity and strength. Positioning of concrete block must be done before the mortar stiffens or the bond strength is destroyed and cracks result. As each block is laid, cut off the excess mortar with the trowel on a slight angle, Figure 19-6. If the work is progressing at an efficient pace, this mortar may be used for the next head joint. Never allow mortar that has been cut off to fall to the ground, since this mortar cannot be used again.

ADJUSTING BLOCKS

If a block is set unevenly on the mortar, first check to see if a pebble or other foreign matter has become lodged between the mortar and the block. If this is the case, remove the block from the mortar bed and lay it in fresh mortar. If the block simply requires readjustment in the mortar, it is permissible to tap it into place, Figure 19-7. The block should not be tapped with the trowel extended over its face, as the face of the block could be chipped or mortar on the trowel could be splattered over the wall. Although it is not recommended to

Fig. 19-7 Tapping a concrete block into place. The trowel must remain centered on the block to avoid chipping the face of the block.

tap the block with the handle of the trowel, it is permitted if done sparingly. If it is done frequently, the handle becomes rough and worn, which may be bothersome to the mason. The mason's hammer should be used if adjustment cannot be easily made with the trowel.

INSTALLING THE CLOSURE BLOCK

The closure block, the last block laid in the wall, is most prone of all the block to leakage. It is imperative that the mortar joints are strong and that the spacing of the block is correct. The amount of space allowed for a standard block is the length of the block plus room for 2 mortar head joints. For a standard concrete block, this measurement is a total of $16\frac{3}{8}''$, $15\frac{5}{8}''$ for the length of the block, and $\frac{3}{8}''$ for each mortar joint.

Before placing the closure block in the wall, apply mortar to form joints on all four edges of the block which is already in place. Next, apply mortar to all four edges of the closure block. Lower the closure block into the space without disturbing the other block,

Fig. 19-6 The trowel should be held at the proper angle when cutting off excess mortar.

Figures 19-8 and 19-9. Immediately fill any holes that are left in the joint by picking up mortar on the back edge of the trowel and pressing it into the head joints. It is extremely important to apply mortar to all head joints when laying the closure block.

After the closure block has been laid, erect the line for the next course and continue building the wall to the specified height. Some masons stand a number of block on end to apply the head joints rather than applying mortar on one block at a time, Figure 19-10. This practice speeds the process, but the mason must guard against the danger of mortar drying before the block is laid.

Fig. 19-10 Applying mortar to form head joints on several blocks at one time can speed up the block-laying process.

REINFORCING BED JOINTS OF CONCRETE BLOCK

When walls require additional strength, wire reinforcements can be laid in the mortar joints, Figure 19-11. The wire is laid on the wall with the mortar spread on top of and fully surrounding it. In places where two pieces

Fig. 19-8 Mortar head joints are applied to both ends of adjoining blocks to help form a more full joint for closure block.

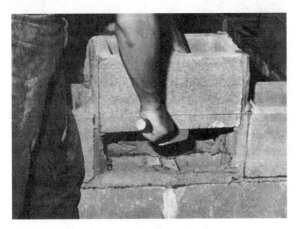

Fig. 19-9 Both ends of closure block are also buttered with mortar and laid gently in position to line.

Fig. 19-11 Wire reinforcement track laid in mortar joint for additional strength.

of wire meet, there should be a minimum overlap of 6″. The average distance between vertical courses of wire track is 16″, with 24″ as the maximum allowed. Job specifications should always be consulted before wire reinforcements are installed.

TOOLING JOINTS

Joints on concrete blocks are usually concave joints, with the V-joint ranking second in popularity. A jointer slightly larger than the mortar joint should be used to form the joint.

The sled runner jointer, preferably 16″ long, should always be used when striking joints on concrete block work because it strikes a straighter joint, Figure 19-12.

Strike the head joint first and the bed joints last. This will help to eliminate tear drops of mortar showing in finished bed joints. Enough force should be used to press the mortar tightly against each side of the joint so that the surface is smooth, water resistant, and free of cracks. When striking, fill any cracks or holes with fresh mortar. Never pick up old or dead mortar to use

for jointing. At the completion of the striking process, allow sufficient time for the mortar to dry before brushing the wall. If it is brushed too soon, the mortar joint may be destroyed.

CUTTING CONCRETE BLOCK

Concrete block can be cut with the use of various tools. The masonry saw provides the most precise cut. However, the apprentice first learns to cut block with the hammer and blocking chisel (or *brick set*) since a masonry saw may not always be available. Proper eye protection must be worn when cutting block.

Lay the block on its side and mark where the cut is to be made with a pencil and ruler or straightedge. Score one side of the block lightly with the blocking chisel. Turn the block over carefully and score the other side. Set the block with the bed side down and with the holes in a horizontal position. With the brick hammer, lightly tap the top of the block above the point at which the scoring was done, Figure 19-13. The block piece should break evenly. If it does not, rescore the block again.

Fig. 19-12 Striking a joint on a block wall with a sled runner jointer.

Fig. 19-13 Cutting a concrete block with the brick set and hammer.

Even when it is cut correctly, the block may not always be perfectly straight. If it is not straight enough, grasp the web of the block nearest the cut end and even the edge with a hammer. Cut downward in a vertical position so that the shell of the block does not break. When it is not necessary that the block be cut accurately, the brick hammer may be used alone.

Do not cut concrete block on a hard surface such as a concrete slab. The hammer or brick set combined with the hard surface may cause the block to shatter or break irregularly.

SAFETY PRACTICES ON THE JOB

Masons should be especially careful when lifting concrete block because of the heavy weight. When lifting the block, keep the feet close together and the block close to the body. If uneven ground or ditches are present in the work area, the chance of a shift in body weight is greater than under normal conditions. A fall or slip with a concrete block in hand could cause a severe injury to the mason.

Cutting concrete block should always be done with consideration for the safety of other workers. Flying pieces of block can cause serious injury to the eyes. Cut the block downward and away from yourself and others. Holding the hand high on the chisel will result in fewer injuries to the hands. Burred chisels should be ground off immediately, as flying pieces of metal can cause injury. Eye protection should be worn when doing any cutting. Cutting concrete block calls for special attention to safety rules.

REPOINTING THE WALL

If patching or repointing must be done after the mortar has hardened, the mortar must be chiseled out with a joint chisel to a depth of at least $1/2''$. All loose mortar should be removed with a brush. The joint should be wet with water. To do this, use a brush or splash water directly on the cut portion. The joint is then repointed with fresh mortar using a steel slicker tool. Wetting the joint delays the setting time and, therefore, produces a better bond. Line pin and nail holes should be repointed immediately after the pin or nail is removed from the wall. Be especially careful that the adjoining units are

not chipped or cracked by the cutting operation when cutting out joints.

CARE AND PROTECTION OF THE WORK

Walls that are constructed above grade line (or ground level) are subject to wind pressure. At times, the pressure can amount to more than 20 pounds per square foot (lb/sq ft) of wall area. While the walls are being built, they should be protected from the wind and given some type of temporary support. This is especially important in the early stages of construction before the mortar has hardened to full strength. As a rule, walls higher than 8' to 10' should be braced. This is done by *shoring* (or propping up) timber or framing lumber about every 8' to 10' of wall length on both sides of the wall, Figure 19-14. After the *joists,* or floor supports are laid in place, the bracing is removed. Bracing similar to this is also used to prevent walls from caving in when a bulldozer backfills a foundation.

When the work is complete for the day, it should be protected from the elements. Walls that are rain-soaked may take months to dry and efflorescence may

Fig. 19-14 Concrete block wall held in position with wooden braces. After the joists are constructed, the braces are removed.

develop. Sheets of plastic or tarpaulin should be used to cover the top of the walls to prevent moisture from entering the core of the blocks. The covering should cover and hang over the wall at least 2′. Loose boards or planks laid on top of the walls are not sufficient protection.

INSULATED CONCRETE BLOCK

A new energy saver, the insulated concrete block, is appearing in commercial and residential masonry construction. A 3″-thick polystyrene insert placed in the cores of the block provides the insulation, Figure 19-15. Although insulated concrete block costs more than noninsulated block, the initial costs are offset by savings in energy costs.

Insulating properties of building materials are usually stated in terms of an R value. Ordinary block produced from a mixture weighing 95 pounds per cubic foot (lb/cu ft) has an energy rating of R-2.86. In comparison, insulated block weighs 80 lb/cu ft and has an energy rating of R-9.52, which is within 1.50 points of the recommended R-11 value for walls. Insulated

concrete blocks can be built in single walls or in double walls with an air cavity between for additional insulation.

Building Insulated Block Walls

In an insulated concrete block wall, a center bed joint must be applied to seal off the joint. This joint is applied directly over the insulation. Wire joint reinforcement is then laid in the mortar. The wire in Figure 19-16 protrudes to tie in the other cavity wall, which will be built later.

The insulated block is then laid and pressed into position to the line. Take care that the joint reinforcement is pressed into the mortar joint, Figure 19-17.

High energy costs are making insulated block very popular with builders. The apprentice mason should be familiar not only with the basic block-laying process, but with the concepts behind the demand for energy-efficient masonry walls as well.

Fig. 19-15 Insulated 8″ × 8″ × 16″ concrete block. Note how the polystyrene fits in the cell of the block.

Fig. 19-16 Mortar bed joint and wire reinforcement in place on insulated block wall.

Fig. 19-17 Laying insulated block to the line. All excess mortar is cut off with the trowel.

ACHIEVEMENT REVIEW

The column on the left contains questions on concrete block and their production. The column on the right lists terms. Select the correct term from the right-hand list to answer each question on the left.

1. What is the term for spreading mortar only on the outside web of a concrete block?

2. Concrete blocks have a top and bottom edge. Which thickness of web should be facing up when they are laid in a wall?

3. A solid joint should be spread on the base when starting the first course. What is the main reason for this?

4. What are ends of stretcher blocks where the mortar head joints are applied known as?

5. If mortar is spread too far ahead on a wall, it dries too quickly. What type of wall strength does this affect?

6. When adjusting a block to the line, where on the block should the mason tap?

7. What is the last block that is laid in the wall called?

a. Brick set
b. Bat
c. Closure block
d. Ears
e. Concave
f. Header
g. Wider web
h. Rake out
i. 24″
j. 16″
k. 8″
l. Face shell bedding
m. Center
n. Jointing
o. Bond
p. To waterproof
q. Jamb
r. Butt edge

8. What is the standard height at which wire reinforcement should be installed in a bed joint?

9. What is the most common joint finish on concrete block construction?

10. When an accurate cut is needed and a masonry saw is not available, what is the best tool to use?

s. Brick hammer
t. Furrowing
u. For design
v. Narrower web

MATH CHECKPOINT

1. A mason is building a block foundation that has 1200 blocks in it. Type S masonry cement mortar is being used. If one bag of mortar will lay 30 blocks, how many bags will it take to build the foundation?

2. Type S masonry cement costs $4.90 a bag. How much will it cost for the masonry cement to build the foundation in problem 1?

3. A mason is building a block wall 50′ long and 12 courses high. The plan calls for metal joint reinforcement every other course in the bed joints. How many linear feet of reinforcement will be needed?

4. Every 8 bags of mortar will require 1 ton of sand. How many tons of sand will it take to lay 1200 block?

5. A mason lays 896 blocks on a job. If he lays 28 per hour, how many hours did he work?

6. A foundation is being built of concrete block. There are six window openings in the job. Each one is 4′ wide. Each lintel must bear (rest) on both sides of the opening 8″. How long must the lintel be for one opening? What would be the total length of all the lintels needed?

PROJECT 13: LAYING A CONCRETE BLOCK WALL

OBJECTIVE

- The masonry student will be able to lay an $8' \times 8' \times 16''$ concrete block wall in the running bond with the use of a corner pole, Figure 19-18.

    Note: This concrete block wall is typical of a block wall which surrounds a stairway in an apartment building.

EQUIPMENT, TOOLS, AND SUPPLIES

2 mortar pans or boards
Trowel
Plumb rule
Chalk line
Line blocks
Modular rule
Convex sled runner
Brick set chisel
Ball of nylon line

Brick hammer
Brush
2 corner poles
Mixing tools
60, $8'' \times 8'' \times 16''$ concrete blocks
   (4 are allowed for breakage)
2 bags of mason's lime (or carotex), 32
   shovels of sand
Supply of clean water

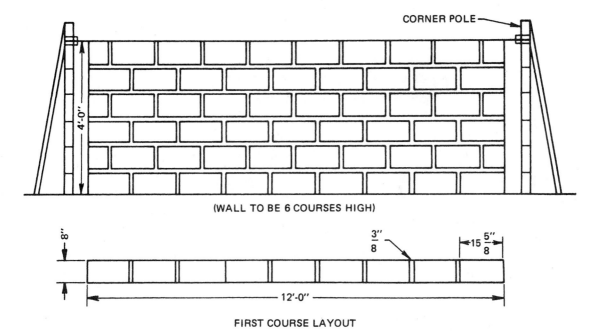

(WALL TO BE 6 COURSES HIGH)

FIRST COURSE LAYOUT

Fig. 19-18 Project 13: Laying a concrete block wall.

SUGGESTIONS

- Use only face shell bedding for the bed joints.
- Spread the mortar on the concrete block with the front of the trowel blade.
- Check the jambs every course to be sure they are plumb.
- Return all mortar which is removed from the wall for retempering in the pan.
- Be sure that the wider web of the block faces up on each course.
- Do not splash the wall with mortar.
- Be certain that each block laid is spaced the correct distance from the line ($1/16''$).
- Use eye protection when cutting the blocks.
- Strike joints when the mortar is thumbprint hard.
- Follow proper safety rules at all times.

PROCEDURE

1. Stock the work area with concrete block and space the mortar pans the correct distance from the work to allow for sufficient working room.
2. Mix the mortar to the desired consistency.
3. Lay out the first course using a rule to check the bond.
4. Set up the corner poles on the wall line and erect the line for the first course.

5. Mark the corner poles in 8″ divisions.
6. Cut the necessary half blocks as shown on the plan.
7. Move the line as needed and erect the wall to a height of 4′-0″ or 6 courses.
8. Strike the mortar joints with a convex jointer as required.
9. Brush and recheck the wall with the plumb rule before inspection.

UNIT 20
Laying the Block Corner

————————————————— OBJECTIVES ——————————————

After studying this unit, the student will be able to

■ lay out the first course of block on a corner.

■ install wire reinforcement in bed joints.

■ build a concrete block corner to a specified height.

BUILDING WITH BLOCK VERSUS BUILDING WITH BRICKS

A concrete block corner and a brick corner are constructed in similar ways. The main differences are that concrete blocks are larger and heavier than brick, and mortar is applied in a slightly different manner. Because blocks are heavy, hollow, and somewhat awkward to handle, care must be taken to prevent them from chipping (also called *spalling*). In addition, since they are larger and heavier than bricks, blocks must be laid gently in the mortar bed to prevent sinking. Sinking would require relaying the unit.

Because concrete blocks are greater in length and height than brick, they may be more difficult to keep level and plumb. Adjustment of the blocks after they are laid may also be more difficult. For this reason, it is very important that the first course be the correct height. Excessive bedding or squeezing of mortar joints is not an acceptable method of adjusting block.

Another important consideration is the length of time needed to lay the units in the wall. Laying a corner using concrete block requires approximately 40 minutes. Laying a corner with bricks requires about 3 hours. Both of these figures are based on normal working conditions.

The volume of bricks and block also relates to efficiency in construction. For example, to build a corner to scaffold height with concrete block requires 7 courses. To construct the same corner with bricks requires 21 courses. When comparing volume, 12 bricks are required to replace 1, 8″ concrete block in a wall. It is not possible for any mason to lay 12 bricks as

rapidly as another mason could lay 1, 8″ concrete block. The time and cost of materials and labor saved through the use of concrete block rather than bricks is the principal reason for the popularity of concrete block walls.

CONSTRUCTING THE CORNER

The corner is the key to a straight, plumb wall. It is therefore important for the beginning mason to perfect a step-by-step procedure in corner construction. Figures 20-1A through 20-1G show the steps involved in the construction of a concrete block corner.

Preparing the First Course

As is the case in preparing to lay out a brick wall, it is extremely important that a good bond between the first course and the base be established. The corner should never be laid out on mud or loose dirt. It should be laid out on a clean concrete base or footing. In warm, dry months it is good practice to dampen the footing with water to prevent too rapid absorption of moisture before the mortar is spread. Use a brush and a bucket of water to dampen the area where the blocks will be laid. The footing should not be saturated with water. Never soak the block, as this causes cracks from shrinkage in the mortar joints. In the wintertime, it is just as important not to spread mortar over ice or snow.

Locate the point of the corner and strike a chalk line across the wall to the opposite corner. Repeat the procedure on the other side of the corner. If it is not

223

Fig. 20-1A Leveling the block. Block should always be leveled by length and width.

Fig. 20-1B Spreading the mortar bed for the first course.

Fig. 20-1C Leveling the first course.

Fig. 20-1D Checking the first course on the corner to be sure that it is plumb.

Fig. 20-1E Checking the first course along the top of the block to be sure that it is straight.

Fig. 20-1G Measuring the corner for the proper height. The modular rule should read number two at the top of the corner.

Fig. 20-1F Swiping mortar on the block with the trowel. Notice that this technique differs from the one used in brick construction.

convenient to strike a chalk line, a ball of nylon line can be unrolled along the side where the corner is to be built. This line aligns the corner which is to be built with the opposite corner. This practice ensures the construction of a true, straight wall. Check the accuracy of the chalk line with a steel square before laying any block on the corner.

Laying the First Course

The first course must be laid with care to ensure that it is level, plumb, and properly aligned. It is good practice to lay the plumb rule on the footing or base before any mortar is spread to determine if the base is level. Spread a solid bed of mortar with the trowel. Do not *furrow* (or groove) the mortar as this may detract from the strength of the bond. A solid joint produces more waterproofing qualities.

It is essential that the first course be perfectly level to the specified height. This is usually determined by measuring from an established mark located close to the corner with a rule. The mason should attempt to correct any differences in height on the first course. The block may require reducing or bedding with split-block or brick to correct a faulty footing. Footings are not always poured perfectly level. Any variance of $\frac{3}{4}''$ or under usually can be corrected in the mortar joints without any noticeable effects.

Lay the corner block first and align it with the wall line. Pull the line tight on the corner block to the opposite corner and lay out the correct number of block needed to reach the specified height. (The scaffold height is usually about 4' or 5'.) As in the construction of brick corners, the sum of the block in the first course equals the number of courses in the height of the corner.

It is a good practice to level the block lengthwise and crosswise on the first course. As concrete block is very square, it can be assumed that if the block is level in these two ways, it will also be plumb on the face. Do not remove excess mortar that has been squeezed from the block immediately, as it could cause the block to settle unevenly. This is usually done after the second course has been laid.

When a block corner consisting of 1 unit in thickness is being built, check only one side of the block to be sure the corner is plumb. Walls consisting of 2 units laid back to back must be plumbed on both sides and aligned. This is done because although concrete blocks are reasonably square and straight, they are not always exactly the same width. Select the side that is the most important as far as appearance is concerned (usually the side facing out). Plumb the point at which every block corner returns. Then plumb the end of the lead and check the complete course with a plumb rule. The line should be attached to the side which was plumbed. If necessary, both sides can be struck or tooled.

BUILDING THE CORNER
TO THE SPECIFIED HEIGHT

Each course should be aligned before proceeding to the next course. Since the corner is racked back one-half

block on each course, stop spreading mortar one-half block from the end of the course.

Each block laid on the corner should be checked for the proper height and leveled from the point. This prevents the possibility of having to re-lay the course in fresh mortar.

After completing each course, cut off excess mortar with the trowel and return it to the mortar board or pan. Temper the mortar as needed to retain its plasticity and workability.

INSTALLING WIRE REINFORCEMENT IN JOINTS

If joint reinforcement is called for in the building specifications, install it at the proper height. When turning a corner, cut one side of the wire and bend the wire on a 90° angle rather than cutting the wire in half, Figure 20-2A. A special prefabricated wire for corners is also available, Figure 20-2B.

Be sure that the wire is bedded solidly in the mortar joint. Lap the wire at least 6'' over the next piece if more than 1 length of wire is needed. This lends adequate strength to the joint if lateral pressure is encountered.

STRIKING JOINTS

Several types of joints may be used with concrete block, Figure 20-3. As previously mentioned, the concave and V-joints are most commonly used. There are times,

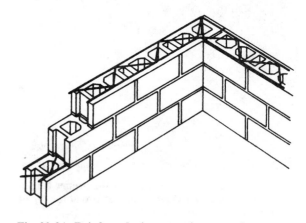

Fig. 20-2A Reinforced wire returning around corner.

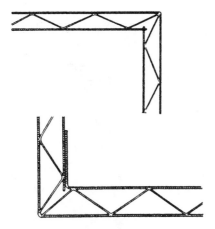

Fig. 20-2B Prefabricated reinforced wire corners for concrete block.

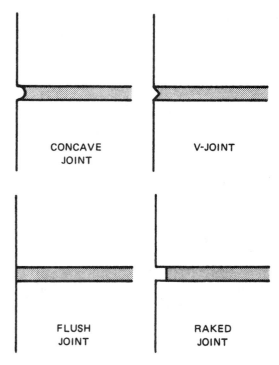

Fig. 20-3 Various types of joint finishes used on concrete block.

however, when a flush joint is required. In a flush joint, the joint is even with the face of the concrete block. This is usually done when a wall is to be painted and the mortar joints should not be evident.

There are two different ways of finishing a flush joint. The first involves carefully cutting off mortar and pointing any holes. This is the simpler method. The second and most often used method is to cut off the mortar and rub the joints with a piece of flat material to seal all holes or voids, Figure 20-4. Any flat piece of material, such as a broken block, scrap of wood, rubber heel, or rubber ball split in half may be used. All masons working on the same wall should use the same type of rubbing material so that the overall appearance is the same. Whatever material is used, it is important that all holes are filled and that the mortar is rubbed flat to meet the surface of the block. Guard against excessive rubbing and smearing of the block, since paint does not completely hide the joints but only covers them.

The raked joint can be used to achieve a special effect but is usually not recommended for concrete block. If a raked joint is to be used, do not rake deeper than $3/8''$, since most block is laid with face shell bedding, and the joint will be weakened and subject to water penetration. Excessive raking out of the mortar weakens the wall. Outside concrete block walls should not be raked out, as there is a greater possibility of the mortar joints leaking.

When the corner is completed, brush it off as soon as the mortar has sufficiently set. Lightly restriking the joints after brushing sharpens the edges of joints and removes any small particles of mortar which remain.

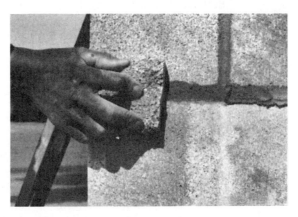

Fig. 20-4 Finishing a flush joint by rubbing with a piece of block.

This is required practice on a job where the work must have a very clean appearance. Be sure to check the end of the lead before leaving the corner, Figure 20-5.

PROGRESS CHECK

At this point, masonry students should have acquired certain basic skills. They must practice these skills, as every mason must have them. Students should demonstrate for their instructor each of the following skills. Any skills not mastered should be practiced until they can be demonstrated correctly.

- Spread mortar on a concrete block without excessive waste.
- Level a concrete block with a plumb rule.
- Plumb a concrete block with a plumb rule.
- Cut a concrete block with a hammer or brick set.
- Lay concrete block correctly to the line.
- Install a closure block correctly while retaining solid joints.
- Strike or tool mortar joints using the joints described in this section.
- Build a concrete block corner to a specified height using the modular rule as a guide.

Fig. 20-5 The last step in the construction of a block corner is to check to be certain that the ends of the block are in line.

ACHIEVEMENT REVIEW

Select the best answer from the choices offered to complete the statement. List your choice by letter identification.

1. Concrete block must be handled very carefully to prevent spalling. *Spalling* is
   a. shrinkage of the masonry unit.
   b. chipping or flaking of the unit.
   c. cracking of the unit.
   d. discoloring of the unit.

2. Concrete block should never be soaked with water before it is laid because
   a. it discolors the unit.
   b. the strength of the block is affected.
   c. shrinkage cracks may occur in the joints.
   d. a good bond cannot be established with the mortar.

3. Furrowing a mortar joint on the first course is not recommended because
   a. the compressive strength of the wall is affected.
   b. the joint would allow water penetration.
   c. good suction between the block and the mortar is not necessary.
   d. only face shell bedding is used.

4. A range line is used for
    a. the layout of the wall.
    b. checking the elevation of the finished grade.
    c. lining up one corner with the opposite corner.
    d. checking the corner for the finished height.

5. When more than one piece of wire reinforcement is to be used, it is best to lap each piece over the next piece at least
    a. 12".   c. 8".
    b. 16".   d. 6".

6. If a concrete wall is to be painted and it is important that the mortar joints not be noticeable, the best mortar joint finish to use is the
    a. concave joint.   c. V-joint.
    b. flush and rubbed joint.   d. raked joint.

7. A joint finish which should not be used on outside concrete block walls that are exposed to weather is the
    a. concave joint.   c. V-joint.
    b. raked joint.   d. flush joint.

8. The correct number on the modular rule when marking off the height for concrete block is
    a. 6.   c. 2.
    b. 3.   d. 5.

9. If a concrete block corner is being built to a height of 56", the number of courses in height is
    a. 4.   c. 6.
    b. 5.   d. 7.

10. Raked joints should not have a depth greater than
    a. ¼".   c. 6".
    b. ⅜".   d. 7".

PROJECT 14: LAYING AN 8″ × 8″ × 16″ CONCRETE BLOCK CORNER IN THE RUNNING BOND

OBJECTIVE

- The masonry student will be able to build a concrete block corner 7 courses high in the running bond, Figure 20-6.

  Note: A block corner such as this is used for interior partitions and corridors.

EQUIPMENT, TOOLS, AND SUPPLIES

Mortar pan or board

Mixing tools

Mason's trowel

Brick hammer

Plumb rule

Chalk box

Brush

28, 8″ × 8″ × 16″ concrete blocks
(7 of these must be corner block)

1 bag of lime or masonry cement (Lime may take the place of cement in a training situation as it may be reused.)

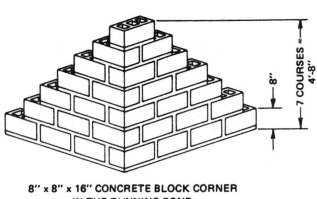

**8″ x 8″ x 16″ CONCRETE BLOCK CORNER
IN THE RUNNING BOND**

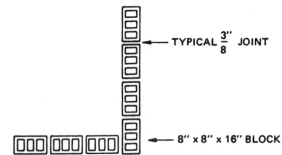

PLAN VIEW OF FIRST COURSE LAYOUT

Fig. 20-6 Project 14: Laying an 8″ × 8″ × 16″ concrete block corner in the running bond.

2′ framing square

Convex sled-runner jointer

Mason's modular rule

18 shovels of sand

Supply of water

SUGGESTIONS

- Space the mortar pan or board approximately 2′ away from the corner to allow sufficient working room.
- Snap a chalk line longer than the one actually needed to serve as a reference point.
- Tie the block together on the stock pile by reversing every other course.
- Mark the distance the block will be laid on the base to prevent spreading the mortar too far.
- Measure the height of every course laid.
- Remove excess mortar from the corner and return it to the mortar pan. Temper it as needed.
- Strike the mortar joints when they are thumbprint hard. Do not wait until the entire corner is built.
- Observe safe working practices at all times.

PROCEDURE

1. Mix the mortar.
2. Prepare the work area by stocking the specified amount of concrete block and placing the mortar in the mortar pan.
3. Brush off the area where the corner is to be built.
4. Square a corner with the framing square.
5. With the plumb rule and chalk box, extend the squared line far enough to build the corner.
6. Spread a solid bed of mortar. Lay the first block, being sure that it is level and plumb.
7. Lay the rest of the first course, being sure that it is level and plumb and that the height measures 8″ on the modular rule. (Scale 2 is equal to 8″ on the modular rule.)
8. Continue building the corner using these procedures until the specified height is reached.
9. Strike the corner on both sides with a convex sled-runner jointer. Brush the work when the striking is finished.
10. Recheck the end of the lead and the corner for the correct height with the plumb rule.

PROJECT 15: RECOMMENDED OPTIONAL BLOCK CORNER PROJECTS

A $4'' \times 8'' \times 16''$ and a $6'' \times 8'' \times 16''$ block corner can be built as practice so the students can familiarize themselves with various sizes of blocks frequently used on the job. Each of these corners will require a cut block on every corner of every course to create the half lap. The $8'' \times 8'' \times 16''$ block is the only block that returns on the corner and creates a half lap bond, because $8''$ is half of $16''$. The illustrations below show the correct layout for the two sizes of block corners. All other procedures and materials are the same as for the $8'' \times 8'' \times 16''$ block corner. If there are any questions, review Unit 15, Bonding Concrete Block and Rules For Bonding.

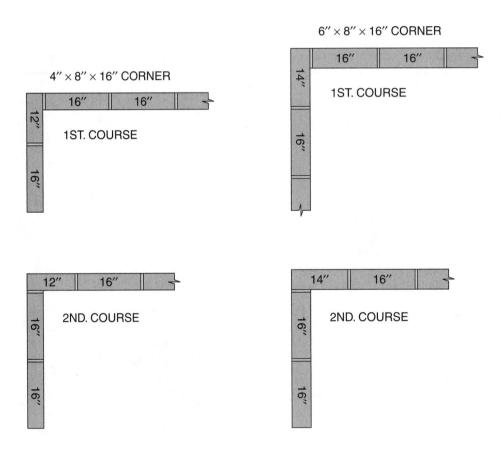

UNIT 21
Estimating Concrete Block by Rule of Thumb

―――――――――――――――――― OBJECTIVES ――――――――――――――

After studying this unit, the student will be able to

■ estimate concrete block by using the rule of thumb method.

■ estimate mortar and sand for jobs which require concrete block.

As is the case in estimating bricks for masonry, a measuring system that uses modular concrete masonry units can be used to simplify estimating. For accuracy, the square foot wall area method is considered the best method for large jobs. When using the wall area method, determine the amount of square feet in the wall and deduct all openings from this figure. A percentage of the figure is included to account for waste and breakage. The total figure is calculated by use of a table giving the number of block in each square foot. Additional tables state the correct amount of sand and masonry cement needed to lay the concrete block.

Rule of thumb estimating is usually adequate for a small-sized or average-sized job. When it is necessary to estimate large quantities of materials, tables should be consulted. There are also slide rules that speed the estimating process. Tables and slide rules should be used by masons only after they learn the rule of thumb method.

DETERMINING THE NUMBER OF UNITS

This method involves adding all of the wall lengths around the *perimeter* (outside) of the building and figuring the total lineal feet. The number of concrete block needed to lay 1 complete course around the building is determined. The plans are studied to determine the height of the wall, and the number of courses necessary to build the wall to the specified height are figured. The number of block needed to build 1 course is multiplied by the number of courses in the height of the

wall. The result is the total number of concrete block needed to build the wall or project. Deduct the number of block in all the openings in the building.

ESTIMATING MATERIALS FOR A FOUNDATION BY RULE OF THUMB

Estimating Block

Assume that the foundation is of average size, measuring 50′ (length) × 32′ (width) × 8′ (height). It is to be constructed of block measuring 8″ × 8″ × 16″.

The first step is to determine the total lineal feet around the outside of the foundation.

1. Determine the length and width of the foundation.

   50′ (length of foundation wall)
   32′ (width of foundation wall)

2. Double these 2 figures since there are 2 walls on each side of the foundation.

   50′ × 2 = 100′ (total length of foundation)
   32′ × 2 = 64′ (total width of foundation)

3. Add the total length and width together.

   100′ + 64′ = 164′ (total lineal feet in foundation)

Three concrete blocks are required to lay 4′ of length in a concrete block wall. The ratio of blocks to feet of length is $\frac{3}{4}$ to 1, or 0.75 to 1.00.

233

4. Multiply the total lineal feet in the foundation by 0.75 to find the number of concrete block required to lay 1 course around the foundation.

   Note: Be sure to place the decimal point in the correct position.

   164 (total lineal feet of foundation wall)
   $\times 0.75$
   820
   1148
   123.00 (Total number of concrete block for 1 course)

5. Find the number of courses in the height of the wall. Since 1 concrete block measures 8″ in height including the mortar joint, divide 8 into the height expressed in inches to find the number of block of which the height consists.

   12″ (number of inches in 1 foot)
   $\times 8$ (number of feet in foundation)
   $= 96″$ (height in inches)
   $96″ \div 8 = 12$ (courses high)

6. Find the total number of concrete block in the wall by multiplying the number of block in each course by the number of courses in the height of the wall.

   Multiply: 123
               $\times\ \ 12$
                246
                123
               1476 (total number of block required for foundation)

   Note: There are no openings, so no deductions are made.

Allowing for Openings

When openings are specified in a concrete block structure, a certain amount of block should be deducted before estimating the number of block for the job.

A concrete block building which will have 2 windows and 2 doors is to be constructed. The doors measure 32″ × 6′8″. The windows measure 2′ × 4′. The number of concrete block to be deducted is estimated as follows:

1. Add the width of the 2 doors.

   $32″ + 32″ = 64″$ (total width of doors)

2. Determine the number of concrete block that would be included in that space. Divide 64″ by the length of 1 block (16″).

   $64″ \div 16″ = 4$ (number of blocks per course)

3. Express the height of the doors in inches.

   $6′ = 72″$
   $72″ + 8″ = 80″$

4. Determine the number of courses by dividing the height of the doors by 8″.

   $80″ \div 8″ = 10$ (total number of courses for both doors)

5. Multiply the number of courses of block in the doors by the number of block on each course.

   $10 \times 4 = 40$ (concrete block to deduct for doors)

The space for windows is estimated in the same manner.

1. Add the width of both windows.

   $24″ + 24″ = 48″$ (total width of windows)

2. Divide 48″ by the length of 1 concrete block (16″).

   $48″ \div 16″ = 3$ (number of block per course)

3. Each window is 4′, or 48″ in height.
4. Divide 48″ by 8″ (height of 1 concrete block).

   $48″ \div 8″ = 6$ (courses of block in window space)

5. Multiply the number of courses by the number of block per course.

   $6 \times 3 = 18$ (total number of block to deduct for windows)

To find the total number of block to be deducted, add the block for door space to the block allowed for window space.

$$40 + 18 = 58 \text{ (total block to be deducted)}$$

To find the total number of block needed, subtract the number of block deducted for openings from the total number of blocks needed for the wall.

$$1476 - 58 = 1418 \text{ (total number of block needed)}$$

Estimating Masonry Cement

To estimate masonry cement by the rule of thumb method, assume that 1 bag of masonry cement is used to cover a certain number of units. Face shell bedding is used for most block work. Therefore, no extra allowance must be made for block of different widths. The exception to this rule is the $6''$ concrete block, since the web is thinner and more mortar is forced into the holes between the webs of the block.

One bag of masonry cement is sufficient for 30 concrete block. This figure applies to the standard 70 lb bag. If portland cement and lime are used, a table must be consulted.

To estimate the masonry cement for the foundation in question, divide 1418 by 30 (the number of block that can be laid per bag). The answer gives the number of bags of masonry cement needed to build the foundation.

$$1418 \div 30 = 48 \text{ (bags of masonry cement)}$$

Estimating Sand

One ton of sand is needed for each 8 bags of masonry cement. If more than 3 tons of sand are needed, $\frac{1}{2}$ ton should be allowed for waste.

1. Multiply 30 (concrete block) $\times$ 8 (bags of masonry cement per ton) to find the number of blocks for each ton of sand.

   $$30 \times 8 = 240 \text{ blocks per ton of sand}$$

2. Divide 1418 by 240 to find the necessary sand.

   $$1418 \div 240 = 6 \text{ tons of sand}$$

3. Add 6 tons and the $\frac{1}{2}$ ton for waste to find the sand needed.

   $$6 \text{ tons} + \frac{1}{2} \text{ ton} = 6\frac{1}{2} \text{ tons of sand}$$

TOTALING MATERIALS

The materials needed for the job in question would be listed as such:

> 1418 concrete block
> 48 bags of masonry cement
> $6\frac{1}{2}$ tons of sand

These materials are adequate to build the foundation and allow a reasonable amount for waste.

ESTIMATING LABOR COSTS

After materials have been estimated, labor costs are estimated. Labor costs for block work are affected by many of the same factors that affect labor costs for brickwork.

To estimate labor costs, the contractor will want to know the average production rate for masons laying concrete block. The International Union of Bricklayers and Allied Craftsmen supplies these figures for the amounts of block that a journeyman mason can lay in an 8-hour day:

- 225, $8'' \times 8'' \times 16''$ blocks
- 210, $6'' \times 8'' \times 16''$ blocks
- 225, $4'' \times 8'' \times 16''$ blocks
- 200, $10''$ or $12'' \times 8'' \times 16''$ blocks

As with bricklaying, these figures will vary depending on specific job conditions.

POINTS TO REMEMBER

Rule of thumb estimating is the most commonly used method for figuring materials for the average or small job. The important points in the method are:

- Multiply the lineal feet of the wall by 0.75 to determine the number of concrete block for one course.

- Divide the height of the wall expressed in inches by 8 inches to determine the number of courses in the height of the wall.
- One bag of masonry cement is the recommended amount for thirty concrete block. (For 6-inch block, estimate 25 block per bag of masonry cement.)
- One ton of sand is needed for every eight bags of masonry cement. Waste is about $\frac{1}{2}$ ton if the quantity required exceeds 3 tons. Be liberal when estimating sand, as it is usually the least expensive material used by the mason.
- Water does not require estimating as a material. As a rule, 5 gallons of water are needed for each bag of masonry cement. Note: The allowance for water depends on the moisture content of the sand being used.

Remember, the skills involved in estimating by rule of thumb are learned only through experience and on-the-job training.

TWO HANDY TIME SAVERS TO USE WHEN ESTIMATING CONCRETE BLOCK

Figure 21-1 shows a handy table to use when matching a given height with the corresponding number of courses of a standard concrete block.

Figure 21-2 shows a handy Concrete Block Slide Rule Calculator which is available free for the asking from many concrete block manufacturers or masonry building supply firms, that will show the number of concrete blocks for a given height and length of wall.

(Height of unit 7-5/8″)
(Joint thickness 3/8″)

Height	No. of courses
8″	1
1′4″	2
2′0″	3
2′8″	4
3′4″	5
4′0″	6
4′8″	7
5′4″	8
6′0″	9
6′8″	10
7′4″	11
8′0″	12
8′8″	13
9′4″	14
10′0″	15
10′8″	16
11′4″	17
12′0″	18
12′8″	19
13′4″	20
16′8″	25
20′0″	30
23′4″	35
26′8″	40
30′0″	45
33′4″	50

Fig. 21-1 A handy table to use when matching a given height with the corresponding number of courses of a standard concrete block. (Courtesy National Concrete Masonry Association)

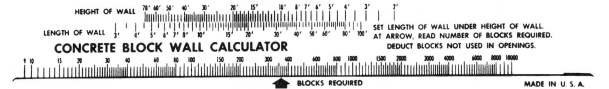

Fig. 21-2 A handy concrete block slide rule calculator which is available free for the asking from many concrete block manufacturers or masonry building supply firms. It will show the number of concrete blocks for a given height and length of wall.

ACHIEVEMENT REVIEW

Solve each of the following problems using the information presented in this unit. Show all of your work.

1. A foundation measures $70' \times 40'$.
   a. What is the number of lineal feet in the foundation?
   b. Estimate the number of $8'' \times 8'' \times 16''$ concrete block needed to lay 1 course completely around the foundation.
   c. The foundation is to be built $8'$ high. Estimate the number of courses for this height.
   d. Estimate the total number of concrete block needed to build the foundation. There will be no windows, doors, or other openings to consider.
   e. Estimate the amount of masonry cement needed to build the foundation.
   f. Estimate the amount of sand needed to build the foundation.
   g. If masonry cement costs $4.90 per bag, what is the total cost of masonry cement for the job?
   h. If concrete block measuring $8'' \times 8'' \times 16''$ cost $0.60 each, what is the total cost of the concrete block for the job?
   i. If sand costs $20.00 a ton, what is the total cost of sand for the job?
   j. What is the total cost of materials for the job?

2. A basement without any openings is to be built from $8'' \times 8'' \times 16''$ concrete block. The outside dimensions are $60' \times 32'$. How many blocks are needed for 1 course around the basement?

3. A wall is to be built of concrete block measuring $12'' \times 8'' \times 16''$. How many courses are needed if the wall is $10'$ high?

4. A concrete block garage is to be built of $8'' \times 8'' \times 16''$ block. The outside dimensions are $20' \times 24'$. The total height of the wall is $10'8''$ from the footings to the top of the wall. There are two large garage door openings, each measuring $8' \times 8'$, and one standard door opening measuring $32'' \times 6'8''$. Estimate the number of concrete block needed to build the garage. Be sure to deduct for all openings.

5. A motor repair shop $20' \times 40'$ is to be built of $10'' \times 8'' \times 16''$ concrete block. The walls are to be $10'$ high. There are six windows measuring $40'' \times 48''$, and one large door measuring $8' \times 8'$. Estimate concrete block, masonry cement, and sand needed to build the shop. Estimate the total cost of materials using the following prices.

$10'' \times 8'' \times 16''$ concrete block	$0.86 each
Sand	$20.00 per ton
Masonry cement	$4.90 per bag

6. Based on the rate of 225, $8'' \times 8'' \times 16''$ blocks laid per 8-hour day, how many days are required for one mason to lay 4500 blocks?

7. A mason lays 175 concrete blocks the first day, 200 blocks the second day, and 250 blocks the third day. What is the total number of blocks that are laid?

8. A mason's contractor has 40 workers on one job, 25 on another, and 38 on a third. Because they are taking their annual vacation, 16 workers are not on the job. How many are left?

9. If one mason lays 25 blocks per hour, how many will be laid in five 8-hour days?

10. Three masons work a total of 120 hours on the job. Each works the same number of hours. How many hours did each work?

UNIT 22
Building a Composite Wall with Brick and Concrete Block

──────────────── OBJECTIVES ────────────────

After studying this unit, the student will be able to

■ describe the different methods of bonding walls with brick headers and metal wall ties.

■ install these ties in a composite wall project.

A *composite wall* is formed when two separate *tiers* or *wythes* (single vertical thicknesses of walls) of masonry units are combined to form a single wall. This requires that each individual tier be bonded together vertically at a specified height so that the wall retains its strength. The generally accepted height at which the tiers are attached is 16″.

There are two different methods of bonding walls together. One method is by the use of a brick header laid across the wall. The second method is by installing metal wall ties. The use of the brick header is an older, more established method. Brick headers provide more strength than other methods. This method is often used in the construction of solid masonry walls. Metal ties and heavy-gauge reinforcement wire have been greatly improved in recent years and are acceptable for most jobs. This is the most common method of tying tiers. Either type of wall tie can be used to effectively join bricks and block together.

TYING TIERS TOGETHER WITH HEADERS

Problems Presented by Headers

The installation of brick headers in walls may present special problems if they are not laid correctly. Extreme care must be used when applying the *cross joint* (a mortar joint that is applied across the full length of a brick)

on the header. If extreme care is not taken, moisture penetration may cause leakage through the wall, Figure 22-1. Single head joints applied only to the ends of the header are not acceptable.

Another problem presented by headers is the difficulty in being sure that the set of the mortar is not

Fig. 22-1 Mason applying a cross joint to a header brick. Note the solid joint that guards against moisture penetration.

239

broken when tying bricks and heavy block, Figure 22-2. Sometimes, the concrete back-up wall is laid ahead of the brick wall. This only occurs when a mason is working on the outside of the brick wall. In this case, brick headers are laid across the block and bricks when the face brick wall reaches the proper height. This is done so that the tiers may be bonded together. This height is usually 16″. Immediately following the laying of the brick headers, a heavy concrete block course is laid in back of the header course. This block course rests on the back half of the header.

With most types of concrete block used as backup, the mortar is spread by the face shell method or by spreading it only on the webs of the block. This is approximately a 1½″ strip of mortar. If the block is laid before the mortar under the header stiffens, the header brick settles unevenly and a crack may develop in the mortar joint under the header. This presents the possibility of moisture penetration. It also detracts from the beauty of the building, since the header lies twisted in the face of the wall.

Expansion and contraction caused by moisture and heat also create problems since the rate of expansion and contraction differs significantly for each material. Movement of the block due to shrinkage creates pres-

sure on the brick header. This may result in hairline cracks developing in the mortar joints surrounding the header course. As masonry units become more complex and new materials are developed, some flexibility in wall ties is essential if materials with different properties are to be used. This is one of the prime reasons the metal tie has become popular in recent years. It lends more flexibility to the wall.

There is also the matter of economy to consider. A header course takes a great deal of time to install as compared with the fairly simple task of laying metal ties or wire joint reinforcement in walls. The materials used to lay header courses are also more expensive than metal reinforcements, as a full header course requires twice as many bricks as a stretcher course.

Laying Headers in the Wall

Remember that it is important to build the brickwork to the height of the header course before laying any back-up block. Block work can be *humored* (gradually adjusted so that a difference in height is not evident) to remain in line with the brickwork. If back parging of the brickwork is specified, it is best to apply it to the bricks before the concrete block is laid since they would interfere with the work. The block must also be laid exactly level in height to match the outside brickwork or the header will not lay level across the two tiers.

Building a 12″ Wall with Headers. When building a 12″ wall using brick headers, there are two different methods of tying the header to the back-up tier. The first involves using a special concrete block called a *header block* (or *shoe block*). Header blocks are 8″ high on one side and the height of 2 courses of standard brick, including the mortar joint, on the other side. The side of the block which receives the header course is approximately 4″ in width, leaving a remainder of about 4″ on the opposite side. The brick wall is laid 5 courses high. The concrete block backing wall is laid 2 courses high with the header block laid on the second course. The header course is then laid across the header block. This method of installing header block is used in cases in which the brick header should not be seen from the inside of the structure, Figure 22-3. A regular 8″ block now can be laid on top of the header block,

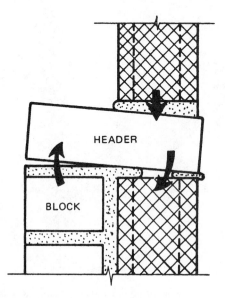

Fig. 22-2 Effect of a heavy back-up block on brick header. Notice that the set of the mortar has been broken.

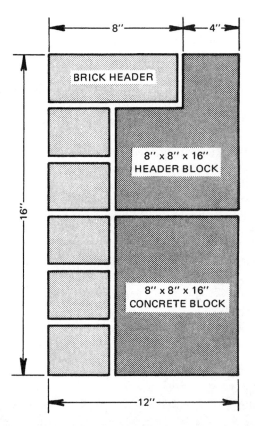

Fig. 22-3 Wall section showing header used in conjunction with header block.

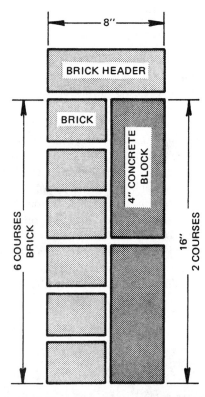

Fig. 22-4 Wall section showing use of brick header on 8″ wall.

which will bond the brick and block together. The building of the wall to the next header level is now resumed.

The brick headers may also be installed on the sixth course by turning the header block upside down. Either method is acceptable and depends on the architect's specification of the height of the headers. Consider that an 8″ wall consisting of 4″ of brick and 4″ of concrete block is built and a header is used to tie the tiers together. Unless the mason uses snap headers, there is no way to avoid having the header course exposed on the back side of the wall, Figure 22-4.

Standard bricks do not actually measure 4″ in width (they measure $3\frac{3}{4}$″) and an 8″ concrete block is approximately $7\frac{5}{8}$″ wide. Therefore, a 12″ wall will always have a $\frac{5}{8}$″ space between the two tiers or wythes of units. This space provides the mason with finger room when laying the back-up units. This space be-

tween the two tiers of masonry units is called a *collar joint.* After the back-up units have been laid and are level with the front tier of the masonry units, the collar joint may be slushed solid with mortar or left hollow. *Slushing* is the trade term for filling in the collar joint in a masonry wall with mortar. If slushing of the wall is done too vigorously, it may cause the wall to be pushed out of line. Be sure that the masonry work has taken an initial set before slushing. Specifications of the job dictate when to slush walls.

If the inside face of the concrete block is to be plastered or is otherwise not to be exposed when completed, the header block can be replaced by a course of stretcher bricks located behind the brick headers. This allows for the full 12″ thickness of the wall. It is more costly to do this, however, as it requires the use of an additional course of bricks.

USING METAL TIES

Prefabricated Metal Joint Reinforcement

The most popular type of metal tie used today is the *prefabricated metal joint reinforcement* or rod joint reinforcement. The truss and ladder type design is designed to be installed in the bed joints, Figure 22-5. The metal wire tie consists of two parallel wire rods connected to a single diagonal cross rod, which is welded to the outside parallel rods wherever they touch. The truss design that is formed provides great strength. Wire reinforcement ties the masonry work together. It also minimizes shrinkage in the mortar joints and helps prevent cracks due to settlement. For a composite wall such as one constructed of bricks and block, the wire should usually be installed no more than 16″ on center (vertically). This is an approved method for a brick header providing the collar joints are filled completely with mortar.

Reinforced wire is usually manufactured in 10′ or 12′ lengths. It should overlap at least 6″ when two pieces are used together. This ensures that the maximum tensile strength will not be interrupted throughout the length of the bed joint. It is extremely important that wire track be embedded and covered completely with mortar before any masonry unit is laid on top of it, Figure 22-6.

When using reinforced wire to tie bricks to concrete block, be certain that the two separate walls are perfectly level with one another since wire reinforcement does not bend easily. Never allow the wire to extend past the ends of the wall farther than necessary, Figure 22-7. Extended wire is not only a safety hazard but is also difficult to work around.

Continuous horizontal metal joint reinforcement is typically made from Number 8, 10, or 11, $\frac{3}{16}$″ diameter gauge wire, the thickness of which allows it to be imbedded completely in the mortar bed joint when the brick is laid over it and still maintain the proper course height. The most common configurations are the lad-

Fig. 22-5 Brick and block tiers tied together with reinforced wire track.

Fig. 22-6 Mason spreading mortar over joint reinforcement. The wire must be completely covered with mortar to be effective.

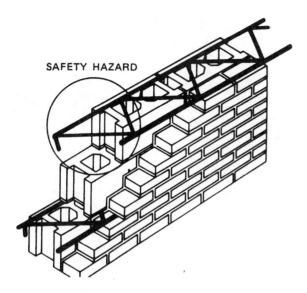

Fig. 22-7 Prefabricated metal joint reinforcement extending past walls could cause serious injury.

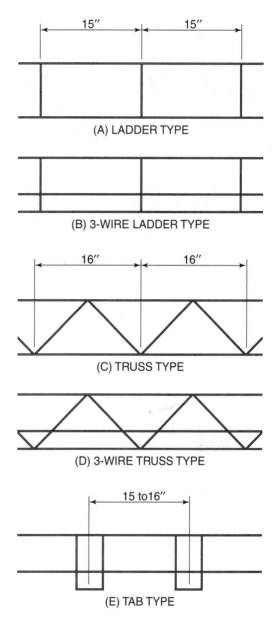

Fig 22-8 Continuous joint reinforcement. (Courtesy Brick Industry Association)

der type, 3-wire ladder type, truss type, 3-wire truss type, and tab type, which are shown in Figure 22.8.

The Z Tie

The term *Z tie* comes from the appearance of the tie when it is laid in the wall, Figure 22-9. It is made from copper-coated or zinc-coated steel that is $\frac{3}{16}''$ in diameter. One tie is usually installed every $16''$ in height and $16''$ apart lengthwise. The Z tie is available with a drip crimp in the center so that water does not cross the tie (used mainly in cavity walls). It is also available without the drip crimp (used in solid masonry walls).

The Rectangular Tie

The *rectangular tie* is rectangular in shape. It is constructed from the same materials and is the same thickness as the Z tie. It is sometimes called a *box anchor.* The rectangular tie is recommended for use in walls where both masonry units used are hollow. It can be obtained with or without a drip crimp in the center. For best results, one tie should be installed every $4\frac{1}{2}$ sq ft of wall surface.

The Dovetail Anchor

Building a masonry wall next to a concrete wall or column requires a special tie. The *dovetail anchor,* used for this purpose, takes its name from the resemblance

Fig. 22-9 Z tie used in cavity wall construction. Notice the drip crimp in the center of the tie which allows drippage of water before it crosses the tie.

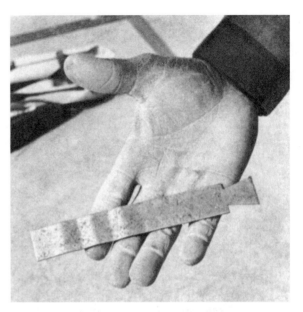

Fig. 22-10 Dovetail anchors are used to tie masonry work to concrete walls or columns.

to the tail of a dove, Figure 22-10. The anchor fits into a slot in the concrete work. The slot is metal with a fiberglass filler. This prevents the slot from being filled with concrete when the wall is poured. The mason removes the filler material and inserts the tapered end of the dovetail anchor into the slot. The anchor is then turned over much the same as a key is turned in a lock, until the flat side is facing up. The slot in the concrete wall, tapered to match the anchor, holds it in place. Minimum dimensions of a dovetail anchor are $\frac{7}{8}''$ in width with a 16-gauge thickness.

The Veneer Tie

Veneer ties are constructed of a corrugated metal, Figure 22-11. The wrinkle in the metal aids in bonding the mortar to the steel, which increases the holding power of the tie.

Veneer ties are usually used to tie masonry veneer to wood frame construction. They are sometimes used to tie two masonry walls together. These ties are not recommended for use in load-bearing walls.

The ties used to attach masonry veneer and wood frame construction should be made of galvanized steel

not less than $\frac{7}{8}''$ wide with a 22-gauge thickness. If the tie is to bond two masonry walls, be certain that a tie of a much heavier gauge is used.

Regardless of the type of metal tie used, it should never be placed closer than $1''$ from the outside surface

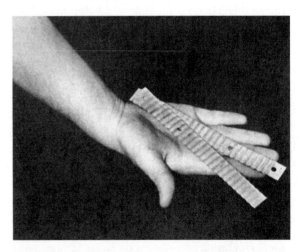

Fig. 22-11 Corrugated metal veneer ties. The holes provide spaces for nails when the ties are used with wood construction.

of the wall, Figure 22-12. If the tie were placed any closer, the joint would not be of the proper thickness and would have a tendency to pop out, exposing the tie. Each method of tying masonry walls has certain advantages over the other. The choice of tie depends upon the requirements of the job. Metal ties are being used more frequently in masonry because they provide adequate strength, economy, better control of shrinkage and expansion in mortar joints, and are faster for the mason to install. All methods discussed have been widely tested and accepted throughout the masonry industry.

Figure 22-13 illustrates unit tie details.

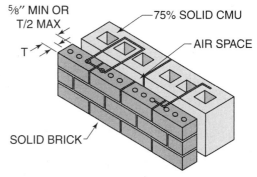

(A) SOLID MASONRY BACKUP

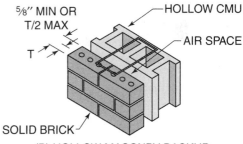

(B) HOLLOW MASONRY BACKUP

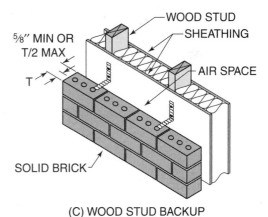

(C) WOOD STUD BACKUP

Fig. 22-12 Mason laying a brick and block concrete wall. Reinforcement bonding wall lies more than 1″ from the outside of the wall.

Fig. 22-13 Typical installation details of various types of unit wall ties. (Courtesy Brick Industry Association)

ACHIEVEMENT REVIEW

Select the best answer from the choices offered to complete the statement. List your choice by letter identification.

1. When two different types of masonry units are laid together to form a single wall, the structure is called a
   a. tier.
   b. wythe.
   c. composite wall.
   d. veneer wall.

2. The principal reason for leakage in composite walls is the use of
   a. metal ties.
   b. collar joints.
   c. hollow blocks.
   d. poorly formed cross joints.

3. The space in the center of a composite wall which is filled with mortar is called a
   a. head joint.
   b. collar joint.
   c. expansion joint.
   d. cross joint.

4. Slushing collar joints should not be done to a great extent because
   a. it may cause the wall to be pushed out of line.
   b. it may cause the mortar joints to lighten in color.
   c. it decreases the strength of the wall.
   d. it is a very wasteful practice.

5. The most popular type of metal tie used today is the
   a. Z tie.
   b. rectangular or box tie.
   c. dovetail anchor.
   d. veneer tie.
   e. wire reinforcement.

6. The principal reason for using metal ties is because they
   a. are very strong.
   b. reduce cracks due to shrinkage.
   c. are easier to install.
   d. are very inexpensive.

7. For best results, wall ties should be installed vertically every
   a. 12″.
   b. 16″.
   c. 24″.
   d. 32″.

8. The dovetail anchor is designed mainly for tying
   a. brick to block.
   b. brick to tile.
   c. masonry units to concrete.
   d. brick to stone.

9. Veneer ties are not recommended solely for tying two tiers of masonry because
   a. they do not bend easily.
   b. they are not as strong as other wall ties.
   c. they do not bond well to the mortar joint.
   d. they are too costly to use on composite walls.

MATH CHECKPOINT

1. A composite wall built of regular brick and backed up with 4″ block is being built. The wall is 6′ high and 40′ long. Using seven bricks per square foot of wall, how many bricks would it take to do the brickwork?

2. How many blocks will it take for the wall in problem 1?

3. One bag of masonry cement will lay 125 bricks. How many bags will it take for the brickwork? Round off your answer to the next full bag needed.

4. One bag of masonry cement will lay 30 blocks. How many bags will be needed to lay the block?

5. What is the total number of bags of masonry cement needed to do the job?

6. If 8″-wide masonry wall reinforcement is used to bond or tie the brick and block together, and the wall is to be tied on four individual courses, how many feet of reinforcement will be needed to do the job?

7. How much sand will be needed to do the job?

PROJECT 16: LAYING A BRICK CORNER FOR A 12″ COMPOSITE WALL IN THE COMMON BOND BACKED WITH 8″ × 8″ × 16″ CONCRETE BLOCK

OBJECTIVE

• The student will be able to lay out and build a brick corner 8 courses high in the common bond by tying the work together with a full header on the first and seventh courses to form a composite wall, Figure 22-14. This project will give the student practice in combining basic materials to form a single wall and practice in laying header courses.

Note: Corners such as these are used in buildings where load-bearing solid walls are required, such as warehouses, office buildings, and factories.

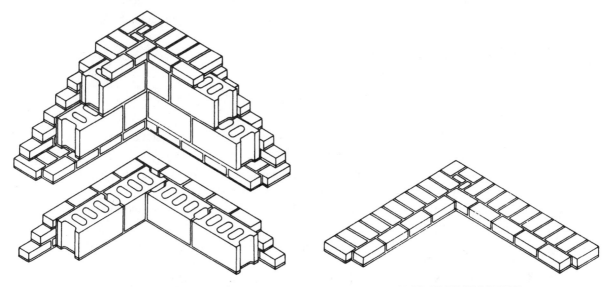

PLAN OF FIRST COURSE OF BLOCK   PLAN OF FIRST COURSE

Fig. 22-14 Project 16: Laying a brick corner for a 12″ composite wall in the common bond backed with 8″ × 8″ × 16″ concrete block.

EQUIPMENT, TOOLS, AND SUPPLIES

Mortar pan or board	Modular rule
Mixing tools	Brush
Mason's trowel	Pencil
Brick hammer	Clean water
Plumb rule	1, $8'' \times 8'' \times 16''$ concrete block in
2'' square	which to set plumb rule
V-sled runner jointer	

Note: The student will estimate the number of bricks and/or block needed by consulting the plans. Based on the number of bricks and block, the student will estimate the needed mortar (lime and sand) by the rule of thumb method.

SUGGESTIONS

- When establishing a reference line for the corner, be sure to extend the square line longer than is actually necessary.
- Select straight, unchipped bricks for the corner point.
- Sight down the corner as each brick is laid.
- Form solid head joints to prevent leakage.
- Cut off all excess mortar from the back of the bricks with the trowel and return it to the mortar pan before attempting to build the backup with concrete block.
- Do not lay concrete block until the brickwork has been constructed to header height.
- Do not lift concrete block over the brickwork; lay them from the inside of the wall.
- Level and plumb the corner on both sides.
- Keep all scraps and tools which are not being used away from the work area.
- Use good safety practices at all times.

PROCEDURE

1. Stock materials approximately 2′ from the work area and set up the mortar pan.
2. Mix the mortar and load it in the pan.
3. Lay out the corner on the floor using a steel square and pencil. Strike a chalk line longer than the actual corner length will be over the top of the pencil line.
4. Measure back from the outside corner line 12″ on each end of the corner and strike the inside line of the corner.
5. The first course to be laid is the header course. Cut 2, 6″ pieces to be used as starter pieces for the corner. Lay out the first course as shown on the plan.
6. Lay the inside stretcher course, being sure that when completed, the width of the wall measures 12″.
7. Recheck the corner with the steel square. Lay the 6 courses of bricks as shown on the plan. Use the number 6 on the modular rule as a reference for course spacing.

8. Back up the corner with 8″ concrete block as shown on the plan. Be sure that the top of each course of block is level with the top of each 3 courses of bricks. Level and plumb all work.

9. Lay the next header course as shown on the plan, using all solid cross joints between each header.

10. Recheck the corner for the proper height, using the number 6 on the modular rule. Strike the brickwork with a V-jointer. Brush the corner. Do not strike the concrete block work since it does not show.

11. Tail the ends of the corner before having the work inspected.

PROJECT 17: LAYING A 12″ BRICK AND CONCRETE BLOCK COMPOSITE PANEL WALL BONDED WITH MASONRY WIRE REINFORCEMENT

OBJECTIVE

- The student will be able to lay out and build a 12″ brick and concrete block panel wall using metal wire reinforcement to tie the wall every 16″ in height, Figure 22-15.

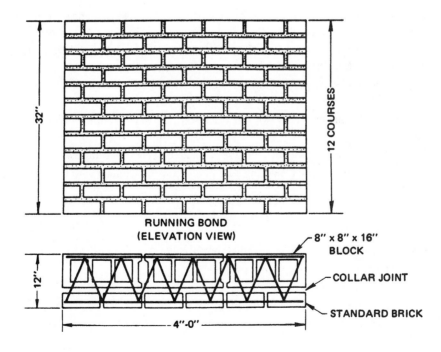

Fig. 22-15 Project 17: Laying a 12″ brick and concrete block composite panel wall bonded with masonry wire reinforcement.

Note: Many of the composite masonry walls built today use masonry wire reinforcement. This project is typical of a panel built between a series of windows on an apartment building.

EQUIPMENT, TOOLS, AND SUPPLIES

Mortar pan or board	Convex sled-runner striking tool
Mixing tools	Modular rule
Mason's trowel	Pencil
Brick hammer	Brush
Plumb rule	Heavy-duty wire or bolt cutters
Brick set chisel	Supply of clean water
Small square	1 length of masonry wire reinforcement
Chalk box	(minimum 4′)

Note: The student will estimate bricks, concrete block, and mortar from the plan.

SUGGESTIONS

- Select straight, unchipped bricks for the end of the jambs.
- Cut all bats needed with a brick set before starting the brickwork. (The same holds true for concrete block halves that are needed unless factory halves are available.)
- Be sure to clean all excess mortar from back of the brickwork before laying the backing block. Mortar which is not cleaned may push the wall out of line.
- The mortar may be tempered so that it is thinner for the brickwork than for laying the block.
- Do not lift the concrete block over the brick wall; lay them from the other side of the wall.
- Keep all scraps and tools away from the work area to prevent injuries.
- Be especially careful when handling wire reinforcement. Contact with the sharp edges could cause injury.

PROCEDURE

1. Stock materials approximately 2′ from the work area.
2. Mix the mortar and load the mortar pan.
3. Strike a chalk line for the project.
4. Dry bond the bricks as shown on the plan.
5. Lay the first course in mortar, leveling and plumbing the course. Square the ends of the wall with a small square.
6. Check the height of the brickwork as it is laid with a modular rule (number 6 on rule).
7. Lay 6 courses of brickwork, which will be the height to which reinforcement wire is installed.
8. Back up the brickwork with concrete block in the running bond to tie height.

9. Install wire track in the wall, being certain that it is $\frac{3}{8}''$ to $\frac{1}{2}''$ from the outside of all face work.

10. Continue laying the brickwork to the height specified on the plans. Strike the work as needed with the convex jointer.

11. Back up the brickwork with the remaining concrete block. Strike the block with the convex jointer.

12. Brush the work and recheck the wall with a plumb rule before having it inspected.

UNIT 23
Cavity and Reinforced Masonry Walls

―――――――――――――――― OBJECTIVES ――――――――――――

After studying this unit, the student will be able to

- describe techniques in building a cavity wall.
- discuss construction details of a masonry reinforced wall.
- build a cavity wall from given plans.

The durability of masonry is dependent for the most part on its resistance to the penetration of moisture. The source of the moisture may be weather or conditions that exist inside the wall. Differences in humidity inside and outside the building cause moisture to condense and collect on the inside.

There are two basic methods used to prevent penetration of moisture into the wall from the outside. One is to provide cavities between the outside and inside wythes of the masonry. The structure formed is called a *cavity wall*. The moisture that collects in the wall is removed from the cavity by weep holes. *Weep holes* are small holes in the bed joints at the bottom of head joints of a cavity wall that permit the leakage of excess moisture.

A second method is to provide a barrier in back of the exterior wythe. Barrier walls may be metal-tied, reinforced masonry, or bonded masonry. Reinforced masonry has steel rods between the two wythes, and sometimes in the cores of the wall. Concrete or *grout* (mortar or concrete that is mixed so that it is liquid enough to flow into the cavity of the wall, completely filling the void around the reinforcing steel) is poured in the cavity and puddled or vibrated. *Puddling* is the process of forcing an object into the grout so that it consolidates and is compact around the reinforcing steel. It is a process similar to vibrating concrete in a form. Both cavity and reinforced walls, when constructed correctly, have a high resistance to rain penetration and will re-

main reasonably dry. The design of any wall should be based on the degree of moisture and the elements to which it will be exposed, and the climate of the area in which it is built.

The degree of exposure varies greatly throughout the United States, from severe on the Atlantic seaboard and Gulf Coast, to moderate in the midwest, and slight in the more arid sections of the west.

CAVITY WALLS

The BIA recommends the construction of cavity walls where there is severe weather exposure, or where a maximum resistance to rain penetration is desired, Figure 23-1. Cavity walls have been built for many years in England and, during the past 40 years, have become popular in the United States.

The cavity wall consists of two tiers or wythes of masonry separated by a continuous air space not less than 2″ wide. The air space acts as an insulator. The cavity wall may be composed of any two types of masonry units but as a rule is constructed of bricks and concrete block. Metal ties are used to connect the two wythes of masonry. Cavity walls can also be masonry bonded if the tying unit does not completely close off the cavity.

For exterior walls, the facing tier is usually of the minimum thickness permitted by local building codes. Most codes define this as a 4″ wall, making the smallest

252

Since the primary function of the air space is to serve as a barrier against moisture, anything that would reduce the effectiveness of the moisture barrier should be avoided. The selection of the insulating material should in no way impair the cavity. In view of the continued depletion of natural energy resources, moisture-proof insulation in cavity walls is a great asset. The new rigid insulation boards and plastic-formed insulation are especially effective. When cavity walls are insulated, the insulation must permit the moisture to drain from the wall so that it is not carried across the cavity to affect the inside wall.

Flashing in Cavity Walls

Flashing is installed in masonry walls to divert moisture, which may enter the masonry at floorlines, rooflines, where chimneys come through the roof, and above windows or doors, Figure 23-2. Flashing should extend through the outer face of the wall and be turned down so that moisture drips off it. Install flashing under brick windowsills as well, since water tends to collect on the sills before finally running off. The dryer a masonry wall is, the dryer the interior wall will be.

Copper is the most popular flashing material for masonry construction, although zinc, aluminum, lead, and plastic flashings are also available.

Masons must provide *weep holes* above the flashing to allow water to drain from the wall. Weep holes are located in the head joints of the first course which lays on the flashing, Figure 23-3. Weep holes should be

Fig. 23-1 Typical cavity wall constructed of brick and concrete block. Note the Z ties which bond the block and bricks together.

cavity wall 10″ thick. When constructed of brick masonry, 2, 3″ units in addition to the air space are acceptable, resulting in an 8″ cavity wall. This practice is limited to 1 story construction.

The features of the cavity wall include excellent resistance to rain penetration, good insulation due to the air space located in the center of the wall, and ease of construction. Remember, however, that a certain amount of moisture is always present in cavity walls.

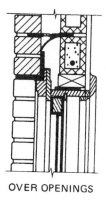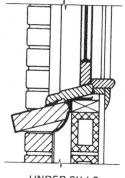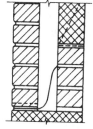

OVER OPENINGS   UNDER SILLS   AT BOTTOM OF CAVITY

Fig. 23-2 Installation of flashing.

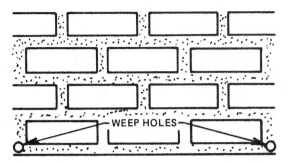

Fig. 23-3 Location of weep holes in masonry wall.

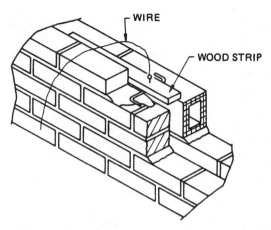

Fig. 23-4 A wood strip inserted on the wall ties prevents mortar from dropping into the cavity of the wall.

provided at intervals of 16″ to a maximum of 24″. Spacing depends on the severity of the moisture problem as determined by the architect. The most common material used to form weep holes is a piece of sash cord or rubber tube about $3/8''$ in diameter. These materials should be soaked in oil to aid in removal from the wall. After the wall is built scaffold high and before it completely hardens, the mason pulls the weep rope or tube from the wall, leaving a clean hole. Sometimes metal screening or fibrous glass is placed in the open weep hole and left in place to act as a wick that draws moisture from the cavity.

When weep holes are installed, they should never be located where they will be covered with ground when the building is completed. For this reason, weep holes are usually installed on top of the third course of bricks above ground level.

Keeping the Cavity Free of Mortar

If mortar falls into the cavity of the wall, it may form *bridges* upon which moisture may cross or it may drop to the flashing, blocking drainage of the weep holes. To prevent this, the mason should use a trowel to flatten any mortar projections on the inner face of the cavity.

After the first set of wall ties has been installed, a drip strip may be laid on the ties to catch any mortar that may drop, Figure 23-4. A *furring strip* (wood strip) is an effective drip strip. A heavy wire or string can be attached to the wood strip to make it easy to remove the drip strip from the cavity. After the wall is built to the next level of ties, the wood strip is carefully removed so that no mortar falls into the cavity.

Another popular method of removing the mortar droppings is to lay every third brick dry (without mortar) when laying out the first course. After the wall is completed, the mason removes the dry brick and cleans out the mortar droppings from the wall. The dry bricks are then relaid in fresh mortar, sealing the wall, and the weep holes added. The spaces made when the dry bricks are removed are known as *clean-outs* in cavity or reinforced wall construction.

Tie Placement

In a properly constructed cavity wall, both wythes of materials must be adequately and properly tied together. The ties must be firmly embedded in and bonded to the mortar. To achieve this, the two wythes surrounding the cavity must be laid with a completely filled mortar bed, and the ties must be placed firmly and solidly in the bed. Extensive tests prove that wall ties have excellent tying capacity if they are well bedded in the mortar joints.

Most building codes require at least 1, $3/16''$ diameter steel wall tie or equivalent in every $4\frac{1}{2}$ sq ft of wall area. The most common type of wall tie for brick construction is the Z tie. In addition, there are rectangular and U-shaped ties for use with hollow back-up units with vertical cores.

The most important factors to consider when using wall ties in cavity walls are the following:

- Corrosion resistance
- Proper spacing
- Placing the wall tie firmly in the mortar bed joint to assure a good bond
- Using cavity wall ties with preformed crimps in the center. The crimps collect moisture which then drops into the cavity.

Prefabricated horizontal joint reinforcement is acceptable for tying cavity walls when used with rectangular ties. The rectangular tie ties the wall together and the joint reinforcement gives added stability. Horizontal joint reinforcement is not usually required in brick masonry walls since they are not subject to shrinkage stresses. However, using them makes it more likely that the mason will remember to lay the cavity wall ties.

The Importance of Good Mortar in Cavity Wall Construction

Mortar has an important bearing on the strength of cavity walls so it is critical that masons follow the recommendations of the architect for the proper mortar mix.

Tests by the BIA indicate that Type M or Type S portland cement-lime mortars, under ASTM C270 or BIA Designation M1-72, provide maximum bond between masonry units and mortar. When wind velocities in excess of 80 miles per hour (mph) are expected, Type S mortar is recommended for cavity wall construction. Under average conditions, Type M may be used. In any case, the designer selects the lowest strength mortar that is still compatible with the structural requirements of the building.

Insulation of the Cavity Wall

In addition to their ability to protect interior walls from moisture, cavity walls also serve as sound and temperature insulators. These features have made cavity walls even more popular.

While the air space between the two wythes of masonry acts as an insulating layer, placing insulating materials inside the cavity increases insulating capacity considerably.

Some important requirements for cavity wall insulation are the following:

- Insulation must permit moisture drainage within the cavity.
- Insulation must not hold moisture so that it loses its thermal insulating efficiency.
- Insulation must support its own weight without settling.
- Insulation must be resistant to rot, fire, and vermin.
- Granular fill insulation is usually poured directly from the bag into the cavity or from a hopper placed on top of the wall. The insulation should be protected from moisture during installation.

Types of Insulation. Suitable types of insulation are granular fills and rigid boards. Two types of granular fill insulation have been tested by the BIA and found to meet the above requirements. They are water-repellent vermiculite masonry fill and silicone-treated perlite loose fill.

Composition

Vermiculite is an inert, lightweight, granular insulating material manufactured by expanding an aluminum magnesium silicate mineral, which is a form of mica. The raw mineral is made of approximately one million separate layers per inch, with a minute amount of water between each layer. When particles of the mineral are suddenly exposed to temperatures in the range of 1800°F to 2000°F (980°C to 1100°C) in a specially designed furnace, the water changes to steam, causing the vermiculite to expand into cellular granules of vermiculite insulation about 15 times their original size.

Perlite is a white, inert, lightweight granular insulating material made from volcanic siliceous rock. When the crushed stone is heated to approximately 1800°F (980°C), it expands or pops much like popcorn as the combined water (2% to 6%) vaporizes and creates countless, tiny bubbles in the heat-softened, glassy particles. Perlite can be expanded up to 20 times its original volume. Specifications for water repellent vermiculite and silicone-treated perlite are published by the Vermiculite Association and the Perlite Institute, Inc., respectively.

Rigid board insulation can be molded polystyrene, expanded polyurethane, rigid urethane, cellular glass, preformed fiberglass, or perlite board. The rigid board insulation fits flush against the cavity face of the internal wythe. There should be at least 1″ of air space between the cavity face of the external wythe and the insulation board, Figure 23-5. The 1″ space supplies room for drainage of the cavity to weep holes at the base of the wall. All boards must fit between ties and abut against the ties. An adhesive holds the insulation in place.

Insulated cavity walls, whether filled with granular fill or rigid board insulation, are still popular. They offer resistance to moisture and, equally important in light of high energy costs, they can greatly decrease the amount of energy needed to heat and cool buildings.

REINFORCED MASONRY WALLS

A reinforced masonry wall is built like a cavity wall with an air space between the wythes. However, this air space is later filled with reinforcing steel rods and grout. For even greater strength, steel rods can also be placed in the hollow cores of the masonry units. Grout is poured into the cavity and puddled or vibrated to consolidate it around the reinforcement steel. Reinforced

masonry walls are designed to supply greater strength and to withstand greater than normal stresses.

Steel Reinforcement Rods

The steel rods in reinforced masonry are the same type as rods used in reinforced concrete. Projections on the outside surface, called *deformations,* help the grout adhere to the rods. When placing rods in a reinforced wall, masons usually first place short dowel rods into the concrete footing before it hardens. Two wythes of masonry are then built, and the long reinforcements are wired securely to the short dowel rods to form continuous steel rods inside the cavity. As the wall increases in height, the vertical rods are periodically wired to horizontal rods for extra strength, Figure 23-6. The stress the wall will bear determines the size of the rods used.

Mortar Joints in Reinforced Masonry

It is extremely important to use solid mortar joints in reinforced masonry walls. A nonfurrowed bed joint is preferred, with the inside of the mortar joint angled with the edge of the trowel blade, Figure 23-7. This taper helps to prevent mortar from protruding beyond the bricks but still allows full bed joints. If mortar does ex-

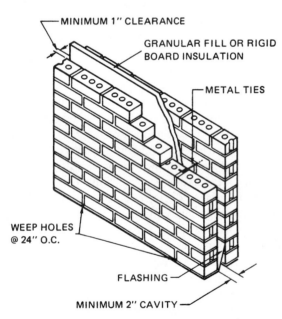

Fig. 23-5 Typical, insulated brick cavity wall.

Fig. 23-6 Vertical steel reinforcement rods are tied to the horizontal rods to attain greater lateral strength in the wall. Grout will eventually be poured between the rods.

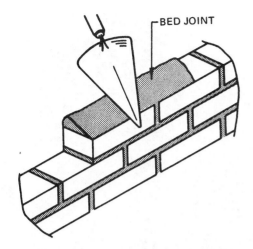

Fig. 23-7 Beveling the bed joint with the blade of the trowel.

tend into the cavity, carefully cut off the excess mortar. Be careful not to drop any in the space where the reinforcement is to be placed. If mortar does drop into this space, the grout will not fill in solidly around the reinforcing steel.

The head joints between each masonry unit must be filled completely with mortar. This is done by completely buttering the ends of each unit and shoving them forcefully into place. The joint formed is known as a *shove joint.* When this method is used by a competent mason, it produces a solid section of wall with no voids.

Cleanout holes are provided during construction by leaving out some masonry units along the base of one side of the wall. These holes can be used not only to clean out the cavity but also to position and tie the steel rods to the dowels in the footings. Mortar that projects more than $\frac{1}{2}''$ into the cavity should be removed so that the grout fully fills the cavity. Many contractors place a layer of polyethylene film or a layer of sand on the bottom of the cavity. Mortar dropped on such layers is easily removed.

Use of Grout in Reinforced Masonry

Grout may be fine or coarse. Fine grout is used when the grout space does not exceed $2\frac{1}{2}''$ in thickness. Coarse grout is used for spaces which measure more than $2\frac{1}{2}''$. The aggregate used in coarse grout should be

about $\frac{3}{8}''$ in size. The thickness of the grout between the masonry units and the reinforcement rods should be at least the diameter of the bars. This ensures that grout will surround each rod.

The grout must contain sufficient water so that it flows easily between the walls. One means of measuring the consistency of the grout is by using a 12″ high slump cone. (A *slump cone* is a metal truncated cone. A *truncated cone* is a cone with the top cut off.) The cone is placed on a flat surface with the large end facing down and is filled with grout. The cone is then removed. The distance the grout falls below the top of the cone, known as the slump, shows the consistency of the grout. Grout for use in reinforced brick masonry should have a slump of 9″, using a 12″ slump cone.

Grouting Methods

There are two accepted methods of placing grout in the wall, the low lift method and the high lift method. In low lift grouting, one wythe of wall is laid about 12″ high. The other wythe is then laid 1 course higher before any grout is poured into the cavity. Fifteen minutes should be allowed between each pouring of grout. Masons should not work any closer than 10′ to 15′ apart to avoid the possibility of blowouts. In a *blowout,* the grout swells the wall and leaks through to the outside surface. If the wall bows out, or if blowouts occur, the work must be rebuilt since the bond strength of the bed joint is destroyed. Such repairs are costly and time consuming.

In low lift grouting, pour the grout in the wall from a container equipped with a pouring lip so that the grout does not run onto the face of the masonry work. The best procedure is for the masons to work in pairs with one pouring the grout in the wall and the other puddling with a wooden stick to vibrate and pack the grout into the cavity. Puddle sticks cut from $\frac{3}{4}''$ to 1″ wood are the most practical size for this purpose. Sticks which are too large are usually clumsy and may cause blowouts.

If operations on the wall are to be suspended for a period of more than 1 hour, it is best to build both wythes of the wall to the same height and pour the grout to within $\frac{3}{4}''$ of the top. This gives the next pouring of grout a better chance of bonding to the section of wall underneath.

High lift grouting involves building the wall to the specified height (not greater than 9′) and then pouring the grout into the cavity. The two wythes of masonry are tied together with metal ties similar to those used in a cavity wall. When the wall is built to the full height, the grout is poured with either a concrete bucket or pumping equipment. After the grout stiffens, pouring resumes. Grouting should not be poured more than 4′ high at any one time. Puddling or vibrating should always be done before the plasticity of the grout is lost.

Remember that the cavity should not be filled with grout until the masonry is strong enough to prevent blowouts from the pressure of the grouting procedure. Code requirements often determine the minimum curing time. Under ordinary conditions a cure of at least 3 days is required before any grout can be poured. Only after inspection of the masonry is the grout poured.

Reinforced masonry walls can be constructed to combine beauty with strength, Figure 23-8. Figure 23-9 shows an example of testing the strength of a reinforced masonry wall section. Another advantage of masonry reinforced walls is that they are highly resistant to moisture penetration, penetration of heat and cold, and expansion. The use of reinforced masonry is highly recommended in areas where earthquakes frequently occur. All these features have led to the in-

Fig. 23-9 The strength of a reinforced masonry wall section is tested by stacking a large amount of bricks on the section. Notice how little the wall is bending under the strain. As a rule, a compressive load is not applied to a wall horizontally.

Fig. 23-8 Reinforced masonry piers arranged on an angle to create an unusual appearance.

creased demand in recent years for reinforced masonry construction, Figure 23-10.

PROGRESS CHECK

At this point, masonry students should have acquired certain basic skills. They must practice these skills as every mason must have them. Students should demonstrate for their instructor each of the following skills. Any skills not mastered should be practiced until they can be demonstrated correctly.

- Proportion and mix mortar ingredients to suit specific job conditions.
- Stock and place materials.
- Fill chalk lines. Hold chalk lines properly when striking a line.
- Cut, spread, and furrow mortar.

Fig. 23-10 This modern brick building is a perfect example of design and superior strength derived by using reinforcement and high strength mortar.

- Form mortar joints.
- Dry bond unit.
- Square jambs or returns of structures.
- Read the plumb rule and level structures with it.
- Lay units to the line. Use line blocks, line pins, and nails.
- Strike joints.
- Brush and clean work at the finish of the job.

ACHIEVEMENT REVIEW

Select the best answer from the choices offered to complete the statement. List your choice by letter identification.

1. The wall type that best ensures interior dryness is the
   a. cavity wall.
   b. solid masonry wall.
   c. veneer wall.
   d. reinforced wall.

2. The weep holes in a brick cavity wall should be spaced the same distance apart to be effective. The proper distance is
   a. 36″.
   b. 24″.
   c. 8″.
   d. 4″.

3. The term *grout* is used frequently when discussing masonry reinforced walls. Grout is
   a. caulking used to seal ends of walls.
   b. steel reinforcement used in the cavity of a cavity wall.
   c. special mortar mixed in liquid form.
   d. wall ties used to tie cavity walls together.

4. An 8″ cavity wall is recommended for buildings
   a. 1 story in height.
   b. 2 stories in height.
   c. 3 stories in height.
   d. 4 stories in height.

5. If the cavity wall is to be effective in draining moisture, the weep holes should be installed over the
   a. header course.
   c. flashing course.
   b. metal wire course.
   d. floor level course.

6. The most popular wall tie used to tie cavity walls together is the
   a. veneer tie.
   c. Z tie.
   b. masonry reinforcement wire.
   d. strap anchor.

7. The consistency of grout can be measured accurately by using a
   a. cylinder compression test.
   b. slump cone.
   c. pouring rate scale.
   d. table which lists exact amounts of water to be used in the mix.

8. Steel reinforcement rods have rough projections and designs on the outside surface. These are there for
   a. the appearance of the steel rod.
   b. easier handling by the person installing the rods.
   c. better bonding of the rods to the grout.
   d. easier bending or cutting when joining one or more rods.

9. The principal reason for using a shove joint is
   a. to ensure a more solid head joint.
   b. for greater speed in laying a reinforced wall.
   c. to increase the compressive strength of the joint.
   d. to allow for expansion of the mortar joint.

PROJECT 18: BUILDING A 10″ CAVITY WALL IN THE RUNNING BOND

OBJECTIVE

- The student will be able to lay a 10″ cavity wall in the running bond, and install metal Z ties and weep holes at designated heights, Figure 23-11.

  Note: Cavity walls such as this are used in the construction of many school buildings since the interior walls remain extremely dry.

EQUIPMENT, TOOLS, AND SUPPLIES

2 mortar pans or boards	Ball of nylon line and line pins
Mixing tools	Plumb rule
Mason's trowel	Convex sled-runner striker
Brick hammer	Chalk box
2′ square	Brush
Modular rule	Clean water
8, 2″ ties with drip crimps	
3 pieces of $\frac{3}{8}$″ rope, each 14″ in length	

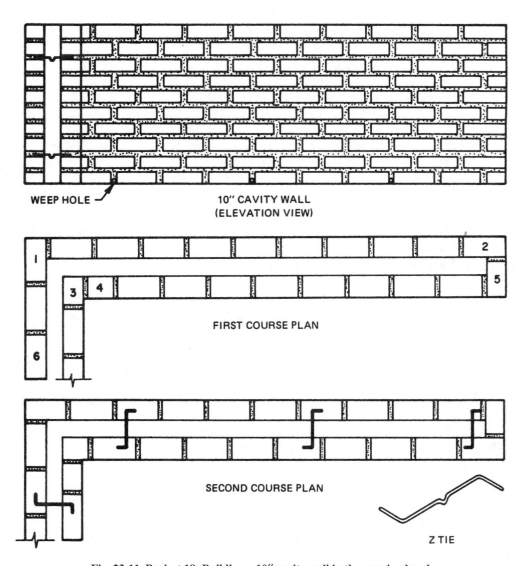

WEEP HOLE

10" CAVITY WALL
(ELEVATION VIEW)

FIRST COURSE PLAN

SECOND COURSE PLAN

Z TIE

Fig. 23-11 Project 18: Building a 10″ cavity wall in the running bond.

2 drip sticks $1\frac{3}{4}''$ wide and $\frac{3}{4}''$ thick;
    one about 76″ long and the other about
    16″ long

Note: The student will estimate the required number of bricks, lime, and sand from the drawing.

SUGGESTIONS

- Brush off the base carefully to remove any dust or dirt.
- Strike chalk lines longer than are needed.

- Angle the back of the bed joint with a trowel to prevent mortar from falling in the cavity.
- Be sure that the drip stick lies flat on the Z ties to ensure that no mortar falls into the cavity.
- Install the Z ties with the drip facing down.
- Keep all materials and tools away from the immediate work area to avoid accidents.
- Wear eye protection when cutting bricks or mixing mortar.
- Be sure to embed all metal Z ties solidly in mortar.
- Remove the drip stick as soon as you complete the wall.

PROCEDURE

1. Mix the mortar and stock the work area with mortar and bricks. Set up 1 mortar board on each side of the wall.
2. Brush off the base and strike chalk lines.
3. Lay out the entire first course dry as shown on the plan.
4. Lay bricks #1 and #2 in mortar to the layout line. Be sure that they are level and plumb. Brick courses are laid to number 6 on the modular rule. Put up the line and lay the course. Lay brick #6 and fill in between #1 and #6, using the plumb rule as a guide. Square the returns.
5. Lay brick #5 ($\frac{3}{4}''$) and the adjoining brick. Be sure they are to the line with the plumb rule. Measure 12″ from the course already laid and lay bricks #3 and #4 ($\frac{3}{4}''$) true. Double-check to be sure that all returns are square with the main wall. Fasten the line and lay the balance of the course in mortar, installing weep holes at the proper locations as shown on the plan. Tail out several bricks from brick #3. Using the plumb rule, be sure these bricks are true to the line. Lay a brick in each corner for the second course and move the line up. Lay the entire course.
6. Install wall ties as shown. Be sure that the front and rear tiers of masonry are level before laying wall ties across the cavity.
7. Lay the drip stick on top of the Z ties and continue laying the project until the wall is 8 courses high. Remove the drip stick and install the second set of Z ties over the eighth course.
8. Continue building the project to the height shown on the plan, being sure all courses are laid to the number 6 on the modular rule.
9. Strike the mortar joints as needed with the convex sled-runner striker and brush the wall.
10. Clean as much mortar as possible from the cavity.
11. Pull the weep hole ropes from the wall.
12. Recheck the wall to be sure that it is level and plumb before having it inspected.

PROJECT 19: BUILDING A 12″ BRICK AND BLOCK CAVITY WALL

OBJECTIVE

- The student will be able to lay out and build a 12″ brick and block cavity wall in running bond, install weep holes with rope, and install metal Z ties as shown on the plan, Figure 23-12.

  Note: It is very important to keep all mortar droppings out of the cavity!

EQUIPMENT, TOOLS AND SUPPLIES

2 mortar pans or boards

Mixing tools

Mason's trowel

Brick hammer

Brick set chisel

2′ square and pencil

Masons modular scale rule

10 metal Z ties

Chalk box

4′ Plumb rule

Ball of nylon line and line pins

Convex sled-runner striker

Brush

4 pieces of 3/8″ cotton clothes line rope 14″ in length for the weeps

1 wood drip stick $3\frac{3}{4}'''$ wide by $\frac{3}{4}'''$ thick as long as the project to catch any mortar that falls into the cavity.

Note: The student will estimate the required number of brick, $6'' \times 8'' \times 16''$ concrete block and mortar from the plan.

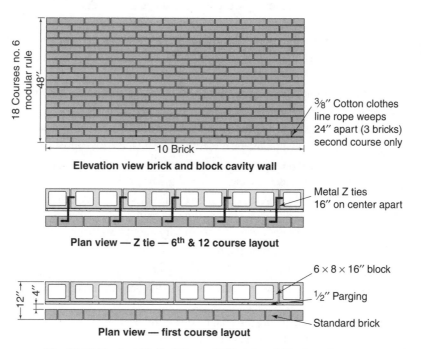

Fig. 23-12 Project 19: Building a 12″ brick and block cavity wall.

SUGGESTIONS

- Brush off the base to remove any dust or dirt.
- Strike the chalk lines longer than needed as a reference line.
- Angle the back of the bed joints with the trowel to help prevent mortar from falling into the cavity.
- Be sure that the wood drip stick lies flat so it will catch the mortar and prevent it from falling in the cavity.
- Install the Z ties with the indented drip down.
- Keep tools out from under your feet to avoid accidents.
- Wear eye protection when cutting brick or mixing mortar.
- Be sure to embed all metal Z ties solidly in the mortar bed joints.
- Remove the drip stick when necessary to remove excess mortar.

PROCEDURE

1. Mix the mortar and stock the work area with mortar and brick. Set up one mortar board on each side of the wall line.
2. Study the plan, brush off the base, and strike the chalk line.
3. Dry bond the first course out on the layout lines.
4. Lay the first course of brick level, plumb, to the correct height, and to a line.
5. Lay the second course of brick to the line and install the weep ropes as shown on the plan. (Review the information on weep holes in the unit if needed.)
6. Lay the six-inch concrete block work backing up two courses high ahead of the brickwork and back-parge with mortar the inside face of the block to make it waterproof.
7. Clean out any mortar dropping in the cavity space and lay the drip sticks in the cavity to catch any mortar.
8. Continue building the brickwork until it is 6 courses high or level with the top of the 6-inch block work backing courses.
9. Remove the drip sticks and any mortar in the cavity area.
10. Install the Z ties as shown on the plan view of the 6th and 12th course layout.
11. Lay the drip stick back on top of the Z ties spanning the cavity area and continue laying up two more of the block courses ahead of the brickwork.
12. Lay the brick courses six more courses high to the height shown on the plan until they are level or even with the concrete block backing wall. Tool or strike the mortar joints as needed. Brush the masonry work when it is dry enough so that it will not smear.
13. Install the second set of Z ties on top of the 12th course as shown on the plan.
14. Move the drip stick up and lay it on the Z ties again.
15. Lay the remaining two courses of 6-inch block to the project height shown on the plan for a total of 6 courses or 48 inches.
16. Complete the project by laying the final 6 courses of brick to the project height shown on the plan for a total of 18 courses of brick or 48 inches.
17. Tool the mortar joints and brush the project.

18. Remove the drip sticks and all mortar droppings from the cavity area.
19. Recheck the project to make sure it is level, plumb, and to the correct height before having it inspected.

SUMMARY, SECTION 5

- Laying bricks and concrete block is more productive with the aid of a line.
- The corner pole eliminates the need for building a corner before erecting a wall and makes the mason more competitive in the building market.
- The masonry corner serves as a guide for the wall and must be built true.
- The mason learns the skills involved in building a corner only by constant practice and strict adherence to basic tool skills.
- When learning to build the corner, do not attempt to work very quickly. Try instead to build the corner as perfectly as possible.
- Pulling and attaching lines to the walls demands safe work practices to avoid injury to the mason and fellow workers.
- Masons spend most of their time laying bricks or concrete block to the line. Using time productively is the key to being a successful mason.
- The mason should be able to estimate the necessary materials for a small brick job by applying the rule of thumb method of estimating. The rule of thumb method of estimating allows for a reasonable amount of waste and is accurate enough for a small job.
- Methods of applying mortar to bricks and concrete block for bed joints are different due to the size of the units.
- Only face shell bedding is necessary when laying concrete block unless otherwise specified.
- Once the masonry unit takes its initial set, it should not be shifted or moved unless fresh mortar is applied and the unit is relaid.
- Mix mortar to the correct consistency for the masonry unit to set properly. As a rule, concrete block require a slightly stiffer mortar than bricks.
- Reinforced wire greatly increases the strength of the bed joint in a masonry wall.
- The best joint finish for bricks or concrete block is the concave joint.
- Always wear eye protection when cutting masonry units.
- When laying concrete block, observe proper lifting practices due to the weight of the units.
- You can erect a structure consisting of concrete block more rapidly than a brick structure since the units are larger.
- Plumb single units of concrete block only on one side.
- When a block wall is to be painted and the mortar joints are not to be evident, use a flat, rubbed joint.
- Rule of thumb estimating for concrete block is done by calculating lineal feet around the structure and converting that figure to block. Multiply the number of block on one course by the number of courses in height for the total number of block needed for the job. Deduct block for all openings.

- When two separate tiers of masonry units are built to form a single wall, the structure is known as a composite wall.
- Tie a composite wall together with masonry headers or metal wall ties.
- Metal ties of different types are now being used more than brick headers to tie walls since they allow for more expansion and contraction in the mortar joints.
- Regardless of the type of metal tie selected for a job, make sure to embed the ties solidly in mortar where they rest on the wall.
- To obtain maximum resistance to rain penetration, the best choice of wall type is the cavity wall.
- Weep holes are necessary to allow moisture to drain from cavity walls.
- Cavity walls may be insulated for greater efficiency.
- Keep cavity walls free from mortar droppings so they will drain properly.
- Wall ties for cavity walls must have a drip crimp in the center so that moisture is not carried to the walls.
- Reinforced masonry walls are built when superior strength is required.
- The middle of reinforced walls is constructed of steel rods and mortar or cement grout.
- Use solid mortar joints when building a reinforced masonry wall to avoid blowouts.
- Never attempt grouting of a reinforced wall until the wall has had time to set.
- Reinforced walls offer the strength of concrete and pleasing appearance of masonry work.

SUMMARY ACHIEVEMENT REVIEW, SECTION 5

Complete the following statements referring to material found in Section 5.

1. Most masonry units laid by masons are laid to the _____.
2. As the mason faces the wall, the left-hand lead where the line is attached is called the _____. The right-hand lead where the line is pulled and fastened is called the _____.
3. If a long wall is being built and the center of the wall sags, the sag can be corrected by setting a _____.
4. Masonry units (brick or concrete block) should always be laid so that the top edge of the unit is even with the _____ and _____ inch (inches) from the line.
5. Masonry corners are also known as _____ in the trade.
6. One of the most important points in building a corner is looking down over the work when laying the unit to be sure it is in close alignment. This technique, which reduces the amount of necessary plumbing, is known as _____.
7. Masonry walls are sometimes plastered on the back or inside of the wall to prevent moisture penetration. This practice is known as _____.
8. Estimating brick masonry for small jobs is done by the rule of thumb method. To determine the number of bricks in a 4″ wall, multiply _____.

9. When laying concrete block to the line, mortar is applied only on the outside web. This method of applying mortar is called _____.

10. The most effective and popular joint finish for concrete block is the _____.

11. The major difference in laying bricks and concrete block is because of the _____ and _____ of the units.

12. When cutting bricks or concrete block, the mason should always wear _____ for safety.

13. Additional strength can be added to mortar bed joints by installing _____.

14. When a masonry wall is built of 2 different tiers of masonry units and bonded together by some type of tie, it is known as a _____.

15. The composite wall may be tied in two different manners, by use of either a _____ or a _____.

16. Concrete block are estimated by the rule of thumb method by determining the number of lineal feet around the total structure and multiplying by _____.

17. The number of concrete block required to lay 1 course around a structure is multiplied by _____ to determine the total number of block needed to build the structure.

18. Before figuring the total amount of materials for a job and their costs, all _____ must be deducted.

19. A masonry wall that consists of 2 tiers of masonry with an air space in the center is called a _____.

20. A masonry wall that has 2 tiers of masonry with steel reinforcement and grout in the center is known as a _____.

21. Masonry walls should be built to suit specific climates or weather conditions. In a very wet, humid area, the best wall type to use is a _____.

22. Grout used in reinforced masonry walls must be more _____ than regular mortar to fill all voids and holes.

23. For wall ties in a cavity wall to remain free of moisture, they must have a _____.

24. Aside from the architectural beauty of reinforced masonry, the most important feature is _____.

25. Under ordinary conditions, before placing grout in a reinforced masonry wall, the period of time that must pass is _____.

SECTION SIX
MASONRY PRACTICES AND DETAILS OF CONSTRUCTION

UNIT 24
Masonry Supports, Chases, and Bearings

---------- OBJECTIVES ----------

After studying this unit, the student will be able to

- describe how lintels are installed in a masonry wall.
- construct a pier or pilaster using given plans.
- improve tool skills by building various masonry projects.

After studying and learning the fundamentals of basic masonry, the student progresses to the details of laying bricks and concrete block. These processes are not difficult, but demand the application of the information presented thus far and mastering of tool skills to a greater degree.

LINTELS

A *lintel* is a horizontal member or beam support placed over a wall opening to carry the weight of the masonry to be laid over it. The design of the particular building and load to be placed on the lintel are the deciding factors in the type of lintel selected.

Steel Lintels

One of the most common types of lintels in use today is the *structural steel lintel,* or *angle iron.* The steel

angle iron, which is shaped like the letter L, should have a thickness not less than $\frac{1}{4}''$ and a width not less than $3\frac{1}{2}''$ to support the standard 4″ masonry unit. Angle irons such as this one may be installed back to back when building an 8″ wall. Three angle irons are required to build a wall 12″ in thickness, Figure 24-1.

There are some important items to remember when installing steel lintels over openings. The mason should consult the specifications on the job to determine the *lintel bearing* (the load which the lintel must bear). The lintel should bear a minimum of 4″ on each side of the opening on houses or small commercial buildings.

Most angle irons have been cut from a larger piece of iron with a burning torch. This burning causes small beads of steel to form on the surface of the bottom and top portions of the angle iron. The beads should be removed from the irons with the brick hammer before installing. The masonry units will not lay level over these

269

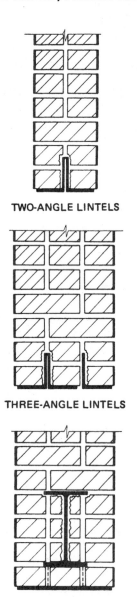

TWO-ANGLE LINTELS

THREE-ANGLE LINTELS

**STEEL I-BEAM
AND SUSPENDED PLATE**

Fig. 24-1 Typical lintel assemblies used in masonry walls. Top to bottom: an 8″ wall requires two-angle irons; a 12″ wall requires three-angle irons; the steel I-beam and suspended plate combination is used for heavier walls.

rough welds. This is especially important if flashing is to be installed over the lintels, as the rough beads of steel may puncture the flashing.

The lintel must be set plumb and level so that the unit which is laid on it is also in the proper position. It is also important that a steel lintel has rust protected paint. The steel lintel must be set back $\frac{3}{8}″$ to $\frac{1}{2}″$ from the face of the wall so that a mortar joint may be formed in front of the steel without cracking, Figure 24-2. Never set the angle iron on the wall until the masonry unit under it has set sufficiently to avoid sinking.

Concrete Block Lintels

Concrete block lintels are made of the same materials and are the same texture as concrete block. These lintels have steel rods running throughout for extra strength.

Fig. 24-2 Installing a steel angle iron over a window opening. The mason is positioning the lintel about $\frac{1}{2}″$ from the face of the wall to allow room for a mortar joint.

They are available in various thicknesses to suit individual concrete block walls. They are available in lengths to fit most openings. A minimum bearing of 8″ should be provided on both sides of the openings for concrete block lintels.

When installing concrete block lintels, caution should be used since they are bulky and heavy. It is recommended that two masons lift and place the lintels in position. Very heavy lintels should be installed with the aid of power-driven lifting equipment.

Bond Beam Lintels

The *bond beam lintel* is used if the lintel must match the rest of the surface of the wall or if other types of lintels are not readily available. To form a bond beam lintel (also called a *reinforced lintel*), the inside of the brick is sawed out or a special bond beam block is purchased. Steel reinforcement rods are placed in the cavity on a layer of concrete and then the cavity is filled with concrete. The bottom of the reinforced lintel must be braced with lumber until the lintel cures, Figure 24-3.

Architects sometimes specify that a course of reinforced lintel run completely around a building. Many missile sites use this type of construction for extra strength.

Fig. 24-3 Bond beam lintel. This type of lintel is typical of reinforced masonry walls which cross over an opening.

PIERS

Piers, Figure 24-4, are vertical columns of masonry that are not bonded to a masonry wall. They can be used as supports for beams, arches, porches, or wherever a free-standing masonry column is needed. The difference between a pier and a column is that columns are usually much higher in proportion to their thickness.

The size and type of construction of a pier is determined by the load the pier will support. Its horizontal length should not exceed four times its thickness. If a pier is hollow, excess mortar must not fall inside the center of the pier since this could cause the pier to swell and be pushed out of position. The centers of solid piers should not be filled immediately after the outside course of bricks is laid. Instead, the pier should be laid to a height of 6 courses and then the bricks should be laid in the center.

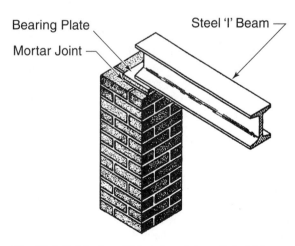

Fig. 24-4 Brick pier with bearing plate and steel I-beam.

Materials must be dry if the pier is to be constructed level and plumb. Because piers are relatively small, they are laid more quickly than most masonry projects. If the masonry units are laid wet and the mortar is too fluid, the units do not set properly, making the task of leveling and plumbing almost impossible. Piers should also be squared frequently and rechecked to be sure that their positions remain the same throughout the building process.

STEEL BEARING PLATES

Steel bearing plates are placed on top of piers to provide a solid bed for steel beams or girders. They also serve to distribute weight over a greater area. Steel plates are available in various thicknesses, but should be thick enough so that they are not bent out of shape under load.

The job foreman should consult individual job specifications for the required size of the plates. The height of the plates are shown on the plans and should be strictly followed if the steel beam is to be set at the correct height.

Installing Plates

A full bed of mortar should be spread with a trowel and the plate should be laid on the mortar bed. It is recommended to bed bearing plates in portland cement mortar for greater strength. Always spread more mortar than is actually needed. Then, when the plate is tapped into position with the handle of the hammer, no voids will exist under the plate. Be sure that the bearing plate is level to ensure that the beam sets level on the plate. When steel anchor bolts are installed in piers to draw the plates against them, they must be built into the masonry work beforehand.

PILASTERS

A *pilaster,* Figure 24-5, is much like a pier except that a pilaster is tied into the wall of a structure. Pilasters may be tied to a wall by use of wall ties or by bonding the masonry units into the wall. Some building codes require that pilasters be tied into the masonry wall at

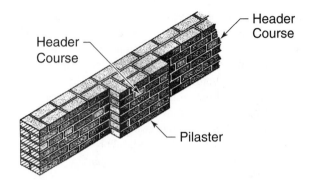

Fig. 24-5   Pilaster extending 4″ from an 8″ brick wall.

least 4″ and extend out from the wall 4″ to 8″. Local regulations will determine this.

Pilasters should be laid from the footing to a predetermined height. If possible, pilasters should never be laid so that units must be cut. For example, a pilaster that extends from a brick wall 6″ would involve cutting 2″ pieces of masonry units to fill the space behind the pilaster.

Pilasters are used for supporting loads, for ornamental design, and as vertical supports for strengthening masonry walls. When pilasters are used for ornamental purposes or for vertical support, they are also known as *buttresses.* Buttresses are usually finished at the top by a sloping face.

In reinforced concrete masonry structures and large buildings, pilasters are also sometimes used to enclose heating and air-conditioning ducts. Steel columns can be hidden from view and given some protection from fire by constructing pilasters around them. When a pilaster is used for this purpose, the wall is usually only 4″ thick since it is acting only as a screen and is not bearing a load.

CHASES AND RECESSES

A *chase,* Figure 24-6, is a vertical or horizontal recess left in a wall to conceal certain items such as plumbing pipes, electrical wires, or heating accessories. Many jobs, especially the commercial type, require that these mechanical items be installed behind the finished surface of the wall. If chases are not left in at the beginning

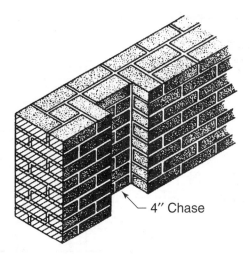

Fig. 24-6 Chase constructed in 12″ wall.

of the construction, the workers must cut them out with a chisel, causing much greater expense and delay in the job.

When buildings are constructed with chases, wall thickness and wall strength are always reduced. This must always be considered when designing a structure. Deep chases should never be built where the wall is to receive a heavy beam or steel bar joists over it.

Some general rules apply to the installation of chases in masonry walls. Do not build chases that are deeper than one-third the thickness of the wall. If the wall is load bearing, install lintels over the chase to sufficiently support the load. Check local building codes if in doubt, but remember, chases that will weaken the wall should never be built.

Masons build the chases as they lay the wall. The plans should specify exactly where the chase will be built. If it is not shown on the plan, do not build a chase in a bearing wall without first consulting the architect or the job engineer. Chases are usually on the inside face of the wall and may be 4″, 8″, or 12″ or more wide. The masonry units forming the chase should be bonded into the rest of the wall. Stack joints are not acceptable for bonding chases, since they are weak and may separate when the load of the structure increases. A chase should be plumb with all excess mortar cleanly cut off so that it does not interfere with the installation of pipes or wires.

SOLID MASONRY-BEARING WALLS

A basement- or foundation-bearing wall (a wall that supports joists, concrete slabs, or steel) often must meet special requirements. Many building codes require that the top 8″ of such walls be built of either 3 courses of brick or a solid course of concrete blocks, usually 4″ or 6″ thick. This helps distribute the weight of the structure on the foundation. Holes in the course of the regular concrete block beneath this solid course are filled with mortar. In addition, face shell bedding such as is used on a standard block wall is not acceptable underneath a solid bearing course.

The bearing course, whether it is brick or concrete block, is the last course of the wall that is laid. Joists or concrete slabs rest on the bearing course. It is, therefore, essential that the bearing course be perfectly level and plumb if the joints are to lay straight on the wall.

LAYING UNITS AROUND DOOR AND WINDOW FRAMES

The mason must often lay masonry units to, against, and around various types of door and window frames. Door frames may be tied to masonry work in different ways. Wooden door frames can be attached to the masonry work by nailing a special tempered nail through the frame and into the unit or by nailing into a wooden block *(nailing block)* that is laid in the mortar against the door frame. When using nailing blocks, 3 blocks are used on each side of the door; one at the bottom, middle, and top of the door, Figure 24-7.

Metal doors are in great demand for commercial construction. Metal ties are used to tie masonry work to metal doors, Figure 24-8. The process involves inserting the tie into the jamb of the door, which is filled with mortar to create a solid door jamb. The mason should use caution when filling the door jamb with mortar so that the mortar does not push the door out of alignment. The door should be rechecked to be certain it is plumb after filling it and before leaving for the day. A wooden spreader installed at the bottom of the door helps prevent the door jamb from being pushed out of alignment.

Windows are built in walls much the same way as doors are. On brick veneer walls, the bricks are butted

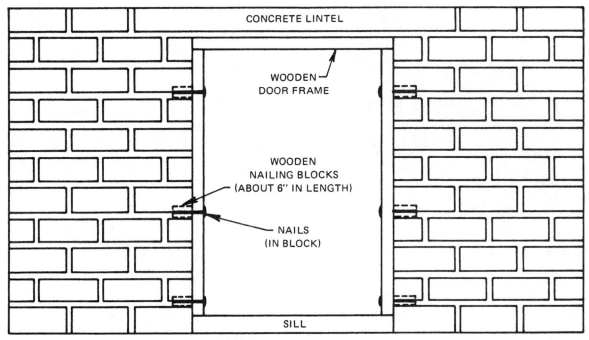

CONCRETE LINTEL

WOODEN
DOOR FRAME

WOODEN
NAILING BLOCKS
(ABOUT 6" IN LENGTH)

NAILS
(IN BLOCK)

SILL

Fig. 24-7 Concrete block wall with wooden door frame and nailing blocks.

flush against the wooden window jambs and the window frame is nailed into the framing of the wall. Metal window frames may be tied with metal ties in the same way as metal door frames. Solid composite walls that have wooden window frames can be secured by nailing

Fig. 24-8 Mason installing metal ties in a metal door frame. Notice that the mortar is filled in solid against the door frame.

a tempered nail through the wooden frame into the masonry unit.

Regardless of how frames are fastened to the masonry wall, the mason must remember that the frame should be positioned the proper distance back from the face of the wall. The position depends on the thickness of the wall; that portion of masonry work which extends from the window frame to the face of the masonry wall is known as the *reveal*. The frame must be kept level on the sill and plumb with the face of the wall. The carpenter has the responsibility of setting the frame. However, the mason should always check the frame to be sure that it is level and plumb before finishing the wall.

Brick Veneer/Steel Stud Walls

For many years brick veneer over a wood stud frame was a standard method of brick masonry and still is with many builders, especially in residential construction. Back in the 1960s the brick veneer/steel stud system was introduced and it has evolved into a wide variety of commercial, industrial, and institutional structures, which include building types such as schools, churches, hospitals, office buildings, etc. Unlike residential (house) construction, they, as a rule, are not designed and built with overhangs, eaves, or gutters to protect the brick veneer from rain water or the elements. For this reason, it is important that they be designed and constructed properly to ensure that they perform to the level required.

The brick veneer/steel stud wall system is considered to be an anchored veneer wall. An anchored veneer wall is a brick wythe (single thickness) secured to the backing—steel studs in this case—with metal anchor ties and supported vertically by the foundation or other structural elements. Brick veneer can be, but is not usually designed to be, a load-bearing wall, but only to carry its own weight and enclose the structure's framing, whatever it may be. Therefore, a brick veneer wall with steel stud backing with anchors or ties usually consists of a nominal 3- or 4-inch-thick exterior brick mechanically attached to a steel stud tied with corrosive-resistant metal ties with a minimum air space of 2 inches between them. See Figure 24-9.

Advantages of Brick Veneer/Steel Stud Wall

The biggest advantage of a brick veneer/steel stud system is that since it is not supporting or carrying the load of the building, the interior structure of a building can be constructed prior to laying the brick veneer work without any delay. This allows the building to be closed in independently of the brick work and placed under roof more quickly. Thus, the time-consuming interior work that the other building trades have to perform can go ahead on schedule, instead of having to wait on the brick masonry. This also enables the masons to take advantage of the best weather and temperatures to lay up the brick in mortar without worrying about the possibility of freezing. This makes the general contractor, subcontractors, and owners of the building very happy!

Another big advantage of this type of wall is that it is highly resistant to moisture because the cavity between the brick and steel studs can be drained efficiently by the use of properly placed flashing and weep holes in the brick work. The cavity between the two walls can also greatly reduce the heat gain or loss through the overall wall. The air space provides a thermal separation between the brick and the steel studs. Brickwork has a

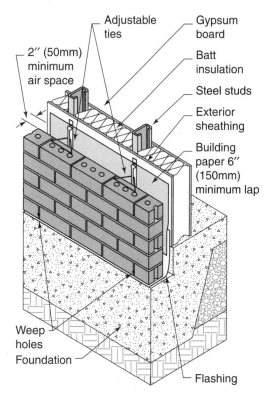

Fig. 24-9 Brick veneer/steel stud wall. (Courtesy Brick Industry Association).

high thermal mass, giving it the ability to store and slowly release heat over time. This effect, according to current energy codes, provides a higher R-value for a wall of this type. Rigid board closed-cell type insulation can also be placed inside the cavity area to prevent additional thermal loss.

Foundations for a Brick Veneer/Steel Stud Wall

Although some building codes permit the support of brick veneer on wood foundations, it is highly recommended that the wall be supported by concrete or masonry foundations. The brick work may extend below finished grade if it is built properly to minimize water penetration. A typical detail wall section view of a foundation is shown in Figure 24-10.

Notice that the base flashing and weep holes are located a minimum of at least 6 inches above finished grade line so that they do not clog up. Brickwork below the flashing should be considered to be a barrier wall by completely filling the cavity with mortar to again minimize water penetration. Metal ties should be spaced across the cavity area the same as in a normal brick veneer wall above grade.

Ties

Specially designed metal ties should be used to anchor or tie the brick veneer to the steel studs. Regular corrugated metal ties are not permitted when brick veneer is tied to metal studs. They should be spaced vertically every 16 inches in height and be 32 inches apart horizontally. They should also be well-imbedded in the mortar bed joint. It is very important that the back assembly of the tie that holds it be securely attached to the steel stud itself and not just the sheathing, so that they do not pull out! Well-filled mortar joints and good workmanship are of particular importance in brick veneer/steel stud type walls if good results are expected. Figure 24-11 shows a variety of tie assemblies that attach ties to steel studs.

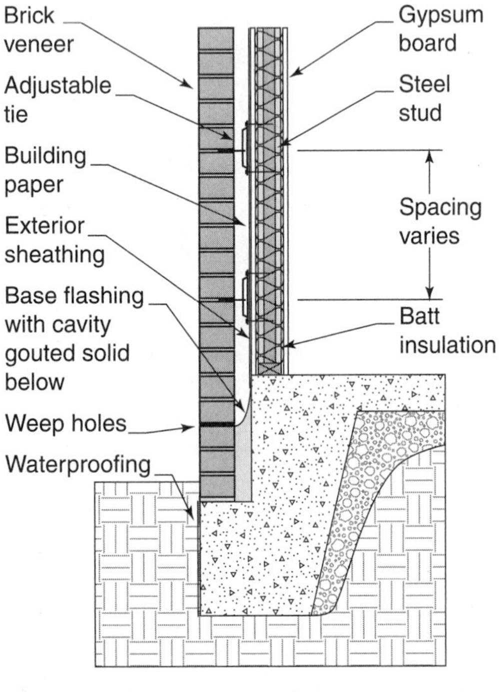

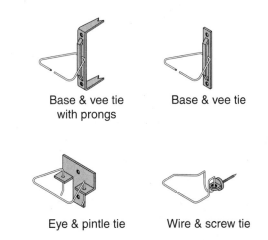

Base & vee tie with prongs

Base & vee tie

Eye & pintle tie

Wire & screw tie

Fig. 24-11 Tie assemblies. (Courtesy Brick Industry Association)

Fig. 24-10 Wall section at foundation. (Courtesy Brick Industry Association).

ACHIEVEMENT REVIEW

Select the best answer from the choices offered to complete the statement or answer the question. List your choice by letter identification.

1. The primary purpose of a lintel is to
   a. provide a means of expansion for masonry work.
   b. support a masonry unit over an opening.
   c. tie a masonry wall together.
   d. provide an extra measure of strength in a bearing wall.

2. If a 12″ brick wall is being constructed and lintels are required for a window, how many lintels should be used if angle irons are specified?
   a. 1 lintel.   c. 3 lintels.
   b. 2 lintels.   d. 4 lintels.

3. When a course of masonry units has the inside cut out and steel rods and concrete placed within, the resulting structure is a (an)
   a. angle iron.   c. bearing plate.
   b. hanging plate.   d. bond beam.

4. The main difference between a pier and a pilaster is that
   a. the pier is always wider than the pilaster.
   b. piers do not carry any weight and are only built for decorative purposes.
   c. a pilaster is tied to the main wall.
   d. the pilaster is free standing.

5. It is extremely important to have dry bricks when constructing a pier because
   a. the pier is built more quickly than a regular wall and, therefore, may not set with wet bricks.
   b. the wet bricks stain and cause an unsightly appearance.
   c. the mortar joints do not attain full strength.
   d. the mason's fingers become sore with the result that productivity is seriously curtailed.

6. Steel plates are laid on piers in mortar for the purpose of
   a. fireproofing the area where the beam sets on the plate.
   b. distributing the weight of the beam more uniformly.
   c. spot welding the beam to the steel plate.

7. On many jobs involving the installation of pipes and wires in the wall, a recess is built by the mason. This recess is called a
   a. reveal.   c. pier.
   b. pilaster.   d. chase.

8. Bearing walls which carry joists or concrete slabs should have at least
   a. 6 courses of bricks or 2 courses of solid concrete block at the top of the wall.
   b. 3 courses of bricks or 1 course of solid block.
   c. 9 courses of bricks or 3 courses of solid block.

9. When laying a concrete block wall against a wooden door frame, nailing blocks are installed in the wall. How many nailing blocks are needed for 1 door, including both sides of the door?
   a. 6   c. 2
   b. 3   d. 8

MATH CHECKPOINT

1. A brick apartment building is being built. It has 48 windows in the walls and each one measures 44″ wide. Concrete block lintels 4″ wide are being used over the head of the windows that are to bear on each side of the opening 8″. What is the total length of the lintel?
2. If each lintel costs $10.40, what would the total cost of all lintels be?
3. In addition to the cost of the lintels, a 5% sales tax is added to the bill. What is the total cost of the bill then?
4. A brick pier is built that has nine bricks on each course. How many bricks will it take to build the pier 6′ high? Assume that each 24″ in height equals nine courses of brick including the mortar joints.
5. Assuming that each bag of masonry cement will lay 125 bricks, how many bags of masonry cement will it take to build the pier?

PROJECT 20: BUILDING A 12″ × 16″ AND A 20″ × 24″ HOLLOW BRICK PIER

OBJECTIVE

- The student will be able to lay out, square, build, level, and plumb a hollow brick pier, Figure 24-12.

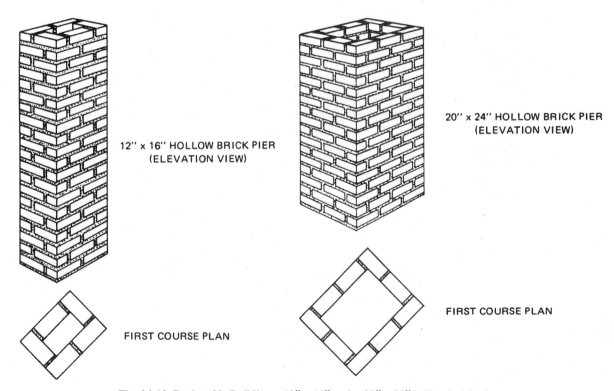

Fig. 24-12 Project 20: Building a 12″ × 16″ and a 20″ × 24″ hollow brick pier.

Note: These piers are typical of those built to support a porch or steel beam in a foundation.

EQUIPMENT, TOOLS, AND SUPPLIES

Mortar pan or board

Mixing tools

Mason's trowel

Brick hammer

$4'$ plumb rule and $2'$ level

Convex striking tool

$2'$ square

Chalk box and pencil

Mason's modular rule

Brush

$1, 8'' \times 8'' \times 16''$ concrete block in which to set plumb rule

The student will estimate the amount of standard face bricks, lime, sand, and water needed to build the project based on the rule of thumb method.

SUGGESTIONS

- Strike the chalk line longer than the actual size of the pier. This gives a reference line with which to check the work as the project is being built.
- When laying out the first course, select the highest point on the base and start the project there. The rest of the course can then be bedded with mortar to the highest brick.
- Use only single head joints on the immediate corner bricks until the entire course has been leveled and plumbed. This reduces the possibility of the corner brick being pushed out of alignment when the next brick is laid against it. Be certain to fill all joints after the course is completed.
- Use the $2'$ level for the first 9 courses since it is easier to handle.
- Keep the levels stored in the block when they are not in use. The $8''$ block is not upset as easily as a smaller block is.
- Prevent the mortar from dropping into the center of the pier by cutting it off with the trowel, as an excessive amount may cause the pier to be pushed out of alignment.
- Do not beat on the pier excessively, as this may cause the brickwork to settle out of alignment.
- Keep the work area free from materials. Always use safe working practices. Always consider those working nearby.

PROCEDURE

1. Set the mortar pan or board and bricks approximately $2'$ from the project.
2. Mix the mortar. Mix only as much mortar as can be used in the time provided. (Generally, never mix more mortar than can be used in a 2-hour period.) Fill the mortar pan with mortar. Clean tools immediately after mixing. Strike a chalk line on one side only.
3. Lay out the first course dry and square it. Mark the layout course with a pencil according to the plan. Remove the layout course and strike a chalk line over the pencil lines, extending the lines past the actual size of the pier for reference lines.

4. Bed the corner brick and check the height with the number 6 on the modular rule. Level and plumb. Spread mortar for the rest of the course and lay the bricks. Be certain that the leveling is done from the original corner brick and completely around the first course. Plumb the course on its corner points and align the pier with a straightedge on all four sides. Do not remove mortar that has been squeezed from under the bricks at this time, as they may settle unevenly. Remove the mortar after 3 courses have been laid.

5. Check the project with the steel square and make any necessary corrections.

6. Repeat the operations involved in laying each course of bricks until the pier has been built the specified number of courses. All brick courses are laid in height to the number 6 on the modular rule.

7. Strike the mortar joints as needed with a convex jointer. Remember to apply the thumbprint test. Brush the work.

8. Recheck the pier to be certain it is level, plumb, square, and the proper size before having it inspected.

PROJECT 21: CONSTRUCTING AN 8″, METAL-TIED BRICK WALL WITH TWO 4″ × 12″ PILASTERS

OBJECTIVE

- The student will be able to lay out and construct an 8″ brick wall with two 4″ × 12″ pilasters. The wall will be laid up to number 6 on the modular rule and tied together with metal Z ties, Figure 24-13.

  Note: The project is typical of garden walls which surround a residence or divide parking lots in an urban area.

  Note: This project can be done by two students working as a team.

EQUIPMENT, TOOLS, AND SUPPLIES

2 mortar pans or boards	Line pin and nail
Mixing tools	Chalk box and a pencil
Mason's trowel and small pointing trowel	Mason's modular rule
Brick hammer	Brush
4′ plumb rule and 2′ level	Supply of clean water
Convex striking tool	One 8″ × 8″ × 16″ concrete block
2′ square	in which to set levels
Ball of nylon line	6 Z ties

The student will estimate bricks, sand, and lime by studying the plan.

SUGGESTIONS

- Strike the chalk line longer than is actually needed to serve as a reference line.
- Use the 2′ level to plumb the first 6 courses of the pilasters as it is easier to handle.
- Lay the bricks to the line on the straight section of the wall before laying the bricks on the pilaster. This prevents the pilaster from interfering with the line.

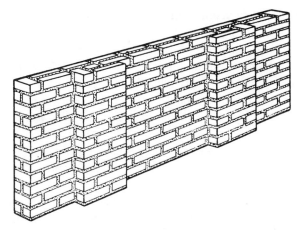

**8" METAL-TIED WALL WITH TWO 4" x 12" PILASTERS
(ELEVATION VIEW)**

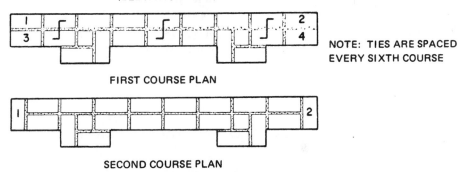

FIRST COURSE PLAN

NOTE: TIES ARE SPACED
EVERY SIXTH COURSE

SECOND COURSE PLAN

**Fig. 24-13 Project 21: Constructing an 8", metal-tied brick wall
with two 4" × 12" pilasters.**

- Square the pilaster as often as is necessary.
- Wear eye protection when cutting bats for the pilaster.
- Cut off excess mortar on the inside of the pilaster and return it to the mortar pan. Temper the mortar for reuse in the wall.
- Be certain that the inside angle joint is pointed solidly and neatly. Use the small pointing trowel.
- Keep the work area free from scraps of materials and follow good safety practices.
- Do not fill in the center of the wall (collar joint) with mortar as this may cause the wall to be pushed out of alignment.

PROCEDURE

1. Mix the mortar and arrange the necessary materials in the work area.
2. Strike a chalk line for the wall. Lay out the project dry as shown on the first course of plan.

3. Bed bricks #1 and #2. Stretch a line from brick #1 to brick #2 and tighten with the line pin.

4. Lay the rest of the course in mortar. Repeat the procedure for bricks #3 and #4. Lay the pilaster bricks in mortar, being sure to square off of the main wall. Pilaster bricks should project $4\frac{1}{2}''$ from the face of the wall, including the head joint.

5. All brick courses are laid to number 6 on the modular rule. On the second course, bed up bricks #1 and #2 being sure they are true with the level. Install the metal Z ties as shown on the plan.

6. Lay up a small 2-brick lead on both ends of the project. Attach the line and lay the courses in mortar. Alternate the line on each side of the wall as the work proceeds. Strike mortar joints as needed with the convex jointer.

7. Repeat this procedure until the wall is 6 courses high from where the last Z ties were installed. Install the second set of Z ties, being sure that the ties are not directly in line with the ties on the second course.

8. After each course of bricks is laid in the wall, lay the bricks for the pilaster level and plumb. Be certain that the pilaster is kept level with the wall. Recheck for squareness.

9. Continue constructing the wall until the number of courses shown on the plan (12 courses) is reached. Strike the joints, fill any holes, and recheck the wall before having it inspected.

PROJECT 22: LAYING A BRICK WALL WITH A PIPE CHASE

OBJECTIVE

- The student will be able to lay out and build a 12" brick wall with a vertical pipe chase in the common bond, Figure 24-14.

  Note: Pipe chases such as this one are constructed in many commercial buildings to conceal drainpipes descending from roofs.

EQUIPMENT, TOOLS, AND SUPPLIES

1 mortar pan or board	2 line blocks
Mixing tools	Chalk box and pencil
Mason's trowel	Mason's modular rule
Brick hammer	Brush
4' plumb rule and 2' level	Pointing trowel
Convex jointer or V-jointer	Supply of clean water
Small square (12" maximum size)	1, 8" × 8" × 16" concrete block
Ball of nylon line	in which to set levels

The student will estimate bricks, lime or cement, and sand from the plan.

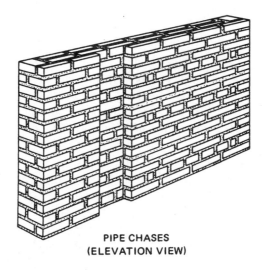

PIPE CHASES
(ELEVATION VIEW)

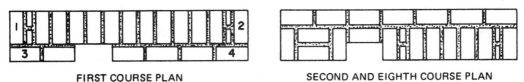

FIRST COURSE PLAN   SECOND AND EIGHTH COURSE PLAN

Fig. 24-14 Project 22: Laying a brick wall with a pipe chase.

SUGGESTIONS

- Select unchipped bricks for all corners.
- Do not fill in the collar joint with mortar, as it may cause the wall to bulge or be pushed out of alignment.
- Start the header course either with a 2″ piece as shown on the plan or with a 6″ piece. In this particular project, the 2″ starter piece is the stronger tie since 1 extra header could be used in the wall. As a rule, walls with pipe chases do not require striking. The wall in this project, however, requires striking because it is exposed.
- Frequently recheck all corners and chase with the square while the project is being built.
- Practice sighting down the corners as the bricks are laid. This practice helps reduce the amount of plumbing needed.
- Observe safety practices at all times.
- Form solid, well-filled joints between all bricks.
- Cut excess mortar from the backs of the bricks as soon as possible and return it to the mortar pan for reuse. Retemper the mortar as often as needed.

PROCEDURE

1. Prepare the work area and mix the mortar. Stock the materials at least 2′ away from the project to allow sufficient working room.
2. Strike a chalk line for the project as shown on the plan.
3. Lay out the first course dry. Lay bricks #1 and #2 in mortar. Erect the line with line blocks and lay the balance of the course. All courses are laid to the number 6 on the modular rule. The first course on the side of the wall that does not have a chase is a header course.
4. Repeat the procedure on the second course. Notice that the header course of the second course is on the same side of the wall as the chase. This is done in 12″ walls so that the headers occur over each other on succeeding courses.
5. Build a small lead on each end of the wall (2 or 3 courses high) and resume laying bricks to the line until the next header height is reached (6 courses high).
6. Lay the header course and continue building the wall until the total number of courses shown on the plan is reached.
7. Clean out excess mortar from the chase and point with mortar any remaining holes.
8. Check the project for the correct height with the modular rule. Recheck to be sure it is square, level, and plumb.
9. Strike the project with the convex or V-jointer as needed. Brush work before having it inspected.

UNIT 25
Small, One-Flue Chimneys

─────────────────── OBJECTIVES ───────────────────

After studying this unit, the student will be able to

■ describe the various components of a chimney.

■ explain how a one-flue chimney is built.

■ build a one-flue chimney from given plans.

■ list general steps in the installation of a wood-burning stove.

Due to the energy crisis and the high cost of conventional types of fuel, the heating of homes throughout North America with wood stoves has made a dramatic return. Although many of the homes that are converting to wood stove heat utilize the fireplace, there is also a great demand for new one-flue chimneys to serve wood stoves.

Local building authorities require permits and inspections of these chimneys while they are being constructed and after the wood stove has been installed in the home. Brick or concrete chimney block are acceptable for building one-flue chimneys for wood stoves. It is important not only that the chimney function correctly but also that there is ample brick laid around all combustible areas to prevent the danger of fire.

The primary function of a chimney is to produce the draft necessary to carry smoke or gases to the outside of the structure without endangering the structure or its occupants. A *draft* is defined as the movement of flue gases or air through the chimney. If a good draft is not present in the chimney, it cannot perform correctly.

A natural draft is influenced by the differences of temperature in the chimney and the outside atmosphere. Air travels down the chimney and returns to form a draft. When the draft is supplied by a fan in the furnace, the draft is said to be forced or induced. Most small chimneys operate on the forced draft principle.

Formerly, chimneys did not have protective linings to prevent the chimney from burning out. Almost all chimneys built today have a fired clay or terra-cotta lining to make them more fireproof and long lasting.

Most areas have building codes that require flue linings to be installed in all chimneys. They are obtainable in various widths and in a standard 2′0″ length. The clay from which the flue lining is made is fireproof and gives many years of service under extreme heat. It is recommended that flue linings be installed in all chimneys.

This unit discusses the one-flue chimney that is normally used in furnace and stovepipe hookups.

THE CHIMNEY BASE

Chimneys should be built on solid concrete footings that have been poured below the *frost line*. Any ground above the frost line freezes when the temperature falls sufficiently low. The depth at which frost lines occur differs in various parts of the country. Local building codes should be consulted to determine the frost line of a particular area. All masonry work should be started on a base that is below the frost line or deterioration of mortar joints may occur.

Footings at the base of chimneys require more concrete than the average footing of a home. This is because of the greater height and weight that is concentrated in a small area. Footings should be a minimum of 12″ deep and twice as wide as the chimney to properly distribute the load over the base area, Figure 25-1.

Chimneys should always be started from footings, never from wood platforms or metal supports. Local building codes should be consulted for proper footing sizes. As a rule, metal reinforcement is used only on

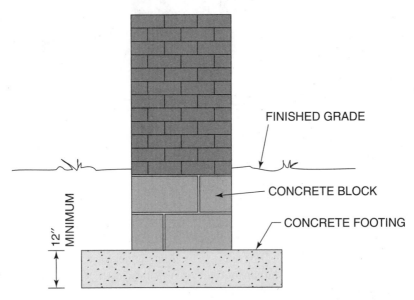

Fig. 25-1 Sectional view of concrete footing and chimney. The base of the chimney should be twice as wide as the chimney itself for proper weight distribution.

large commercial chimneys; it is not necessary for the average-sized or small-sized house chimney.

MORTAR FOR THE CHIMNEY

The same type of mortar that is used to lay the masonry units of a structure can be used to set the flue liners. Standard masonry cement (Type N) is generally accepted for house chimneys. The average house chimney normally does not necessitate the need for a special high-strength mortar. Whether the masonry unit is a brick or concrete block, the same mortar serves equally well. If extra strength is desired, Type M or S (portland cement and lime mortar) can be used. It is important to use solid mortar joints at all times in chimney construction.

UNITS FOR THE CHIMNEY

Any hard, fired masonry unit can be used to build a chimney. However, bricks are selected most of the time. Old or used soft bricks should not be used for the outside face of the chimney or near the hot part of the flue as they may not be able to withstand the dampness or heat.

Concrete masonry units may also be used for chimneys. They may be the solid type or a specially manufactured chimney block with or without a section of flue lining built into the block. If the block does not come with a flue lining, the lining can be inserted as the chimney is built by the same method used on brick chimneys. Generally, the same type of mortar used for building brick chimneys can be used on concrete masonry units.

INSTALLING THE CLEANOUT DOOR

After the chimney has been laid out and construction has started, the mason must determine how to deal with soot from the chimney. All chimneys require periodic cleaning. For easy access to the chimney, cleanout doors are installed near the bottom of the chimney, Figure 25-2.

Cleanout doors should be installed over the third course on a brick structure above grade or the first course on a block structure. They are always located above the level of the finished grade, Figure 25-3. They should never be installed lower than the first course on a brick building. When they are this low, there may be a problem since soot builds up rapidly at this point and moisture may prematurely deteriorate the metal door.

Fig. 25-2 Metal cleanout for soot removal in chimney.

Cleanout doors should always be installed tightly so that they do not interfere with the draft of the chimney. Inserting a wire or nail in the holes on the flange of the door frame and into the mortar joint helps to hold the door in place while it is being constructed. Cast-iron cleanouts are recommended since they last longer than the more lightweight metals.

Fig. 25-3 Cleanout door properly installed above grade.

THE FLUE LINING

The lining in the chimney serves two purposes. The lining prevents the extreme heat from burning into the masonry work. Also, it adds to the efficiency of the chimney, since its smooth surface prevents soot from clinging to it as readily as to masonry units. Flues may be square or round in design. Round flues are more effective than the rectangular type, as gases tend to travel up the flue in a spiral motion. The corners of rectangular flues contain dead spaces. Therefore, a rectangular flue must be larger to obtain the same results.

Nothing is gained by using a flue lining that is larger than the size the manufacturer of the heating system recommends for the chimney. In fact, when a flue lining of the correct capacity is used, the chimney performs better and is more economical and safe. Using linings larger than necessary is a common mistake in the building trades. The $8'' \times 12''$ flue lining is recommended for use with the standard heating unit.

Flue linings that curve in the chimney are as efficient as straight linings provided that there are no abrupt turns or completely flat places. The curvature must be smooth, causing no interference with the passage of air or gases leaving the chimney. Nothing should project within the chimney or block the flue. Any offsets of the flue should not exceed 30° from a vertical position. Mortar projecting from the bed joint between the flues should be scraped off cleanly and all holes filled to prevent the heat from penetrating into the chimney walls.

BEGINNING INSTALLATION

The first section of flue lining should be installed at least 1' below the point at which the thimble is to be inserted. The first section should be located approximately 2' from the point at which the joists will rest on the wall. Since the flue liner is 2' high, only one section is installed below the joist level.

A hole must be cut into the flue lining to receive the terra-cotta thimble, or *flue ring,* in order to make a fireproof connection. The stovepipe or furnace pipe fits into the thimble. This is done by first filling and packing the flue lining with sand to prevent it from cracking while the hole is being cut, Figure 25-4.

Fig. 25-4 Packing flue lining tightly with sand prior to cutting.

A circle is drawn on the flue lining exactly where the hole is to be cut. A small hole is cut in the center with a point chisel. Then a small, light brick hammer or tile hammer is used to gently tap the hole through following the line drawn on the flue. The trick is not to cut too much with any one tap. Figure 25-5 shows a mason cutting the hole in the flue lining.

Care should be taken to see that the flue ring does not project any further inside the chimney than the inside face of the flue lining or the draft may be affected.

Fig. 25-5 Cutting hole for thimble with small tile hammer.

Installing the thimble requires good workmanship since it must fit snugly in the flue. It is mortared in to prevent air leakage.

Sometimes, flue linings are started by driving a nail into the chimney wall where the flue starts and setting the flue on the nail. This should never be done on any chimney. The nails could become dislodged, causing the flue to settle in the chimney, thereby shearing off the thimble and cracking the joints. The proper method is to project a header course at the height of the bottom of the flue lining flush with the inside of the flue and rest the flue liner on the headers, Figure 25-6.

INSTALLING WOOD FRAMING AROUND CHIMNEYS

All wood or other flammable materials should be kept at least 2″ from chimneys to prevent any chance of fire. This space may be filled with some type of loose fireproof material to act as a fire stop. As a rule, carpenters on such jobs box the opening around the chimney with framing lumber to maintain the 2″ distance.

The Brick Industry Association states that fireplace and chimney assemblies must be isolated from combustible materials. General requirements incorporated into many buildings are: (1) that all spaces between masonry fireplaces, chimneys, wood and other combustible materials should be firestopped by placing 1 inch (25 mm) of noncombustible materials in such spaces; (2) that in the plane parallel to the front wall of a fireplace, combustible materials should not be placed within 6 inches (150 mm) of a fireplace opening; and (3) that combustible material within 12 inches (300 mm) of the fireplace opening should not project more than $\frac{1}{8}$ inch (3.2 mm) for each inch distance from the opening. This is to protect items such as wood trim or mantels

The BIA recommends the use of non-combustible bat-type fiberglass or mineral wool insulation without any paper lining, available from any building supply dealer, for most situations. They prefer not to use loose particle noncombustible insulation because most building codes require that it be contained by metal strips, which could build up heat. They also recommend that you contact your local building code authority and check their specific requirements before insulating around a chimney or fireplace.

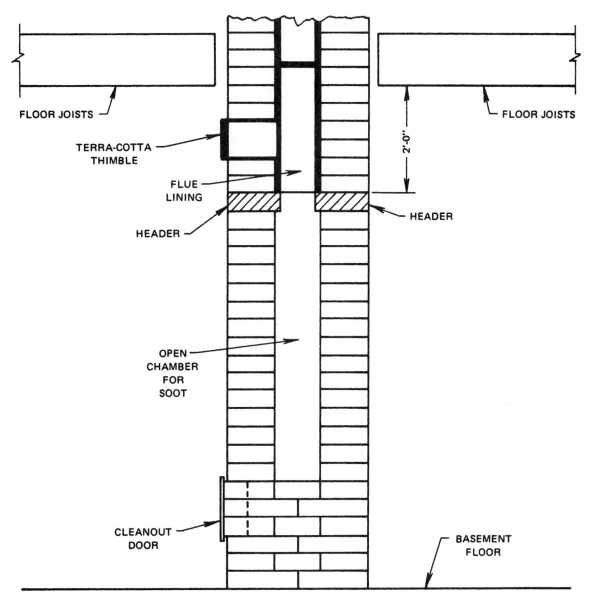

Fig. 25-6 Cross section of chimney showing how the flue lining rests on brick headers.

INSTALLING FLASHING AT THE ROOF

The point at which the chimney passes through the roof must be weatherproofed by the insertion of some type of flashing and counter flashing into the chimney. *Flashing* is a piece of metal such as lead, copper, or alu-

minum that covers the point at which the chimney and roof meet to prevent the penetration of moisture, Figure 25-7. If the mason does not install the flashing, he or she must be sure to rake out the bed joint sufficiently so that the carpenter can install it at a later time. The

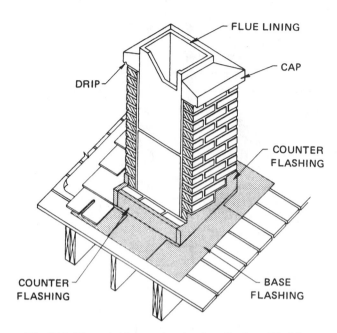

Fig. 25-7 View of chimney showing installation of flashing.

recommended method is to mortar the flashing into the chimney rather than to rake out the mortar bed joint, followed by installation of the flashing.

Note: If the joint must be raked out, it should be to a depth of at least ½″.

CAPPING THE CHIMNEY

The top of all chimneys should be at least 2′ to 3′ above the ridge or peak of the roof to assure a good draft in the flue. The presence of large hills or taller buildings nearby may necessitate building higher than this to prevent downdrafts.

The top or cap of the chimney may be finished in different ways. Various courses of bricks may be projected and recessed to accomplish different patterns and designs. Precast concrete or stone may also be used for caps.

Years ago, the trend was to make the cap of the chimney one of the important architectural features of the home. Very intricate designs were used to accomplish this. The trend now is to keep the cap as simple as possible. The fewer projections that are on the chimney

cap, the less chance of deterioration there is, since snow and water that collect on caps cause the bricks to eventually loosen.

The most frequent method of capping a chimney is to apply a wash of mortar on an angle by approximately 45°, Figure 25-8. The wash is applied from the face of the last course laid to the side of the flue lining with the trowel. A wash coat serves two purposes, to direct air currents upward and to prevent moisture from eroding the chimney top. Regular mortar (Type N) should not be used to apply a wash coat as it is not durable enough for such a project. Instead, a richer portland cement mortar, such as Type M, should be used. The flue lining should project 6″ above the last course of masonry units laid on the chimney.

CHECKING FOR A GOOD DRAFT

When the chimney is completed, the mason should check to make sure that the chimney creates a good draft, Figure 25-9. A simple method of doing this is to roll a piece of paper into a long cylindrical shape and light it with a match. Hold the lighted paper near the

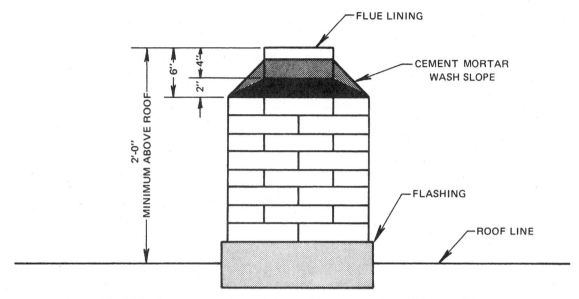

Fig. 25-8 The mortar wash coat helps to direct currents toward the top of the flue and prevents water from entering the chimney top.

flue ring in the chimney. The flame should draw into the ring strongly. It may take a short time for this to happen, as the air in the flue will have to warm slightly. If the flame and smoke do not draw into the ring, it will

Fig. 25-9 Checking chimney draft with a lighted piece of paper. Flame and smoke should be drawn strongly into the thimble and up the chimney.

be necessary either to extend the chimney top or to check for an obstruction in the chimney.

THE IMPORTANCE OF GOOD WORKMANSHIP

Care must be taken to lay bricks and block level and plumb on chimneys. Because of the height and width of the chimney, mistakes are especially noticeable. All mortar joints must be solid and uniform in size. Any work built near wood framing of structures must be free of the wood to prevent the possibility of fire. Tooling and striking of the mortar joints should be done neatly. Excessive smearing of mortar on the face of the work must be avoided.

Flue lining should be cut with a masonry saw, if possible, or packed (reinforced) with sand and cut neatly with a chisel. Broken sections of flue lining should never be used in the chimney.

All materials used in the chimney should be free from dirt and foreign matter. Scaffolding should be removed as soon as possible after the chimney is built to prevent dirt from splashing onto the surface of the chimney.

Note: The chimney is a true test of the mason's skill because the major portion of it is done with the use of the plumb rule.

BUILDING CODE REQUIREMENTS

Building code requirements for residential chimneys can vary from one locality to another. The following information as stated by The Brick Industry Association is accepted nearly everywhere. The requirements state that:

1. Chimney wall thickness should be a nominal 4 inches (100 mm) unless no flue liner is used, in which case a nominal 8 inches (200 mm) is required.
2. Neither chimney nor flue liner may change size or shape within 6 inches (150 mm) of either floor components, ceiling components or rafters.
3. The minimum chimney height for fire safety is the greater of 3 ft (1.0 m) above the highest point where the chimney penetrates the roofline, or 2 ft (600 m) higher than any portion of the structure or adjoining structures within 10 ft (3.0) of the chimney.
4. Chimney clearance from combustible materials is a minimum of 2 in (50 mm), except where the chimney is located outside the structure, in which case 1 in (25 mm) is acceptable
5. The spaces between a chimney and combustible material should be firestopped using a minimum of 1 in (25 mm) thick noncombustible material.
6. All exterior spaces between the chimney and adjacent components should be sealed. This is most commonly accomplished by flashing and caulking.
7. Masonry chimneys should not be corbeled more than 6 in (150 mm) from a wall or foundation nor should a chimney be corbeled from a wall or foundation which is less than 12 in (300 mm) in thickness unless it projects equally on each side of the wall, except that on the second story of two-story dwellings corbeling of chimneys or the exterior of the enclosing walls may equal the wall thickness. Corbeling may not exceed 1 in (25 mm) projection for each course of brick projected.

SEALANTS

Caulking is often used as a means of correcting or hiding poor workmanship, rather than as an integral part of the construction process. It should be detailed and installed with the same care as any other part of the structure. In all cases, a good grade of either polysufide, butyl, or silicone sealant should be used. Oil-based sealants should not be used. Regardless of the sealant used, proper priming and backing rope should be used first.

WOOD STOVE INSTALLATION

The one-flue chimney has been in great demand in the last several years due to the high cost of oil, natural gas, and electric heat. The principal use of one-flue chimneys is to service wood-burning stoves, Figure 25-10. Wood-burning stoves are as safe as any other heating system if sensible safety rules are followed by the homeowner.

Building one-flue chimneys for wood stoves has become a valuable source of work for masons. Therefore, they must be familiar with some basic rules of installa-

Fig. 25-10 A wood stove installed in a one-flue chimney. The brick hearth and wall make this a very safe job. The metal hood causes heat to rise to the floor through the grate.

tion. It is the mason's responsibility to build a fireproof chimney in the home and to make recommendations to the homeowner for safe installation of the stove. Following are some of the more important rules to observe when hooking up a wood stove to a one-flue chimney.

- It is necessary to have a building permit and inspection to build a chimney and install a wood stove. This is governed by local authorities and codes.
- All masonry chimneys with walls less than 8″ thick should have a flue lining. Some areas of the country require a flue lining regardless of thickness. Check local codes.
- Cleanout doors should be provided to remove soot. They should be located approximately 2′ below the inlet (thimble). The door will last longer if it is cast iron. Cleanout doors are available at building trades supply houses.
- An exterior chimney may touch the house siding or sheathing if 8″ of solid masonry is provided on the face between the flue lining and house.
- Masonry chimneys with flue linings built on the exterior of the house must be placed a minimum of 1″ from any combustible materials.
- The size of the thimble (connector) being installed should not be smaller than the stove flue collar as heat may build up at the point where the connector passes through the house wall.
- The terra-cotta thimble, when installed, should be surrounded by no less than 8″ of solid brickwork or fireproof materials. Check local codes for the specific measurement in your area.
- The thimble should be installed flush with the inside face of the flue lining and mortared in with fire clay or high-strength mortar to prevent burning out.
- The thimble should have an elevated pitch where it meets the chimney. The general rule is that there should be a horizontal position of not less than $\frac{1}{4}$″ to the linear foot, so that the chimney connection point is higher than the stove connection point.
- Flue linings are available in either modular, nonmodular, or both sizes depending upon the availability in certain areas of the country.

Modular flue lining is manufactured on multiples of the 4″ grid (module), which is the common unit of measurement used in the construction industry. Nonmodular flue lining is not manufactured to the 4″ grid and is a carryover from the days when the modular system was not used as a construction industry standard. Today there are still several companies that make the nonmodular-size flues. An example of a nonmodular-size flue lining is 9″ × 13″, and a modular-size flue lining of approximately the same size is the 8″ × 12″ flue.

When determining what size flue lining to use in a chimney for a stove, the cross-sectional area of the flue lining should be approximately 25% more than the cross-sectional area of the thimble (connector). This will promote good draft and will decrease the creosote buildup in the flue. This rule will work for open-face or close-face stoves.

For example, a stove is being connected to a chimney in which the stove outlet the stove pipe fits on is 6″ in diameter. Multiplying 6″ × 6″ equals 36″. Therefore, an 8″ × 8″ flue lining is needed because 8″ × 8″ equals 64″. Subtracting 36″ from 64″ equals a total of 28″, which is easily more than the 25% required for good draft. There is nothing to be gained by installing a flue lining that is much more than the 25%. This will only cause most of the heat to go up the chimney and be wasted. However, the closest possible flue lining size that will work out should be selected. Many times it is not possible to match the size perfectly. When this happens, the next larger size should be used. The chimney should also be built higher than the standard 2′ above the peak of the roof. It is good practice when building a chimney to look over the area and take note of other houses. If they have a higher than normal chimney or extension, then you will need one also.

Modular and nonmodular flue linings are available in a wide variety of sizes to fit building needs. Check with local building supply dealers in your area to find out what is available.

- The chimney should be tied to the building with some type of approved wall ties as stated in local building codes.
- The top of the chimney should be a minimum of 2′ above the peak of the roof for a good draft.

- Solid brick masonry walls cannot burn but will conduct heat, so combustible materials should be kept out of contact with masonry walls. It is recommended to set the stove on firebrick or a nonburning masonry base at least 18″ from the masonry or noncombustible wall. Stovepipe should always be a minimum of 24″ from the ceiling.

- The masonry base or hearth should extend a minimum of 12″ on each side of the stove and 16″ in front of the stove to protect the floor from hot embers. If the area is carpeted, allow 16″ around the stove.

- There is a simple test to see if there is enough clearance for a hot stove. Place your hand on the surface closest to the stove. If you can keep your hand there comfortably, the location probably is acceptable.

- Before hooking up a stove on existing masonry chimneys with flue linings, a visual inspection should be performed to be sure that the chimney is not blocked. A strong light held in the chimney area or a large mirror reflected down the chimney from the top usually provides enough light for a check.

A smoke test indicates the workability of an existing chimney. This is performed by lighting a small, smokey fire or setting off a smoke bomb in a connected stove or fireplace, and then partially closing off the top of the chimney. If any leaks or cracks are in the chimney, the mason will see smoke coming out of them. Repairs can be made based on the results of the test. Stains on masonry or loose mortar joints must also be repaired. An old chimney should be inspected carefully before ever attempting to install a wood stove.

CHIMNEYS BUILT TO SERVE APPLIANCES

Unlike fireplace and wood stove chimneys that should have a separate or individual flue for each heat source,

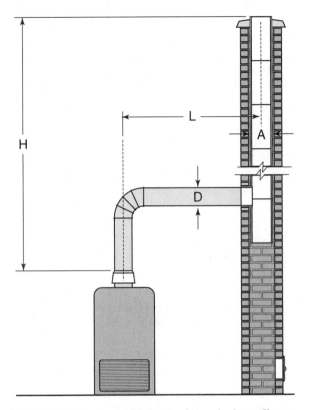

Fig. 25-11 Masonry chimney serving a single appliance. (Courtesy Brick Industry Association) (See Table 2)

an appliance chimney can serve more than one heat source if it is designed and built correctly, because it utilizes a forced draft. According to the Brick Industry Association, appliance chimneys are divided into two types: those venting one appliance, see Figure 25-11, and those venting two or more appliances, see Figure 25-13. The two variables that must be known to the designers are the input rating and configuration of the system. Typical design criteria are shown in Figure 25-12 (Table 2) and Figure 25-14 (Table 3). Since both fireplace and appliance chimneys have identical functions, their construction methods and materials are also identical.

**Capacity of Masonry Chimneys
Serving a Single Appliance**[a]
(Combined Appliance Input Rating,
Thousands of Btu/h)

Height H, Feet	Lateral L, Feet	Minimum Internal Area of Chimney, A, *Square Feet*				
		0.26	0.35	0.47	0.66	0.92
6	2	130	180	247	400	580
	5	118	164	230	375	560
8	2	145	197	265	445	650
	5	133	182	246	422	638
	10	123	169	233	400	598
10	2	161	220	297	490	722
	5	147	203	276	465	710
	10	137	189	261	441	665
	15	125	175	246	421	634
15	2	178	249	335	560	840
	5	163	230	312	531	825
	10	151	214	294	504	774
	15	138	198	278	481	738
	20	128	184	261	459	706
20	2	200	273	374	625	950
	5	183	252	348	594	930
	10	170	235	330	562	875
	15	156	217	311	536	835
	20	144	202	292	510	800
30	2	215	302	420	715	1110
	5	196	279	391	680	1090
	10	182	260	370	644	1020
	15	168	240	349	615	975
	20	155	223	327	585	932
	30	NR[b]	182	281	544	865
50	2	250	350	475	810	1240
	5	228	321	442	770	1220
	10	212	301	420	728	1140
	15	195	278	395	695	1090
	20	180	258	370	660	1040
	30	NR[b]	NR[b]	318	610	970
		6	7	8	10	21
		Single-Wall Vent Connector Diameter, D, Inches				

[a]SI conversions: W = Btu/h × 0.293; m = ft × 0.3048; mm = in. × 25.4; mm² = in.² × 645

[b]Not recommended.

Fig. 25-12 Table 2. Capacity of masonry chimneys serving a single appliance. (Courtesy Brick Industry Association)

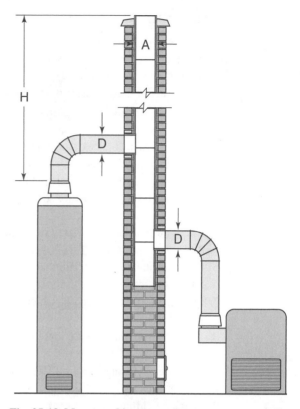

Fig. 25-13 Masonry chimney serving two or more appliances. (Courtesy Brick Industry Association) (See Table 3)

**Capacity of Masonry Chimneys
Serving Two or More Appliances**[a]
(Combined Appliance Input Rating,
Thousands of Btu/h)

Total Vent Height H, Feet	Minimum Internal Area of Chimney, A, Square Feet			
	0.26	0.35	0.54	0.79
6	102	142	245	NR
8	118	162	277	405
10	129	175	300	450
15	150	210	360	540
20	170	240	415	640
30	195	275	490	740
50	NR[b]	325	600	910

[a]SI conversions: W = Btu/h × 0.293; m = ft × 0.3048; mm = in. × 25.4

[b]Not recommended.

Fig. 25-14 Table 3. Capacity of a masonry chimney serving two or more appliances. (Courtesy Brick Industry Association)

ACHIEVEMENT REVIEW

The column on the left contains a statement associated with the construction of chimneys. The column on the right lists terms. Select the correct term from the right-hand list and match it with the proper statement on the left.

1. Raw material used to make flue linings
2. Passage of air through the chimney
3. Piece of metal such as lead, tin, or copper which is laid in the bed joint of the chimney at the roof height to prevent moisture penetration
4. Mortar applied to the top of a chimney at a 45° angle to prevent moisture from entering the chimney
5. Fireproof terra-cotta flue ring into which furnace pipe fits
6. Object installed in the chimney through which soot is removed
7. Safe depth in the ground at which the footing is poured
8. Distance from house (exterior) to chimney if flue lining is used in chimney
9. Distance from masonry wall to back of wood stove inside house
10. Height of chimney above peak of roof

a. Cleanout door
b. Wash
c. Frost line
d. Draft
e. Flue lining
f. Flashing
g. Thimble
h. Fire clay
i. 2′
j. Hood
k. 1″
l. 18″

MATH CHECKPOINT

1. You are going to build a one-flue chimney. Each course will have 11 bricks per course. How many bricks will be needed to build the chimney 24′ high?
2. Each flue lining is 2′ in length. The first 4′ of the chimney does not require any flue lining, as it is near the base and below the heat intake. Allowing for this, how many pieces of flue lining will be needed for the job?
3. How many bags of masonry cement are needed? Round off your answer to the nearest higher bag.
4. How much sand will be needed to build the chimney? Allow $\frac{1}{4}$ ton for waste on the ground. Sand costs $20.00 a ton.
5. If the cost of each brick is $.28, how much will the brick cost for the job?
6. If the cost of the masonry cement is $4.75 per bag, how much will the masonry cement cost? Do not figure any sales tax on these costs.
7. If the cost of each 9″ × 24″ flue lining is $5.50, what is the total cost of the flue lining needed for the job? Do not allow sales tax on this.
8. What is the total cost of all materials for the job?
9. Add 5% sales tax to the above figure. What is the total cost for the job?

10. The mason is earning $15.00 per hour and the mason's helper $8.50 per hour. If it takes 24 working hours for each person to build the chimney, what is the total cost of the labor for the job?

PROJECT 23: BUILDING A ONE-FLUE CHIMNEY

OBJECTIVE

- The student will lay out and build a one-flue chimney of bricks. The student will also install various parts of the chimney, such as cleanout door, thimble, and flue lining, Figure 25-15.

Note: Chimneys such as this are built to service furnaces and small heating plants.

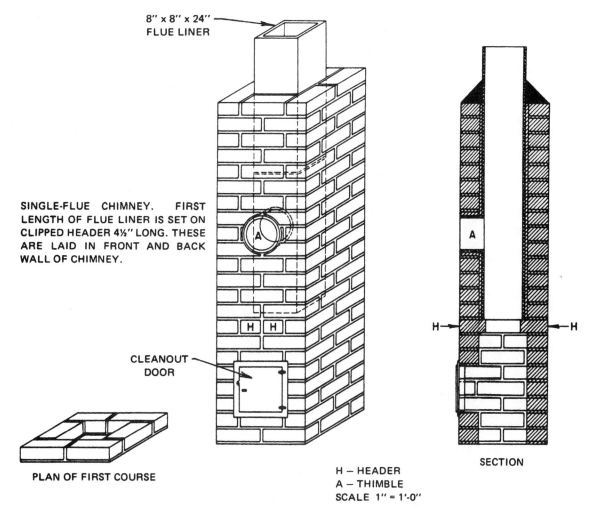

8" x 8" x 24" FLUE LINER

SINGLE-FLUE CHIMNEY. FIRST LENGTH OF FLUE LINER IS SET ON CLIPPED HEADER 4½" LONG. THESE ARE LAID IN FRONT AND BACK WALL OF CHIMNEY.

CLEANOUT DOOR

PLAN OF FIRST COURSE

H — HEADER
A — THIMBLE
SCALE 1" = 1'-0"

SECTION

Fig. 25-15 Project 23: Building a one-flue chimney.

EQUIPMENT, TOOLS, AND SUPPLIES

Mortar pan or board

Mixing tools

Mason's trowel

Brick hammer

Plumb rule

Striking tool (convex or V-jointer)

2′ square and pencil

Modular rule

Brush

Point chisel

1 cleanout door

1 terra-cotta thimble

2 sections of $8'' \times 8'' \times 24''$ flue lining (Because of the fragility of terra-cotta flue liners, a wooden model may be used in a shop situation.)

The student will estimate bricks and mortar from the plan. (For training purposes, it is suggested that only lime and sand mortar be used for the work at the top of the chimney, not portland cement.)

SUGGESTIONS

- Due to variance in bricks and flue liners, it is possible that the flue liner will not fit into the chimney as shown. In this event, increase the size of the chimney by one-half brick in each direction. On some flue linings, the $8'' \times 8''$ dimension is taken from the inside rather than the outside. This would require extremely large head joints in the chimney to accommodate the flue liners inside the brickwork. These points must be determined when the chimney is laid out.
- The space between the liners and brickwork should not be filled with mortar or droppings. Only enough mortar should be used to form a good joint and to hold the liners in a level and plumb position. The proper air space allows the flue to expand without cracking the brickwork.
- Use well-filled mortar joints at all times.
- Be careful when handling flue linings as they have sharp edges.
- If flue linings must be cut, do not do the cutting around other workers because the cutting process can result in flying chips. Wear eye protection whenever cutting flue linings.

PROCEDURE

1. Mix the mortar and stock the work area with the needed materials.
2. Dry bond the first course to obtain the correct spacing.
3. Lay the first course in mortar. Level and square all sides. Be sure to check the inside space to determine if the flue lining will fit, allowing room for expansion. Bricks will be laid to the number 6 on the modular rule.
4. Lay the second course, installing the cleanout door as shown. Anchor the frame to the brickwork using a metal clip nail or tie.
5. Lay 5 more courses, walling in the cleanout door. Recheck the door to be sure that it is level and plumb. Be sure that each course is level and plumb as it is laid.

6. Cut 4 long bats, each measuring $4\frac{1}{2}''$ in length. Lay these in mortar on the eighth course, allowing the rough end to project inside the flue space. The projecting edge provides a bearing on which the first flue lining rests. (See plan.)

7. Strike joints as needed.

8. Lay 2 or 3 additional courses to anchor the header course.

9. Before installing the first flue liner, carry it to the sand pile and pack it tightly with sand. Mark the point at which the thimble will be cut in the flue. With a steel-pointed chisel, carefully cut a hole in the center of the marked area. Enlarge the hole with the head of the brick hammer by chipping until the thimble fits neatly. Remove the sand from the flue liner and install it in the chimney as shown on the plan.

10. Lay the brickwork, installing the thimble at the required height. (See plan.) Solidly point up around the thimble and flue lining.

11. Continue laying brickwork until the top of the lining is reached. Set the second section of the flue liner, being sure it is level and plumb. Cut off any projecting mortar on the inside of the flue lining.

12. Build the chimney to the height shown on the plans. Strike remaining joints and brush the work.

13. Apply a wash coat of mortar on a slope. It should be located on top of the last course against the flue liner. Smooth the work with the trowel.

UNIT 26
Movement Joints, Intersecting Walls, and Uses of Scale Rules

—————————————— OBJECTIVES ——————————————

After studying this unit, the student will be able to

■ mark courses of various masonry units with the modular rule and the spacing rule.

■ explain the term *hog in the wall* and what steps must be taken to correct it.

■ describe an expansion joint and explain how it is installed in a masonry wall.

■ lay out and construct a concrete block intersecting wall.

THE MASON'S RULES

All masonry units in a wall must be laid to a predetermined, specified height. Close attention must be given to the height of such items as windows, doors, beam pockets, and joist seats. Since mortar joints and unit size may vary, the mason uses two different rules to be certain that mortar joints are divided equally under any circumstances. The two principal rules with which all masons must acquaint themselves are the *spacing rule* and the *modular rule*.

Building a wall to the same height on both ends of the building, but ending with one more course on one end than the other, is a very costly error. With the correct starting point (bench mark) established, proper use of the mason's rules and story pole, and adherence to rules of construction, the mason can be assured that the work will be constructed to the correct level.

The Modular Rule

The modular rule, Figure 26-1, has a standard 72″ marked on one side. The reverse side of the rule has various numbers to match the different sizes of masonry units which are manufactured under the modular manufacturing system. These masonry units may be

reviewed by referring to the section in Unit 1 on modular bricks.

It is important to remember that all modular units are bonded together at a 16″ increment. This height may be expressed in terms of a multiple of 16″, such as 32″, 48″, or 64″. Examine the various scales on the rule in Figure 26-1. The last scale, scale 2, accommodates either a concrete block or a masonry unit which equals two courses reaching 16″ in height. Concrete block is usually laid off in increments of 8″ instead of a set number on the scale.

Reading the modular rule requires no learning of new techniques, Figure 26-2. All that is necessary is a basic understanding of how the modular system coordinates with building materials and how masonry units fit into the 4″ modular grid.

The Spacing Rule

The spacing rule was designed before the modular system was adopted. Its purpose at that time was to divide mortar joints evenly in brickwork. It was also used to lay out courses of units since frames and doors at that time were not designed and manufactured in a standard size. The spacing rule is still used today in gauging mortar joints that are not modular.

300

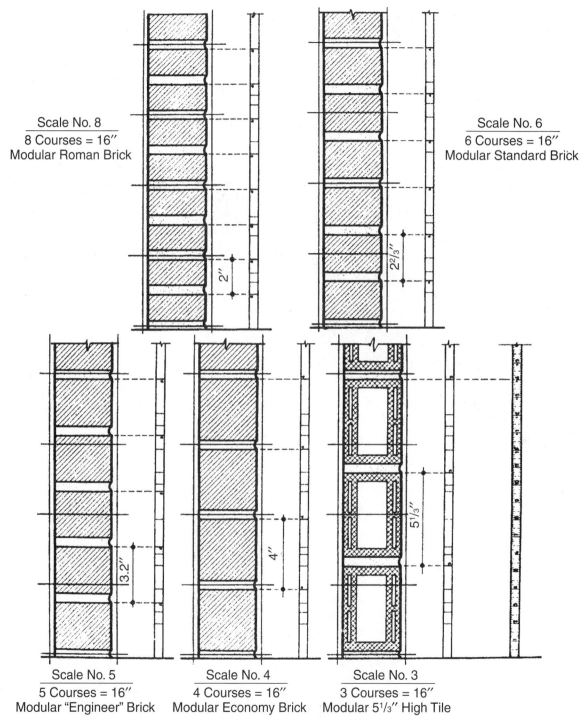

Scale No. 8
8 Courses = 16″
Modular Roman Brick

Scale No. 6
6 Courses = 16″
Modular Standard Brick

Scale No. 5
5 Courses = 16″
Modular "Engineer" Brick

Scale No. 4
4 Courses = 16″
Modular Economy Brick

Scale No. 3
3 Courses = 16″
Modular 5⅓″ High Tile

Fig. 26-1 Various scales of the modular rule applied to masonry units based on
the modular system of measurement.

Fig. 26-2 Brick wall laid to number six on the modular scale rule. Notice course number six includes mortar between joints.

One side of the spacing rule is marked off in increments of $\frac{1}{16}''$ and is read as an ordinary 6' rule. The other side of the spacing rule is marked off in several groups of numbers from 1 to 0, Figure 26-3. (The 0 actually represents 10.) As the mortar bed joint increases in size, so does the number on the rule representing that course of bricks. The smaller the number, the tighter or thinner the mortar joint is. For example, number 6 on the rule equals 4 courses of bricks and 4 mortar joints which will measure $11''$ in height. Number 7 on the spacing rule represents 4 courses of bricks laid to $11\frac{1}{4}''$ in height. By using various numbers, the mortar joint size can be increased or decreased as needed.

Another important feature of the spacing rule is that the total number of courses shown on the scale side from the bottom of the rule to the top is shown in red. The height of any wall can be checked and, at the same time, the number of courses needed to reach the specific height is known. For example, to lay a standard brick wall to the height of $71\frac{1}{2}''$ using the number 6 on the spacing rule, 26 courses of bricks including the mortar joint are required. Do not confuse number 6 on the spacing rule with number 6 on the modular rule, as they represent entirely different measurements.

Compare application of the two rules in this example involving installation of a brick windowsill. The bricks are laid in a rowlock position, but the spacing is figured the same as if the course were in a vertical position. The brick sill is to be laid for a window $36''$ long.

Checking the distance with the modular rule, the space coincides with the number 4, which will not fit into the modular grid. On the other hand, using the spacing rule, the number 6 is only $\frac{1}{4}''$ from coinciding perfectly with the $36''$ mark. Therefore, the number 6 on the spacing rule should be used to gauge the distance, gaining the $\frac{1}{4}''$ in the mortar joints. Compare the two rules in Figure 26-4, in which $36''$ is marked.

The mason will seldom use a scale larger than the number 7 on the spacing rule since the thickness of the mortar joint would be too great in most cases. By the same token, the number 4 on the rule generally forms the tightest joint possible. Regardless of which rule is being used, always read from the top of one brick to the top of the next brick.

Hog in the Wall

The term *hog in the wall* is a trade term that indicates that two opposite corners or leads have reached the same height but do not contain the same number of courses, Figure 26-5. This usually occurs when the story pole being used does not have each course numbered from the bottom of the pole to the top. A hog in the wall can be caused by the following:

- A footing that is not level
- A very long wall between doors or windows
- Structures in which the wall is built over a series of concrete footers
- One mason laying a thicker bed joint under the bricks than another mason is using

Correcting this situation is extremely difficult, especially if the wall has been partially constructed.

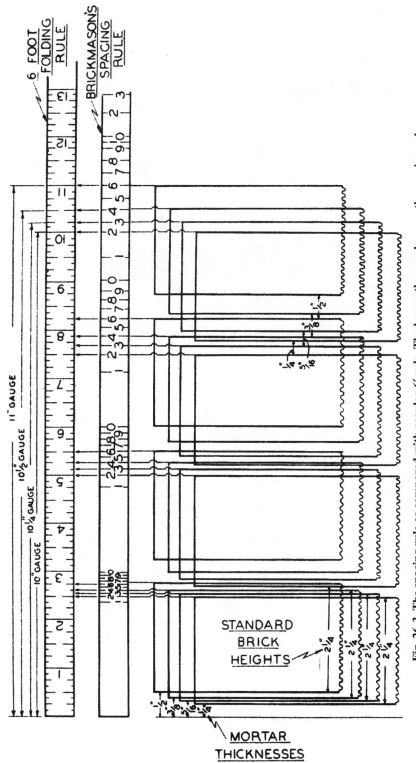

Fig. 26-3 The spacing rule as compared with regular 6' rule. The larger the numbers on the spacing rule are, the bigger the mortar joint being formed is.

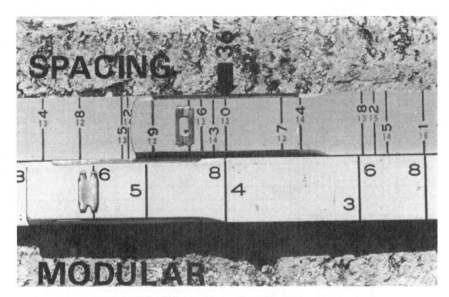

Fig. 26-4 Comparing the spacing and modular rules. The arrow indicates the 36″ measurement on both rules. Notice the difference in size of the two rules.

Sometimes it can be corrected by increasing the mortar joint thickness at one end and decreasing the joint thickness at the opposite end of the wall. This action equalizes the number of courses but usually can be detected in the wall, as there are bigger joints on one end than on the other. This is not considered good building practice and should be avoided. Usually, the best method is to tear down the portion of the wall containing the hog and rebuild it correctly. This is a costly procedure since the work must be torn down and rebuilt, and there are no materials that can be reused. It also delays completion of the job.

Note: It is recommended that all measurements be checked with a story pole or gauge rod from the bench mark when a job is started, and that the correct number of courses are laid to the specified height when the corner is completed.

A hog in the wall is associated with the laying of bricks rather than concrete block because of the size of the units. It is possible, however, to have a hog in a concrete block wall if the footings are extremely unlevel.

Movement Joints In Masonry Buildings

Structures are constantly moving. The movement may be caused by the settling of the building. However, it is more often caused by expansion and contraction of the different building materials from temperature and moisture changes. This movement may be small by comparison, but if no allowances are made by the designers, it will likely cause cracks at some points in the masonry walls. According to the Brick Industry Association (BIA), the technical authority on brick masonry in the United States, there are four main types or classifications of movement joints generally designed and used in buildings. For many years most movement joints were mistakenly classified by many builders and tradesmen as simply expansion joints, when in reality expansion joints are only one of the movement joints in a typical building. The four main types of movement joints are expansion joints (which are by far the most common ones a mason comes in contact with), control joints, building expansion (isolation) joints, and construction joints (cold joints). Each type of these movement joints is designed to perform a specific task and

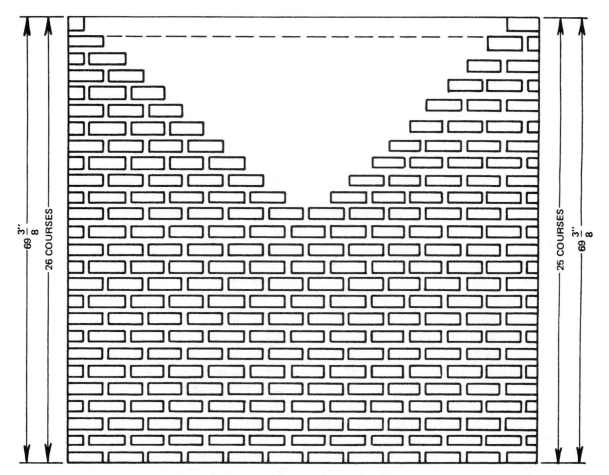

Fig. 26-5 Brick wall demonstrating a "hog in the wall."

they should not be used interchangeably. It is important to remember that, in general, concrete and concrete masonry shrink, whereas brick and brick masonry expand.

An *expansion joint* is used to separate brick masonry into segments and to allow a controlled minor movement of the masonry to prevent cracking due to changes in temperature, moisture expansion, elastic bending or flexing due to loads imposed on them, and creeping, which is defined as a slight movement that occurs when materials are under a heavy load or stress. Creeping occurs more in concrete work than brickwork and then only in the mortar joints to a very minor extent. It is

also important to remember that expansion joints can be either horizontal or vertical. The joints are formed of highly elastic materials that are placed in a continuous, unobstructed opening through the brick wall. This allows the joint to close as a result of an increase in size of the brickwork. For example, the BIA has determined in past studies that a brick wall 100′ long expands or contracts approximately 7/16″ for every 100 degrees of change in temperature. The design of expansion joints is the responsibility of the architect and structural engineers and should never weaken the structural strength of the building. The skill of the mason plays a very important role in the installation of a successful expansion joint. If the joint is clogged with mortar or foreign matter, it most likely will not function properly. Figure 26-6 shows typical expansion joints located in various positions.

A good example of a horizontal expansion joint is when brickwork is supported on steel angle irons that are bolted to the main structure. Horizontal expansion joints are needed to provide space beneath the angle for movement to occur. High-rise frame structures typically have horizontal expansion joints located at every floor level or every other floor level. Figure 26-7 shows a typical detail of a horizontal expansion joint on a brick veneer building. Notice that a clear space or highly compressible material is placed beneath the angle and that a backer rod and sealant are used at the toe of the angle to seal the joint.

A *control joint* is used in concrete masonry work to create a plane of weakness which, in conjunction with joint reinforcement, "controls" the location of cracks due to dimensional changes resulting from the shrinkage and creep properties of concrete masonry. A control joint is usually vertical in nature. A control joint is designed to control and regulate the development of high stress in a building at a given point. Reinforcement should not be allowed to bridge a control joint or it will not work as designed. The width of a control joint is usually not critical. The placement and spacing of control joints are determined by the type of joint materials used, location of the wall, number of openings, height of the wall, and the spacing of the joint reinforcement or bond beams, Figure 26-8.

There is an excellent rubber control joint material, Figure 26-9, which is sometimes installed in the slot on the end of adjoining concrete window sash blocks in a concrete block wall. This makes a very neat job. Neoprene compound flanges act as a seal to the outside edge of the block face, making caulking unnecessary. Copper water stops have also been used very successfully for many years. They consist of short pieces of copper sheets that overlap at the joint. Pre-molded compressible (rubber or plastic) fillers may be inserted in back of the copper to seal the joint.

A *building expansion (isolation) joint* is used to separate a building into sections so that stress that builds up in one section will not affect the integrity of the entire structure. The isolation joint is a complete through-the-building joint.

A *construction joint (cold joint)* is used primarily in concrete construction in which construction work is interrupted. Construction joints are located where they will least impair the strength of the concrete structure.

INTERSECTING WALLS

On many jobs, there is a need to build intersecting walls that branch off the main wall at a 90° angle. This can be accomplished in two different ways. The intersecting wall may be tied into the main wall by bonding the masonry unit into a hole that has been left in the main wall, Figure 26-10. Another method is to butt one wall against the other and use metal ties to bond the two walls together.

The first method mentioned, which makes use of a physical tie, was used for many years as the only method of tying walls together. In this method, one masonry unit is usually left out at the point at which the walls are to be tied together. At a later time, the intersecting wall is built. Every other course ties into the various holes to strengthen the wall.

If there is a settlement or movement between the walls, a crack may develop at the point of intersection or the intersecting masonry unit may break completely. Because of this, the other method mentioned was developed. In this method, the intersecting wall is butted against the main wall with a mortar joint at the point of contact. This joint is usually called a *stack joint*. The mortar is then raked approximately $\frac{3}{8}''$ and filled with a caulking compound until it is flush with the unit. When a crack occurs, all that is necessary is recaulking of the joint.

Intersecting walls that carry the load of floors and roofs of buildings must also be able to stretch or flex. The wall, however, must be tied together very securely for lateral support. Heavy metal tie bars are a very good method of tying these walls together, Figure 26-11. The tie bars are usually heavy-gauge steel, which is $\frac{1}{4}''$ thick, $1\frac{1}{2}''$ wide, and approximately $28''$ long. They usually have $2''$ right-angle bends on each end to lock into the masonry unit. The end of the tie bar should be

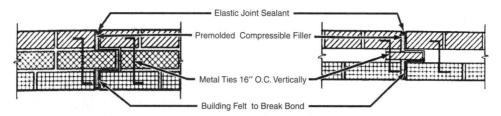

Expansion Joints in Straight Walls

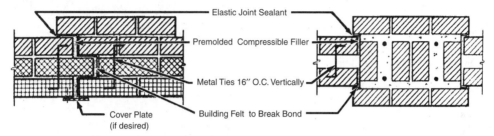

Expansion Joints at Pilasters

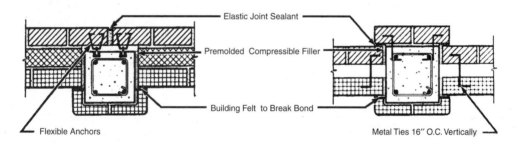

Expansion Joints at Concealed Column

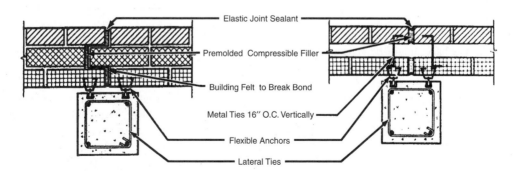

Expansion Joints at Exposed Column

Fig. 26-6 Expansion joints located in various positions.

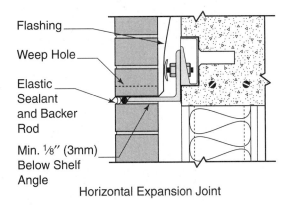

Horizontal Expansion Joint

Fig. 26-7 Horizontal expansion joint on a brick veneer building. (Courtesy Brick Industry Association)

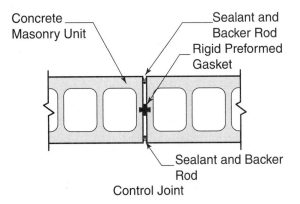

Control Joint

Fig. 26-8 Typical control joint in a concrete masonry unit wall. (Courtesy Brick Industry Association)

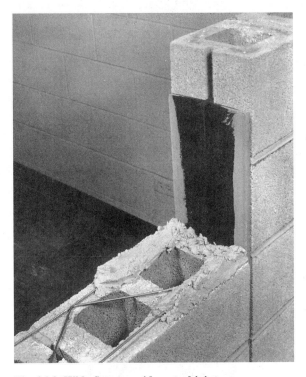

Fig. 26-9 Wide flange rapid control joint.

filled completely with mortar before laying the next course over it. Tie bars should be spaced vertically every 24″.

Metal reinforcement wire also can be used to tie intersecting walls. The best type is the reinforcement wire that is made especially for this purpose. It is welded in the shape of a T and is the same diameter as standard reinforcing wire, Figure 26-12.

Intersecting walls that do not bear loads, such as partitions, can be tied together with strips of metal lath or hardware cloth, Figure 26-13. This is ¼″ wire mesh, which is galvanized, and is placed across the joint between the two walls. When one wall is built first, the

hardware cloth is allowed to extend at least 8″. Metal lath or hardware cloth should be placed in the wall every other course for a secure tie. When using hardware cloth, it is recommended that a caulking joint be applied against the point at which the two walls meet.

> **Caution:** Always fold the hardware cloth down flat against the wall to avoid the possibility of someone being injured by the wire.

Later the exposed portion of the wire is tied onto the mortar joint of the wall being built (intersecting wall) and mortared in securely. Wire mesh can be cut with ordinary snippers.

All tying methods discussed in this unit are effective when used properly and may be used with any type of masonry unit. At no time should an intersecting wall be built without some type of tie. All ties should be mortared in securely.

Fig. 26-10 Hole left in concrete block wall to receive an intersecting wall. The tie used is known as a physical tie.

Fig. 26-12 T-shaped metal reinforcement wire used to tie intersecting concrete block walls together.

Fig. 26-11 Section of metal tie bar used to tie a concrete block bearing wall to the main wall.

Fig. 26-13 Intersecting block walls tied together by use of metal lath. Notice that the mortar joint has been raked out to receive caulking compound.

ACHIEVEMENT REVIEW

Select the best answer from the choices offered to complete each statement. List your choice by letter identification.

1. The term *hog in the wall* describes
   a. a wall which is not level.
   b. a wall which has a bulge in the brickwork.
   c. a wall which is level but contains one more course on one end than the other.
   d. a wall which is very poorly built and on which the mortar joints are not struck properly.

2. Scale number 8 on the modular rule represents
   a. standard brick courses.
   b. tile courses.
   c. concrete block courses.
   d. Roman brick courses.

3. All courses of the various masonry materials found on the modular rule tie together every
   a. 8".
   b. 12".
   c. 16".
   d. 24".

4. The mason's spacing rule differs from the modular rule in that it
   a. has more listings of masonry units on the rule.
   b. allows the adjustment of standard brick by the application of a larger mortar joint or a thinner mortar joint.
   c. is much easier for the mason to use and understand.
   d. is used only to mark off doors and windows.

5. The spacing rule offers the added advantage of
   a. larger numbers, which make the rule easier to read.
   b. wood construction, so that it does not rust or deteriorate.
   c. courses numbered in red from the bottom of the rule to the top.
   d. the fact that it will fold and fit into the mason's pocket.

6. The major reason that the expansion joint does not function correctly is that
   a. masonry debris sometimes remains in the joint.
   b. a copper water stop is selected instead of a rubber strip for filler material.
   c. it is not installed at the proper location in the wall.
   d. it is not built perfectly plumb with the wall.

7. Fiberboard is a poor choice as a filler for an expansion joint because
   a. it is not thick enough to fill the joint.
   b. when it compresses, it will not return to its original size.
   c. it absorbs moisture and resists movement of the wall.
   d. it is very costly and difficult to install in the wall.

8. When an intersecting bearing wall is being built, the strongest type of metal wall tie to install is
   a. the reinforced masonry T wire.
   b. hardware cloth.
   c. the veneer tie.
   d. the T bar.

9. When building intersecting walls that specify a physical tie, it is recommended that
   a. the walls be tied together by interlocking the masonry units of the intersecting wall into the main wall.

    b. all mortar joints which butt against the main wall be pointed with a high-strength mortar.

    c. the mortar joint against the main wall be raked and filled with caulking compound.

    d. no joint be formed where the intersecting wall and main wall meet.

10. When raking out a mortar joint at the intersection of a wall to receive caulking, the correct depth is

    a. $\frac{1}{8}''$.   c. $\frac{3}{4}''$.

    b. $\frac{3}{8}''$.   d. $1''$.

MATH CHECKPOINT

1. How many courses of standard brick are there in a wall built $64''$ high?

2. How many courses of Roman brick are there in a wall built $48''$ high?

3. How many rowlock bricks, including the mortar head joint, would it take for a brick sill of $40''$? (Use number six on modular rule for spacing.)

4. How many courses of modular $5\frac{1}{3}''$ tile are needed for a height of $48''$?

5. You are building a wall in which a vent is to be installed. The height of the vent is $45\frac{3}{4}''$, which is not a modular measurement. What scale number would you select on the spacing rule to work this out in full courses? How many courses of brick would it take?

6. How many courses of concrete block would it take to build a wall 8' high? Which scale number would you use on the modular rule to lay them off?

7. How many courses of standard brick laid to scale number 6 on the modular rule will it take to equal 8 feet in height?

8. How many courses of standard brick laid to no. 5 on the course counter spacing rule will it take to equal $56\frac{7}{16}''$?

9. Using scale number 5 on the modular rule, how many courses of brick will it take to equal 4 feet in height?

10. How many courses of regular concrete block will it take to back up a brick wall 6 ft. high?

PROJECT 24: BUILDING AN $8'' \times 8'' \times 16''$ CONCRETE BLOCK WALL WITH A ROWLOCK BRICK WINDOWSILL

OBJECTIVE

- The student will be able to lay an $8'' \times 8'' \times 16''$ concrete block wall 5 courses high and install a brick rowlock sill. The student will use the modular rule when constructing the sill, Figure 26-14.

  Note: Windowsills such as this are typical of windowsills on wooden or metal windows in a concrete block garage.

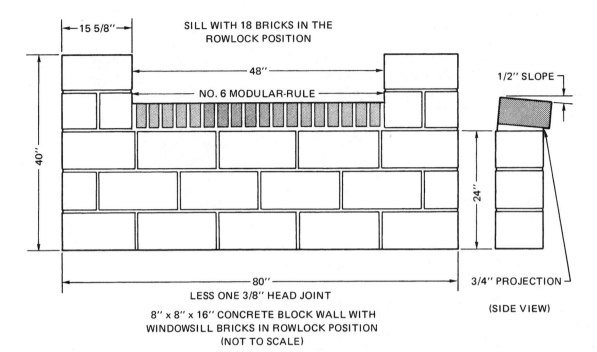

SILL WITH 18 BRICKS IN THE
ROWLOCK POSITION

15 5/8"

48"

NO. 6 MODULAR-RULE

40"

24"

1/2" SLOPE

80"
LESS ONE 3/8" HEAD JOINT

3/4" PROJECTION

(SIDE VIEW)

8" x 8" x 16" CONCRETE BLOCK WALL WITH
WINDOWSILL BRICKS IN ROWLOCK POSITION
(NOT TO SCALE)

Fig. 26-14 Project 24: Building an 8″ × 8″ × 16″ concrete block wall with a rowlock brick windowsill.

EQUIPMENT, TOOLS, AND SUPPLIES

1 mortar pan or board	Modular rule
Mixing tools	Chalk box
Mason's trowel	Brick set chisel
Brick hammer	Brush
Ball of nylon line	Supply of clean water
2 line blocks or line pin and nail	1 piece of wire mesh or hardware cloth approximately 48″ long and 4½″ wide
Plumb rule	
Convex striker	
Flat slicker jointer	

The student will estimate the required number of bricks, block, lime (masonry cement), and sand according to the plans.

SUGGESTIONS

- The concrete block wall should be completed according to the plans before any of the sill bricks are laid.
- Holes in the cells of the concrete block directly underneath the point at which the sill is to be laid should be filled with masonry scraps, or wire mesh should be laid across the holes to assure a solid joint under the sill.

- Select dry bricks for the sill, as they set more quickly.
- Select bricks of approximately the same height, width, and length for the sill.
- Spot check the brick sill periodically with the modular rule as it is laid to be sure that no brick pieces will have to be used.
- Level the back of the brick sill every 3 bricks to be sure that it is level.
- The last brick laid in the sill will require a buttered mortar joint on each side of the brick and the adjoining bricks. Be sure to allow one extra joint for the closure brick when the sill is checked with the modular rule.

PROCEDURE

1. Stock the necessary materials in the work area and mix the mortar.
2. Strike a chalk line and lay out the first course as shown on the plans.
3. Lay the concrete block wall to a height of 8″ per course or to number 2 on the modular rule.
4. Make all concrete half-block with the brick saw if manufactured halves are not available. Wear eye protection.
5. Lay the concrete block wall up to 24″ or 3 courses high. Lay out the brick sill with the rule according to the plan.
6. Using square-ended block for the jamb, continue building the block wall to the height shown on the plan.
7. Strike all block work with the convex striker and brush before installing the sill.
8. Fill in any holes in the block underneath the sill with masonry scraps, hardware cloth, and mortar.
9. Butter a mortar joint on the side of a brick and lay it in the rowlock position against the jamb of the opening. Slope the brick to the front of the wall a minimum of ½″ to drain water. Project the bottom of the sill bricks ¾″ from the outside face of the wall. Set a brick in the rowlock position on the opposite end of the sill in the same way.
10. Attach a line to the wall with the aid of a line block or a line pin and nail about 1″ above the sill far enough back so it will not fall out. Lay a dry half-brick on each of the preset sill bricks, pinning the line to the face of the brick sill.
11. Lay the brick sill, working from both ends to the center. Check with the modular rule as often as necessary to be sure the sill will accommodate whole brick.
12. Form a mortar joint on both sides of the closure brick and the adjoining brick to prevent penetration of moisture and to form a sound joint. Check the back of the sill with the level for proper alignment and parge to ensure a watertight joint.
13. Strike vertical head joints on the front of the sill bricks with a convex jointer. Strike joints on the top of the sill with a flat slicker jointer. (The flat joint helps to prevent water from remaining on the sill.) Brush the work.
14. Recheck the sill to be sure the frame is level before having the work inspected.

$$\text{PROJECT 25: TYING AN INTERSECTING } 8'' \times 8'' \times 16'' \text{ CONCRETE}$$
$$\text{BLOCK WALL AND A } 6'' \times 8'' \times 16'' \text{ CONCRETE BLOCK WALL}$$
$$\text{WITH WIRE MESH EVERY 2 COURSES HIGH}$$

OBJECTIVE

- The student will be able to lay out and build an $8'' \times 8'' \times 16''$ block wall and lay a $6'' \times 8'' \times 16''$ stub wall on a 90° angle off the main wall. The student will rake out the mortar joint in the intersecting wall angle and tie the wall with wire mesh every 2 courses, Figure 26-15.

  Note: Walls such as this are found in many office buildings where partitions which are not load bearing are used.

EQUIPMENT, TOOLS, AND SUPPLIES

1 mortar pan or board	2' square
Mixing tools	Tin snips
Mason's trowel	Modular rule
Brick hammer	Short section of nylon line
Brick set chisel	2 line blocks
Convex striker and brush	Supply of clean water
Plumb rule	Length of wire mesh
Chalk box	

The student will estimate the number of concrete block and mortar materials from the given plan.

SUGGESTIONS

- Lay the 8″ concrete block wall to the height shown on the plans before laying out or building any part of the 6″ concrete block wall.
- Project the wire mesh out from the wall over the second course and bend it down to avoid an accident.
- The wire mesh should be mortared in solidly to ensure maximum strength.
- Cut all halves needed with the brick set chisel before laying the project.
- Do not cut the wire mesh to the full width of the block, as it may project from the block, which would result in a poor striking job.
- Apply a solid head joint to the square end of the block where the two walls intersect. (It will be raked out for caulking.)
- Practice good safety rules when cutting the block and wire.

PROCEDURE

1. Mix mortar and stock the work area.
2. Strike a chalk line for the project.
3. Lay the first course of 8″ block, using a line and line block as shown on the plan.

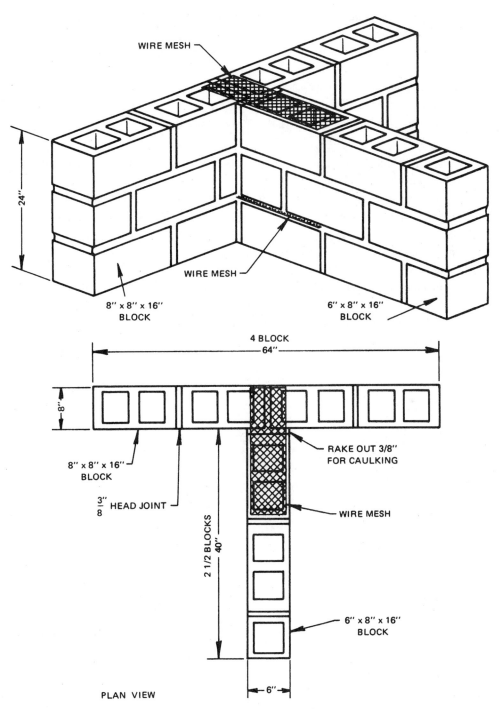

Fig. 26-15 Project 25: Tying an intersecting $8'' \times 8'' \times 16''$ concrete block wall and a $6'' \times 8'' \times 16''$ concrete block wall with wire mesh every two courses high.

4. Cut a piece of wire mesh, being sure that it ties into the 8″ wall at least 4″ and into the 6″ wall at least 8″.

5. Cut the halves needed for the project. (2, 8″ block halves and 3, 6″ block halves will be required.)

6. Lay the second and third courses according to the plan.

7. Strike and brush the 8″ wall. Use the convex jointer.

8. Locate the center of the 8″ wall. Mark off, from both sides of the center, the 6″ block wall.

   Note: The 6″ block wall will actually measure $5\frac{5}{8}$″. This must be remembered when working from the centerline.

9. Square off the 8″ wall and lay out the 6″ block stub wall with a pencil or chalk line.

10. Lay the first course of 6″ block to the layout lines. Be sure it is level and plumb.

11. Bend the wire mesh flat on top of the first course and lay the next course of 6″ block on the wall. Be sure to enclose the wire solidly in mortar.

12. Continue laying the wall as shown on the plans, periodically checking the height and maintaining plumb and level corners.

13. Strike the joints in the wall. Brush and rake out the mortar joint in the angle to a depth of $\frac{3}{8}$″ to receive caulking. (Caulking will not be done on this project as it is usually the work of the carpenter.) Recheck the work before having it inspected.

UNIT 27
Installing Anchor Bolts, Brick Corbeling, and Wall Copings

OBJECTIVES

After studying this unit, the student will be able to

- set an anchor bolt to receive a wood plate or a steel beam in a masonry wall.
- describe the process of corbeling bricks for a masonry wall.
- list the different types of copings used on masonry walls.

The art of masonry is not confined to building corners and laying units to the line. There are many fine details and arrangements of bricks and concrete block that apprentices will learn in the course of their daily work. Many of these detailed practices may be defined as basic masonry construction, such as installing anchor bolts, brick corbeling, and wall copings.

SETTING ANCHOR BOLTS

Anchor bolts are used to fasten a member of a structure, such as a wooden plate, to the structure itself. Sometimes the carpenter installs the anchor bolts, but it is usually the job of the mason. Nailing wood plates to a wall is not an acceptable method of attaching it to a structure. An anchor bolt must be used. After anchor bolts are installed, they should be mortared into place so that they will not loosen when the nut is tightened against the plate.

Anchor bolts are available in different sizes to suit the requirements of specific jobs. To hold down the wood plate on an average-sized job, a bolt ½″ in diameter and 12″ long should be used. An L-shaped bend on the bottom of the bolt helps to hold the bolt in the mortar, Figure 27-1. Bolts should be spaced according to plans or drawings.

The bolt should normally be set in the center of the wall unless otherwise specified. The position of the top of the bolt should be determined before the wall is completed, since it may be necessary to cut a notch out of the bricks to allow the L-shaped section of the bolt to fit snugly in the brickwork. The carpenter on the job should be consulted for the correct height of the bolts as it may vary from job to job. If more than 1 bolt is to be set in the wall, be sure they are all spaced the same distance from the outside face of the wall. This can be done easily by using a line.

Form a base under the bolt either with mortar or scraps of brick or block to ensure a solid base. Fill mortar around the bolt, ramming it tight with a wooden stick or trowel blade to be sure a solid joint is formed. The bolt should be installed plumb as it will be easier for the carpenter to fit the wood plate over the bolt. Try to keep as much mortar as possible off the threaded part of the bolt where the nut will be installed. Figure 27-2 shows a bolt set in a masonry wall.

The mason should never stuff empty cement bags into the hollow concrete block cell on the last course and then fill around the bolt with mortar. This is poor workmanship and may eventually cause the bolt to loosen in the wall.

Installing bolts that anchor steel beams is very critical, as the holes are predrilled in the steel, so there is no

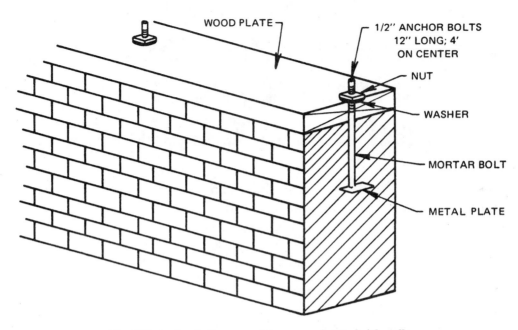

Fig. 27-1 Anchor bolts are used to secure plate to brick walls.

room for error. In most cases, the carpenter or building engineer is on hand to help the mason install them. The responsibility for correct placement of the bolts for steel erection is usually the responsibility of the carpenter.

Fig. 27-2 Bolt installed in masonry wall for garage door jamb.

FASTENERS FOR BRICK MASONRY THAT CAN BE INSTALLED DURING CONSTRUCTION

Wooden nailing blocks and wall plugs can be placed and mortared in the mortar joints as the brick are being laid. Wooden blocks are not used as much today as they once were, but are still a good method of providing an acceptable way of attaching an object, particularly a wood door buck that needs to be held securely to a masonry wall. Any wood blocks used should be of a seasoned and preferably a treated soft wood to prevent shrinkage, such as a piece of $2'' \times 4''$ framing lumber, and so on.

Metal wall plugs are made of galvanized metal and may contain wooden or fiber inserts. They can be walled in either the mortar head or bed joints. When you can make them fit the situation, metal plugs are better than wood blocks, since problems with shrinkage and decay do not occur with them. It is always important to work with the carpenter or tradesman on the job who will be utilizing them to determine their exact locations or their use will be a wasted effort. Metal wall plugs are used frequently in school construction for attaching chalk boards, baseboards, and chair rails, Figure 27-3.

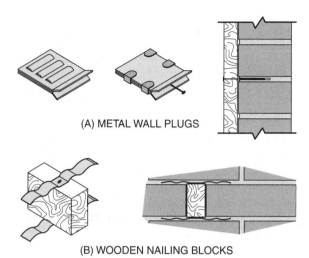

(A) METAL WALL PLUGS

(B) WOODEN NAILING BLOCKS

Fig. 27-3 Fasteners installed during construction. (Courtesy Brick Industry Association)

OTHER TYPES OF FASTENERS TO ATTACH TO A MASONRY WALL AFTER IT HAS BEEN BUILT

Screw Shields and Plugs

There are a variety of screw shields and plugs made of plastic, fiber, rubber, nylon, and lead that are available to insert into masonry work, see Figure 27-4. These fas-

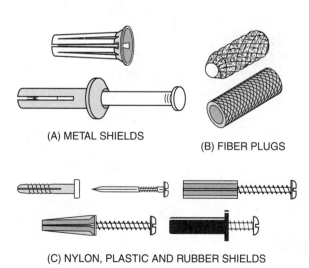

(A) METAL SHIELDS

(B) FIBER PLUGS

(C) NYLON, PLASTIC AND RUBBER SHIELDS

Fig. 27-4 Shields and plugs. (Courtesy Brick Industry Association)

teners are generally used for lightweight attachment purposes and are typically installed in the mortar joints or by drilling and then inserting them into the brick itself.

Bolt and Screw Attachments

A number of bolts and screws are available for use in solid and hollow masonry units. As a rule, these fasteners are used to attach and hold medium- to heavyweight fixtures. Toggle bolts (made of steel or plastic), hollow wall screws, small diameter sleeve anchors and screws are used to attach fixtures to hollow masonry units, see Figure 27-5. They can be inserted into holes in the mortar joints or through the hollow face section of the brick and into the hollow cell area.

Fasteners for Solid Masonry Units

Smaller diameter sleeve anchors, wedge anchors, screws, and lag bolt shields are shown in Figure 27-6.

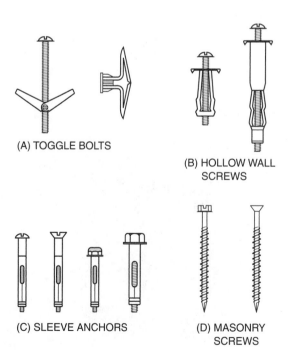

(A) TOGGLE BOLTS

(B) HOLLOW WALL SCREWS

(C) SLEEVE ANCHORS

(D) MASONRY SCREWS

Fig. 27-5 Fasteners for hollow masonry units. (Courtesy Brick Industry Association)

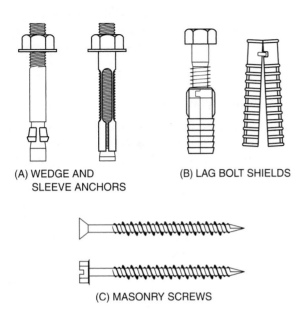

(A) WEDGE AND
SLEEVE ANCHORS

(B) LAG BOLT SHIELDS

(C) MASONRY SCREWS

Fig. 27-6 Fasteners for solid masonry units. (Courtesy Brick Industry Association)

They are used to attach fixtures to solid masonry such as a solid brick wall. Holes are usually drilled and then they are installed in the mortar joints. There are also power-driven hardened steel pins that are driven into masonry by using a special power-actuated tool called a "gun." The power for this gun tool is provided by a powder charge typically ranging from a .22 caliber to .38 shell with varying charges, depending on the material and required penetration depth. Power-driven penetration tools are used frequently on commercial or industrial projects where large quantities of fasteners are required. Strict safety procedures and training are required to operate these tools and in many localities a certificate or license to use them is mandatory!

BRICK CORBELING

Corbeling is the projection of masonry units to form a shelf or ledge, Figure 27-7. When corbeling, the mason must especially follow good bonding practices and use well-fitted mortar joints to ensure that the projection

Fig. 27-7 Corbeling bricks to form a decorative cornice. Notice that on corbeling, all bricks are in a rowlock position to give it greater strength.

will remain stable. Corbeling can be strictly decorative or used to thicken a wall to suit a particular job condition or requirement.

In years past, the tops of many brick buildings were identified by the masonry trim work designed by corbeling bricks in intricate patterns. This type of fancy masonry work was called *gingerbread work* and is now very costly due to high labor rates. The trend now is to simplify work to keep the costs down as much as possible. However, on some public buildings and restoration work, this type of work is still done. Inner walls inside chimneys and fireplaces still use corbeling to a great degree.

Headers are used extensively in corbeling as they lend the wall greater strength by tying further into the wall than they extend beyond it. The maximum projection should not exceed one-half the unit's height or one-third the bed height.

As a rule, the total projection of the corbeling should not exceed more than one-half the thickness of the wall. If metal supports or angle irons are used, the figure may be adjusted.

Bricks should be dry and free from chips or cracks, as most edges are exposed to view. A dry brick sets more quickly, which eases the operation for the mason. The bricks must be level and plumb. Well-filled mortar joints give maximum strength and reduce the amount of striking and jointing required. Corbeling should not be done so quickly that the mortar and bricks cannot take a firm initial set and remain in position.

Masonry work that is projected past the face of the wall is more susceptible to moisture penetration than regular walls. Once the mortar begins to deteriorate, it proceeds at a very rapid rate. Masonry work, as a rule, should last at least 40 years before any repair needs to be considered.

Excessive speed in this type of work may defeat the purpose of the job and result in shoddy work. For this reason, the contractor estimating a job that includes corbeling work usually allows extra time and money for it.

COPINGS

Coping is a material used to cap or cover a masonry wall to prevent moisture penetration. Coping is available in many shapes and forms, depending on the desired effect and the materials with which it must blend on the job. Coping can be an interesting architectural high point of a building.

Types of Coping

A course of bricks in the rowlock position can be classified as coping if it projects from both sides of the wall. Limestone coping has a very attractive appearance and sets fairly quickly. White Indiana limestone, when used to trim the top of walls, adds a variation of color and texture to a brick structure, Figure 27-8. Solid concrete blocks are sometimes used as coping but do not add much in the way of design or beauty to the building.

Terra-cotta tile, Figure 27-9, is used when economy is important. Each section of tile has a projecting bell-shaped edge on one end which fits over the end of the tile laid next to it. The tile is very hard and, therefore, resists the effects of weather very well. Builders of large buildings such as plants and industries many times select this type of coping since it is quick to install and is long lasting. A piece of burlap bag can be used to clean the terra-cotta.

Long sections of walls that are equipped with coping should have some room for expansion. Pliable filler

Fig. 27-8 White Indiana limestone coping on a brick building. The coping projects past the ends of the structure so that water may drop off. Otherwise a leak in the coping could allow the water to run onto the masonry work, possibly resulting in efflorescence and deterioration.

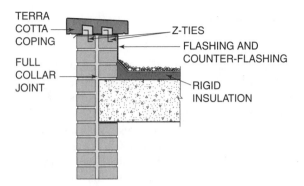

Fig. 27-10 Coping on a solid brick masonry wall above the roof line. (Courtesy Brick Industry Association)

Fig. 27-9 Terra-cotta coping before installation. Each section shown is eventually cut on the seam with a sharp mason's hammer to form two pieces.

materials can be inserted in the head joints of the coping and caulked much as a vertical expansion joint is installed in a masonry wall.

Regardless of the type of coping selected, it should be laid in mortar as other masonry is laid. Lay a piece of coping in mortar on each end of the wall, being sure that it is level and plumb. Attach a line and lay the remainder of the coping to the line. Fill any holes that may have formed under the projecting edge of the coping.

Installing Coping

Coping that is improperly designed or set may result in moisture damage to the building. Always use a full mortar joint under the coping. A good practice is to caulk the mortar joint that is exposed to the weather with some type of material that expands and contracts with changes in temperature. Flashing can also be used to be sure that no moisture penetrates by laying the flashing on the wall and then bedding the mortar over the flashing, Figure 27-11. Weep holes should be installed on flashing to drain water.

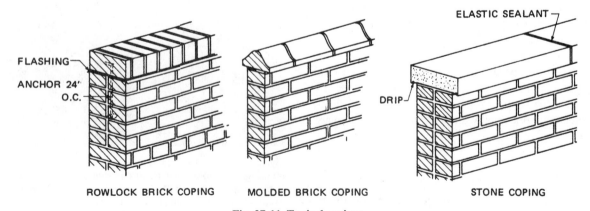

ROWLOCK BRICK COPING   MOLDED BRICK COPING   STONE COPING

Fig. 27-11 Typical copings.

ACHIEVEMENT REVIEW

Use basic math to answer the following questions. Show all of your work.

1. A foundation measuring $60' \times 20'$ is to be built. Anchor bolts are to be spaced $4'$ apart to anchor the wooden plates. How many bolts are needed? (Add the lengths and widths of the foundation for a total distance.)

2. A decorative corbeling ledge is planned for a building. The wall is $12''$ thick before any corbeling is done. The contract for the job specifies that the corbeling total $6'$. If each corbeling course projects $\frac{3}{4}''$, how many courses of corbeling are required to reach the total projection of $6''$.

3. Limestone coping is to be laid on a wall measuring $110'$ in length. If each section of coping is $2'$ long, including the mortar joint, how many sections of coping are required for the wall?

4. Coping projects over walls on both sides to protect them from moisture penetration. If a wall measures $12''$ in width, how wide must the coping be to project $2''$ on each side of the wall?

5. A large industrial plant is being built. Plans include limestone coping installed on top of the walls. The building measures $300' \times 100'$. The specifications require that there be an expansion joint every $50'$ in the coping. How many expansion joints are required?

PROJECT 26: CORBELING THREE COURSES ON A 12″ WALL

OBJECTIVE

- The student will be able to build a $12''$ brick wall, corbel a ledge, and return to the original wall line to complete the project, Figure 27-12.

  Note: This type of corbeling is typical of decorative work at the top of some buildings.

EQUIPMENT, TOOLS, AND SUPPLIES

1 mortar pan or board	Mason's modular rule
Mixing tools	Brush
Mason's trowel	Slicker striker
Brick hammer	Ball of nylon line
Plumb rule and 2′ level	Nail and line pin
Convex striker	Supply of clean water
12″ square	1, 8″ × 8″ × 16″ concrete block
Chalk box and pencil	on which to set level
Brick set chisel	

The student will estimate the number of bricks needed from the given plans.

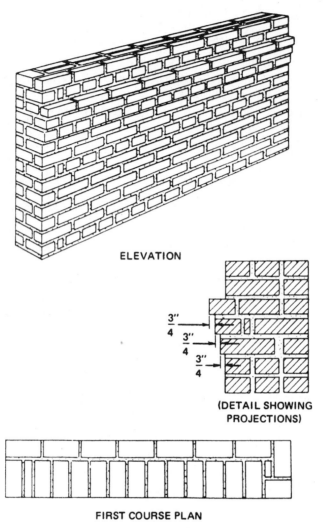

ELEVATION

(DETAIL SHOWING PROJECTIONS)

FIRST COURSE PLAN

Fig. 27-12 Project 26: Corbeling three courses on a 12″ wall.

SUGGESTIONS

- Strike the chalk line a little longer than is actually needed.
- The plan shows a 2″ piece to be laid against the header course as a starter piece. If this is too difficult, cut a 6″ or three-quarter brick piece to use in place of the 2″ piece and the header next to it.
- Use solid joints between all bricks, being especially careful on the header courses.
- Strike all the joints at the start of the corbeling since the mortar joints under the corbeling may harden too much to strike later.

- In all projecting courses, the bottom of each course should be checked to be sure it is level since it is very noticeable.
- Wear eye protection when cutting any material.
- Select straight, unchipped bricks for the corbeling.

PROCEDURE

1. Space the mortar pan or board and bricks approximately 2' from the project.
2. Mix the mortar and fill the pan.
3. Strike a chalk line for the front line of the project. Dry bond the project.
4. Mark off the returns with the square and pencil. Lay a brick in mortar on each end, being sure it is level and plumb with the layout line.
5. Attach a line and lay the first course as shown on the drawing. All brick courses will be laid to number 6 on the modular rule.
6. Lay a short lead on each end of the project. Continue to fasten the line and lay bricks until the corbeling height is reached (9 courses from base). Strike the work with a convex striking tool.
7. The first corbeling course is a header course projecting $3/4''$ from the face of the wall. Lay the corner bricks, attach the line, and fill in the course. Be sure to level the bottom of the course as well as the top. See the detail of the projection on the plan.
8. Lay the second corbeling course. The second corbeling course is a stretcher which also projects $3/4''$ beyond the first corbeling course, or $1 1/2''$ beyond the original face of the wall. Back up the course with a header course.
9. Lay the third corbeling course as a stretcher projecting $3/4''$ beyond the second corbeling course, or a total projection of $2 1/4''$ beyond the original face of the wall.
10. Lay the last 2 courses by returning to the original wall line as shown on the plan.
11. Carefully fill all holes under the projecting bricks. Strike the mortar joints on the last projecting course with a flat slicker so that the wall can better resist water penetration. Brush and recheck all work before having the work inspected.

PROJECT 27: CORBELING A DENTIL AND SAWTOOTH CORNICE

OBJECTIVE

- The student will be able to lay out and build a brick dentil and sawtooth cornice (Figure 27-13) utilizing corbeling.

  Note: This project is typical of the decorative cornices at the top of fine brick buildings such as banks, libraries, and government buildings, and will test the student's skill.

EQUIPMENT, TOOLS AND SUPPLIES

1 mortar pan or board	Mason's modular rule
Mixing tools	Brush
Mason's trowel	Slicker striker

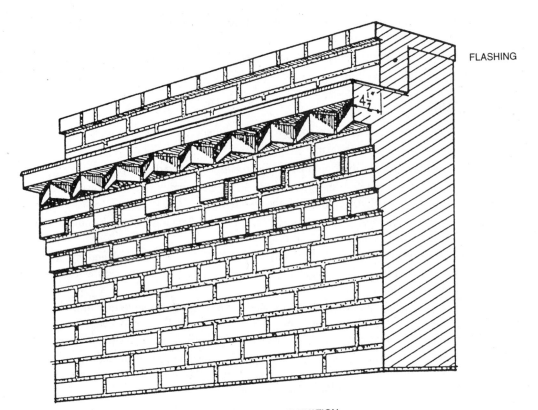

FLASHING

FINISHED FRONT ELEVATION

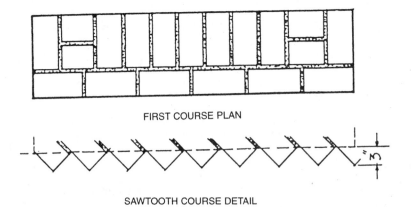

FIRST COURSE PLAN

SAWTOOTH COURSE DETAIL

Fig. 27-13 Project 27: Corbeling a Dentil and Sawtooth Cornice.

Brick hammer

4′ and 2′ level

Convex sled runner

Square

Chalk box and pencil

Brick set chisel

Nylon line

Nail and line pin

4′ length of 16″ aluminum flashing (actual job construction would use a rigid copper flashing)

1, 8″ × 8″ × 16″ concrete block to set level in when not using

The student will estimate the numbers of brick needed from the plan.

SUGGESTIONS

- Strike the chalk line a little longer than needed.
- Select unchipped brick for the ends and especially the corbeling and sawtooths.
- Use solid well-filled mortar joints.
- Strike (tool) all the mortar joints in the corbeling work as soon as they are thumb print hard to help press the mortar firmly in and hold the brick in place better.
- Be careful not to let the corbeled course droop down on the bottom edge. This may require hand picking these brick for size.
- Wear eye protection when cutting brick.
- Be sure to line up with a line the front of all corbeled and sawtooth work.
- Clean out neatly around the corbeled and sawtooth work with a flat slicker tool.
- Ends of the project do not have to be struck or tooled as the project represents a section of wall.

PROCEDURE

1. Space the mortar board or pan and brick approximately 2′ from the project.
2. Mix the mortar in a 1:3 proportion.
3. Strike a chalk line on the front line of the project.
4. Square off the end returns and strike a chalk line for the back side of the project.
5. Dry bond the first course of brick out to the layout line.
6. Lay a brick on one end of the project to number 6 on the modular rule and level across.
7. Lay up a short lead on each end of the project, making sure each course is laid to number 6 on the modular rule.
8. Attach a line and lay out the first course.
9. The bond used for the project is the Common Bond with a full header on the 6th course. See plan view.
10. Build the wall to the height of the first corbel shown on the plan.
11. The first projected corbeled course is a header and the second course is a stretcher course.
12. Notice that the dentil course means that every other header is placed ½″ beyond the face of the lower course. One header is flush with the course below it and the next one projects ½″.

13. Line up the next stretcher course flush with the projected dentil course.

14. The sawtooth point is laid on a diagonal $3''$ beyond the last projecting course, beneath it. See sawtooth detail.

15. The last corbeled course is laid flush with the projected points of the sawtooth course. On top of this course, crimp and install the flashing as shown on the elevation plan.

16. Complete the brickwork by laying the last header course, projecting it $\frac{1}{2}''$ as shown on the elevation plan. This will tie in or bond all of the projected courses.

17. Header course should also be laid on the back side of the project as needed to tie the project together. Alternate it one course higher or lower than the front header so they will not interfere with each other.

18. Tool the mortar joints with a convex sled runner; use a flat slicker tool for filling in around the corbeled brickwork. It is also recommended to strike a flat head joint on top of the last header course. It is not necessary to strike the back side of the project; just make sure that all of the mortar joints are filled well.

19. Clean out carefully all protruding mortar from around the dentil and sawtooth courses so it does not interfere with the appearance of the completed work.

20. Brush the finished brickwork when it has set enough that it will not smear and repoint any holes.

UNIT 28
Glass Block Masonry

OBJECTIVES

After studying this unit, the student will be able to

- list the principal types, sizes, and advantages of glass block.
- describe the three basic methods of installing glass blocks.
- describe how to clean glass block masonry properly.

INTRODUCTION

Glass block masonry is not new; it has been around for many years. The use of glass block in commercial and residential structures has increased dramatically in recent years, due to an ever more effective technology, and the availability of a larger variety of glass block and accessories with many desirable features and advantages that appeal to architects, builders, and home owners. It is also in great demand in remodeling projects. Many masons who have been trained mainly to lay brick and block can branch out and include glass block masonry work in their expanded list of saleable skills. These skills can be applied to the replacement of existing masonry, the installation of windows, or new construction. Working with glass block does require the development of a higher degree of skills than does normal masonry work because the material is a very hard non-absorbent glass, but any proficient mason with the proper training and skills can master laying glass block in a relatively short time. The most important benefits for a mason are the increased job opportunities and the work that result.

What Are Glass Block?

The leader in the field of development and manufacture of glass block in the United States is Pittsburgh Corning Corporation, to whom I am indebted for their assistance in preparing this unit. Probably no other masonry material possesses glass block's special combination of timeless beauty, practicality, flexibility of design and ability to transmit light, and yet still provides privacy to the area that it surrounds.

According to Pittsburgh Corning Corporation, glass block are described as hollow glass units that have been computer modeled using the finest raw materials, including pure silica sand. After it has been fired at 2300°F in a kiln, each block undergoes a slow strengthening process to ensure consistent color and the ultimate in reliability. It is partially evacuated of air in the manufacturing process, which creates a partial vacuum. The block is made of clear, colorless glass of the highest quality with a polyvinyl butyral edge coating. This edge coating has a twofold purpose: to create a better bond for the mortar to stick to and to provide for the expansion/contraction mechanism for each block once it has been laid in mortar in a wall or panel.

It is very important to remember that glass block are designated as *non-load bearing units*, which means that they are to support only their own weight and none of the weight of the structure overhead. Therefore, adequate provisions have to be made for the support of any construction above them on the job! It is the responsibility of the architect and structural engineer to specify this information in the job specifications and on the plans. As a point of information, the compressive strength of a regular hollow series glass block is technically 400 to 600 psi.

Figure 28-1 shows a beautiful example of glass block in the front of the Metropolitan Toronto Police Headquarters Building in Canada.

Fig. 28-1 Metropolitan Toronto Police headquarters building in Canada. (Courtesy Pittsburgh Corning Corp.)

Types or Series of Glass Blocks

Glass block are available from Pittsburgh Corning Corporation in two basic series: Premiere™ and Thinline™. Both of these series are available in various assorted patterns and sizes. In addition, each series has different types of finishing units available to enhance or enrich almost any application. The Premiere Series are 4″ thick (nominal size) or $3\frac{7}{8}$″ thick (actual size), which are preferred by architects and building professionals because they have a higher insulation value, sound transmission, and fire resistance rating. They are recommended for curved or angled panels and shower walls. They are also recommended in residential work for interior walls up to 250 square feet and exterior walls up to 144 square feet. Premiere glass block are available in the following nominal face sizes:

$4″ \times 8″$

$6″ \times 6″$

$6″ \times 8″$

$8″ \times 8″$ (the most popular size)

$12″ \times 12″$

The Premiere Series Finishing Units (DECORA™ and Ice Scapes® patterns only) are available in a variety of sizes and shapes to accommodate special needs presented by a project. There are special shapes such as the EndBlock $6″ \times 8″$ and $8″ \times 8″$ for finishing horizontal or vertical edges of interior panels, and the ENCURVE® 8″ glass block with adjacent arched soft edges to round out any design options or finish panels. Used with an $8″ \times 8″$ EndBlock Finishing Unit for a stepped panel, HEDRON® Corner Blocks 8″ high are hexagonal corner units that allow you to form 90 degree corners that result in a gently rounded continuous glass face. They can also be used in combination with other finishing units. TRIDRON 45 Degree Block® Units 8″ high, create angles or configurations from 45 to 360 degrees for columns, alcoves, and faceted panels of glass block. ARQUE® Block 8″ high are radius designed to allow construction of smooth, tightly curved walls with consistent widths of mortar joints.

The Thinline Series are 3″ thick (nominal), which includes the thickness of the mortar joint, and are popular for residential or home use. They are more economical and are ideal for windows and non-curved walls. The Thinline Series can be used for silicone or mortar installations. If installed in mortar, they are recommended for interior walls up to 150 square feet and exterior walls up to 85 square feet. The Thinline Series glass block are available in the following sizes:

$4″ \times 8″$

$6″ \times 6″$

$6″ \times 8″$

$8″ \times 8″$

The Thinline Series Finishing Units also have a specially made Endblock $4″ \times 8″$ for finishing horizontal or vertical edges of interior panels in the DECORA pattern only.

Figure 28-2 shows the variety of glass blocks that are available in the Premiere and Thinline Series along with special order glass block. Figure 28-3 shows a typical glass block display in a building materials show room.

Premiere Series

04270/PIT
BuyLine 0980

ARGUS® Pattern
Rounded perpendicular flutes. Maximum light transmission/medium degree of privacy.

ARGUS® Parallel Fluted Pattern
Rounded parallel flutes on each face complements SPYRA® Pattern. Maximum light transmission/medium privacy.

DECORA® Pattern
Distinctive wavy undulations. Maximum light transmission/subtle distortion.

DECORA® Pattern "LX" Filter
Fibrous glass insert adds significant thermal and light characteristics. Maximum privacy.

ESSEX® AA Pattern
A fine grid of closely spaced ridges. Moderate light transmission/maximum privacy.

IceScapes® Pattern
Non-directional pattern lets light in without sacrificing privacy. Maximum light transmission/medium to maximum privacy.

SPYRA® Pattern
Complements a variety of glass block patterns. Many options for decorative designs. Maximum light transmission/minimal privacy.

VUE® Pattern
Smooth, undistorted faces allow maximum light transmission and ultimate visibility.

THICKSET® Block
Cutaways show the three-times greater face thickness of the THICKSET® Block vs. the standard units shown above.

THICKSET® Block, DECORA® Pattern
Thicker-faced version of this popular pattern with its distinctive wavy undulations. Maximum light transmission/subtle distortions.

THICKSET® Block, ENDURA® Pattern
Ideal where thick-faced, heavier glass block is needed. Narrow flutes provide moderate light transmission/maximum privacy.

THICKSET® Block, VUE® Pattern
Modest distortion, thicker-faced block. Maximum light transmission/good visibility.

VISTABRIK® Solid Glass Block
With 3" of solid glass, high impact strength provides enhanced vandal resistance. Excellent light transmission/good visibility.

VISTABRIK® Solid Glass Block, Stipple Finish
Solid 3 inch thickness of glass with a stippled finish to add privacy. Same characteristics as the VISTABRIK® Solid Glass Block, VUE® Pattern. Good light transmission/medium privacy.

EndBlock™ Finishing Units, DECORA® Pattern
6" x 8" and 8" x 8" for finishing horizontal or vertical edges of panels. Also available in the IceScapes® pattern.

TRIDRON 45° Block® Units, DECORA® Pattern
Create angles or configurations from 45-360° for columns, alcoves, and undulating panels. Also available in the IceScapes® pattern.

HEDRON® Corner Block, DECORA® Pattern
Hexagonal corner unit allows you to form 90° corners resulting in a gently rounded continuous glass face. The block can be used in combination with other finishing units. Also available in the IceScapes® pattern.

ENCURVE® Finishing Unit, DECORA® Pattern
With arched, soft edges to round out your design options or finish panels. Use with 8" x 8" EndBlock™ Finishing Units for a stepped panel.

ARQUE® Block, DECORA® Pattern
Allows for construction of smooth, tightly curved walls with consistent mortar joints. Also available in the IceScapes® pattern.

Thinline™ Series

SPECIAL ORDER

DECORA® Pattern
Distinctive wavy undulations. Maximum light transmission/subtle distortion.

DELPHI® Pattern
Prismatic diamond design offers moderate light transmission/maximum privacy.

IceScapes® Pattern
Non-directional pattern lets light in without sacrificing privacy. Maximum light transmission/medium to maximum privacy.

EndBlock™ Finishing Units, DECORA® Pattern
For finishing horizontal or vertical edges of panels.

PC® Custom Signature Block
Custom manufactured with your corporate logo or other design pressed into one or both inside surfaces of an 8" square standard unit.

Edge Coating
Choose standard brown (shown) or black.

VISTABRIK® Solid Glass 45°/90° Corner Block
Slight radius for corners or angles.

Fig. 28-2 Glass blocks that are available in the Premiere™ and Thinline™ Series along with special order glass block. (Courtesy Pittsburgh Corning Corp.)

Fig. 28-3 Typical glass block display in a building materials show room.

Years ago glass block were laid in mortar much the same as regular brickwork. The general practice was to use a richer mortar, allowing for expansion on both sides and the top of the panel and a predetermined space that permitted whole units of glass block to be installed without any cutting. This work was usually done by highly skilled masons. Reflecting the recent technological advances in the glass block industry, Pittsburgh Corning Corporation has led the way in developing, testing, and marketing many accessories that make the planning and installation of glass block not only more accurate and easier but also less labor intensive and costly to install. This does not change the fact that exercising special care and following specified procedures are still required when installing glass block. Because of the advances in technology, the use of glass block is now in much greater demand in residential as well as in commercial buildings.

THREE BASIC INSTALLATION METHODS

There are three basic methods of installing glass block that have been developed and recommended by Pittsburgh Corning Corporation: Mortar I System, Mortar II System, and KWiK'N EZ® RIGID TRACK Silicone System. All of the materials and information needed to install glass block using these three systems are available from Pittsburgh Corning Corporation or building supply houses that stock Pittsburgh Corning Corporation glass block products. See Figure 28-4.

Determining the Size of a Rough Opening to Fit Whole Units of Glass Block

Glass block cannot be cut; therefore, openings must be planned to allow whole units of glass block. To determine the rough opening width required for a Mortar I System installation, multiply the number of glass block horizontally (level) you have selected, times the nominal width of the glass blocks plus $\frac{1}{4}''$. The extra $\frac{1}{4}''$ takes into account the thickness of the Perimeter Channel. To determine the height of the rough opening, multiply the number of glass block of the size you want to use vertically, times the nominal height of the glass block, plus $\frac{1}{4}''$. This is the rough size of the opening that will be required for the installation of glass block. Example: using an $8'' \times 8''$ glass block with a panel height of $48'' \frac{1}{4}''$ and a width of $32'' \frac{1}{4}''$, height = $48'' \div 8'' = 6$ glass block, width = $32 \div 8 = 4$ glass block. Multiply the number of block needed for the height by the number needed for the width, $6 \times 4 = 24$ glass block total for this opening. *Note:* In a Mortar II System, you have to add $\frac{1}{2}''$ to the height and to the width instead of $\frac{1}{4}''$! Otherwise, the number of glass block needed is figured the same way.

MORTAR NEEDS

You can buy White Premixed Glass Block Mortar or mix your own. For small amounts, it is more economical to buy bagged dry premixed mortar than to buy regular bags of white portland cement, hydrated lime, and white sand and having to combine them. For a large commercial application, it definitely would be more economical to mix your own. The table in

1. Mortar I System

VeriTru® Spacers
One-piece, all-plastic VeriTru® Spacer speeds construction, assures uniform placement and helps keep panels flush.

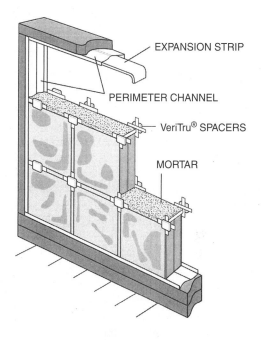

EXPANSION STRIP

PERIMETER CHANNEL

VeriTru® SPACERS

MORTAR

2. Mortar II System

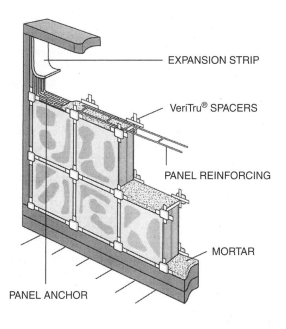

EXPANSION STRIP

VeriTru® SPACERS

PANEL REINFORCING

MORTAR

PANEL ANCHOR

3. KWiK'N EZ® RIGID TRACK
Silicone System

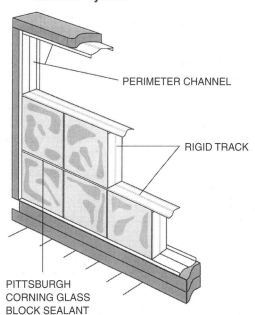

PERIMETER CHANNEL

RIGID TRACK

PITTSBURGH CORNING GLASS BLOCK SEALANT

Fig. 28-4 Basic installation methods. (Courtesy Pittsburgh Corning Corp.)

Figure 28-5 gives you an idea of the number of blocks that can be installed per 50 pound bag of premixed mortar.

Mixing Your Own Mortar

Pittsburgh Corning Corporation recommends that Type S or N ASTM C270 Mortar be used to install glass block. Type S Mortar is defined as 1 part waterproof portland cement to $\frac{1}{2}$ part Type S masons hydrated lime, and sand equal to $2\frac{1}{4}$ to 3 times the amount of cementitious materials (cement plus lime), all measured by volume, not with a shovel. Type N Mortar is defined as 1 part portland cement to 1 part masons hydrated lime and $2\frac{1}{4}$ to 3 times sand, which is a more user-friendly mortar mix. Either mortar mix is acceptable to use for glass block mortar. *For exterior mortar, an integral type water proofer should be added to the mortar mix. Never add antifreeze compounds or accelerators.* If mixing white mortar use a white quartzite or silica sand. Not all glass block installations use a white mortar, but as a rule it is greatly preferred for the majority of glass block installations.

Mixing mortar that will be used for installing glass block properly is half the secret of a successful job. Add the water slowly and in small amounts while mixing, as it should be the consistency or stiffness that will allow it to stick to the edges of the glass block when the block is held on a 90 degree angle. The stiffness should be such that you can spread it with a trowel, and yet it will not be crumbling or separating. A good example of the desired stiffness would be the consistency of "peanut butter." Since glass block are non-absorbent, the mortar has to be just right to support the unit and at the same time adhere to the edges. You do not want any water bleeding out of the mortar joints after the glass

block is in place, because that might cause it to slump or slide out of position. You should not mix any more mortar than can be used in a one-hour period.

Another cardinal rule is to never furrow the mortar joints, as you do not want any voids that might allow moisture to seep in later. Do not shift or move glass block in the mortar bed after initial set has occurred or the bond of the mortar to the block's edges will be broken, which can cause moisture to penetrate into the joints. Glass block will not stand tapping or beating on with a hard object such as a hammer, so it is a good idea to have a rubber crutch tip on your trowel handle when setting or adjusting the block into the mortar bed. It is also recommended that you not temper glass block mortar (add water later) after it has stiffened past initial set, or it will seriously affect the strength and bond of the mortar.

MAIN DIFFERENCES BETWEEN MORTAR I AND MORTAR II INSTALLATIONS

Mortar I System is designed for small residential panels or openings that are 25 square feet or less. It is not to be used in curved glass block walls. Interior and exterior panels must be framed on all four sides.

Mortar II System is designed for panels larger than 25 square feet that are anchored on three sides. Mortar II System does not use the perimeter channels but instead uses expansion strips, mortar joint reinforcement, and panel anchors. Joint reinforcing is imbedded in the mortar bed joints and used for the entire width of the opening in every other horizontal mortar bed joint. Joint reinforcing is available in 48″ and 96″ lengths. There are recommended glass block panel size limitations. The Premiere Series interior installation should be 250 square feet maximum and the exterior limitations 144 square feet maximum. The Thinline Series interior installation should be 150 square feet maximum and exterior 85 square feet maximum panel size.

The expansion strips should cover both jambs and the head of a panel and are available in 2-foot lengths. Panel anchors are installed in every other course up the jambs and across the head of the panel. Each panel anchor can be cut in half to create two 12″ long anchors.

Block Sizes	$4'' \times 8''$	$6'' \times 6''$	$6'' \times 8''$	$8'' \times 8''$	$12'' \times 12''$
(Nominal)	34	34	30	26	18
No. of Thinline Series (3″ thick)	42	42	36	32	N/A

(Based on using $\frac{1}{4}''$ mortar joints)

Fig. 28-5 Block requirements.

VeriTru Spacers can be used throughout the panel in the mortar joints on a ratio of 1.5 per glass block.

PANEL ANCHOR DETAILS

Panel anchors provide lateral support for Pittsburgh Corning Corporation glass block panels. In lieu of panel anchors, channel-type construction can be used. All panel anchors must be bent within the expansion joint and are generally placed 16″ apart occurring in the same joint as the metal joint reinforcing. They must be imbedded completely in the mortar bed joint beneath the glass block and extend a minimum of 12″ into the joint. Figure 28-6 shows typical construction details of a panel anchor and channel-type restraint construction detail.

CLEANING GLASS BLOCK THAT ARE INSTALLED IN MORTAR

Cleaning glass block that are laid in mortar is a relatively simple task. All surplus mortar is removed from the faces of the joints at the time they are tooled with inexpensive polyfoam paint brushes available at any hardware or paint store, which works great to fill any holes. It is recommended that you not remove excess mortar with a regular brick trowel for small or residential installations, as it could scratch the surface of the glass. After the mortar has dried enough so that it won't smear or cause holes in the joints, wipe the surface of the panel with a damp sponge to smooth out the mortar joints and close any remaining holes.

Never use harsh cleaners such as acid solutions, alkaline materials, or abrasive materials that might scratch the glass. Never use any type of wire brush to remove stubborn stains, as this is a sure-fire way to scratch the surface of the glass. Let the panel dry again until it shows a white dry haze, and wipe it off once more with a clean damp sponge. Rinse frequently with clean water to remove any particles that may scratch the glass surface. Allow any remaining film to dry again to a thin smear or powder. If any organic sealants, caulking, and so on have been used, you can remove them safely with commercial solvents such as xylene, toluene, mineral spirits, or naptha, followed by a wash and rinse of water.

Be careful not to scrub too hard with the cleaning agents as they might damage the caulking.

Do the final cleaning of the glass block panels after they have dried sufficiently when exposed to direct sunlight, by starting at the top and washing down with a sponge with generous amounts of clean water. Use a rubber-bladed squeegee to remove excess water from the face of the glass block and let it dry. Do the final wiping and cleaning with a soft cloth such as an old towel, changing it as often as necessary. If there is still a stubborn powder or stain present, remove it with a very fine steel wool (grades from 000 or 0000 work well). I suggest that you try this in a corner near the bottom first to make sure that it does not present a problem. Cleaning procedures for glass block are very similar to cleaning ceramic tile in a bathroom.

Figure 28-7 shows an example of a beautiful glass block window installed in a bathroom. The textured design of the face of this type of glass block not only allows natural light to pass through, but also insures complete privacy.

Although the KWiK'N EZ RIGID TRACK Silicone System previously mentioned was designed mostly for the home owner or handy man to utilize, it does offer additional work for the mason who wishes to take advantage of installing it, as not all home owners may want to try it. It is an ideal solution when renovating an existing bathroom to provide a small window with a minimum of effort. Complete instructions and materials are available from Pittsburgh Corning Corporation or building suppliers that sell Pittsburgh Corning Corporation glass block products.

INSTALLING A GLASS BLOCK WALK-IN SHOWER

Pittsburgh Corning Corporation has developed several glass block shower system kits that come complete with DECORA Pattern or Ice Scapes Pattern glass block and all installation materials and accessories. By utilizing these well-designed kits, the mason is relieved of a lot of the planning and work of installing a shower. The kits do not include glass block mortar, plumbing fixtures, ceramic tile or finished tile boards for the backing. I especially like the simplicity and appeal of the walk-in shower design and have selected it to use here

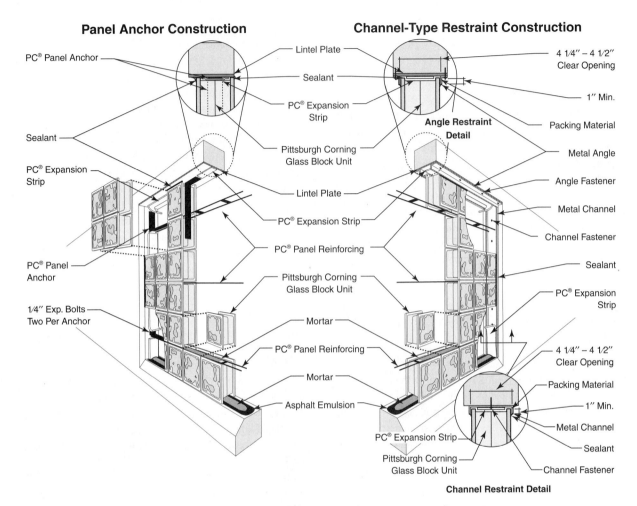

Fig. 28-6 Typical construction details of a panel anchor and channel-type restraint construction detail. (Courtesy Pittsburgh Corning Corp.)

to illustrate a typical installation of a glass block shower in mortar.

Setting and Preparing the Shower Base

The glass block are installed in mortar on a walk-in acrylic shower base developed by Pittsburgh Corning Corporation that measures $72'' \times 51''$ and is available in selected colors and entry. Although installing the shower base is not considered to be masonry work, you may be called on to install it, so it is essential to know how to do this. Mortar or plaster mix is mandated to reinforce and provide full bottom support under the shower base. Pittsburgh Corning Corporation recommends that you spread the mortar with a trowel on the sub floor or base $2''$ thick the full width and length of the shower base. This includes under the curb area. Coordinate setting the base with a plumber so that drains are lined up properly and any future problems are alleviated. Avoid standing in or loading the shower base with weight until the mortar has set. Using a plaster mix in place of mortar will result in the mix getting hard quicker.

Before laying any glass block in mortar, sand with an 80 grit abrasive paper over the textured surface of the curb where the blocks will be laid to remove the gloss

Fig. 28-7 A beautiful glass block window installed in a bathroom. (Courtesy Pittsburgh Corning Corp.)

of the acrylic and provide a rough surface for the mortar to bond to. After the mortar has been mixed, prepare a solution of Weldbond® primer and 1 part clean water. Use a brush and apply a primer coat to the sandpapered area. Let it set until it is tacky to the touch (3 to 5 minutes). You should work on one glass block wall at a time and make sure that the surface area is still tacky before applying the mortar and panel anchors.

Installing Horizontal Anchors

Figure 28-8   Installing vertical panel anchors with a screw drill in the curb base at the proper spacing locations using No. $12 \times 2''$ screws through predrilled holes. Apply a daub of sealant over the head of the screw so they won't be visible or leak later.

Figure 28-9   A length of expansion strip is placed against the wall and between the panel anchors. The first course of glass block is then laid out in mortar on the base. It is essential that full mortar head joints be

Fig. 28-8 Installing vertical panel anchors.

used and that you do not furrow the bed joint. The standing vertical panel anchors should be fully imbedded in the mortar head joints.

Figure 28-10   Laying the curved end block.

Figure 28-11   Installing the VeriTru Spacers in the top of each mortar head joint between each glass block.

Figure 28-12   Laying the metal reinforcement in the mortar bed joint on top of the second course of glass block. The reinforcement is installed every 16″ in height horizontally across the bed joints throughout the walls.

Figure 28-13   Attaching the wall anchors with a screw; drill every 16″ in height.

Fig. 28-9 Placing a length of expansion strip.

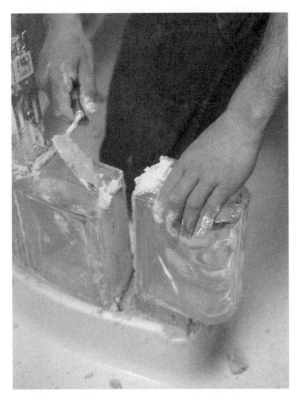

Fig. 28-10 Laying the curved end block.

Fig. 28-12 Laying the metal reinforcement in the mortar bed joint.

Fig. 28-11 Installing the VeriTru® spacers.

Fig. 28-13 Attaching the wall anchors.

Fig. 28-14 Laying the glass block panels.

Fig. 28-16 Tooling the mortar joints.

Fig. 28-15 Checking the wall for plumb.

Figure 28-14 Laying the glass block panels up to the prescribed height of the shower.

Figure 28-15 Checking the wall for plumb with a level.

Figure 28-16 Tooling the mortar joints at the completion of laying the glass block.

Figure 28-17 Caulking the edges with sealant.

After the mortar has dried enough, the final step is to clean the glass block as previously described. (All photographs of installation of shower are courtesy of Pittsburgh Corning Corporation.)

Figure 28-18 Completed glass block walk-in shower.

COMMERCIAL INSTALLATIONS OF GLASS BLOCK

The major differences between installing glass block on larger commercial jobs and on residential projects

Fig. 28-17 Caulking the edges.

Fig. 28-18 Completed glass block walk-in shower. (Courtesy Pittsburgh Corning Corp.)

are mostly related to the techniques and methods used to increase production and profit for the masonry contractor. The high standards that apply to installing glass block still remain the same. Obviously, longer walls or panels of glass blocks have to be laid to a line in mortar, as this is the only practical and accurate method of building them. Masons will use a brick trowel in place of the smaller margin trowel that is used in smaller residential installations. The selection of glass block for a commercial job is still determined by the job's particular needs and if the installation is interior or exterior.

Larger installations of glass block will require that they be divided into a series of individual panels with a vertical steel column type divider between them to withstand wind pressure stress against the panel. Designing and specifying the size and details of these types of glass block installation panels are the responsibility of the architect and structural engineer on the job and must meet local building codes. You still need

the accessories such as expansion strips, reinforcing, vertical and horizontal anchors in the panels and so forth. Masons who install glass block in commercial structures usually mix their own white portland cement mortar following the specifications described previously in this unit, as small prepackaged mixes would be too expensive for large jobs. Figure 28-19 shows a first typical layout course of glass block on a commercial job with expansion strips on the jambs and horizontal anchors attached to the jamb.

The sill area on which the glass block will rest needs to be coated with a heavy coat of asphalt emulsion and allowed to dry at least two hours before placing any mortar on it. Pittsburgh Corning Corporation recommends the use of a water-based asphalt emulsion made by Karnak Chemical Corp (Karnak 100 or equal. Phone 1-800-526-4236). Pittsburgh Corning Corporation also recommends the use of a sealant for caulking that is

Fig. 28-19 A typical first layout course of glass block on a commercial job. (Courtesy George Moehrle Masonry Co.)

Fig. 28-20 Checking the end face for plumb. (Courtesy George Moehrle Masonry Co.)

non-staining, waterproof mastic, silicone or urethane type.

You should always store unopened cartons of glass block in a clean, dry area off the ground and, if possible, in a relatively warm place, to keep frost and dampness out. They should be kept covered with a waterproof covering such as plastic. The accepted rule for installing glass block is that they should not be laid in mortar when the temperature is 40°F (4°C) and falling. This same temperature has to be maintained in cold weather for glass block installation in mortar for the first 48 hours after installation by using heaters if necessary, to allow the mortar to set properly. As stated previously, you are not allowed to add any antifreeze or accelerators to the mortar mix! It is a good idea when laying out the first course of glass block to check the end and inside face of each glass block for plumb with a small level to make sure they are right. See Figures 28-20 and 28-21. The edge of the base that the glass block are laid on should be protected with a layer of construction-grade plastic during the installation process, as shown in Figure 28-22.

CONCLUSION

Glass block masonry has been a highly desirable, unique type of masonry construction since its begin-

ning, and now offers not only a more superior product developed through continuing technology, but free technical support through the Pittsburgh Corning Glass Block Resource Center in the United States and Canada. They can be contacted anytime at 1-800-624-2120 or by visiting their computer website at www.pittsburgh corning.com.

Fig. 28-21 Checking the inside face for plumb. (Courtesy George Moehrle Masonry Co.)

Fig. 28-22 Protect the edge of the base with plastic during installation.

Fig. 28-23 Glass block masonry used to transmit light and also ensure privacy. (Courtesy Pittsburgh Corning Corp.)

The author wishes to express his thanks for the cooperation and assistance in preparing this unit to the Pittsburgh Corning Corporation, Glass Block Division, its staff and especially Mr. Nicholas T. Loomis, P.E., Senior System Engineer, Pittsburgh Corning Corporation. Figure 28-23 shows a beautiful example of utilizing glass block masonry to transmit light and at the same time provide privacy in a modern office building.

ACHIEVEMENT REVIEW

Select the best answer from the choices offered to complete each statement. List your choice by letter identification.

1. Glass block have a high insulating value because they
   a. they are filled with air.
   b. they have a clear insulation material inside.
   c. they have a partial vacuum inside.
   d. the design of the glass block surface causes heat to build up inside.

2. The recommended masonry cleaner to clean glass block masonry is
   a. clear water.
   b. muriatic acid.
   c. scouring powder.
   d. mineral spirits.

3. The preferred mortar joint thickness for glass block masonry is
   a. ¼″.
   b. ½″.
   c. ¾″.
   d. 1″.

4. The recommended stiffness of mortar for glass block should be like the stiffness of
   a. heavy paint.
   b. cake dough.
   c. whipped cream.
   d. peanut butter.

5. Glass block mortar that has been mixed with water should not be used after sitting
   a. 1 hour.
   b. 2 hours.
   c. 3 hours.
   d. 4 hours.

6. The most popular size of glass block is
   a. $4'' \times 8''$.
   b. $6'' \times 12''$.
   c. $8'' \times 8''$.
   d. $12'' \times 12''$.

7. Mortar I System is designed for panels of glass block that are
   a. 35 sq. ft. or less.
   b. 25 sq. ft. or less.
   c. 50 sq. ft. or less.
   d. 75 sq. ft. or less.

8. Panel metal anchors for glass block should be installed vertically
   a. every course.
   b. every other course.
   c. every third course.
   d. every fourth course.

9. VeriTru™ spacers for glass block masonry should be estimated on a ratio of
   a. 1.5 per glass block.
   b. 2 per glass block.
   c. 1 per glass block.
   d. 4 per glass block.

10. What is the accepted rule for laying glass block in mortar to prevent freezing?
    a. 32°F and falling
    b. 40°F and falling
    c. 50°F and falling
    d. 60°F and falling

SUMMARY, SECTION 6

- Masonry units over openings are supported by lintels. The most common types of lintels are the steel angle iron and the reinforced concrete block.
- Lintels should always be set level and plumb over any opening.
- Extreme care should be taken when lifting heavy lintels or injury may occur. When in doubt as to whether the weight can be lifted safely, get help.
- A pier is a free-standing vertical mass of masonry and is used mainly to carry beams or support parts of a structure.
- Piers differ from columns in that columns are usually higher in proportion to their thickness.
- Pilasters are similar to piers but are tied to the wall of the structure. They can be used to carry beams or for decorative purposes.
- Steel-bearing plates are set in a high-strength mortar over walls and piers to help distribute the weight of beams.
- Chases should be carefully designed so that they do not reduce the strength of the wall.
- Solid masonry bearings are important at the point where joists or concrete rests on the masonry work. The purpose of a bearing is to distribute the weight of a structure on a wall underneath more evenly.
- The proper tie should be installed when installing doors and window frames.
- Chimneys carry smoke and fumes safely away from buildings.
- Chimneys should have the flue lining installed in the smoke chamber to prevent the heat from burning through to the masonry work and to form a more efficient chimney.
- The mason should be careful about installing the proper-sized footing for a chimney since the weight is concentrated in a small area.
- Hard, burned brick should be selected for a brick chimney.
- The chimney should be built to a minimum of 2′ above the roof to assure a good draft and discourage any downdraft from the roof.
- The flashing should be built securely into the chimney to avoid water penetration.
- Use the size recommended by the manufacturer when installing a furnace flue in a chimney.
- All wood or other burnable materials should be kept at least 2″ from the chimney for fire protection.
- Use portland cement mortar (Type M or S) for the wash coat on the chimney.
- The two rules used by the mason are the modular rule and the spacing rule.
- The spacing rule is used to evenly divide mortar joints when the modular scale or grid will not accommodate them.
- The trade term *hog in the wall* indicates that one end of a wall has one more course of bricks than the other, but both are built to the same height. This is usually caused by larger joints being formed on one end.
- A *hog in the wall* can usually be prevented by checking the corner carefully with a story rod from a level point or from the benchmark.

- Expansion joints are built into masonry work to permit some movement of the wall without cracking.
- The full width of the expansion joint must be kept clean of mortar and foreign materials if the joint is to function properly.
- Workmanship is very important in the installation of expansion joints. If the joint is not perfectly plumb, it presents a very poor appearance.
- Intersecting walls can be tied by bonding the masonry units into each other or by using various types of metal ties.
- If a straight (stack) joint is being used against the intersecting wall, the mortar should be raked out to a depth of $\frac{3}{8}''$ and filled with a pliable caulking material.
- Anchor bolts should be set plumb in the masonry wall and to the correct height specified with mortar surrounding them.
- It is important that well-fitted mortar joints be used in corbeling and that the bottom of the course as well as the top be level.
- The total projection of work which is corbeled should not exceed more than one-half of the thickness of the wall.
- When installing coping on exceptionally long walls, the mason should allow for expansion joints to relieve the stress.
- Glass block are designated as non-load bearing units.
- Type S Mortar is recommended for installing glass block.
- Glass block cannot be cut, therefore, any openings they are installed in must be roughed in so that full blocks can be accommodated.
- Installing glass block with silicone instead of mortar for small openings is increasing in residential work.
- Never clean glass block with an acid solution or wire brush.
- Glass block have an excellent insulating rating due to the partial vacuum inside the unit.
- Glass block should not be laid in mortar if the temperature is 40°F or falling.
- The use of glass block in residential and commercial construction is increasing due to their esthetic appeal, low maintenance, ability to transmit light and provide privacy, and their insulating value.

SUMMARY ACHIEVEMENT REVIEW, SECTION 6

Complete each of the following statements referring to materials found in Section 6.

1. The deciding factor in selecting a lintel for a particular job is _____.
2. Steel lintel bearings for residential construction should be a minimum of _____.
3. Concrete block lintels should bear a minimum of _____ on each side of the opening.
4. The main difference between a pier and a column is _____.
5. Steel-bearing plates should be bedded in _____.

6. Pilasters differ from piers in that _____.

7. A vertical or horizontal slot left in a masonry wall to provide space for wires or pipes is called a _____.

8. The last 8″ of masonry work in many basement walls is required to be brick or solid block. This is done to _____.

9. Nailing blocks are installed in masonry walls for the purpose of _____.

10. There are two major types of drafts in a chimney. They are _____ and _____ drafts.

11. In order to prevent burning out of masonry work, tile or terra-cotta _____ is installed in chimneys.

12. When the carpenter is installing wood framing around a chimney, a space of _____ is recommended.

13. The point where the chimney goes through the roof must be made watertight by installing _____.

14. The wash coat which is applied to the top of a chimney to prevent water from seeping in is composed of _____.

15. The mason uses two different types of rules in his work, the _____ and the _____.

16. If a wall is built which is level on each end but with one end containing one more course, it is said to have a (an) _____.

17. To allow for movement in a building which is caused by temperature changes, the mason can install a (an) _____.

18. When using wall ties to tie an intersecting wall and a main wall together, it is advisable to rake out the mortar joint in the place where the two walls meet and fill the space with _____.

19. Wood plates located over masonry units are held securely in place by the use of _____.

20. The ledge formed by the projection of one unit over the unit underneath it is called a (an) _____.

21. The type of work mentioned in the previous question must not extend more than one-half _____.

22. The capping material laid on masonry walls to prevent moisture penetration and prolong the life of the wall is called _____.

23. Three major types of coping in use today are _____ , _____ , and _____.

24. Coping is always wider than the masonry wall on which it rests because _____.

SECTION SEVEN
SCAFFOLDING AND
CLEANING MASONRY WORK

UNIT 29
Types of Scaffolding

━━━━━━━━━━━━━━━ OBJECTIVES ━━━━━━━━━━━━━━━

After studying this unit, the student will be able to

■ erect a section of tubular steel sectional scaffold.

■ discuss the advantages offered by adjustable scaffolding.

■ explain how suspended scaffolding is erected and when and how it is used.

When masonry work has been built as high as the mason can comfortably reach, scaffolding must be erected to complete the work. *Scaffolds* are platforms that set the worker a certain distance off the ground or surface.

For many years, wooden scaffolding was the only type of scaffolding available. Wooden scaffolding is not in common use today for several reasons. The high cost of lumber, the time required to erect it, and the superior strength of steel scaffolding have limited its use.

Today, almost all scaffolding is constructed of some form of steel. The most commonly used is the steel sectional scaffold. This type of scaffold is used on most residential work and on moderately sized buildings. The main reason that steel scaffolding is preferred over scaffolding of other materials, such as wood, is the safety factor. Steel does not decay or split, and the load capacity of steel scaffolding is extremely high. It is possible to overload a steel scaffold to the breaking point. However, the section that actually breaks is the wooden planking rather than the steel framework.

New designs in scaffolding have been developed to satisfy modern-day construction needs, speed the work process, and reduce strain on the mason.

This unit deals with the three major types of scaffolding used in masonry work. All types discussed are of steel construction, although some types are also available in wood.

TUBULAR STEEL SECTIONAL SCAFFOLDING

Sectional steel scaffolding is assembled by the mason from steel frames. The most popular size of tubular steel scaffolding is 4' to 5' wide by 5' high. Sections that are 4' high are available but require more sections for use. Steel supporting braces, welded horizontally between the top and bottom of a section, can be used as ladder rungs if all are kept in a straight vertical line when constructing the scaffold, Figure 29-1. The frames are tied together by braces. These braces fit on threaded bolts, pins, or slip locks and are welded into the tubular frame of the scaffold. Each section of scaffolding is set

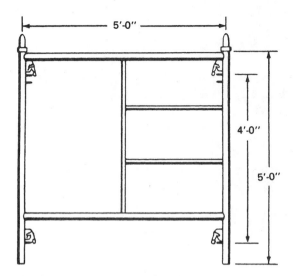

Fig. 29-1 Steel sectional scaffolding.

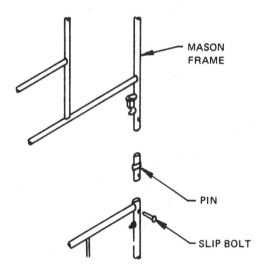

Fig. 29-2 Sections of steel scaffolding held together by pins.

on the next by slipping the hollow post on top of a steel pin, or sprocket, which fits into the hollow tubing of the section underneath, Figure 29-2. In the trade, these pins are known as *nipples*. It is important to keep the tubular sections into which the pin fits free from mortar and dirt to speed up the process of erecting the scaffolding.

By law, scaffolding must have certain safety features to protect masons and other people underneath the scaffolding from injury, Figure 29-3. These guidelines were established by the Occupational Safety and Health Act, passed by Congress in 1970. The purpose of this congressional act (commonly referred to as OSHA) is to establish safe working practices in all places of employment in the United States.

Two of the most important safety items on scaffolding are guardrails and toe boards. Guardrails are placed around the outside edge of the scaffolding to prevent the worker from falling off the scaffolding. Toe boards are installed at the bottom outside edge to prevent masonry materials or scraps from dropping off the scaffolding and injuring persons below.

A tubular steel scaffold with more headroom has been developed to allow a walking area for the mason and laborer. It has become increasingly popular since it permits the laborer to walk or push wheelbarrows along the scaffold without having to bend over. This type of scaffold is usually 5′ wide and approximately 6½′ high. Since the structure is taller than other scaffolds, the mason does not have to stop to erect new scaffolding as often.

Platform extenders (more commonly known as *scaffold hangers* or *brackets*), are fastened on the front of

Fig. 29-3 Scaffolding section showing features required by the Occupational Safety and Health Act. Features shown include the guardrail, toe boards, toe board clips, and climbing ladder with 12″ spacing of steps.

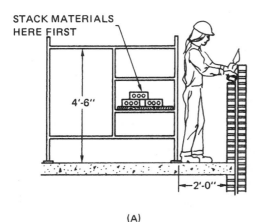

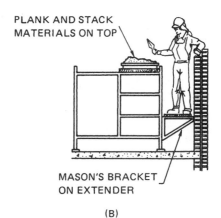

(A)

(B)

Fig. 29-4 The bracket and extender allow the mason to work at the most convenient height. Note: the mason should never place materials on the bracket. Until a wall reaches a height of about 4′ above floor level, masons can lay brick overhand standing on the floor (A). Above the 4′ height, brackets can be attached to tubular scaffolding, creating a platform on which masons can stand (B).

the scaffolding frame, allowing the mason to work at the best possible height, Figure 29-4.

> **Caution:** The mason should always check the scaffold before beginning work to be sure that the hangers are fastened securely and that the planks are lapped well over one another. Scaffolding that holds materials should always be stocked with all necessary equipment and supplies before the mason starts work, since the weight of the material counterbalances the weight of the mason. Ground settlement should be checked periodically after wet weather.

A masonry building may be completely encased in scaffolding as the work progresses. Power equipment, such as hydraulic lifts on a portable powered vehicle, can be used to lift materials as high as 4 stories to place the materials on the scaffolding, Figure 29-5.

Many accessories for steel sectional scaffolding make it adaptable to many situations. Wheels may be inserted so that the scaffolding becomes a *rolling* scaffold, Figure 29-6. This type of scaffold is acceptable for striking and pointing joints but is not recommended to be used as a base from which to lay units. When the

scaffold is loaded with materials, there is too great a possibility of the scaffold moving, thereby possibly causing the mason to fall and be seriously injured. The rolling scaffold is popular with workers who are working on a job that does not require the permanent, conventional scaffold. Screw jacks with baseplates are also available to level the sectional steel scaffold after it has been built.

Fig. 29-5 Mechanical tractor lifting material onto steel sectional scaffolding. Note: the guardrails are put into place after the scaffold is loaded with materials.

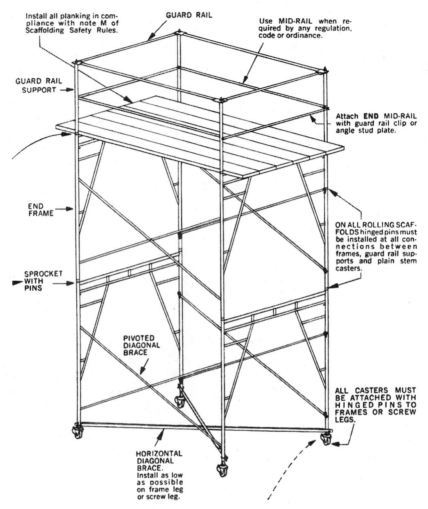

Fig. 29-6 The rolling scaffold. Note: the wheels on this type of scaffold should always be locked into position while the scaffold is in use.

The Advantages of Steel Scaffolding

Ease of Erecting. There are several advantages to using steel scaffolding. The option of interchanging parts and frames is a great time saver. Fatigue is minimized by the placement of the masonry materials at the best working level. No special tools are needed to erect the scaffold. Built-in coupling pins and wing nuts for the vertical frame assembly and an improved slip lock fastener for attaching cross braces speed the process of erection. Standard parts are available from suppliers to fit the major types of scaffolds with a minimum of delay. If more scaffolding is needed, it is usually rented.

Durability. Steel scaffold lasts indefinitely with reasonable care. Scaffolding should not be tossed to the ground from high heights. The mason should not use unnecessary rough treatment when disassembling the sections after the job is completed. Pins or scaffold nipples should be tied together in bundles with a piece of wire and stored for use on the next job. If wing nuts are being used, the threaded bolt and the nut should be coated with grease. This is done so that the person erecting the scaffold can assemble and disassemble it without a wrench. Wing nuts should not be tightened or loosened with a blow of a heavy hammer as the nut or bolt may be permanently damaged. Steel scaffolding is the property of the contractor and is costly to replace. It should be treated as if it were personal equipment.

Strength. The superior strength of sectional steel scaffolding is a major advantage of its use. The strength and complete protection from fire give the mason a feeling of security while working on the scaffold, Figure 29-7.

ADJUSTABLE TOWER SCAFFOLDING

The *adjustable tower scaffold* consists of vertical metal towers braced to the structure with alternating bays of metal scaffolding between each tower section. The main feature of this type of scaffold is the winch mechanism that raises the working platform as the work progresses.

Fig. 29-7 Masons walking on a sectional steel scaffold. The masons attempt to stock materials to a level higher than themselves for convenience.

In this way, the mason can always be at his or her best working position.

Scaffolding of this type is a relatively new concept in the masonry business. One of the best adjustable scaffolds, designed by a mason, is the Hoist-O-Matic trademark sold by Patent Scaffolding Company. In this scaffold, 6′ wide base panels are erected perpendicular to the proposed wall line. A sliding, locking cross arm is then installed as the load-bearing member of the platform to support materials, equipment, and workers. Additional scaffold height can be obtained from the base unit by adding posts and braces. This is done as the mason continues working, thus eliminating the loss of time and labor usually involved in rescaffolding.

As the wall is being built, workers raise the platform using a portable mechanical winch, which is attached to the posts. An operator moves the winch, which elevates the scaffold as necessary to keep the mason at the best working height.

The Advantages of Adjustable Scaffolding

Scaffolding of this type (Figure 29-8) offers many advantages to the mason and the contractor. There is no need for the mason to relocate several times a day while the scaffold is being rebuilt or restocked with materials. Material platforms are wide enough to enable movement of materials without hindering the mason. An overhead canopy can be easily built for wintertime enclosure, thereby allowing the job to continue. The mason is always working at the most convenient height, reducing fatigue and increasing production.

SUSPENDED OR SWINGING SCAFFOLDING

Suspended scaffolding is usually the best choice for very tall, multistory buildings, Figure 29-9. Steel beams are usually embedded and bolted into the concrete framework of the roof with steel cables dropped to the ground. Affixed to these cables are heavy-duty drum mechanisms with crank handles. Stop bolts at the end of the steel outrigger I-beams prevent the shackles from slipping off.

The planking of the scaffold is inserted in the metal frame below the drum mechanism and overlapped with lengths of scaffolding planks reaching to the next drum.

Fig. 29-8 Example of adjustable tower scaffolding.

Fig. 29-9 Suspended scaffolding extends around the entire structure. Notice the planking above the scaffolding to protect the workers.

A heavy steel frame is installed underneath the wood planking on each set of suspended hangers. This frame makes the scaffolding extremely strong and safe. Guardrails and toe boards are standard equipment. An overhead canopy attachment provides protection from the elements or objects falling from overhead.

Scaffolding machines provide a platform approximately 5' wide, which is adaptable for use on buildings of normal height.

As the work progresses the scaffold is cranked up, keeping the mason at the best working level. Each drum mechanism has 2 steel locking devices that engage in a large steel gear or sprocket to prevent the wire cable from unwinding or slipping.

Suspended scaffolding is the most economical way to build masonry work on a very tall building. It is one of the safest types of scaffolds that can be used since it is highly unlikely that the steel outrigger beams would become disattached, causing the scaffolding to collapse. Also, masons are constantly facing their work, which means easier working conditions for the masons and improved quality of the work being done, Figure 29-10.

Fig. 29-10 Masons working on suspended scaffolding. Safety rails, protective wire screen, and overhead planking all add to the mason's safety. Notice the winch with which the scaffolding is raised.

Accidents are less likely to occur with suspended scaffolding because guardrails, toe boards, and overhead planking are used for the workers' protection. Suspended scaffolding is usually rented and installed by scaffolding specialists.

COMMON SENSE SCAFFOLDING HAZARDS

Scaffolding-related accidents are one of the major causes of accidents in the construction industry and most of them could be avoided if the masons would observe and follow good safety practices that are mostly common sense. Although it is the responsibility of the employer to make sure the work place and scaffolding are safe, it is also up to the mason, particularly when working on scaffolding, to ensure that it is erected correctly and safely. It is a matter of self-preservation. Statistics in the construction trades reveal that every year more than 60 workers are killed in scaffolding mishaps. Problems involving poor planking practices, poorly secured or missing guardrails, and just plain carelessness working on and around scaffolding cause the majority of accidents. Masons are especially susceptible because they spend a large percentage of time working on some type of scaffolding.

Hazards You Should Be Looking for Before Getting on the Scaffold

- Not just anyone working on the job should be allowed to erect scaffolding. Most masonry jobs I have worked on have regular scaffolding workers who have been trained for that purpose. Scaffolding should always be inspected by the Masonry Supervisor before assigning masons to work on it. This means that the supervisor should get on the working surface of the scaffolding and check it out!

- OSHA states that a supported scaffold must be able to support its own weight and at least four times the maximum intended load.

- Keep electrical power lines at least 10 feet or more away from scaffolding or 3 feet away if the lines are less than 300 volts, unless you are positively sure the lines are dead.

- Do not work on scaffolding during periods of high winds or storms, or when it is coated with snow or ice.
- If a scaffold is more that 2 feet above or below a level, there must be a way to get on or off, such as by using a ramp, ladder, or approved hoist. It cannot be more than 14 inches horizontally from the scaffold to the structure to prevent falling through.
- A standing scaffold's legs have to be on a firm foundation, which means there must be metal base plates attached to the legs. Another example would be one piece of solid, firm wood under both legs that extends at least 1 foot past each leg so it does not work off.
- Upright sections of scaffold must be as vertically true as possible and braced to prevent swaying and the working platforms must be as level as possible.
- A scaffold that is more than four times as high as its base is wide must be tied or fastened securely to stable supports in the structure so that it cannot work loose.
- Scaffold working platforms and walkways have to be at least 18 inches wide and kept clear of materials or obstructions at all times. OSHA requires that if the work area is less than 18 inches, personal fall-arrest gear must be used by the workman.

- On most scaffolds, guardrails must be securely in place on all open sides and ends. OSHA also requires that on suspension scaffolds, a safety harness be worn for maximum safety protection.
- Scaffold walkways should have no more than a $9\frac{1}{2}''$ gap between the edges of the planks and a guardrail.
- Persons beneath or near a scaffold also have to be protected. There must be a $3\frac{1}{2}''$ high toe board to prevent objects from falling off a scaffold. If objects taller than the height of the toe board are on the scaffold, debris nets, screen wire, or other restraining devices can be erected to prevent objects or workers from falling off or though.
- Never throw tools, materials, or other objects that could possibly cause harm off of scaffolding.
- Loose clothing such as torn overalls or straps are a hazard to the worker, as are shoe strings that are not tied.
- Horseplay can never be tolerated on the job and can be especially deadly when engaged in by those working on scaffolding.
- The final recommendation concerning scaffolding safety hazards, regardless of the type, is that if you feel the scaffolding is not safe to work on, report it immediately to the supervisor on the job and refuse to work on it until it is corrected.

ACHIEVEMENT REVIEW

Select the best answer from the choices offered to complete each statement. List your choice by letter identification.

1. The most important factor to consider when erecting any steel scaffolding is the
   a. required amount of time to build the scaffolding.
   b. safety of the masons involved.
   c. weight of the scaffolding.
   d. requirements for the job.
2. Adjustable scaffolding was developed to
   a. speed the process of constructing the scaffold.
   b. reduce the costs of constructing scaffolding.
   c. give the mason the opportunity to work at the best possible height.

3. Steel sectional tubular scaffolding is fastened together vertically by using
   a. steel braces.
   b. bolts.
   c. steel pins or nipples.
   d. wire cables.

4. A major safety law was passed in 1970 by the United States Congress. The law is referred to as
   a. FHA.
   b. VA.
   c. HEW.
   d. OSHA.

5. One of the following scaffolds is *not* recommended for the mason to use when laying masonry units. This scaffold is the
   a. steel sectional scaffold.
   b. rolling scaffold.
   c. suspended scaffold.
   d. adjustable tower scaffold.

6. Scaffolds that are built higher than a few feet should be equipped with two major safety devices. They are
   a. wire screening and safety ties.
   b. toe boards and guardrails.
   c. reflector pads and warning buzzers.

7. Suspended scaffolding is recommended for very tall buildings because
   a. it is more economical to construct.
   b. the height can be continually adjusted.
   c. more materials can be stocked on the scaffold.
   d. it is quicker to erect.

8. Scaffolding can be of various heights but should be a standard width of
   a. 4'.
   b. 5'.
   c. 6'.
   d. 8'.

UNIT 30
Safety Rules for Erecting and Using Scaffolding

—————————— OBJECTIVES ——————————

After studying this unit, the student will be able to

■ list safety rules for erecting and working on scaffolding.

■ explain how to safely construct a foot board.

■ describe in detail how a steel sectional scaffold is built.

Figure 30-1 shows the advantages of using an adjustable tower scaffolding for masons working on a high concrete block wall. The scaffolding is adjusted up as the building of the wall progresses and keeps the masons always working at their most productive height.

A considerable amount of the mason's time is spent working on scaffolding. The danger of an accident resulting from poor scaffolding erection or faulty safety practices while working on the scaffold is higher than that of other practices in masonry work. Masons must be constantly aware of other workers while working on scaffolding. Materials should never be thrown to the ground from a scaffold, where they might cause injury.

Regardless of the type being used, scaffolds must be built properly to support the load placed on them. The National Safety Council recommends that scaffolds be built to support at least four times the anticipated weight of the materials and workers placed on them. This measure gives the mason protection in case additional loads are placed on the scaffold.

Scaffolding should be periodically inspected after it has been erected. Heavy rains and other severe weather conditions may cause sections of the scaffold to become disconnected from the wall. The mason should always inspect every scaffold before beginning work. If the mason is not satisfied, work should not begin on the scaffolding until the problem is corrected. Contractors are bound by federal law to provide a safe working environment for their employees. Immedi-

Fig. 30-1 Adjustable tower scaffolding.

356

ately report any unsafe conditions to the employer or safety engineer.

SAFETY RULES

Figure 30-2 shows a listing of safety rules established by the Scaffolding and Shoring Institute for persons erecting and working on scaffolds.

SAFETY IN STRUCTURAL COMPONENTS

Footing

The mason should make sure there is good footing under the scaffold. The board on which the scaffolding rests is known as the *scaffold sill*. The mason should be sure that it is level and on solid ground before placing any scaffolding on it.

> **Caution:** Never rest a scaffold leg on a hollow block. The weight of the scaffold after it has been loaded could cause the block to break through its web, possibly causing the scaffold to collapse.

Metal base plates equipped with screws are available for the legs of the scaffold, Figure 30-3. The screws are tightened or loosened to adjust the scaffold until it is level. The metal plates assure that the scaffolding rests firmly on the scaffold sill.

Foot Boards

Foot boards (also known as *blocking boards*), Figure 30-4, are erected to give the mason more height. Technically, however, they are not classified as scaffolding. When building a wall to the first level of the scaffold, it is sometimes necessary to lay a wooden plank on several block or bricks so that the mason is able to reach high enough to build. This is especially true when 5′ scaffolding is being used.

There are safety rules that must be followed when erecting a foot board. Select a wood plank that is free from splints and breaks, and one that is not excessively

warped or twisted, to ensure a stable footing. An insecure footing could slip out from under the masons' feet while they are lifting a heavy load and cause severe injury. The idea that masons must work high on the scaffold to be injured is false. Many injuries occur at ground level.

Divide the supporting materials under the foot board evenly along the length of the plank to avoid a sag in the middle. Be sure that the blocking material is placed level on the ground. In places where two boards meet, lay each board halfway over the block.

> **Caution:** A foot board should never be allowed to extend over the end block. This constitutes a safety hazard.

Do not build foot boards very high, as they will not be stable support. As a rule, foot boards are usually built 8″ from ground level and are never higher than 16″ from the ground. Space the foot board approximately 2″ from the wall to allow sufficient working room and room for mortar droppings. Do not attempt to make a scaffold out of foot boards. Their purpose is to allow only a small amount of additional height for the mason to prevent undue strain.

Putlogs

On occasion, there may be a ditch near the wall being constructed; therefore, there is no base on which to set the scaffold frame. In these cases, putlogs are used, Figure 30-5. *Putlogs* are crosspieces of heavy timber or steel beam, one end of which rests in a hole left in the wall, with the other end resting on solid ground or on a suitable base. This provides a strong footing on which to set the scaffold frame. After the masonry work has been completed and the scaffold disassembled, the putlog is withdrawn from the hole, which is filled by the mason.

Braces

All braces must be fastened securely. Extra braces may be added if necessary. If the nuts are missing from the bolt where the braces fit, the brace should not be wired;

SCAFFOLDING SAFETY RULES

as Recommended by

SCAFFOLDING AND SHORING INSTITUTE

(SEE SEPARATE SHORING SAFETY RULES)

Following are some common sense rules designed to promote safety in the use of steel scaffolding. These rules are illustrative and suggestive only, and are intended to deal only with some of the many practices and conditions encountered in the use of scaffolding. The rules do not purport to be all-inclusive or to supplant or replace other additional safety and precautionary measures to cover usual or unusual conditions. They are not intended to conflict with, or supersede, any state, local, or federal statute or regulation; reference to such specific provisions should be made by the user. (See Rule II.)

I. **POST THESE SCAFFOLDING SAFETY RULES** in a conspicuous place and be sure that all persons who erect, dismantle or use scaffolding are aware of them.

II. **FOLLOW ALL STATE, LOCAL AND FEDERAL CODES, ORDINANCES AND REGULATIONS** pertaining to scaffolding.

III. **INSPECT ALL EQUIPMENT BEFORE USING**—Never use any equipment that is damaged or deteriorated in any way.

IV. **KEEP ALL EQUIPMENT IN GOOD REPAIR.** Avoid using rusted equipment—the strength of rusted equipment is not known.

V. **INSPECT ERECTED SCAFFOLDS REGULARLY** to be sure that they are maintained in safe condition.

VI. **CONSULT YOUR SCAFFOLDING SUPPLIER WHEN IN DOUBT**—scaffolding is his business, NEVER TAKE CHANCES.

A. **PROVIDE ADEQUATE SILLS** for scaffold posts and use base plates.

B. **USE ADJUSTING SCREWS** instead of blocking to adjust to uneven grade conditions.

C. **PLUMB AND LEVEL ALL SCAFFOLDS** as the erection proceeds. Do not force braces to fit—level the scaffold until proper fit can be made easily.

D. **FASTEN ALL BRACES SECURELY.**

E. **DO NOT CLIMB CROSS BRACES.** An access (climbing) ladder, access steps, frame designed to be climbed or equivalent safe access to the scaffold shall be used.

F. **ON WALL SCAFFOLDS PLACE AND MAINTAIN ANCHORS** securely between structure and scaffold at least every 30' of length and 25' of height.

G. **WHEN SCAFFOLDS ARE TO BE PARTIALLY OR FULLY ENCLOSED,** specific precautions must be taken to assure frequency and adequacy of ties attaching the scaffolding to the building due to increased load conditions resulting from effects of wind and weather. The scaffolding components to which the ties are attached must also be checked for additional loads.

H. **FREE STANDING SCAFFOLD TOWERS MUST BE RESTRAINED FROM TIPPING** by guying or other means.

I. **EQUIP ALL PLANKED OR STAGED AREAS** with proper guardrails, midrails and toeboards along all open sides and ends of scaffold platforms.

J. **POWER LINES NEAR SCAFFOLDS** are dangerous—use caution and consult the power service company for advice.

K. **DO NOT USE** ladders or makeshift devices on top of scaffolds to increase the height.

L. **DO NOT OVERLOAD SCAFFOLDS.**

M. **PLANKING:**
1. Use only lumber that is properly inspected and graded as scaffold plank.
2. Planking shall have at least 12" of overlap and extend 6" beyond center of support, or be cleated at both ends to prevent sliding off supports.
3. Fabricated scaffold planks and platforms unless cleated or restrained by hooks shall extend over their end supports not less than 6 inches nor more than 12 inches.
4. Secure plank to scaffold when necessary.

N. **FOR ROLLING SCAFFOLD THE FOLLOWING ADDITIONAL RULES APPLY:**
1. **DO NOT RIDE ROLLING SCAFFOLDS.**
2. **SECURE OR REMOVE ALL MATERIAL AND EQUIPMENT** from platform before moving scaffold.
3. **CASTER BRAKES MUST BE APPLIED** at all times when scaffolds are not being moved.
4. **CASTERS WITH PLAIN STEMS** shall be attached to the panel or adjustment screw by pins or other suitable means.
5. **DO NOT ATTEMPT TO MOVE A ROLLING SCAFFOLD WITHOUT SUFFICIENT HELP**—watch out for holes in floor and overhead obstructions.
6. **DO NOT EXTEND ADJUSTING SCREWS ON ROLLING SCAFFOLDS MORE THAN 12".**
7. **USE HORIZONTAL DIAGONAL BRACING** near the bottom and at 20' intervals measured from the rolling surface.
8. **DO NOT USE BRACKETS ON ROLLING SCAFFOLDS** without consideration of overturning effects.
9. **THE WORKING PLATFORM HEIGHT OF A ROLLING SCAFFOLD** must not exceed four times the smallest base dimension unless guyed or otherwise stabilized.

O. For "PUTLOGS" and "TRUSSES" the following additional rules apply.
1. **DO NOT CANTILEVER OR EXTEND PUTLOGS/TRUSSES** as side brackets without thorough consideration for loads to be applied.
2. **PUTLOGS/TRUSSES SHOULD EXTEND AT LEAST 6"** beyond point of support.
3. **PLACE PROPER BRACING BETWEEN PUTLOGS/TRUSSES** when the span of putlog/truss is more than 12'.

P. **ALL BRACKETS** shall be seated correctly with side brackets parallel to the frames and end brackets at 90 degrees to the frames. Brackets shall not be bent or twisted from normal position. Brackets (except mobile brackets designed to carry materials) are to be used as work platforms only and shall not be used for storage of material or equipment.

Q. **ALL SCAFFOLDING ACCESSORIES** shall be used and installed in accordance with the manufacturers recommended procedure. Accessories shall not be altered in the field. Scaffolds, frames and their components, manufactured by different companies shall not be intermixed.

Reprinting of this publication does not imply approval of product by the Institute or indicate membership in the Institute.   Printed In U.S.A.

Permission to reproduce in entirety can be obtained from: Scaffolding and Shoring Institute, 2130 Keith Bldg., Cleveland, Ohio 44115   51   9-73—10M

Fig. 30-2 Safety rules of the Scaffolding and Shoring Institute.
(Courtesy Scaffolding and Shoring Institute)

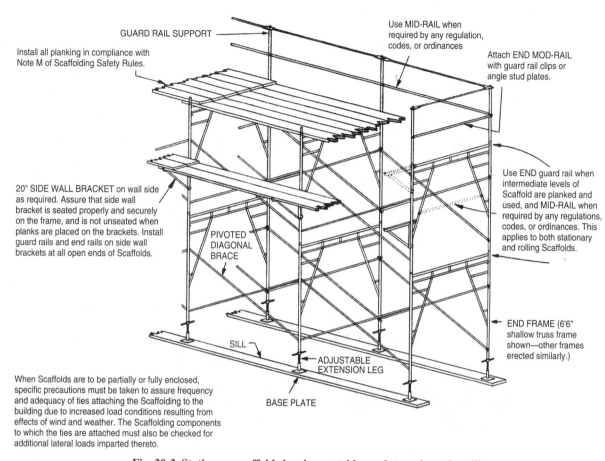

GUARD RAIL SUPPORT

Use MID-RAIL when required by any regulation, codes, or ordinances

Install all planking in compliance with Note M of Scaffolding Safety Rules.

Attach END MOD-RAIL with guard rail clips or angle stud plates.

20" SIDE WALL BRACKET on wall side as required. Assure that side wall bracket is seated properly and securely on the frame, and is not unseated when planks are placed on the brackets. Install guard rails and end rails on side wall brackets at all open ends of Scaffolds.

PIVOTED DIAGONAL BRACE

Use END guard rail when intermediate levels of Scaffold are planked and used, and MID-RAIL when required by any regulations, codes, or ordinances. This applies to both stationary and rolling Scaffolds.

END FRAME (6'6" shallow truss frame shown—other frames erected similarly.)

SILL

ADJUSTABLE EXTENSION LEG

BASE PLATE

When Scaffolds are to be partially or fully enclosed, specific precautions must be taken to assure frequency and adequacy of ties attaching the Scaffolding to the building due to increased load conditions resulting from effects of wind and weather. The Scaffolding components to which the ties are attached must also be checked for additional lateral loads imparted thereto.

Fig. 30-3 Stationary scaffold showing metal base plate and wooden sill.

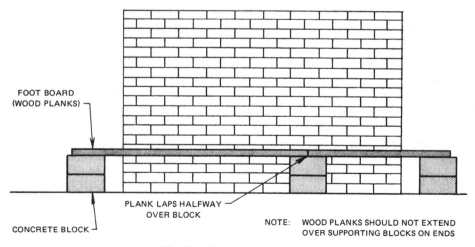

FOOT BOARD (WOOD PLANKS)

PLANK LAPS HALFWAY OVER BLOCK

CONCRETE BLOCK

NOTE: WOOD PLANKS SHOULD NOT EXTEND OVER SUPPORTING BLOCKS ON ENDS

Fig. 30-4 Foot board structure.

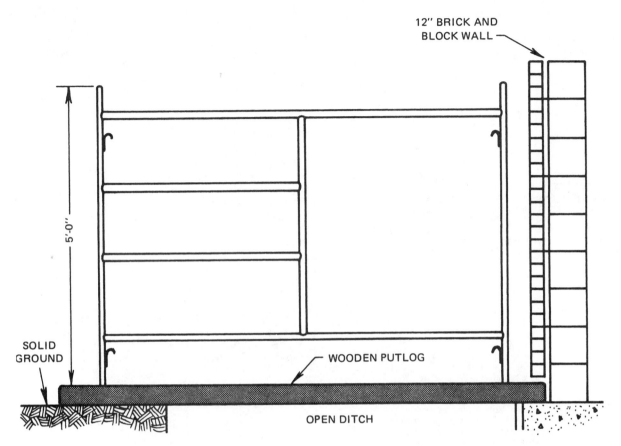

Fig. 30-5 Scaffold frame resting on putlog. After the wall is completed, the putlog is removed and the hole is filled by the mason.

it will not be secure. New nuts should be obtained and installed to hold the braces in position.

A mason should not climb on cross braces; the rungs built into the scaffold or a ladder should be used. Steel ladders that attach to scaffolding are available. Wooden ladders are also acceptable for scaffolds, providing they are wired to the top of the scaffold.

Figure 30-6 shows mason's tubular steel sectional scaffolding being erected around a tower in a shopping center that will be brick veneered. The advantage of this type of scaffolding is that it allows the masons to work completely around the structure once it is in place.

Ties

When scaffolding is higher than 2 sections, it should be tied to the structure at every 26′ of height with a heavy gauge wire. The wire is usually fastened to the frame of the structure and then walled into place. This is accomplished by wrapping the wire around one of the interior columns and laying it in the mortar joint through the masonry wall. The wire then cannot be pulled out. Ties should also be installed every 30′ in length. After the masonry work has been completed and the scaffold is ready to be torn down, the wire can be cut as the scaf-

Fig. 30-6 Tubular steel scaffolding.

fold is disassembled, and the hole can be filled with fresh mortar.

Planking

The scaffolding is no stronger than the planking used on it. Reasonably straight, warp-free boards should be selected. Rough, crooked sawmill lumber should be avoided as it does not provide stable footing. Poplar is the most commonly used wood due to its strength and light weight.

Scaffold planks are usually at least 2″ thick. All planks should be overlapped at least 12″ and extended at least 6″ beyond the center of the support; or, cleats may be installed at both ends to prevent them from sliding off the scaffold. The mason should secure the planks to the scaffold with nails when necessary. All holes in the floor of scaffolding should be repaired before work is begun. Pull all loose nails out of the planking; do not merely bend them over with a hammer.

The planks should be laid as close together as possible to prevent the possibility of materials or workers falling through. The total planking space should be a minimum of 4′ wide to allow sufficient working room. For best results, it is recommended that the scaffold be built 5′ wide. At the completion of the job, all planking is removed from the scaffolding with care. Do not throw it to the ground. Stacking it in a neat pile will prevent excessive warping until it is ready to be used on the next job. Wooden scaffold planking should last for many years with reasonable care.

Planking laid on top of the metal frames provides protection from falling objects. Planking should always be used for overhead protection on swinging scaffolding. Unsafe areas underneath the scaffold should be marked with rope or trestles to protect other workers or people passing by.

Scaffold Brackets

When scaffold brackets are attached to the scaffolding, they must be installed correctly. Brackets should support only the weight of the mason; materials should never be placed on them. Materials are stocked on the main section of the scaffold.

Scaffolding brackets should never touch the wall being built, since the scaffold will probably be moving somewhat as the work is being done. The movement could cause the brackets to knock the masonry work out of alignment. Provisions to install the brackets the correct distance from the wall must be made when the scaffold is begun by proper spacing of the scaffold frames from the wall with the bracket attached. Scaffolding that is not plumb when it is erected will cause the brackets to be knocked either in or out and may necessitate rebuilding or moving the frames before the masonry can be laid.

When building a single section of scaffolding and laying the wooden planks, be sure that they extend at least 8″ farther than the brackets. A cleat should be nailed to the bottom of the planks to prevent them from becoming disconnected from the brackets.

Guardrails, toe boards, and midrails should be on all steel scaffolds at an elevation of 6′ or more. Protect all open sides and ends of a scaffold with guardrails. Protective screen can be fastened on the side that faces a street if the work is being done in a populated area.

SPECIAL PROBLEMS OF ROLLING AND SUSPENDED SCAFFOLDING

Rolling and suspended scaffolding present some unique problems and, therefore, require safety regulations in their use.

As a rule, the mason does not use rolling scaffolds, but they are sometimes used on small repair jobs or when pointing joints. Rolling scaffolds are limited in height to four times their narrowest base dimensions, unless the base is widened by outriggers or more end frames.

Rolling scaffold should have the caster brakes locked in place when it is being worked on. Brackets are never used on rolling scaffolding. Rolling scaffolds should never be parked on a grade or slope without placing blocks against the wheels. A mason should never ride on rolling scaffolding.

Because of the nature of its construction, suspended scaffolding also has particular safety rules. Before using the scaffolding, inspect all wire cables to be sure that the cables are not frayed or worn. Before installation, be sure that the roof of the building will sustain and support the weight of the scaffold and the load to be placed on the scaffold. The working parts of the drum mechanism must be checked frequently to be sure that they operate freely. The working parts must also be kept well greased. Overhang of the outriggers must not exceed the distance recommended by the manufacturer. Be sure that toe boards and guardrails are installed and that all boards are lapped sufficiently over one another.

Scaffolding should be raised in small intervals to keep the entire scaffold reasonably level so that planking does not slip out of the metal frame. Overhead protection should be provided at all times by laying planking on top of the metal frames. Hard hats must be worn at all times if there is any overhead work being done.

SCAFFOLDING GENERAL REQUIREMENTS FROM THE 1995 OSHA SAFETY & HEALTH STANDARDS FOR CONSTRUCTION 29 CFR 1926, UNITED STATES GOVERNMENT CODE OF FEDERAL REGULATIONS

- The footing or anchorage for scaffolds shall be sound, rigid, and capable of carrying the maximum intended load without settlement or displacement. Unstable objects such as barrels, boxes, loose brick, or concrete blocks, shall not be used to support scaffolds or planks.

- No scaffold shall be erected, moved, dismantled, or altered except under the supervision of competent persons.

- Guardrails and toeboards shall be installed on all open sides and ends of platforms more than 10 feet above the ground or floor, except needle beam scaffolding and floats. Scaffolds 4 feet to 10 feet in height, having a minimum horizontal dimension in either direction of less than 45 inches, shall have standard guardrails installed on all open sides and ends of the platform.

- Guardrails shall be 2×4 inches or the equivalent, approximately 42 inches high, with a midrail, when required. Supports shall be at intervals not to exceed 8 feet. Toeboards shall be minimum of 4 inches in height.

- Where persons are required to work or pass under the scaffold, scaffolds shall be provided with a screen between the toeboard and the guardrail, extending along the entire opening, consisting of No. 18 gauge U.S. Standard wire $\frac{1}{2}$ inch mesh, or the equivalent.

- Scaffolds and their components shall be capable of supporting without failure at least 4 times the maximum intended load.

- Any scaffolding, including accessories such as braces, brackets, trusses, screw legs, ladders, etc. damaged or weakened from any cause shall be immediately repaired or replaced.

- All load-carrying timber members or scaffolding framing shall be a minimum of 1,500 fiber (Stress Grade) construction grade lumber. All dimensions are nominal sizes as provided in the American Lumber Standards, except that where rough sizes are noted, only rough or undressed lumber of the size specified will satisfy minimum requirements.

- The maximum permissible span for $1\frac{1}{4} \times 9$ inch or wider plank of full thickness shall be 4 feet with medium duty loading of 50 p.s.f.

- All planking or platforms shall be overlapped (minimum of 12 inches) or secured from movement.

- An access ladder for safe access from scaffolding shall be provided at all times.

- The poles, legs, or uprights of scaffolds shall be plumb, and securely and rigidly braced to prevent swaying or displacement.

- Overhead protection shall be provided for workers on a scaffold exposed to overhead hazards.

- Slippery conditions on scaffolds shall be eliminated as soon as possible after they occur.

- No welding, burning, riveting or open frame work shall be performed on any staging suspended by means of fiber or synthetic rope. Only treated or protected fiber or synthetic ropes shall be used for or near any work involving the use of corrosive substances or chemicals.

- Wires, synthetic, or fiber ropes used for scaffolding suspension shall be capable of supporting at least 6 times the rated load.

- The use off shore or lean-to scaffolding is prohibited.

- Lumber sizes, when used in this subpart, refer to nominal sizes except where otherwise stated.

- Materials being hoisted onto a scaffold shall have a tag line.

- Employees shall not work on scaffolds during storms or high winds.

- Tools, materials and debris shall not be allowed to accumulate in quantities to cause a hazard.

- Metal tubular frame scaffolds, including accessories such as braces, brackets, trusses, screw legs, ladders, etc. shall be designed, constructed and erected to safely support four times the maximum rated load.

- Spacing of the frames shall be consistent with the load imposed.

- Scaffolds shall be properly braced by cross bracing or diagonal bracing, or both for securing vertical members together laterally, and the cross braces shall be of such length as will automatically square and align vertical members so that the erected scaffold is always plumb, square and rigid. All brace connections must be made secure.

- Scaffold legs must be set on adjustable bases or plain bases placed on mud sills or other foundations adequate to support the maximum rate load.

- The frames shall be placed one on top of the other with coupling or stacking pins to provide proper vertical alignment of the legs.

- Drawings or specifications for all frame scaffolds over 125 feet in height above the base plates shall be designed by a registered professional engineer.

GENERAL PRECAUTIONS IN SCAFFOLDING

Electric power lines near scaffolds are potentially very dangerous, since masons working on the scaffolding could receive an electrical shock. Use extreme caution when working around power lines. Call the local power company if there is any question of safety.

Do not prop piles of building materials above the scaffold level as a base from which to complete the work. Makeshift devices can easily tip over and cause a serious accident.

Materials stocked on the scaffold should always be placed directly over the scaffolding frame for greater strength. Stack and interlock the masonry materials to a reasonable height so that they do not topple over. When a mechanical tractor is placing materials, a worker should be on the scaffold to direct their placement. Scaffolding should not be overloaded. Even steel scaffolding has limits as to what it will support. Each scaffold manufacturer states a safe load limit for its scaffolding.

> **Caution:** Any accidents should be reported immediately to the supervisor so that prompt medical attention can be given.

Practical joking can be hazardous in any phase of construction work but is absolutely forbidden when working on or around scaffolding. Scaffolding work requires strict adherence to safety regulations, common sense, and a serious attitude about the work.

ACHIEVEMENT REVIEW

A. Select the best answer from the choices offered to complete each statement. List your choice by letter identification.

1. The most popular type of scaffolding for general masonry work is
   a. suspended scaffolding.
   b. adjustable tower scaffolding.
   c. steel tubular sectional scaffolding.
   d. wooden scaffolding.

2. A building is being constructed with a ditch around the wall, making it impossible to set the scaffold close enough to the wall. This situation can be overcome by placing
   a. a brace against the wall.
   b. a putlog in the wall.
   c. brackets on the scaffold.
   d. filling in the ditch.

3. The purpose of brackets or scaffold hangers is to
   a. strengthen the scaffold.
   b. tie the scaffold to the wall for stability.
   c. prevent the worker from falling from the scaffold.
   d. act as a platform on which the mason works.

4. When erecting scaffolding with one vertical section on top of another, the sections should be tied to the wall every
   a. 10'.
   b. 15'.
   c. 20'.
   d. 26'.

5. In thickness, scaffolding planks should be at least
   a. 4".
   b. 2".
   c. 1".
   d. ¾".

6. The mason can be protected from overhead hazards on scaffolding by using
   a. wire screen fastened to the overhead section of the scaffold.
   b. metal braces tied to the scaffold frame.
   c. wooden planking laid over the metal frame.
   d. canvas on the top of the scaffolding frame.

7. If a mason is working on a scaffold and a crane begins to swing, accidentally knocking objects overhead, the proper procedure for the mason is to
   a. call the supervisor and complain.
   b. call the safety inspector.
   c. leave the scaffold immediately.
   d. cover the scaffolding with planking.

8. If a mason is injured on scaffolding, the proper procedure is to
   a. fill out an accident report.
   b. treat the victim.

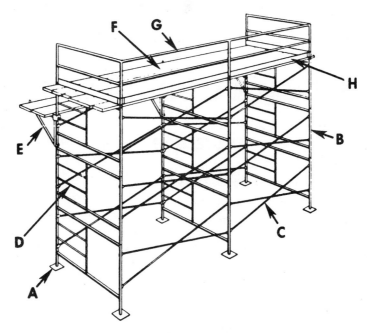

Fig 30-7 Illustration for achievement review, part B.

    c. call the supervisor.

    d. move the victim from the scaffolding.

9. Swing scaffolding should be raised very gradually because

    a. the cables may break under the strain.

    b. stacks of masonry materials on the scaffold may be upset.

    c. it involves too many laborers to raise the scaffolding.

    d. boards on the bottom of the scaffold may slip out of the hangers.

B. Identify each lettered part of the scaffolding illustration, Figure 30-7.

UNIT 31
Cleaning Brick and Concrete Block

OBJECTIVES

After studying this unit, the student will be able to

■ describe methods the mason uses to keep masonry work clean.

■ list poor cleaning techniques which damage masonry work.

■ describe the step-by-step procedure in cleaning brickwork with muriatic acid.

■ explain how concrete block are cleaned.

The finishing touch on all masonry work is the removal of dirt and stains. All brickwork should be thoroughly cleaned to enhance the color of the masonry units. In the construction of masonry walls, a skilled mason generally keeps the surface reasonably clean of mortar and other agents which cause stains. However, even the highly skilled mason may find it very difficult to keep the work perfectly clean. For this reason, the cleaning of masonry work is considered an essential part of every masonry job.

The National Apprenticeship and Training Standards for Bricklaying, established by the United States Department of Labor, requires that all apprentices have work experience totaling 150 hours in pointing, cleaning, and caulking during the course of their apprenticeship. This experience should be gained under the supervision of a competent journeyman mason.

GENERAL INFORMATION

There are different methods involved in the cleaning of bricks and block. The mason should be acquainted with these methods, including the proper tools, supplies, and safety measures. The mason should also practice techniques for keeping the work clean while it is being constructed so that cleaning of the work does not have to be extensive.

The correct clothes are important to protect the mason from injury and effects of the cleaning agent. A

hard hat should be worn on the job at all times; this includes cleaning work. When cleaning, the mason is facing the masonry wall and is subjected to falling objects. Since masons are constantly moving when cleaning, it is impossible that they always be protected from overhead danger. A rubber apron or rubber pants are best to keep the clothing dry and to protect the worker from the acid or cleaning agent. Rubber boots or galoshes keep the feet dry. Rubber gloves long enough to cover a good portion of the arm should be worn. Standard, shorter rubber work gloves do not give enough protection from the cleaning agents.

Some masons do not wear safety goggles when cleaning because of poor vision from the spray of the solution. However, it is recommended that eye protection be worn during all cleaning work.

BRICKWORK

According to the Brick Industry Association, cleaning failures usually fall into one of three categories:

1. Failure to thoroughly saturate (soak) the brick masonry surface with water before and after application of chemical or detergent cleaning solutions: dry masonry allows absorption of the cleaning solution and may result in "mortar smear," "white scum," or the development of efflorescence or "green stain." Saturation of the

366

surface prior to cleaning reduces the absorption rate, permitting the cleaning solution to stay on the surface rather than be absorbed.

2. Failure to properly use chemical solutions: improperly mixed or overly concentrated acid solutions can etch or wash out cementitious materials from the mortar joints. They have a tendency to discolor masonry units, particularly lighter shades, producing an appearance frequently termed "acid burn," and can also promote the development of "green" and "brown" stains.

3. Failure to protect windows, doors, and trim: many cleaning agents, particularly acid solutions, have a corrosive effect on metal. If permitted to come in contact with metal frames, the solutions may cause pitting of metal or staining of the masonry surface and trim materials, such as limestone and cast stone.

When cleaning and washing masonry work with bucket and brush, be sure to do the following:

- Refill the bucket before it is completely empty.
- Replace the cleaning agent when it becomes dirty.
- Follow safety rules at all times.

Keeping Brickwork Clean during Construction

To keep brickwork clean during construction, the mason stocks the mortar pans and boards far enough away from the wall (usually a minimum of 2′) to avoid splashing mortar on the work. Poor tempering practices also contribute to mortar smears on walls.

The mason should also be careful when laying the units in the wall. Cutting off the mortar from the bricks without smearing it with the trowel greatly reduces the amount of washing required. Striking the work too soon causes the masonry units to be smeared unnecessarily. Mortar joints should be tooled when thumbprint hard. After tooling with the striking iron, all particles of mortar remaining are scraped off with the blade of the trowel before brushing. Never brush the masonry work while the mortar joints are still wet, or they will smear. Always avoid any motion that will result in rubbing or pressing mortar particles into the brick face. A good,

medium-bristle brush such as a soft stove brush works well. If the bricks are too wet when they are laid, the mortar does not absorb properly and the joints bleed and smear on the outside surface of the wall.

Mixing the mortar to the right consistency prevents excessive smearing on the face of the wall. Clean materials should be used. When bricks arrive on the job, they should be stored out of the way of concrete, tar, and other materials that may stain them prior to laying them in the wall. Never pile masonry units under the scaffolding while work is in progress.

Additional staining may be caused by rain beating against the scaffold boards and splashing dirt and mortar on the wall. Many times this happens at night after the jobsite has been abandoned. This problem can be prevented to a certain extent by turning the scaffolding boards on their edges with the clean side to the wall at the end of the day's work. Always cover the walls at the end of the day to prevent the mortar joints from washing out and allowing rainwater to enter the wall.

Tools and Equipment

The tools used in cleaning masonry work are few but essential. A wooden paddle and nonmetallic scraper tool or chisel removes the large particles and does not leave rust or stains on the wall. A rubber bucket or heavy-duty plastic pail is needed to hold the cleaning solution. A water hose with a nozzle attached to adjust the spray of the water and a good stiff-bristled brush are also necessary. The handle should be long enough to protect the hands from the cleaning solution. Also useful are a couple of brick pieces with which the mason can rub stubborn particles of mortar loose from the wall.

Testing the Cleaning Agent

The best method of cleaning new brickwork is the least severe method. Acids are necessary on some occasions. However, if they are used when they are not really needed, they can cause severe problems. Similarly, detailed instructions on how to use a given product should always be superceded by the specific manufacturer's recommendation.

Before the actual cleaning of a brick wall begins, all cleaning procedures and solutions should be tested and evaluated on a test wall area of approximately 20 sq. ft. in size. The size of the test area could be larger but never smaller than this. The testing is important because small quantities of certain minerals found in some fired clay units, and materials added to color brick, such as manganese, may react with some solutions and cause staining. Chemical staining solutions are generally more effective when the outdoor temperature is 50°F or above.

Cleaning New Brickwork

Figure 31-1, developed by the BIA, lists cleaning methods for new masonry. It can be used as a general cleaning guide for all new masonry.

The three cleaning methods for new masonry are bucket and brush hand cleaning, high-pressure water cleaning, and sandblasting. When using any of these methods, be sure to wear proper clothing and eye protection, Figure 31-2.

Brick Category	Cleaning Method	Remarks
Red and Red Flashed	Bucket and Brush Hand Cleaning	Hydrochloric acid solutions, proprietary compounds, and emulsifying agents may be used.
	High Pressure Water	*Smooth Texture:* Mortar stains and smears are generearaly easier to remove; less surface area exposed; easier
	Sandblasting	to presoak and rinse; unbroken surface, thus more likely to display poor rinsing, acid staining, poor removal of mortar smears.
		Rough Texture: Mortar and dirt tend to penetrate deep into textures; additional area for water and acid absorption; essential to use pressurized water during rinsing.
Red, Heavy Sand Finish	Bucket and Brush Hand Cleaning	Clean with plain water and scrub brush, or *lightly* applied high pressure and plain water. Excessive mortar stains may require use of cleaning solutions. *Sandblasting is not recommended.*
	High Pressure Water	
Light Colored Units, White, Tan, Buff, Gray, Specks, Pink, Brown and Black	Bucket and Brush Hand Cleaning	*Do not use muriatic acid!!* Clean with plain water, detergents, emulsifying agents, or suitable proprietary compounds. Manganese colored brick units tend to react to muriatic acid solutions and stain. Light colored brick are more susceptible to "acid burn" and stains, compared to darker units.
	High Pressure Water Sandblasting	
Same as Light Colored Units, etc., plus Sand Finish	Bucket and Brush Hand Cleaning	Lightly apply either method. (See notes for light colored units, etc.) *Sandblasting is not recommended.*
	High Pressure Water	
Glazed Brick	Bucket and Brush Hand Cleaning	Wipe glazed surface with soft cloth within a few minutes of laying units. Use soft sponge or brush plus ample water supply for final washing. Use detergents where necessary and acid solutions only for *very difficult mortar* stain. For dilution rate, see Step 1d, *Select the Proper Solution,* under Bucket and Brush Hand Washing. Do not use acid on salt glazed or metallic glazed brick. Do not use abrasive powders.
Colored Mortars	Method is generally controlled by the brick unit	Many manufacturers of colored mortars do not recommend chemical cleaning solutions. Most acids tend to bleach colored mortars. Mild detergent solutions are generally recommended.

Fig. 31-1 Cleaning guide for new masonry.

Fig. 31-2 Proper protective clothing and eye protection for cleaning masonry.

Fig. 31-3 Removing mortar particles with wood scraper.

Bucket and Brush Hand Cleaning. This is by far the most popular of all the methods used for cleaning new brick masonry. This is because of the simple procedures required and the easy availability of muriatic acid and proprietary cleaning compounds. The following example is a recommended step-by-step procedure for cleaning brickwork with muriatic acid solution or proprietary compounds.

1. The cleaning operation should be one of the last phases of the job. Do not start before the mortar is thoroughly set and cured. If a prolonged time period passes, there may be a problem, since after a period of 6 months to 1 year, mortar particles are very difficult to remove. As a rule, 1 week of good weather should pass before the wall is cleaned. This allows the mortar time to cure well.

2. Dry clean the wall first with a wooden paddle. Large particles may require removal with the chisel, wood scraper, or piece of brick, Figure 31-3.

3. Protect any metal, glass, wood, limestone, and cast stone surfaces. Mask or otherwise protect windows, doors, and fancy trim work from the acid solution. Do not allow metal tools to contact acid solutions.

4. Prepare the cleaning solution. Use a clean, stain-free commercial grade of muriatic (hydrochloric) acid. Mix not more than 1 part acid to 9 parts of clean water in a nonmetallic container. Pour the acid into the water; do not pour the water into the acid.

Caution: If any of the acid solution gets in the eye or on bare skin, immediately flush out with plenty of clean water.

5. Presoak the wall to remove loose particles or dirt. Flush the wall with plenty of water with the hose, Figure 31-4. Start from the top of the

Fig. 31-4 Prewetting the wall with water from a hose. The wall is thoroughly saturated before applying cleaning solution.

Fig. 31-5 Scrubbing the wall with a long-handled brush. Eye protection is worn at all times.

wall and work down. A saturated wall does not absorb dissolved mortar particles. The area immediately below also is soaked with water to prevent the cleaning solution from drying into the wall.

6. Scrub with the brush starting at the top. Begin with a small area, not more than approximately 20 sq. ft., Figure 31-5. Heat, direct sunlight, warm masonry, and drying winds affect the drying time and reaction time of the acid solutions.

7. To remove stubborn spots, try using a piece of brick to rub the wall, Figure 31-6. After rubbing the spot with the brick, rescrub with more acid solution to work the particle or stain loose.

8. Rinse the wall thoroughly with water from the hose starting at the top and letting it run to the bottom of the wall. Acid solutions generally lose their strength after 5 to 10 minutes of contact with mortar particles. Failure to completely flush the wall with water and remove the cleaning solution may result in the formation of *white scum,* or the acid may burn the wall. For

stubborn places, repeat the washing action to assure a clean wall.

When an acid solution is not desired as a cleaning agent, detergents or proprietary compounds can be used.

Detergent or soap solutions are usually used to remove dirt, mud, and soil that has gotten on the wall during construction. A suggested solution is ½ cup (c) of trisodium phosphate (such as Calgon®) and ½ cup laundry detergent dissolved in 1 gal of clean water.

Many brick manufacturers now recommend using a proprietary compound in place of muriatic acid. This compound is considered to be milder and less likely to burn or stain the brickwork if used according to directions on the container. Before cleaning any brickwork, be sure to read the manufacturer's recommendation to determine the proper cleaner to use for that particular product.

A very popular cleaning product for brick masonry used by masonry contractors around the nation is called

Fig. 31-6 Rubbing a stubborn spot with a brick piece. The wall is washed afterwards with a cleaning solution.

Fig. 31-7 SURE KLEAN 600™ Detergent Masonry Cleaning Compound.

SURE KLEAN 600™ Detergent Masonry Cleaning Compound, Figure 31-7. It is manufactured by Prosco, Inc., 1040 Parallel Parkway, Kansas City, KS 66104.

The Prosco Company also makes a variety of very effective masonry cleaning products and stain removers for specific uses that are manufactured so that they do not harm the masonry walls they are used on. Some of the more popular Prosco masonry wall cleaners that I have used with a lot of success are shown in Figure 31-8.

SURE KLEAN VANA TROL™ is an excellent Prosco product that can be used for cleaning brick or block that is subject to vanadium, molybdenum, and other metal stains.

SURE KLEAN RESTORATION CLEANER™ is another Prosco cleaning product that is a general-purpose acidic cleaner and is highly recommended for cleaning brick, removing atmospheric stains, and generally brightening up brickwork that has become dull or faded from exposure to the elements or environmental pollutants. It is often used by restoration contractors.

SURE KLEAN ASPHALT & TAR REMOVER™, which is made by Prosco, is excellent for removing asphalt and tar on brickwork. These kinds of stains are

Fig. 31-8 Prosco masonry wall cleaners.

frequently encountered on brick masonry walls when built-up tar roofs are being installed. Because there are so many different types of stains that can show up on a brick or masonry building during the construction process, it is not possible to discuss all of cleaners that can be used here. However, your best bet is to visit a reputable masonry material supplier in your area, because they have a lot of cleaners on display and can usually give you good advice on the proper one to use for a specific cleaning problem.

SURE KLEAN 600 DETERGENT™ is a Prosco product that is very popular with professional masons, builders, and architects for washing down and cleaning brick masonry. It is recommended by many brick manufacturers because it is less harsh than muriatic acid to the colors in brick.

High-Pressure Water Cleaning. One of the major improvements in the cleaning of masonry work is the use of high-pressure water cleaning. However, certain practices should be followed for good results. If these practices are not followed, a sizable increase in the cost of cleaning could occur from the creation of deep-set stains and discolorations that would require costly, job-delaying, remedial cleaning. When pressure-cleaning equipment is being used, the cleaning should not be started until the mortar has cured for a minimum of seven days. There are some mortars that take longer to cure, and this should be taken into consideration by reading the manufacturer's recommendations.

The most important advantage of high-pressure water cleaning is labor savings. A much higher degree of efficiency can be achieved through the use of high pressure application rather than the old garden hose. Using the proper amount of both water pressure and volume will greatly reduce the amount of scraping and scrubbing that normally goes into washing down masonry work. It is important to remember not to set the water pressure and volume so high that the force will cut off all the mortar by itself, as this would in effect cut the mortar joints completely out of the wall or rod them out, damaging the integrity of the wall and causing it to leak.

What is proper pressure? The ProSol Company recommends pressure equipment that will develop from 400 to 800 psi water pressure for the highest degree of efficiency. Desirable results have been obtained with lower pressures in the range of 200 to 300 psi.

The proper water volume is possibly more important than pressure. A minimum flow of water through a gun of three gallons per minute should be obtained, although four to six gallons per minute would be highly desirable. It is better to sacrifice pressure and increase volume. Both pressure and volume efficiency depend greatly on the right spray tip. The common opinion is that the most efficient spray tip is the fan type, stainless steel tip, dispersing a 25° to 50° fan spray, Figure 31-9. It is never recommended to use less than a 15° fan spray tip.

Excellent cleaning results have been obtained by applying products such as SURE KLEAN masonry cleaning product with a masonry cleaning brush in the normal manner and using high-pressure water for the removal of the excess mortar and cleaning materials. This avoids

Fig. 31-9 Spray cleaning brickwork with a high-pressure sprayer. The most efficient spray tip is the fan type which disperses a 25° to 50° fan spray.

driving the cleaning agents into the wall and causing unsightly stains or discolorations that are very difficult or impossible to remove.

Sandblasting. Dry sandblasting has been used for many years on restoration work. It is one method that eliminates the danger of mortar smears, acid burn, and efflorescence. There is the danger, however, of the brick face being damaged or scarred by the wrong use of the equipment.

In this process, air pressure forces the sandblasting materials against the work to remove old paint or finish. The abrasive materials used are mined silica sand, crushed quartz, granite, white urn sand (round particles), crushed nut shells, and other softer abrasives. There are various degrees of cutting or cleaning, and, therefore, many types of abrasive materials. The worker should always direct the abrasives at the brick and not the mortar joints.

Sandblasting is still a very popular method of cleaning and restoring old brickwork. Care must be taken by the cleaning contractor that areas other than the masonry work be protected during the cleaning process, particularly old molding in the building.

Sandblasting should be done by a qualified person, usually one who specializes in the field. Basically, it is done with a portable air compressor, blasting tank, blasting hose, and nozzle. It is very important that the operator wear proper clothing and eye protection.

A suggested procedure for sandblasting is as follows:

1. Select sandblast materials that are clean, dust free and abrasive.
2. Brick masonry should be dry and well cured.
3. Remove all large mortar particles.
4. Protect non-masonry surfaces near cleaning areas. Use plastic sheeting, duct tape or other covering materials.
5. Test clean several areas at varying distances from the wall and several angles that afford the best cleaning without damaging brick and mortar joints. Workers should be instructed to direct abrasive at the brick and not on the mortar joints.

Chemicals and Steam Cleaning. Chemicals and high-pressure steam are used in special cases to re-

move applied coatings, such as paint, from masonry. Like sandblasting, it is a highly specialized field. Frequently, the proper cleaning agent can be determined only after an analysis of the various factors involved in a particular project.

CONCRETE BLOCK

Keeping Block Clean during Construction

Ideally, mortar is not cleaned from concrete masonry units with an acid wash. Therefore, care should be taken to keep the work as clean as possible during construction. The task of cleaning the walls in preparation for painting or finishing is the responsibility of the masonry contractor. The cleaner the work is kept during construction, the less work there is to be done at the completion of the job.

Treat joints carefully during construction to eliminate unnecessary cleaning. When a mortar joint is rubbed flat (when striking a flush joint), there is sometimes a tendency to smear the mortar over too large an area, causing the joint to appear larger than it should. Immediately after joints are rubbed and pointed, they are brushed with a stiff-bristled brush. This should remove most of the excess mortar which clings to the edges. Once mortar smears, it tends to embed itself into the surface of the block, causing noticeable stains. Paint only magnifies the mortar smears, chipped block, and other evidence of poor work.

Unless covered, concrete block should not be stocked near an area where it will be subjected to such things as spillage from tar buckets, paint, and dirt.

Procedures in Cleaning Block

The tools needed to clean mortar from concrete block are fairly simple. A piece of brick, broken block, or rubbing stone can be used to rub the particles of mortar from the wall, Figure 31-10. Larger particles of mortar may be removed with metal scrapers, putty knives, chisels, or the mason's trowel. The work is then brushed. The use of acid solutions is not usually recommended for concrete block. However, if there is a difficult stain on the block, such as efflorescence, an acid wash may be used.

Fig. 31-10 Rubbing a concrete block wall with a piece of block to remove mortar particles.

If a stubborn stain must be removed from concrete block, check with the foreman or superintendent before attempting to remove it chemically. The type of block and material to be removed determine the procedure and cleaning methods to be used. Colored concrete block products could be damaged by using the wrong cleaning agent. Once the wall has been damaged, correcting the problem is usually not a simple procedure. Figure 31-11 shows the various stain-removing materials and the procedure to follow when removing stains from concrete block.

PROCEDURE FOR STAIN REMOVAL		
A	Scrub with brush and chemicals or detergents	H	Apply liquid to surface by brush
A_1	Scrub with brush and water	H_1	Apply liquid to surface by brush at 5 to 10 minute intervals until thoroughly soaked
B	Wash thoroughly with clear water		
C	Cool until brittle. Chip away with chisel	J	Allow stain to disappear by aging
		K	Stir liquids together
D	Mix solids	L	Put paste on trowel. Sprinkle crystals on top of paste. Apply to surface so crystals are in contact with block and paste is on outside.
D_1	Stir solids and liquid to thick paste		
D_2	Apply paste to stain to thickness of 1/8 to 1/4 inch		
D_3	Place heated concrete brick over top to drive liquid in	M	Soak up fast and soon with poultice material until no free oil remains
D_4	Cover to minimize rate of evaporation		
D_5	Let dry as needed for periods up to 24 hours. Scrape off. Use wood scraper if block has tile-like finish	N	Scrape any solidified matter off surface
		O	Let harden. Remove large particles with trowel, putty knife or chisel
		P	Let stand. Remove with scraper and wire brush.
E	Repeat as needed	R	Allow to age three days
E_1	If brown stain remains, treat as for iron stain	S	Absorb with soft cloth or paper towels, then scrub vigorously with paper towels
F	Apply with saturated cloth or cotton batting	T	Provide thorough ventilation
G	Dissolve solid chemical in hot water	V	Follow manufacturer's directions
G_1	Dissolve solid chemicals in water	X	Pour into paste and mix

Fig. 31-11 Materials for stain removal and procedural sequences (continues).

| Stain | Appearance | Materials Needed | | Procedural Sequence (Letters refer to steps listed in Figure 29-8) |
		Chemicals and Detergents*	Poultice Materials	
Aluminum	White deposit	10% Hydrochloric acid	–	A-B
Asphalt	Black	1.* Ice. (Dry ice is not very effective)	–	C
		2.* Scouring powder	–	A
Emulsified Asphalt	Black	Water	–	A_1
Cutback Asphalt	Black	1.* Benzene	Talc or whiting	$D_1 \cdot D_2 \cdot D_3$
		2.* Scouring powder	–	A
		3.* Same	Same	E
Coffee	Tan	Sodium hypochlorite OR Glycerine 1 part Water 4 parts	–	F
Copper, bronze	Green, sometimes brown	Ammonium chloride 1 part Ammonium hydroxide as needed for paste	Talc 4 parts	$D \cdot D_1 \cdot D_2 \cdot D_5 \cdot E$
Creosote	Brown	1.* Benzene	Talc or whiting	$D_1 \cdot D_2 \cdot D_5$
		2.* Scouring powder		A
Ink, ordinary blue	Blue	Sodium perborate Water	Whiting	$G \cdot D_1 \cdot D_2 \cdot D_5 \cdot E \cdot E_1$
Containing Prussian Blue	Blue	Ammonium hydroxide OR Strong soap solution	–	F
Red, green, violet, other bright colors and indelible synthetic dyes	Varies	Sodium perborate Water OR Sodium hypochlorite OR Calcium hypochlorite	Whiting	$G \cdot D_1 \cdot D_2 \cdot D_5 \cdot E$
		OR Ammonium hydroxide OR Sodium hypochlorite OR Calcium hypochlorite	–	F
Indelible, containing silver salt	Black	Ammonium hydroxide	–	F-E
Iodine	Brown	Ethyl alcohol	Whiting or talc	$H_1 \cdot D_1 \cdot D_2 \cdot D_5$
		OR None	–	J
Iron	Brown yellow	Sodium or ammonium citrate 1 part Water (lukewarm) 6 parts Glycerine (lime-free) 7 parts	Whiting or Diatomaceous earth	$G_1 \cdot K \cdot D_1 \cdot D_2 \cdot D_5 \cdot E \cdot B$
		OR Same plus sodium hydrosulfite for step L	Same	$G_1 \cdot K \cdot D_1 \cdot D_2 \cdot L \cdot D_5 \cdot B$
Linseed oil	Dark gray	Trisodium phosphate 1 part Sodium perborate 1 part Liquid green soap or strong soap solution in hot water	Lime, whiting, talc or portland cement	$M \cdot G_1 \cdot D_1 \cdot D_2 \cdot D_5 \cdot B \cdot E$
Lubricating oil or grease that has penetrated	Dark	1.* Trisodium phosphate 1 lb Water 1 gal.	–	$N \cdot G_1 \cdot A$
		2.* Benzene	Talc, lime or whiting	$D_1 \cdot D_2 \cdot D_5 \cdot B$
		OR Amyl acetate (for small areas only)	Asbestos fiber	$D_1 \cdot D_2 \cdot D_3 \cdot D_5$
Mortar smears	Gray	–	–	O
Paint, at least 3 days old	Varies	Trisodium phosphate 2 lb. Water 1 gal OR Commercial paint remover	–	H-P
freshly spilled		Same		S-R-H-P
Perspiration stains from hands, and hair oil stains	Brown or yellow	Trichloroethylene	Talc	$T \cdot D_1 \cdot D_2 \cdot D_4 \cdot D_5$
Plant growth, mold and moss	Green, brown or black	Ammonium sulfamate (from garden supply stores)	–	V, and B if needed
Smoke and fire	Brown to black	Trichloroethylene	Talc	$T \cdot D_1 \cdot D_2 \cdot D_4 \cdot D_5$
Soot and coal smoke	Black	Soap Water Pumice	–	A
Wood tar and smoke	Dense black	1.* Scouring powder Water	–	A
		2.* Sodium hypochlorite	–	F
Tobacco	Brown	1.* Calcium chloride Water	–	D_1
		2.* Trisodium phosphate 2 lb Water 5 qt.	–	$G_1 \cdot X$
		3.* –	Talc	$D_1 \cdot D_2 \cdot D_5 \cdot E \cdot B$
Wood rot	Chocolate	1.* Glycerine 1 part Water 4 parts	–	A
		2.* Trichloroethylene	Talc	$T \cdot D_1 \cdot D_2 \cdot D_4 \cdot D_5$

*Numbers indicate that materials are to be used in sequence.

Fig. 31-11 (continued).

ACHIEVEMENT REVIEW

Select the best answer from the choices offered to complete each statement. List your choice by letter identification.

1. The first procedure in washing a brick wall with muriatic acid is
   a. scrubbing the wall with an acid solution.
   b. soaking the wall with water.
   c. dry cleaning the wall with a wooden paddle or scraper.

2. A metal scraper should not be used on masonry work when an acid solution is being used because
   a. it will scratch and scrape the bricks.
   b. it will catch in the mortar joint and cause permanent damage.
   c. it may cause rust spots or stains.
   d. the mason could be injured by the sharp metal edge on the scraper.

3. A common trade name for muriatic acid is
   a. sulphuric acid.   c. SURE KLEAN.
   b. hydrochloric acid.   d. acetic acid.

4. When mixing muriatic acid for cleaning brick, the correct proportions of mix are
   a. 1 part acid to 3 parts clean water.   c. 1 part acid to 7 parts clean water.
   b. 1 part acid to 5 parts clean water.   d. 1 part acid to 9 parts clean water.

5. To protect the mason from the acid solution when cleaning masonry work, the clothing should be
   a. canvas or heavy twill.   c. plastic.
   b. rubber.   d. light cloth.

6. Temperature and weather conditions have a definite effect on how much of the wall is washed before it is rinsed with water. However, the average area measures
   a. 5 to 10 sq. ft.   c. 20 to 30 sq. ft.
   b. 10 to 20 sq. ft.   d. 30 to 40 sq. ft.

7. Acid solution loses its ability to remove mortar smears after
   a. 5 to 10 minutes.   c. 20 to 25 minutes.
   b. 10 to 20 minutes.   d. 30 minutes.

8. When cleaning rough-textured bricks, it is absolutely essential that
   a. a solution of acid or cleaning agent which is stronger than that for smooth bricks be used.
   b. pressurized water be used.
   c. the bricks not be scraped or rubbed, as the texture may be damaged.
   d. muriatic acid not be used.

9. The major drawback to proprietary compounds is
   a. that the formula is not printed on the container.
   b. their chemical composition.
   c. that they are too slow in removing dirt and mortar from walls.
   d. that there is danger to the mason from the high acid content.

10. Particles of mortar are usually cleaned from concrete block by
    a. washing with an acid solution.   c. rubbing them with a piece of block.
    b. washing only with clean water.   d. scrubbing with a wire brush.

11. The last step in cleaning concrete block is to
    a. fill any holes.
    b. brush the wall.
    c. chisel off any large particles.
    d. rinse the work with water.

UNIT 32
Removing Various Stains

━━━━━━━━━━━━━━━━━━ OBJECTIVES ━━━━━━━━━━━━━━━━━━

After studying this unit, the student will be able to

- identify stains on masonry work.
- select the proper treatment for removing stains.
- explain the importance of the proper use of cleaning agents.
- list protective clothing items to be worn when working with cleaning solutions.

On occasion, there may be dirt and stains other than mortar deposited on masonry work. These must be removed by the mason when the building is cleaned. There are different types of chemicals that are used for this purpose.

> **Caution:** Strict attention should be given to the directions for the use of any chemical, as they have a harmful effect if not used properly.

In many cases, small buildings or buildings that contain hard, smooth surfaces are cleaned successfully with regular soaps or detergents. This method works particularly well on glazed bricks or glazed tiles since mortar or other stains do not cling to this type of surface as easily as they do on a sand or rough-finished brick. The soap or detergent does not burn or damage the face of the masonry unit, as might other cleaning agents discussed in the previous unit. This method is more costly, however, since it is considerably slower. Because of this, regular soaps or detergents cannot be used on very large structures.

Although the mason is responsible for cleaning and washing masonry work, stains such as tar, paint, or welding splatters are the responsibility of the subcontractor and general contractor working on that particular job. The apprentice should be aware of materials that may stain and damage the masonry work, and report them to his or her employer at once so that the proper precautions may be taken. This unit discusses

various stains, originating both internally (within the brick) and externally, and methods that may be used to correct the problem. Figure 32-1 lists various cleaning agents and their sources.

INTERNALLY CAUSED STAINS

Efflorescence

Generally, *efflorescence* refers to a white, powdery substance sometimes seen on masonry wall surfaces. It is composed of one or more water-soluble salts originally present in the masonry materials that have been brought to the surface by water and deposited on the surface by evaporation of the water. In many cases, it can be removed by applying clean water to the wall and then scrubbing with a brush. If this procedure does not remove all of the efflorescence, the surface is scrubbed with a solution of muriatic acid or SURE KLEAN 600 Detergent mixed no stronger than 1 part commercial acid to 12 parts water (by volume). It is very important that the wall be presoaked with water before any washing is done, and that the wall be thoroughly rinsed with water from a hose equipped with a nozzle after it has been washed with the acid solution.

Vanadium Stains

It is generally agreed that greenish stains are caused by salts present in the metallic element known as *vanadium*. While the stain is usually green, it is at times a brownish green and, more rarely, brown. The amount

378

Agent	Supply Source
1. Aluminum Chloride	Pharmacist.
2. Ammonia Water	Supermarket. Household ammonia water.
3. Ammonium Chloride	Pharmacist. Salt-like substance.
4. Ammonium Sulfamate	Nursery and garden stores. Past use was as a base for weed killers. Not now readily available. Substitute any brand weed killer solution.
5. Acetic Acid (80%)	Commercial and scientific chemical supply firms.
6. Hydrochloric Acid	Hardware stores. Muriatic acid is generally available in 18° and 20° Baumé solutions.
7. Hydrogen Peroxide (30 - 35%)	Some commercial and scientific chemical supply firms.
8. Kieselguhr	Commercial, scientific chemical and swimming pool supply firms. Diatomaceous earth.
9. Lime-free glycerine	Drug stores. Used as a hand lotion base.
10. Linseed Oil	Hardware and paint stores.
11. Paraffin Oil	Hardware stores.
12. Powdered Pumice	Hardware stores. A sanding or polishing material.
13. Sodium Citrate	Pharmacist. Appears like enlarged salt granules.
14. Sodium Hydroxide (Caustic Soda)	Supermarket. Available in brand name substances such as Drano.
15. Sodium Hydrosulphite	Pharmacist or photographic stores. A white salt or "hypo" of photographic fixing agent.
16. Talc	Drug stores. Inert powder available as "purified talc." Bathroom talcum powder may be substituted.
17. Trichloroethylene	Commercial/scientific chemical supply firms and possibly some service stations or supermarkets. A highly refined solvent for dry cleaning purposes.
18. Trisodium Phosphate	Paint stores, some hardware stores, supermarkets. Strong base type powdered cleaning material sold under brand names. Also available in brand name substance such as Calgon.
19. Varsol	Service Stations. A refined solvent by the brand name Varsol.
20. Whiting	Paint manufacturers, possibly some large paint stores. A powdered chalk. Substitute kitchen flour, if purchase is difficult.

Fig. 32-1 Sources of cleaning and masking agents.

of vanadium in a brick is very small, about 0.01%. It is not known in what form the vanadium is present in the raw materials, in the fired bricks, or on the surface of the stained bricks. If this could be determined, the problem of removal would be greatly simplified. Research is currently being conducted to identify the sources of the compounds involved.

The following are three facts about the chemistry of vanadium that the mason should know.

• Vanadium salts may be divided into two classes, which include *colorless salts* that crystallize in alkaline or neutral solutions, and *colored salts* that are obtained from an acidic solution. The colorless salts are quickly soluble in water, while the colored salts are slightly soluble.

- The reaction of the colorless salts is practically instantaneous; the colored salts change very slowly.
- Vanadium salts react much more rapidly with acid than they do with alkaline substances.

Green-stained bricks often show no sign of a stain until they are washed in an acid solution, at which time the salts mix with the acid solution and then become evident on the face of the wall as dry, colored salts. It is impossible to determine if vanadium stains are going to appear on masonry. Therefore, it is good practice to test the effect of an acid wash on masonry units by applying it to a sample wall before washing the entire structure.

If green stains appear on the surface of the wall following the acid wash, the following procedure, which provides for neutralization of the acid, should be followed.

1. Flush the wall thoroughly with water.
2. Wash or spray the wall with a solution of potassium or sodium hydroxide, consisting of ½ lb hydroxide to 1 qt water (2 lb per gal). A paint brush may also be used to apply the solution. Allow this to remain on the wall for 2 or 3 days in order to neutralize the acid that causes green staining. An easy way to use sodium hydroxide is in the form of Drano®. The mixture that has been used successfully in testing by the BIA is 1, 12 oz can per quart of water. The sodium hydroxide, or Drano, leaves the white salt that can be washed off with a hose. Various proprietary compounds, such as compounds 5, 6, and 7 in Figure 32-1, have proved successful in some cases.
3. The white salt left on the wall by the hydroxide may be hosed off the wall after 2 or 3 days or allowed to set until the first heavy rain removes it.

To date, research has not developed any single method for removal of green stains that can be recommended as best for all conditions.

Manganese Stains (Brown Stains)

Under certain conditions, manganese stains occur on mortar joints of brickwork made up of units colored with manganese dioxide. It appears as a tan, brown, nearly black, or, sometimes, gray stain. The *brown stain* has an oily look and may streak down the face of the brick. The salts are deposited when the solution reaches the mortar joints and becomes neutralized by the cement or lime.

During the burning process in the manufacture of some brick, the manganese coloring agents experience several chemical changes. This results in compounds that are not water soluble, but are soluble in weak acid solutions. Since brick can absorb acid, such weak acid solutions can prevail in brick washed with muriatic acid. Rainwater is also acidic in some highly industrialized areas.

To solve this problem, do not use muriatic acid solutions on tan, brown, black, or gray brick. There are proprietary cleaning compounds available for cleaning manganese brick. Advice of the brick manufacturer should be followed if there is a problem.

Permanent removal of manganese stains may be difficult. After the first removal, it sometimes returns. The following method has been very effective in removing brown stain and preventing its return.

1. Carefully mix a solution of acetic acid (80% or stronger), hydrogen peroxide (30–35%), and water in the following proportions by volume: 1 part acetic acid, 1 part hydrogen peroxide, and 6 parts water.

> **Caution:** Although this solution is very effective, it is a dangerous solution to mix and use. Consult with your supervisor or instructor before attempting to mix or use this solution. Otherwise, serious injury could result.

2. After wetting the wall, brush and spray the solution on the wall. Do not scrub. The reaction is very rapid and the stain disappears quickly. After the reaction is complete, thoroughly rinse the wall with water.
3. A proprietary compound, Brick Klenz®, is sometimes effective in keeping the stain from reappearing. Brush or spray a solution of 1 part Brick Klenz to 3 parts water by volume. Do

not scrub it; allow it to remain on the brick surface.

An alternate solution suggested for new and light-colored *brown stain* is oxalic acid crystals and water. Mix 1 lb crystals to 1 gal water.

Using the Poultice

A *poultice* is a paste consisting of a solvent and an inert material. The inert material may be talc, whiting, fuller's earth, or bentonite. The solution or solvent used depends upon the stain to be removed. Enough of the solution or solvent is added to a small quantity of the inert material to make a smooth paste. The paste is smeared on the stained area with a trowel or spatula and allowed to dry. It is then scraped off.

The solvent in the poultice dissolves the stain on the bricks. The resulting solution moves to the surface of the poultice where the solvent evaporates. The stain that is left on the loose, powdery residue is then removed. If all of the stain does not come off the first time, the procedure is repeated. The chief advantage of poultices is the way in which they work. Poultices tend to prevent the stain from spreading during treatment and tend to draw the stain out of the pores of the brick.

If the solvent being used to prepare a poultice is an acid, do not use whiting as the inert material. As a carbonate, whiting reacts with acids to produce carbon dioxide. This is not dangerous but is messy and destroys the power of the acid.

> **Caution:** The student should wear eye protection and rubberized clothing when using any of these cleaning agents. If any of the cleaning agents are splashed on the eyes or bare skin, flush with clean water. If any discomfort persists, see a doctor immediately.

Paint Stains

For fresh paint, apply a commercial paint remover or a solution of trisodium phosphate and water (2 lb trisodium phosphate to 1 gal water). Allow the mixture to stand and remove it with a scraper and wire brush. Wash with clean water. For very old, dried paint, or-

ganic solvents similar to the above may not be effective. In these cases, scrubbing with steel wool or sandblasting may be required.

Iron Stains

Iron stains, quite common, sometimes cover entire walls. These stains are easily removed by spraying or brushing with a strong solution of 1 lb oxalic acid to 1 gal water. Ammonium bifluoride added to the solution ($\frac{1}{2}$ lb/gal) speeds the reaction. The ammonium bifluoride generates hydrofluoric acid, which etches the brick. The etching will be evident on very smooth bricks, and, therefore, the solution should be used with caution.

Another method is to mix 7 parts lime-free glycerine with a solution of 1 part sodium citrate and 6 parts lukewarm water, and mix with whiting or kieselguhr to form a thick paste. Apply the paste to the stain with a trowel and scrape it off when it dries. Repeat the process until the stain is removed. Wash well with clean water. A poultice made from a solution of sodium hydrosulphite and an inert powder (such as talc) also has been used for removal of iron rust stains.

Copper and Bronze Stains

Mix together dry 1 part ammonium chloride or sal ammoniac and 4 parts powdered talc. Add ammonia water and stir until a thick paste is formed. Place the mixture over the stain and leave until dry. An old stain may require several applications.

Welding Splatters

When metal is welded too close to a wall or pile of bricks, some of the molten metal may splash onto the bricks and melt the surface. The oxalic acid-ammonium bifluoride mixture, which is recommended for iron stains is particularly effective in removing welding splatters. Scrape as much of the metal from the bricks as possible. The solution is then applied in a poultice. The poultice is removed when it has dried. If the stain has not disappeared, sandpaper is used to remove as much of it as possible and a fresh poultice is applied. For stubborn stains, several applications may be necessary.

Smoke Stains

Smoke stains are usually difficult to remove. A thorough scrubbing with scouring powder (preferably one containing bleach) and a stiff-bristled brush works well. Some alkaline detergents and commercial emulsifying agents may be brushed or sprayed on. These do a good job when they are given time to work. These have the added advantage that they can be used in steam cleaners. For more stubborn stains, a poultice using trichloroethylene usually draws the stains from the pores.

> **Caution:** When using trichloroethylene, make sure the work area is well ventilated since the fumes are harmful.

Oil and Tar Stains

Oil and tar stains are effectively removed by commercial emulsifying agents. For heavy tar stains, the compounds can be mixed with kerosene to remove the tar and then with water to remove the kerosene. Sometimes, a steam-cleaning apparatus is used to remove tar with the use of kerosene. In a small area or in a situation where the job must be very clean, a poultice using benzene, naphtha, or trichloroethylene is more effective in removing oil stains.

Dirt Stains

Dirt can be very difficult to remove from a textured brick. Scouring powder and a stiff-bristled brush are effective if the texture of the unit is not too rough.

Scrubbing with the oxalic acid-ammonium bifluoride solution recommended for iron stains has proved effective on moderately rough textures. High-pressure steam cleaning appears to be the most effective method for cleaning dirt stains.

Straw and Paper Stains

Stains from straw and paper sometimes result from wet materials used to pack bricks for shipment. This stain can be removed by applying household bleach and allowing it to dry. Several applications may be neces-

sary to remove the stain. A solution of oxalic acid-ammonium bifluoride cleans the stain more rapidly.

Plant Growth

Sometimes exterior masonry that is exposed to sunlight and remains constantly damp develops plant growth, such as moss. Applications of ammonium sulfamate (marketed under the manufacturer's brand name and available in gardening supply stores) made according to directions, which come with the compound, have been used successfully to remove such growths.

Stained masonry that is being recleaned after a long period of time may require a more severe method. High-pressure steam and sandblasting are two of the most successful methods employed. The mason seldom uses these methods, as they are usually left to professional cleaning companies.

Ivy

Ivy can damage the mortar joints in a masonry wall as the suckers on the plant enter the mortar joints and cause them to erode. Avoid pulling the vines away from the joints as this may cause more of a problem. Carefully cut a few square feet of the vine away from the mortar joints designated area and examine the joints to see how much the vines are rooted. There will be some deposits left on the surface of the adjacent masonry. These usually are the "suckers" that attached and held the vines. Do not use acids or chemicals to remove the sucker. Leave in place until they dry and turn dark. Then, remove with a stiff brush and detergent.

Egg Splatter

Brick walls that have been vandalized with raw eggs have been successfully cleaned with a solution of oxalic acid crystals dissolved in water. Mix in a non-metallic container and apply with a brush after saturating the surface with water.

White Scum

White scum is a grayish-white haze on the face of a brick. It is sometimes mistaken for efflorescence, but technically is silicic acid scum. This condition results from the failure to saturate the wall before application or failure to thoroughly rinse acid solutions after cleaning. Generally, it is a film of materials that is insoluble in acid solutions except for hydrochloric acid, which is dangerous and not generally recommended for this use. Proprietary compounds formulated to remove this condition may be tested and their effectiveness judged. If removal is too difficult, masking of the haze may be considered. In time, weathering will remove both the mask and the white scum.

Masking solutions may consist of paraffin oil and Varosol, or linseed oil and Varosol, applied by brush to the affected brick work. Linseed oil and Varosol (10–25% linseed oil) or paraffin oil and Varosol (2 to 50% paraffin oil) will darken lighter color brick. Several batches of solutions with various concentrations should be mixed and tested. Generally, solutions of 2 to 25% paraffin oil will be satisfactory. Allow 4 to 5 days of warm drying weather to pass, preferably at 70 degrees Fahrenheit (21 degrees Celsius) minimum, before a judgment is made on the effectiveness of the solutions.

It is always recommended to test any cleaning procedure and chemical cleaning solutions in a small area before attempting to clean the entire job. Cleaning brick masonry still remains a trial-and-error procedure and should be approached in that manner.

SAFETY PRACTICES TO FOLLOW WHEN WORKING WITH MASONRY CLEANERS

Masonry cleaning products can be corrosive and toxic in nature. Common-sense safety practices should be followed at all times when handling or using any type of chemical cleaner. The first rule is to always read the directions on the package or container before using the contents.

Some general recommendations that should be followed are to wear long rubber gloves and safety goggles when using any type of chemical liquid or cleaner, and to wear an approved mask if fumes are present. It also helps to cover areas of skin that may be exposed with a light application of a product such as Vaseline and to make sure there is good cross-ventilation if you are working inside. Try to work downwind as much as possible, especially if using acid type cleaners that emit fumes. Never open a container of chemical cleaner of any type and then take a deep breath of it. It can replace the oxygen in your lungs quickly and cause extreme discomfort and breathing problems.

When mixing an acidic type of cleaning solution in a bucket, it is a good practice to always pour the water in first and then slowly add the cleaner. Don't *"spike"* or add larger doses of cleaner to a solution than the directions call for, in an attempt to remove stubborn spots or stains, as it may not only burn the mortar joints, causing a discoloration, but also actually affect the color of the brick face. If you do splash some cleaner on an exposed surface of your skin or in the eyes, immediately rinse it out with plenty of clear water. If this does not help, seek medical attention immediately.

SOME HANDY CLEANING TIPS THAT I HAVE USED WITH SUCCESS

- If you want to remove a stubborn clump of mortar or stain on a masonry wall, rub it with a small piece of broken masonry material that matches the wall, such as a piece of brick, as it will be less likely to scratch or ruin the face of the unit.

- In some cases, applying dry powered cement or granular cat litter on a grease or oil stain will help lift out the grease stains from masonry. Then, scrub the area with a masonry cleaning detergent and rinse with clear running water.

- If you are trying to remove an asphalt or tar stain (especially if it has not dried), use a cloth saturated with kerosene and wipe over the area. Follow this up by wiping it off with another piece or two of cloth with clean kerosene on it. Rub lightly across the surface and try not to force the stain into the pores of the material being cleaned. Finish up cleaning the area with regular household scouring powder and rinsing with clear water.

- Another technique that will work sometimes is to apply *dry ice* to a lump of tar on the face of a brick or block. It helps to freeze the underside of the lump and make it pop loose. Chewing gum can also be removed rather easily by holding an ice cube against it for a couple of minutes. This will cause it to harden and then it can be scraped off with a putty knife. If a stain still remains, use a little denatured alcohol on a cloth to remove it.

- You can remove wet paint or a stain from a masonry wall by blotting it with a soft cloth or paper towel to absorb the excess. Then scrub the area with a laundry soap detergent and warm water and rinse it with clear water. If it is an oil-base paint, try blotting it with a little turpentine or paint thinner and scrub the area with some household scouring powder, followed by rinsing with water. If the paint has hardened, try using some paint-brush cleaner and rinse afterwards. Usually this will work.

- To kill moss on brickwork, spray with a good weed killer according to directions or use a solu-

tion of 1 part Clorox to 8 parts water. Rinse well afterwards with clear running water from a hose.

- To get rid of mildew on concrete block, scrub with a medium soft brush using a solution of 1 ounce of laundry detergent to 3 ounces of Trisodium Phosphate to 1 quart of chlorine bleach mixed with 3 quarts of water. Rinse well with water from a garden hose.

- To enhance the color of Vermont Slate Flagstone, which has a lot of blue, red and green in it, wash it down with a brick cleaner such as SURE KLEAN 600 or a solution of 1 part muriatic acid to 9 parts water and rinse well with a garden hose with a spray nozzle. After it dries, apply a coat of liquid self-polishing wax to bring out the colors in the stone.

ACHIEVEMENT REVIEW

Select the best answer from the choices offered to complete each statement. List your answer by letter identification.

1. The best method to remove efflorescence is to apply a solution of
   a. sulphuric acid.
   b. potassium.
   c. household cleanser.
   d. muriatic acid.

2. Manganese stains can be removed with
   a. muriatic acid.
   b. a combination of acetic acid, hydrogen peroxide, and water.
   c. whiting.
   d. trichloroethylene.

3. Green stains on masonry walls are caused by
   a. efflorescence.
   b. manganese.
   c. vanadium salts in the bricks.
   d. iron in the bricks reacting to the mortar.

4. The chief advantage of a poultice is that
   a. it is easier to apply than a solution.
   b. it draws the stain to the surface and stops it from spreading.
   c. it presents no danger to the mason who is applying the poultice.
   d. it permanently prevents the stain from reappearing.

5. Oil and tar stains are removed from masonry using certain compounds. Heavy tar stains are removed by adding to these compounds
   a. muriatic acid.
   b. household bleach.
   c. ammonium sulfamate.
   d. kerosene.

6. Plant growth on masonry work is removed by using a solution of
   a. ammonium sulfamate.
   b. muriatic acid.
   c. whiting.
   d. talc.

7. Paint stains are removed with a commercial paint remover or a solution of
   a. muriatic acid and water.
   b. household bleach and water.
   c. trisodium phosphate and water.
   d. oxalic acid and water.

8. When in doubt about the effects of a cleaning compound on a masonry wall, the best practice is
   a. to clean only a small area and observe the results.
   b. to use judgment as to the type of cleaner to be used.
   c. not to attempt to remove the stain.
   d. to apply household bleach.

SUMMARY, SECTION 7

- Steel scaffolding has replaced wooden scaffolding for the most part because of speed of erection, cost of lumber, and requirements of state and federal safety regulations.
- The most commonly used scaffolding is the tubular steel sectional scaffolding.
- The OSHA federal safety laws were established to protect all employees from hazards.
- The employer is responsible for seeing that safe working conditions are maintained at all times on the job.
- Scaffolding must be built to conform to OSHA regulations and requirements.
- Two major types of tubular steel scaffolding are used, the standard frame measuring $5' \times 5'$ and the type with more height which allows the worker to walk freely through the center.
- Platform extenders (more commonly called brackets) and scaffold hangers should carry only the weight of the mason, not the materials.
- Rolling scaffolding is not recommended for general masonry work. It can be used only for quick, small jobs such as pointing joints.
- Wheels must be locked at all times when the mason is working on rolling scaffolding.
- The mason has a responsibility to care properly for steel scaffolding since it is expected to last for a period of years.
- The adjustable tower scaffolding offers the advantage that the mason may remain at the best working height. It also has an advantage in that the mason does not have to move while the scaffold is built higher.
- Suspended or swinging scaffold is used on high-rise structures where the erecting of standard tubular steel sectional scaffolding is not profitable or safe.
- A suspended scaffold is hung from steel cables which are fastened to steel outriggers anchored in the roof of the structure. It is considered a very safe scaffold if built correctly.
- Regardless of the type of scaffolding used, the mason should always inspect the scaffold to see if it is safe before beginning work. If the scaffolding is unsafe, the mason should refuse to work on it until the problem is corrected.
- It is the responsibility of the mason not to throw, kick, or knock anything from scaffolding.

- All scaffolding is limited in terms of the amount of materials which can be loaded on it.
- Scaffolds should be inspected periodically to be sure that they have not moved away from the building during construction and that none of the safety attachments have been removed.
- Foot boards are used to reach scaffold height or work to the top of a wall.
- Foot boards should not be used as a replacement for scaffolding and should never be built over 16″ in height.
- All scaffolding set on ground level should have a wooden sill plate under it to distribute the weight of the scaffolding and to prevent it from sinking into the ground.
- Braces should be attached with nuts or slip fasteners, never wired to the frame.
- Cross braces are not to be climbed upon, but are there only to hold the scaffolding rigid and secure.
- Scaffolding more than 2 sections high should be tied to the structure with heavy wire.
- Wooden planking should be free from splits and twists and lapped 12″ over each other.
- Scaffolds should be spaced far enough from the wall so that no part bumps against the wall when the masons are working on it.
- Federal and state regulations require that guardrails and toe boards be on all scaffolds that rise to a height of 10′ or more.
- Extreme caution should be taken that no contact is made with electrical power lines when they are located near the scaffold or work area.
- Overhead protection in the form of planking laid on steel frames should be used whenever there is a danger of objects falling from overhead.
- Never ride on rolling scaffolding.
- Suspended scaffolding should be inspected frequently for frayed cables, loose boards, or other unsafe conditions.
- Practical joking is potentially dangerous and will not be tolerated on the construction site.
- The purpose of washing brickwork with a cleaning agent is twofold: to remove the mortar stains and to bring out the full color of the bricks.
- Cleaning time and labor can be drastically reduced if care is taken during construction to keep the masonry work clean.
- Damage to the masonry wall can be caused by poor cleaning practices and can result in rebuilding of the work.
- Whenever there is any question of the effect of a cleaning agent on the wall, it is good practice to first clean a small area and observe the results.
- Metal tools should not be used when cleaning with acid solutions since they may leave rust stains on the bricks.
- A rubber, wood, or good-quality plastic container should be used for storage of cleaning solutions.
- The mason should be dressed in waterproof clothing and safety eyeglasses or goggles whenever working with cleaning compounds or solutions.

- When cleaning masonry work on a scaffold, be aware of any person nearby or underneath the scaffold.
- Presoak the wall in water before applying any acid or cleaning solution.
- Immediately after washing a certain section of wall (the size of the section depends on weather conditions), rinse with plenty of water until the water running down the wall is clear.
- Rough-textured brick walls should be rinsed with a pressurized source of water, such as a garden hose with a nozzle, to remove dirt from the deep pores.
- When cleaning brickwork, the mason must be careful not to mix a solution which is too strong, thereby burning the wall, and covering too great an area before rinsing, thereby allowing dirt to be absorbed in the wall.
- Wear a pair of long rubber gloves to cover part of the arm to prevent infections from occurring in open cuts. Rinse any area exposed to a cleaning agent immediately with plenty of clean water.
- Concrete block usually is cleaned without the use of water by rubbing the area with a scrap of block or brick unless there is a special problem. A good brushing usually removes all dirt.
- Filling holes in the wall with fresh mortar usually is done at the same time the concrete block wall is rubbed for stain removal.
- Always read directions before using any cleaning agent.
- Soap and detergents can be used to wash smooth bricks or tile, but this can be a very time-consuming task since they do not destroy mortar the way acid does.
- The way to remove efflorescence from a masonry wall is to apply a solution of water and muriatic acid.
- Green staining is caused by the salts in vanadium and can be removed with sodium hydroxide applied with a paint brush.
- Manganese stains are very difficult to remove. The best method is to brush on an acetic acid-hydrogen peroxide solution and wash it off with plenty of water.
- The chief advantage of a poultice is that it tends to draw the stain out to the surface pores of the brick, where it is easier to treat.
- A fresh paint stain can be removed with a paint remover or a solution of trisodium phosphate and water.
- Iron stains are removed with a combination of ammonium chloride and 4 parts powdered talc.
- A welding splatter can be removed with an oxalic acid-ammonium bifluoride mixture.
- A smoke stain can be removed by using a scouring powder containing bleach.
- Common dirt can be removed by using water and scouring powder.
- Straw and paper stains can be removed with a household bleach, or a solution of oxalic acid and ammonium bifluoride.
- Plant growth can be removed with ammonium sulfamate (sold in garden stores).
- Stubborn stains which cannot be removed with a cleaning agent may require the use of high-pressure steam or sandblasting. These methods are done only as a last resort and are left to professional cleaners rather than to the mason.

SUMMARY ACHIEVEMENT REVIEW, SECTION 7

Complete each of the following statements referring to material found in Section 7.

1. Steel scaffolding has replaced wooden scaffolding because of three main reasons, which are _____ , _____ , and _____ .

2. The most frequently used type of scaffold for masonry work is the _____ .

3. In 1970, Congress passed the Occupational Safety and Health Act, which is commonly referred to as _____ .

4. Scaffolding built in excess of 10′ vertically must have two protective devices, which are _____ and _____ .

5. Brackets are installed on scaffolding to provide _____ .

6. Scaffolding that is equipped with wheels is known as _____ .

7. The outstanding feature of a tower scaffold is _____ .

8. Adjustable tower scaffolding is raised by _____ .

9. Suspended or swinging scaffold is selected for use on high structures because it is _____ .

10. Suspended scaffold is supported by _____ .

11. Overhead protection for suspended scaffold is provided by _____ .

12. Regardless of the type of scaffolding, the first thing the mason should do before beginning work is to _____ .

13. If the mason requires a platform on which to step so that he may build to the point at which the scaffolding begins, a (an) _____ is built.

14. Scaffolding built on the ground should always have a (an) _____ under it to prevent the scaffold from sinking and to permit better distribution of weight.

15. The reason that a scaffold frame is never set on top of a hollow concrete block is because _____ .

16. Scaffolding should be tied to the structure whenever it is higher than _____ .

17. A section of wood or steel that runs the length of a ditch and on which scaffolding rests is known as a (an) _____ .

18. Wood planking selected for scaffolding boards should be free of _____ .

19. If electrical power lines are located so near a scaffold that they interfere with the building of the scaffold, the proper procedure is to _____ .

20. Masonry materials stocked on scaffolding should be placed directly over the _____ .

21. Brickwork is cleaned not only to remove mortar and dirt but to _____ .

22. Before applying any cleaning agents on brick walls, the walls should be _____ .

23. Using too strong an acid solution may cause the wall to _____ .

24. Washing too large a section with a cleaning solution without rinsing the wall may result in _____.

25. When there is a question of how the masonry wall will react to a cleaning agent, it is good practice to _____.

26. Metal tools or containers are not recommended for use around acid or cleaning agents because _____.

27. For best results, acid solutions should be removed from the brick wall by _____.

28. Proper clothing and protective devices worn when cleaning walls consist of _____.

29. The chief disadvantage of proprietary compounds is _____.

30. Concrete masonry products are not usually cleaned with _____ or _____ unless there is a special problem.

31. If an acid solution gets into a cut or into the eyes, the victim should immediately _____.

32. Regardless of the type of cleaning agent being used, the mason should follow _____ as a guide to its use.

33. Efflorescence is removed most easily with a solution of _____.

34. Green staining on masonry walls is usually caused by _____.

35. Manganese stain is usually _____ in color.

36. The main advantage of a poultice is its ability to _____.

37. Paint stains can be removed with a chemical called _____ and water.

38. Iron stains are removed by spraying or brushing the area with a strong solution of _____.

39. Many times, smoke stains are removed from brick walls by use of a common _____.

40. Plant growth is removed from masonry by using a chemical sold at many garden stores called _____.

SECTION EIGHT
FIREPLACES
AND CHIMNEYS

UNIT 33
History, Theory, and Function of the Fireplace and Chimney

―――――――――――――― OBJECTIVES ――――――――――――――

After studying this unit, the student will be able to

■ briefly describe the history of the fireplace and chimney up to modern times.

■ define the meaning of draft and explain how it works in the operation of the fireplace.

■ list the parts of the two- or three-flue chimney and state how each part contributes to chimney operation.

HISTORY OF THE FIREPLACE

Fireplaces and chimneys have been used for many centuries for heating and cooking purposes. The earliest fireplaces in the United States can be traced back to about the year 1600 in the eastern part of the country.

Most early fireplaces had large face openings and flue holes that sucked most of the heat given off in the firebox up the chimney. Since wood was the source of fuel and it was very inexpensive to obtain, not a great deal of thought was given to the efficient design of the fireplace in this time period. In addition, there was no damper on top of the firebox to regulate the draft coming down the chimney, which also contributed to the poor operation of the fireplace.

In the mid-1700s, two people had a dramatic effect on the design and building of fireplaces for hundreds of years to come. They were Benjamin Franklin of Philadelphia, Pennsylvania, and Sir Benjamin Thompson, also known as Count Rumford, of Woburn, Massachusetts. Franklin was interested in what made a fireplace operate and wondered why so many fireplaces smoked into the house and did not draw cleanly. After studying and observing fireplaces in use, Franklin realized that smoke was heavier than air; therefore, the draft that came down the chimney from the outside made the smoke rise up the chimney. The mixing of the hot air and gases that originated from the burning of the fire in the fireplace created the condition that carried the smoke and burning gases back up the chimney. The heat given off by the fire was radiated out into the room and warmed the area.

Franklin also contended that the room in which the fireplace was built should match the size of the fireplace if it was to heat the room efficiently. He also believed that the fireplace needed a supply of air in the room to make it burn properly. Today, we know that a supply of air has to be available in the room or firebox in order for it to burn correctly. Benjamin Franklin published various papers on his thoughts and observations of fireplaces, which are still regarded highly today. The curved back of the firebox common on many American fireplaces is still called the "Franklin Back" and helps promote the smooth passage of the hot gases and smoke up through the damper and into the chimney.

Fireplaces of Franklin's time used heavy wood beams to support the masonry work over the firebox opening. After a period of years the beams often weakened or burned out from the constant heat of the fire. The beams had to be carefully checked and repairs made when necessary. Today, iron or steel lintels are used, thus eliminating the problem. Figure 33-1 shows a modern colonial fireplace that uses a steel lintel over the opening.

Fig. 33-1 Stone fireplace built to resemble the old colonial or early American fireplace. The hearth is recessed from the floor level.

Count Rumford was a native-born American who went to England and met with success working with the British Government in various jobs. He was very interested in the fireplace and spent many years studying how it operated and how it could be improved. While in England, Count Rumford built a reputation in the repairing and correcting of fireplaces that did not operate properly. In the course of doing this, he discovered a number of principles relating to heat and to improving the operation of a fireplace.

The conventional fireplace, which is explained in detail in Unit 34, has a deeper firebox and moderately angled sides with a shorter front opening than a Rumford fireplace. With the Rumford fireplace, the concern was not with the appearance but more its efficiency as a heating source. The Rumford fireplace had a higher, more shallow firebox than a conventional one. Due to the shallow depth of a Rumford fireplace, it is a good idea to build a deeper forehearth in front of the fireplace as there is more possibility of sparks or hot embers popping out on the floor.

The width and height of the fireplace opening could be up to three times the depth of the fireplace. The width of the back wall should be about equal the depth of the firebox. In addition, the back wall should rise vertically from the hearth to a point of about 15 inches and then slant or roll forward gently and uniformly to the front edge of the smoke shelf. The smoke shelf should be about 3 to 4 inches deep, and the centerline of the throat should be directly over the center of the hearth. Both types of fireplaces are built to relating proportions and are considered to be more efficient if they are single face only (only one firebox opening to the room).

There are pros and cons for both types of fireplaces and each should be studied before deciding which type a builder would want to build. Up to this point in time in the United States, however, the majority of fireplaces built have been of the conventional design. Unit 34 shows an illustration of both types of fireplaces with relating proportions. Figures 33-2 and 33-3 illustrate cross-sections of both types described.

Mantels were first built on fireplaces around 1725 and 1750 when plastering replaced the use of wood paneling on the other walls of the room. During this period of time it was common to have very decorative or large mantels or front facings of wood or stone (such as mar-

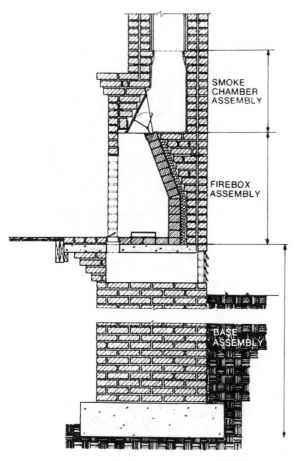

Fig. 33-2 Conventional brick fireplace.

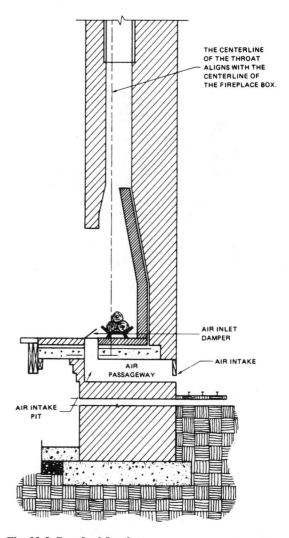

Fig. 33-3 Rumford fireplace.

ble) on the fireplace. When restoring old houses, many of these types of mantles are still encountered.

In today's fireplaces, the trend has been to simplify the mantel and, in many cases, have no mantel at all. Figure 33-4 shows a modern version of the colonial "ring around" fireplace front, in which only one stacked course of brick is seen around the opening of the fireplace. A wood trim piece is fitted against the front of the fireplace to cover where the brickwork joins the finished wall in the room.

BASEMENT FIREPLACES

Basement fireplaces, Figure 33-5, were rare until about 1750. They were large and also served as a support for part of the home. Pockets were built into the base to store the wood ashes for making soap, potash, and lye. Lime mortar was used because portland cement mortar was not in use at this time. Large stone slabs were laid level with the finished first floor. These slabs served as the inside and outside hearth for the fireplace on the first floor. The first-floor fireplace was then started off this base.

The basement fireplace of today serves a different purpose. In clubrooms, recreation areas, or enlarged living spaces for the family, a basement fireplace is used

Fig. 33-4 Modern version of the colonial "ring around" fireplace. It is made by stacking one course of bricks up the jambs and around the opening.

Fig. 33-5 Early American basement fireplace built about 1775.

to provide a pleasant atmosphere as well as heat. Many of today's homes have a fireplace in the basement level as well as the first floor. The additional fireplace increases the value of the home when it is to be sold.

Modern Fireplaces

The modern fireplace is built basically the same as years ago but with a number of improvements, Figure 33-6. Regardless of whether it is a conventional masonry fireplace, Rumford design, or metal heat-form circulating fireplace, the concept of draft and parts of the fireplace remain the same.

Some of the major improvements include the use of dampers to control draft; fireproof, fired-clay tile flue linings; highly improved mortars; firebrick for lining the firebox and hearths; fire screens and glass doors; and a supply of fresh air introduced through the wall of the chimney or hearth into the firebox.

Design wise, the hearth on many fireplaces is raised off the floor to provide a bench seat, architectural design, and a higher level of heat radiation from the fireplace. On the majority of old fireplaces, the hearth was always built approximately level with the finished floor with a hearth strip around it.

Fig. 33-6 Modern stone fireplace with simple wood mantel, fire screen, glass doors and raised brick hearth.

As previously stated, the fireplace is growing in popularity today because of the energy situation and the high cost of fuel. Past traditions influence many people to include a fireplace in the construction of their homes.

It is true that a fireplace is not as efficient as a stove or heating device located in the middle of a room. However, if the dampers are regulated properly, it will radiate much heat. In colonial times, the fireplace was used primarily for heating and cooking purposes. In more recent years the fireplace has served more of a decorative use rather than as a functional source of heat. However, since the cost of energy is rising, there is a growing consideration of fireplaces as an additional source of heat. Real estate brokers state that a fireplace increases the possibility of selling any house on the market place.

Importance of Draft in the Burning Process of the Fireplace

Draft is the current of air created by the variation in pressure resulting from differences in weight between the hot gases in the flue and the cooler air outside the chimney. The movement of draft or intensity depends on the height of the chimney and this difference in temperature. Draft is stronger in the winter than in the summer because of this fact. In addition, a fireplace and chimney built in the center of the home rather than on an outside wall provide a better draft because it remains warmer.

High trees near the chimney or the rise of a nearby hill can influence the proper draft and should be considered when constructing a chimney. Trees can cause downdrafts, which result in puffs of smoke coming from the firebox into the room. They may also cause the fireplace to smoke and not draw at all. The logical solution is to build the chimney top higher above the roof than is regularly done. Under usual conditions, however, the chimney is built 2 to 3 feet above the top of the ridge of the roof, Figure 33-7. Metal extensions to increase the chimney's height should not be used because they rust and burn out in a relatively short period of time. The chimney should be built high enough the first time so that extensions are not necessary.

Two common mistakes are often encountered in chimney design. Sometimes the flue size is not large enough, or else the flue is not high enough to produce a good draft. Either one or both of these mistakes will result in a chimney that does not draw properly, forcing smoke into the house.

Draft Operation in the Fireplace

The actual operation of draft in the fireplace is shown in Figure 33-8. Cool air is drawn down one side of the

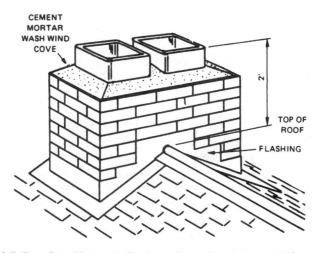

Fig. 33-7 Two-flue chimney built above the roof a minimum of 2′ to prevent down drafts in the chimney. Notice the mortar wash wind cove on top of the chimney to deflect the currents upward.

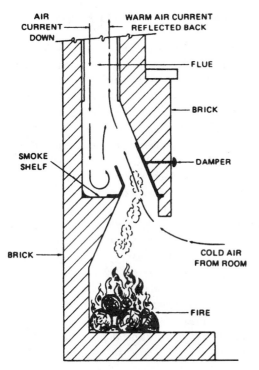

Fig. 33-8 Draft operation in the fireplace. Notice that air comes down one side of the flue, bounces off the smoke shelf, and goes up the other side of the flue. Cold air is sucked in from the room, completing the draft cycle.

flue. The warm air, smoke, and gases return up the other side. When the air coming down the chimney reaches the smoke shelf directly behind the damper, the air is deflected back up the chimney on the opposite side of the flue. Smoke and gases are taken up the chimney in the air. Air is sucked from the room into the firebox as a result of this action. As the air enters the firebox it quickly becomes heated and rises through the throat of the damper. It then meets the colder air bouncing off the smoke shelf.

Note: The *smoke shelf* is a brick ledge laid level with the bottom of the damper. It is built by extending the back firebrick lining forward until it meets the bottom rear flange of the fireplace damper. The mixture of hot and cold air causes the column of air to be drawn rapidly up the chimney. If the fireplace is not built correctly, this action and reaction does not occur.

Caution: The damper of the fireplace must always be checked to make sure it is completely open before building a fire in the fireplace.

Use of Outside Air for Greater Efficiency

The air needed for the combustion process was traditionally drawn from the room in which the fireplace was located. This movement of air occurred by being drawn through any crack around windows or doors, or through a window left slightly open to provide a source of air. This practice is especially wasteful in a time when heating the home is so costly, as much of the air drawn from the room was heated air. Anyone in the room would experience a draft which would tend to make them uncomfortable, if not cold.

When the concept of totally electrically heated homes came along, one of the prime requirements was to fully insulate the house and make it as tight or air resistant as possible. This made the problem of supplying air for the combustion of a fireplace even more difficult as there were no cracks or leaky doors and windows. A vent had to be installed in the wall or a window had to be left open to provide air. Better methods of introducing air into the firebox were sought to combat this problem.

A very successful, rather new development is to use outside air that is admitted into the combustion area by means of a metal duct or tube. This puts the needed air at the correct location in the firebox and prevents the loss of heat and draft in the rooms. Figures 33-9 and 33-10 illustrate an example of a very good manufactured duct and damper arrangement that is now available at most building supply dealers. Regardless of the brand and design, the function is basically the same. The device shown is for the building of a new fireplace and chimney. While the device can be installed in an old chimney, it would be very difficult and costly.

The device works like this: An intake grille or vent is built into the back of the outside chimney wall, above grade line. This vent has a screen-backed louver that can either be opened or closed when not in use to keep out insects or air. A metal tube or duct is attached to the vent and is built in under the hearth. Either concrete or

Fig. 33-9A Outside air kit which is installed in chimney to hearth to supply outside air to firebox. The grille cover for the outside wall and duct is shown in this photograph.

masonry work can be used to support the tubing. The 6-inch telescopic tubing is adjustable to fit any length needed with additional tubing available from the manufacturing company or a local hardware store.

Adequate clearance is provided for the firebrick hearth so the tubing does not interfere with any building of the masonry work. Metal elbows can be obtained if necessary to work around any obstructions.

A standard ash dump door is installed in the center forward section of the hearth. The area directly under the ash dump is formed with masonry into a chase or squared hole that leads directly to the end of the air tube. At the end of the air tube, which is lined up vertically with the back end of the ash dump, is a damper that can be opened or closed to admit air to the firebox through the ash dump door when it is open. The damper rod, which is connected to the end of the air tube and the damper plate, can be adjusted by means of tension to control the amount of air passing through the tubing or duct.

> **Caution:** When the fireplace is burning, this damper must be open at all times.

When the fireplace is not in use, or the fire goes out, the damper can be closed off, preventing the loss of valuable heat from within the house. This results in a very efficiently operating fireplace that does not depend on a supply of air from within the house. In the event that the device described cannot be installed through the back of the chimney, it could be built in between the floor joists of the home and connected to the exterior foundation wall and the firebox. The ideal arrangement, however, is to install the air duct beneath the firebox through the back wall of the chimney.

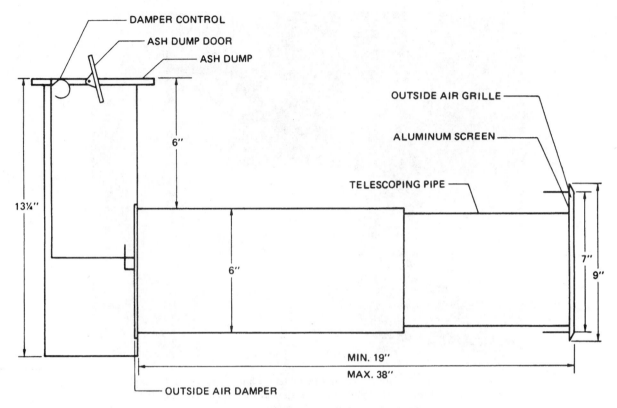

Fig. 33-9B Cross section view of outside air kit.

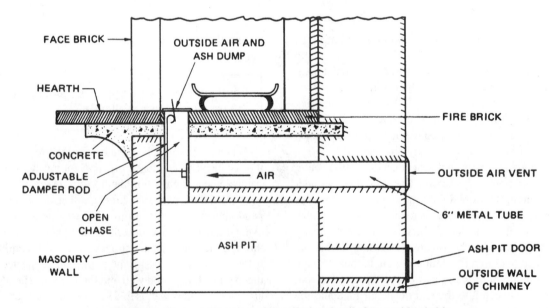

Fig. 33-10 Cross section view of installation details for outside air kit.

In place of the device described, a masonry opening can be formed as the chimney base is built, which would connect the outside air to the combustion area. The damper in this case would be the opening and closing of the ash dump door. Either method works successfully and admits a supply of outside air to the firebox. Figure 33-11 shows a section view of a brick fireplace with a masonry chase under the hearth to the firebox.

Tests have shown that about five times the amount of air required for good ventilation can be drawn into the room by the operation of a fireplace. Further tests also showed that a fireplace is only about one-third as efficient as a good circulating-type heater. Nevertheless, fireplaces still provide a source of heat, add to the

cheerful atmosphere of the home, and are highly desirable by the homeowner.

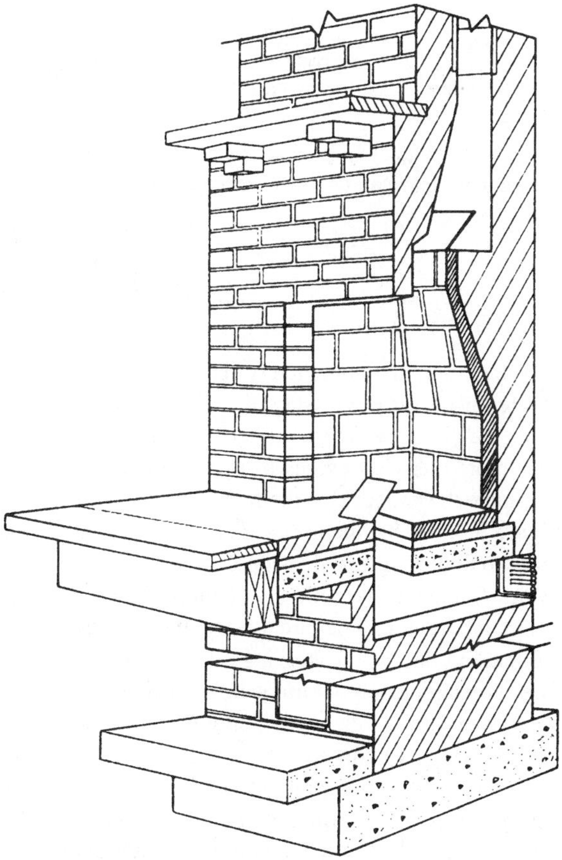

As the fire burns in the firebox, the damper is regulated slowly to the point that it does not smoke; but the fire still burns freely. The maximum amount of heat possible is then radiated into the room discounting the heat, of course, that naturally goes up the chimney. If the damper, which is at the top of the firebox, is left entirely open when the fireplace is not in use, a great amount of heat from the room escapes up the chimney. This is where fire-approved glass doors are helpful on the front of the fireplace. They seal off the heat loss effectively while still allowing radiated heat into the room and provide safety from sparks that could pop out of the firebox.

> **Caution:** When building a fire in a fireplace, keep the fire to the back. Start with a small fire until the flue heats up and then increase the size of the fire.

TWO- (OR MORE) FLUE CHIMNEYS

Solid Masonry

Only masonry materials should be used in the construction of a chimney that serves a fireplace. Hard fired brick is a good choice. If brick is used as the facing material, concrete block or stone can be used inside as the fill unit, except against any area that will be subjected to direct heat. Always use a hard fired brick for the roughing in of the firebox and the smoke chamber of the chimney. Flue liners should also be surrounded by at least 4 inches of hard fired brick for fire protection.

Do not build up the outside walls of the chimney and then fill with trash lying around the job. Using leftover mortar particles, scraps of nonmasonry materials, or sand results in an inferior chimney. Scraps of wood should never be used to fill in around the chimney.

Masonry units should be laid in a strong, fire-resistant mortar such as Type N portland cement-lime or equal strength masonry cement mortar. All of the masonry units should be bonded to the main outside walls either by using metal wall ties or by cutting into the wall and interlocking the chimney into the wall with mortar.

Fig. 33-11 Conventional fireplace with external air supply system.

Footings

The footings must be of concrete. The footing should also be wider on all sides than the actual chimney, for load distribution. Reinforcement steel can be added to the concrete for extra strength.

If the chimney is part of the outside wall, make sure it is built below the frost line to prevent freezing and thawing. The chimney can be located on one of the exterior walls or in the center of the home. It can be built as a separate part completely removed from the outside structural walls. This allows the architect more freedom in designing the structure. Do not locate the chimney and fireplace too close to a doorway or closet.

Installing Flue Linings

As previously mentioned, there must be a separate flue for every fireplace or heat source such as a furnace. Start the flue lining for furnaces approximately 12 inches below the thimble intake, or where the furnace pipe is cut into the chimney. (The *thimble* is a round tile ring used to put furnace pipe through.) The lowest flue should be supported on at least three sides with corbeled brick projecting from the inside wall of the chimney, and flush with the inside wall of the flue lining, set in a bed of mortar and cut off flush with the trowel. Wipe the joint with cloth to make sure it is smooth and full of mortar. Corbeling support eliminates the chance of the flue slipping down into the chimney as the construction proceeds.

Set any additional flues in a bed of mortar not more than $3/8$ inch. Finish off in the same manner with a smooth joint. If possible, set the flues ahead of the brickwork, level and plumb. This results in a better mortar bed joint.

Every flue should be divided by a 4-inch brick partition wall to prevent air in one flue from leaking over and affecting the draft in the flue next to it. Mortar should not, however, be slushed tightly down between the flue and brick because there must be room for expansion due to heat. Full mortar joints are important when building chimneys or fireplaces, Figure 33-12. There should be a soot pocket below the flue lining and a cleanout door to remove the soot.

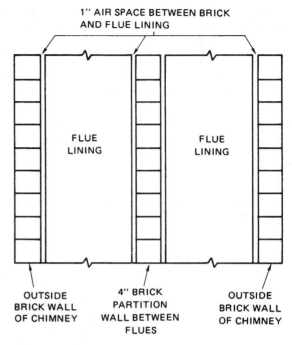

Fig. 33-12 Typical wall section of a two-flue chimney showing air space and 4″ brick partition between flues.

When more than one flue is needed in a chimney, it is often necessary to offset one of the flues for proper fit in the chimney. These offsets or bends should not vary more than 30 degrees from a vertical position to eliminate the chance that the flue might be choked off, Figure 33-13. The flue lining should be cut neatly on a miter to fit the angle of the slope. This can best be done by first filling the flue with sand, tamped tightly. A sharp chisel and light hammer are used to make the cut along the marked line. After working all the way around the flue, it breaks cleanly. Thimble holes are cut in the same way, except the circle size is drawn on the flue and a point chisel is used to cut a hole in the center. The cut is then made from the center out to the edges of the circle by chipping with the hammer head. Figure 33-14 shows a mason cutting a flue lining that has been packed full of sand.

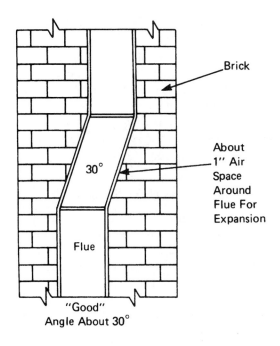

"Good"
Angle About 30°

Brick

About 1" Air Space Around Flue For Expansion

30°

Flue

Fig. 33-14 The mason is using a chisel to cut a thimble hole in the furnace pipe. The flue is packed tightly with sand to support the cutting without breaking.

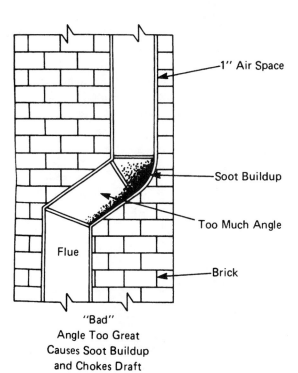

1" Air Space

Soot Buildup

Too Much Angle

Flue

Brick

"Bad"
Angle Too Great
Causes Soot Buildup
and Chokes Draft

Fig. 33-13 Good and bad flue angles in chimney.

Clearance Around Wood

The chimney clearance from wood or combustible materials is a minimum of 2 inches except where the chimney is located outside the structure, in which case 1 inch is acceptable. All exterior spaces between the chimney and other objects should be sealed off outside with flashing, a noncombustible filler material, or caulking, Figure 33-15.

Flashing a Chimney

Flashing around a chimney is installed to prevent water from entering and to divert any water away from the chimney where it meets the roof. Flashing can be copper, aluminum, or heavy-gauge tin. Copper and aluminum seem to do the job better than tin. It should also be remembered that flashing should last for a long time with no repair or upkeep needed.

Flashing around a brick chimney requires a base and a counter flashing. The base flashing is installed first on the faces of the chimney and lapped onto the roof around the perimeter of the chimney. The flashing should extend at least 4 inches up the sides of the

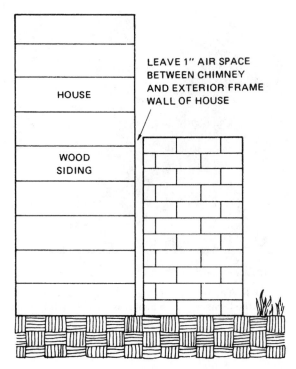

Fig. 33-15 A 1″ air space should be left between the frame wall of the house and the chimney. Later this can be caulked or filled in with fireproof materials.

chimney in order to seal off around the chimney at the point where it comes through the roof. The counter flashing is then installed over the base flashing by squaring it up with the course of brick on the chimney and inserting it into the mortar joint approximately 1 inch. The counter flashing can be built in by the mason as this point is reached in the construction of the chimney, or the joint can be raked out with a line pin or metal tool and the flashing installed after the chimney is topped out. However, it is more effective if the counter flashing is walled into the mortar joint as the chimney is built, making it less likely to pull loose at a later time.

In addition, the counter flashing should lap down over the base flashing by at least 3 inches. If the flashing is installed in sections, the flashing higher up the roofline should lap over the lower flashing a minimum of 2 inches. All joints in the base flashing should be thoroughly sealed. The unexposed side of any bends in the flashing should also be sealed tightly. This can be

done by applying a coat of roofing tar at any point where needed.

Flashing is a task that should be done by a person who has had experience or been trained to do so. On most jobs, this is the responsibility of the carpenter or the roofer, because, if not done correctly, the roof may leak. Figure 33-16 shows a typical section view and flashing details for flashing a one-flue chimney for a fireplace.

Finishing the Chimney above the Roof

A building code requirement, which is accepted nearly everywhere, concerning the height of the chimney in relation to the roof and for fire safety is as follows.

The minimum chimney height for fire safety is the greater of 3 feet above the highest point where the chimney penetrates the roofline, or 2 feet higher than any portion of the structure or adjoining structures within 10 feet of the chimney. This is shown in Figure 33-17.

As previously mentioned, if there is a hill or high trees near the house, in some cases the chimney may have to be built higher than the 3 feet in order to prevent down drafts. Therefore, the mason must be alert to the existing conditions on a particular job. If there is any doubt, the rule is to always build the chimney higher rather than take the chance that the mason would have to return later and rebuild the chimney higher.

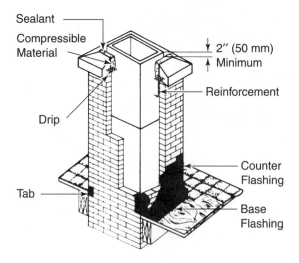

Fig. 33-16 Typical section and flashing detail.

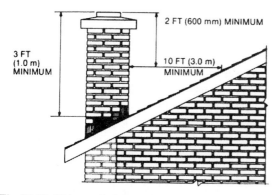

Fig. 33-17 Chimney height in relation to roof height.

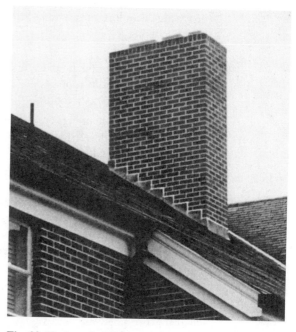

Fig. 33-18 A typical three-flue chimney.

The chimney is capped off with a portland cement-lime mortar wash coat on an angle to form a wind cove and seal off around the flue and bricks. Types M or S are suitable for this job. If using masonry cement, make sure it is a comparable type. The mortar wash can be prevented from drying out too fast by covering it with wet burlap and dampening it periodically with water until it is cured. The sun and wind will crack mortar on the top of a chimney if some dampening with water is not done.

The flue linings should extend at least 4 inches above the top of the finished cove. They are laid level with each other if more than one flue is in the chimney. Figure 33-18 shows a completed three-flue chimney.

Dummy Flues

If the chimney is rather large and only two flues are being used, sometimes a *dummy flue* is added in be-tween the two flues to balance out the top of the chimney. This is done by cutting off a section of flue and building it into the top of the chimney. To prevent moisture or water from lying in the dummy flue, it should be filled to the top with mortar.

The chimney is a column of masonry work. If the work is not plumb and level, it is very obvious to any person viewing it from the ground. Therefore, it is very important that good workmanship practices be followed at all times. Building a chimney tests the skill of the mason.

ACHIEVEMENT REVIEW

Select the best answer from the choices offered to complete each statement. List your choice by letter identification.

1. Fireplaces and chimneys in the United States can be traced back to about the year
   a. 1550.
   b. 1600.
   c. 1776.
   d. 1800.

2. Early fireplaces lost most of their heat up the chimney due to the absence of
   a. a flue lining.
   b. an ash dump.
   c. a wind shelf.
   d. a damper.

3. The main purpose of the fireplace in colonial days was to
   a. provide light for the room.
   b. give warmth and cook food.
   c. provide a cheerful atmosphere.
   d. increase the value of the home.

4. The main reason flue linings are added to chimneys is to
   a. add beauty and design to the chimney top.
   b. provide better draft.
   c. prevent the walls from burning out.
   d. make the chimney stronger.

5. The draft in a chimney is stronger in the winter than in the summer because
   a. the prevailing winds are stronger.
   b. the fireplace is used more.
   c. there is a difference in temperature between the air outside the flue and the air inside the flue.
   d. air currents change in the atmosphere.

6. Air coming down the flue is deflected back up by bouncing off the
   a. damper.
   b. smoke shelf.
   c. hearth.
   d. mantel.

7. Heat from the fireplace is forced into the room by
   a. radiation.
   b. air currents coming down the flue.
   c. forced air.
   d. the draft inside the room sucking the air.

8. The dividing wall that is built around the flue in a two- or three-flue chimney is
   a. stone.
   b. block.
   c. tile.
   d. brick.

9. In order to prevent down drafts, the chimney top is built above the roof
   a. 12 inches.
   b. 2 feet.
   c. 4 feet.
   d. 6 feet.

10. Offsets or bends in a chimney flue should not vary from vertical more than
    a. 30 degrees.
    b. 40 degrees
    c. 45 degrees.
    d. 60 degrees.

11. All woodwork surrounding a chimney in the interior of a house should be kept back
    a. 1 inch.
    b. 2 inches.
    c. 3 inches.
    d. 4 inches.

12. The top of the flue lining projects above the mortar wash a distance of
    a. 4 inches.
    b. 6 inches.
    c. 8 inches.
    d. 12 inches.

PROJECT 28: TWO-FLUE CHIMNEY

OBJECTIVE

- The student will build a two-flue chimney, the type that has a fireplace and furnace flue, Figure 13-19. The basic skills of leveling and plumbing can be developed by the student in this project. A column of masonry, such as a chimney, requires good workmanship.

EQUIPMENT, TOOLS, AND SUPPLIES

Mortar pan or board

Mixing tools

Mortar box or wheelbarrow

Mason's trowel

Brick hammer

Plumb rule

Chalk box

2-foot framing square

Convex sled runner jointer

Mason's modular rule

Brush

Supply of mortar (lime and sand or Carotex if to be reused)

two 13-inch × 13-inch flue liners (wood plywood form can be used instead of flue liners)

two 9-inch × 13-inch flue liners

Standard bricks for project are estimated by counting the number on one course and multiplying it by the number of courses shown on the plan

SUGGESTIONS

- After the first course is laid out according to the plan, set the flue liner in place to make sure it fits correctly.
- All mortar protruding to the inside cutoff should be kept clean to prevent it from building up inside the chimney.
- Use well-filled head and bed joints.
- Plumb and level each course as it is laid.
- Strike the mortar joints as soon as they are thumbprint hard.
- Make sure the first course is laid square since it is the guide for the entire project.
- Lay up one side of the chimney 3 courses high. Rack back and then build up the opposite side.
- Check the height of the chimney every other course with a modular rule.
- Keep all scraps and tools out of the way.
- Set the flue liner in place immediately after the first course is laid out on the base. This prevents the chimney from being shoved out of position by dropping in the flue after the work is up.
- Follow good safety practices at all times.

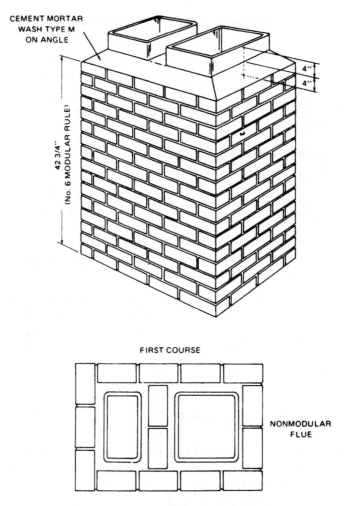

Fig. 33-19 Project 28: Two-flue chimney.

PROCEDURE

1. Mix the mortar.
2. Prepare the work area by stocking the bricks and mortar back away from the project so the mason is not disturbed when working.
3. Lay out the project with the 2-foot framing square and pencil.
4. Dry bond the brick with a ⅜-inch head joint as shown on the plan in Figure 33-19.
5. Lay each course to number 6 on the modular rule.
6. Lay the first course in mortar.
7. Set the first two flue liners in place as shown on the plan.
8. Lay the 4-inch brick partition between the flues.

9. Build the chimney to the height of the first flue liner.
10. Set the second flue in place with mortar.
11. Continue to build the chimney up to the finished height shown on the plan.
12. Strike the mortar joints with the convex jointer.
13. Apply the mortar wash coat to the top of the chimney on a slope as shown on the plan. The flue liner should project 4 inches above the top of the wash.
14. Brush the project when dry enough not to smear the joints.
15. Recheck the project to be sure it is plumb.
16. Have the instructor inspect the work.

UNIT 34
Design and Construction of a Fireplace

─────────────── OBJECTIVES ───────────────

After studying this unit, the student will be able to

- list key factors to consider when planning and building a fireplace.
- identify the parts of a fireplace.
- describe how the flue size is determined for different sizes of fireplaces.
- explain the construction techniques for building a fireplace step by step to the point where the flue lining is set.

FACTORS TO CONSIDER WHEN BUILDING A FIREPLACE

Location

When deciding the location of the fireplace, consider that it is a permanent structure. It is not a good idea to build a fireplace near a doorway because it would be in the way of household traffic. The door could also affect the draft pattern of the fireplace. The fireplace is usually the center attraction in a room. Therefore, thought should be given whether to build it on an outside wall or in the center of the room.

Traditionally, the fireplace was built as part of the outside wall. Now this is not always done because the fireplace and chimney may be a separate unit within the home. They also serve as a column of masonry that strengthens the structure.

The fireplace can be built close to a window if it does not block the light entering the room. Many architects design a window or bookcase on either side of the fireplace using it as a centerpiece. All of these factors are usually considered by the architect when the plans are developed.

Size

The size of the fireplace depends on the size of the room in which it will be located. This is not only for the sake of appearance but also to ensure successful operation.

A fireplace that is too small, even if it burns correctly, will not supply enough heat to the room. If the fireplace is too large, a fire that fills the combustion area (firebox) would be too hot for the room. In addition, a larger fireplace requires a larger chimney flue which draws more air through doors and windows to use for combustion.

A room with 300 square feet of floor area is well served by a fireplace with an opening 30 inches to 36 inches wide. For larger rooms, the width of the fireplace opening is increased. The height of the opening is just as important as the width. The height and width must be in correct proportion for the fireplace to function properly. (See the table in Figure 34-6 for fireplace dimensions.)

A fireplace should also be built in proportion to the wall of which it is a part. The average wall in a modern house is longer than it is high since modern houses no longer have as high ceilings as in older homes. Therefore, a fireplace should also be longer than higher to match the room. A tall, thin fireplace would be completely out of place in this situation. Many fireplaces of today cover the entire wall of a room to display the effect of masonry inside the house. There are no hard and fast rules about this; it is a matter of taste and the owner's preference.

A fireplace that is poorly designed and incorrectly built has many disadvantages: it is unattractive; it may allow leakage of water into the room; it creates constant problems from soot, dirt, or down drafts; and it will not draw properly. Figure 34-1 shows a foundation built to meet requirements of a house.

Offsetting the Fireplace into the Room Past the Wall Line

The front of the fireplace can be built either flush with the wall or offset into the room. Most are built flush with the wall because it is more economical. However, a fireplace that extends into the room adds a feel of depth and character of design.

The major problem to consider when offsetting the fireplace into the room is that the flue lining must be laid on more of an angle than usual at the top of the smoke chamber to attain the proper position. Also, the damper handle may be too short and have to be custom-made. Therefore, the fireplace should not be offset more than eight inches into the room; four inches is suggested.

RECOMMENDED METHOD OF BUILDING A FIREPLACE

A fireplace can be built two different ways. One way is to complete the fireplace as the chimney is built. This includes the firebox and all of the facing work.

Most masons, however, use the second method of building a fireplace, in which the chimney is built completely to the top, leaving a rough opening for the later installation of the finished fireplace. Flue linings that will be installed above the firebox for the fireplace are set at the top of the smoke chamber at the prescribed height on brick corbeling. This corbeling must be built to this height, regardless of which method is used.

Fig. 34-1 Chimney and fireplace foundation squared up to first floor level of house.

The second method of building a fireplace is strongly recommended. If the finished fireplace is built as the chimney is constructed, mortar, chips of masonry units, and other objects may be dropped down the open flue lining causing an obstruction or possibly breaking the brittle cast iron damper. These objects are very difficult to remove or clean out. Also, if the fireplace is installed as the chimney is built, the face can be smeared or coated with mortar from building the chimney. It is possible, too, that the damper rod in the front of the fireplace may be struck with a fallen object breaking the control. This is expensive to repair and may require tearing out the front of the fireplace.

Another reason for using the second method is that most contractors have the masons do the finish work of building a fireplace on bad or rainy days. This also provides a good incentive for the masons.

LAYING OUT THE FOUNDATION

Once the footing is in place and has been approved, the chimney is laid out so it will accommodate the fireplace and other heating units such as the furnace or wood stove flues. This is the factor that determines the measurements of the foundation. As a rule, no more than two fireplaces and one other flue for a furnace or wood stove are in one chimney.

The first step in the actual construction of the chimney is to study the plans and mark off the outline of the chimney on the footing with a steel square and a chalk line. Most of the masonry houses being constructed now are of the brick veneer type. The foundation is built first and then all of the framing is completed, leaving room for the brick veneer that rests on the outside of the block foundation wall. When laying out the chimney wall, the mason must plumb up from the foundation wall to make sure that the chimney will not interfere with any windows or doors that were built in when the framing was done. Now is the time to shift the chimney a little if needed, to compensate for this. Although the architect should have taken this into account, sometimes a problem arises if this is not checked out by the mason before building the chimney foundation.

Brick or concrete block, whichever material is being used in the base, should be dry bonded to work full units without any cutting. If the base is to be concrete block

(due to being below grade), be sure to use brick as the layout material to make sure the bond will work when brick is started at the finish grade line. Dry bonding is done by laying out the units on the base, dry (without mortar), and correctly spaced between the units for the mortar joints to determine the bond. See Figure 34-2.

The brick or block is then pushed aside from the layout line, and mortar is spread on the base. The first course is laid to work full units. This may require a slight adjustment of the size of the chimney, to make the bond work. Always make the chimney slightly larger, not smaller, if this is the case.

It is more convenient and economical to use only one type of mortar for the entire fireplace and chimney. Type N portland cement-lime mortar or equal is recommended as an all-around good average mortar. Type N mortar is 1 part portland cement, 1 part hydrated lime, and 6 parts sand.

Lay the correct number of courses to the height of the cleanout door (usually 4 or 5 courses) for the ash pit. Install the cleanout door in the masonry work as shown in Figure 34-3. The foundation of the chimney is then built to the first floor height and any furnace flues, thimbles, or the like are installed, as shown on the plan in Figure 34-2.

INSTALLING THE ROUGH HEARTH

In the past, when the rough hearth was constructed, a brick arch was laid on a wooden form beneath the hearth. This arch was called a *trimmer arch*. The concrete or rough hearth brick was then placed over the arch and the arch carried the load or weight of the finished hearth.

In present day construction, brick trimmer arches are seldom used because they require more time to install. Instead, plywood or short lengths of lumber are installed under the hearth area and nailed to wood strips located on the headers that surround the hearth area on the floor. Steel reinforcing rods are then placed over the forms and tied together with thin wire, forming squares. Concrete is placed in the forms and vibrated so that it settles around and under the steel rods. The wood forms can be left in place after the concrete hardens, or they may be removed from under the part of the hearth that projects out from the main chimney. As a rule,

Fig. 34-2A Dry bonding bricks for proper mortar head joint.

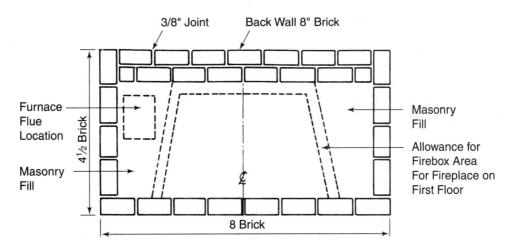

Fig. 34-2B Plan view of dry bonding the first course layout of a chimney and fireplace. This plan is for a fireplace to be installed on the first floor and shows the foundation layout.

however, the builder usually leaves the forms in place. Figure 34-4 shows the method used today to construct a rough hearth for a fireplace.

A form is built at the location in the inner hearth where the ash dump is installed. Use short pieces of wood or sheathing that can be knocked out easily after the concrete has hardened. This stops the concrete from coming through the ash dump hole.

Any flue liners coming from a furnace or basement fireplace should be set so they project through the hearth

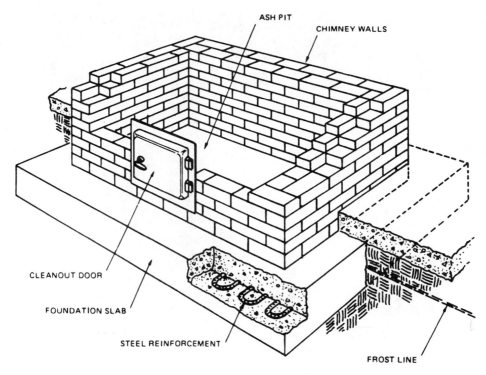

ASH PIT

CHIMNEY WALLS

CLEANOUT DOOR

FOUNDATION SLAB

STEEL REINFORCEMENT

FROST LINE

Fig. 34-3 Installation of cleanout door in the chimney base and ash pit.

slab before the concrete is placed. The part of the hearth that projects out from the actual chimney is called the *outer hearth* (or forehearth). The inside hearth that the fire is built on is called the *inner hearth*. The rough hearth must be strong if it is to support the weight of the finished hearth without cracking. Figure 34-5 shows an overall section view of a typical fireplace and chimney.

THE FINISHED HEARTH AND ROUGH OPENING

The finished outer hearth can be constructed of quarry tile, brick, stone, or other similar materials. This hearth is built flush with the floor or raised off the floor. In many modern fireplaces the hearth is raised up from the floor in order to elevate the fire. The raised hearth also provides a bench seat around the fireplace. Either method is correct; the method chosen depends upon the design and the preference of the architect or home-owner.

Before the hearth can be built, the rough opening of the fireplace must be determined. This is done to make sure that there will be enough room to construct the finished hearth and firebox. (The rough opening should be built at the same time as the chimney.)

Refer to the table in Figure 34-6, Standard Size Fireplace Dimensions in Inches. The finished opening size is given for different fireplaces. When building a rough opening for a fireplace that is to be installed later, always allow 8 inches more than the finished width shown on the table. Allow 19 inches more than the opening shown on the table for rough height. This provides enough room to build the finished fireplace and set the damper in position.

For example, a standard fireplace will have finished opening dimensions of approximately 36 inches by 29 inches. There could be a little difference depending on brick size. Adding 8 inches in width to the finished opening, which is 36 inches, the total rough opening would be 44 inches wide. The height of the rough

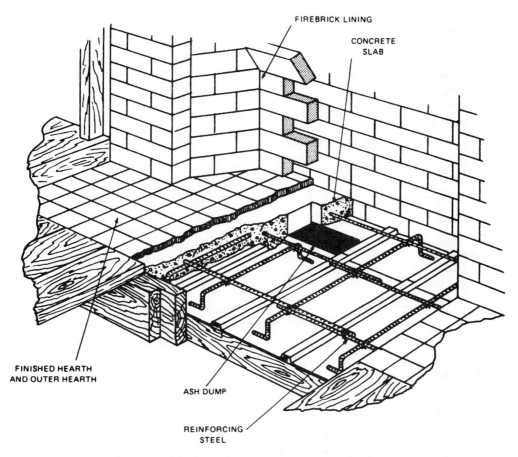

FIREBRICK LINING

CONCRETE SLAB

FINISHED HEARTH AND OUTER HEARTH

ASH DUMP

REINFORCING STEEL

Fig. 34-4 Cutaway view of the installation of a rough hearth. Notice the reinforcing steel in the base of the hearth.

opening should be 19 inches plus the finished height, which is 29 inches, making a total of 48 inches for the rough opening. An angle iron or a brick trimmer arch can be installed over the opening to support the masonry work above. The arch will allow more working room for the mason when the damper is set in position. Either method can be used successfully. Figure 34-7 shows the rough opening for a standard fireplace.

ALLOWANCE IN ROUGH OPENING FOR RAISED HEARTH

If a raised hearth is to be built, the height of the rise must be added to the rough opening to allow for the difference. For example, if the raised hearth is to be 12 inches,

then the rough opening must be 60 inches rather than 48 inches (48 + 12 = 60). This is very important to remember when building a raised hearth fireplace or the fireplace will not operate correctly. Usually a raised hearth is not built higher than 12 inches; if it is built higher, the fireplace appears out of proportion to the room.

To better understand how the various parts of the fireplace are built, see Figure 34-6. Use the table for information as often as necessary throughout the rest of this unit.

LAYING THE HEARTH

The hearth is usually built of firebrick made of special burned clay that is highly resistant to heat. A standard

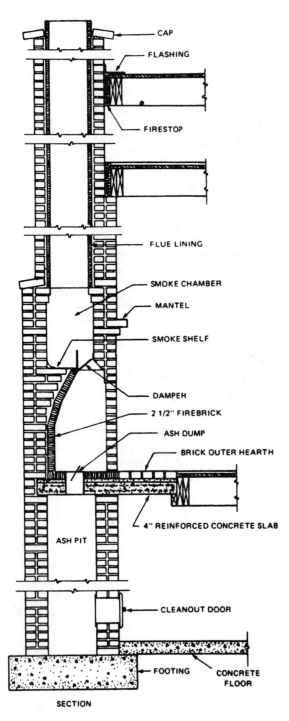

Labels on figure (top to bottom): CAP, FLASHING, FIRESTOP, FLUE LINING, SMOKE CHAMBER, MANTEL, SMOKE SHELF, DAMPER, 2 1/2" FIREBRICK, ASH DUMP, BRICK OUTER HEARTH, 4" REINFORCED CONCRETE SLAB, ASH PIT, CLEANOUT DOOR, FOOTING, CONCRETE FLOOR, SECTION

Fig. 34-5 Section view of a fireplace and chimney. Notice the location of the ash dump in the inner hearth.

firebrick measures 4½ inches × 9 inches thick. Firebricks are buff or an off-white color and can be purchased at any building supply dealer who stocks masonry materials.

For firebrick mortar, it is desirable to use a fire-clay mortar or a high-lime mortar, such as Type O. They have been found to resist the heat in the combustion chamber without cracking better than a high-strength hard mortar. The mortar joint size also affects the ability of the mortar to withstand cracking. A good practice in the firebox area is to limit the size of the mortar joints to ¼ inch. This also makes a neater job.

The hearth is always laid out from the center of the fireplace. After determining the width of the opening to be used from the plans, mark off with a pencil the exact overall distance of the inner hearth. Refer to the illustration in Figure 34-7.

Lay the outside course of firebrick. Make sure it is slightly longer than is needed, since the masonry jamb that is laid on top of it hides the rough edge. Use small (¼ inch to ⅜ inch) head and bed joints and strike with a flat jointer, such as a slicker.

Lay the firebrick so they are half over each other. Continue laying the hearth until it is completed. Install the ash dump in the rough opening left when the rough hearth was built. This is usually slightly off the back wall of the firebox and in the center. The ash dump door is the same size as a firebrick and has a projecting flange that holds it in place. Figure 34-8 shows a completed hearth ready for the walls of the firebox to be laid out. Brush the hearth clean at completion and re-strike all joints.

FIREPLACE OPENINGS

Correct proportioning of the opening greatly affects the operation of the fireplace. Height and width should be in proper proportion to each other. In general, the wider the opening the greater the depth and height of the firebox. The dimensions in Figure 34-6 should be followed as a guide whenever laying out fireplace openings.

If the fireplace is designed mainly to burn hardwood logs, some builders find it desirable to increase the distance from the front to the back of the firebox by 3 to

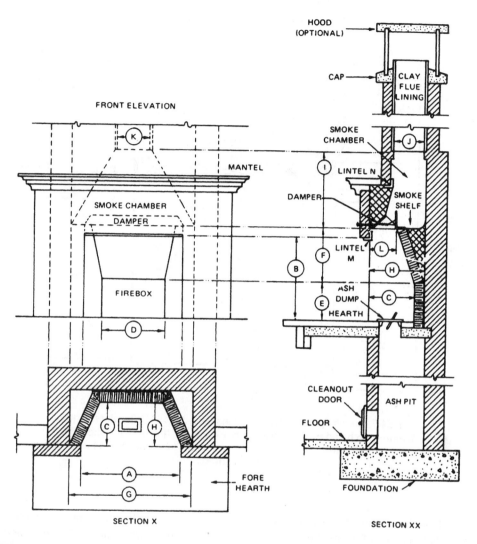

TABLE OF STANDARD SIZE FIREPLACE DIMENSIONS IN INCHES																					
Finished Fireplace Opening						Rough Dimensions			Flue Lining Size		VESTAL MANUFACTURING COMPANY										
									Old Rectangular	New Modular	Damper Model			Damper Model		Damper Model		Ash Dump	Cleanout Door	Lintels	
A	B	C	D	E	F	G	H	I	J x K	J x K	Poker	Rotary	L	Chain	L	Steel	L			M	N
24	24	16	11	14	15	32	20	19	8½ x 13	8 x 12	24	124	10¼	824	9¼	624	9	49	88	36	42
30	29	16	17	14	20	38	20	24	8½ x 13	12 x 12	30	130	10¼	830	9¼	630	9	49	88	42	48
33	29	16	20	14	20	41	20	24	13 x 13	12 x 16	33	133	10¼	833	9¼	633	9	49	108	42	54
36	29	16	23	14	21	44	20	27	13 x 13	12 x 16	36	136	10¼	836	9¼	636	9	49	108	48	54
42	32	16	29	14	23	50	20	32	13 x 13	16 x 16	42	142	10¼	842	9¼	642	9	49	1210	54	60
48	32	18	34	14	23	56	22	37	13 x 18	16 x 16	48	148	10¼	848	9¼	648	9	49	1210	60	66
54	36	20	37	16	26	62	24	44	13 x 18	16 x 20	54	154	10¼					49	1210	66	72
60	38	22	42	16	28	68	26	45	18 x 18	16 x 20	60	160	11¾					49	1212	72	78
72	40	22	54	16	29	80	26	54	20 x 20	20 x 24	72	172	11¾					49	1212	84	90

Fig. 34-6 Illustration and table of fireplace construction principles. The illustration shows how various parts of the fireplace are built.

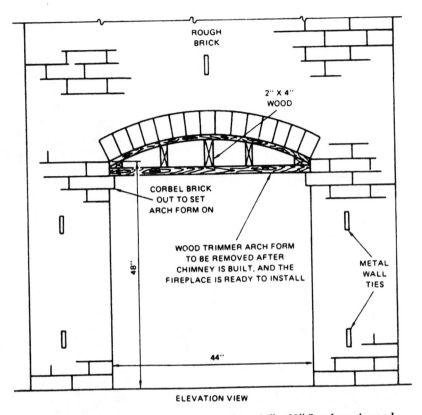

ROUGH BRICK

2" X 4" WOOD

CORBEL BRICK OUT TO SET ARCH FORM ON

WOOD TRIMMER ARCH FORM TO BE REMOVED AFTER CHIMNEY IS BUILT, AND THE FIREPLACE IS READY TO INSTALL

METAL WALL TIES

48"

44"

ELEVATION VIEW

Fig. 34-7 Rough opening measurements for a $36'' \times 29''$ fireplace. A wood trimmer arch and plywood frame are used to lay the arch on. The form is knocked out later after the chimney is completed. The brick arch is better than a metal lintel.

6 inches. This allows the burning of larger logs without much effect on the operation of the firebox.

Wall Thickness

The masonry walls behind the firebrick and on the sides of the firebrick should be at least 8 inches thick to protect the firebox from overheating. If made of stone, the walls should not be less than 12 inches thick. An air space of approximately 1 inch is provided between the firebrick and the other masonry for expansion purposes. This air space can be filled with a compressible, noncombustible material such as fibrous insulation. The purpose of this is only to maintain the air space, not to insulate the wall.

It is especially important for the walls of a fireplace and chimney which are built in the interior of a home to be thick enough since the back of the chimney may also serve as the inside of a closet or project out into a hallway. Metal ties serve to tie the firebox and backing walls together.

LAYING OUT THE FIREBOX

The *firebox* is the part of the fireplace that contains the fire. It is built of firebrick and can either be laid in a stretcher position (as a brick is usually laid) or shiner position. A firebox is generally laid in the shiner position because it is more economical and still provides sufficient fire protection. The shiner position is accom-

Fig. 34-8 A completed firebrick hearth ready for the walls of the firebox to be laid out.

plished by laying the firebrick on its longest surface ($2\frac{1}{4}$ inch bed).

Note: The size of the firebox can be determined, if a plan is not available, by referring to Figure 34-6.

The first step in laying out the firebox is to find the center of the hearth. Hold a 2-foot framing square on that point and draw a straight line from that point to the back of the hearth. All measurements for the wall of the firebox are worked out from this line. Figure 34-9 shows how this is done. This example uses a standard 36-inch width opening.

Examining the drawing in Figure 34-9, it can be seen that from each side of the centerline in the front, 18 inches is measured and marked with a point of a pencil. The total width, therefore, is 36 inches. See point A in Figure 34-9.

Next, the depth of the firebox is marked off on each side of the centerline, from the front to the rear, a distance of 16 inches. See point B on the drawing. Use a pencil line to connect the back wall of the firebox. The 2-foot square can be used as a guide.

Now, measure out from each side of the rear centerline $11\frac{1}{2}$ inches. This totals 23 inches which (referring to the table in Figure 34-6) is correct for a 36-inch fireplace opening. See point C in Figure 34-9.

With the framing square laid flat on the hearth at point C to point A, draw a fine pencil line. This establishes the splayed (inclined or angled) side of the

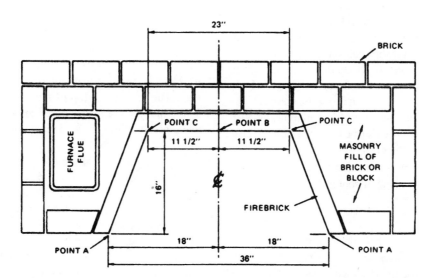

Fig. 34-9 Plan view of a standard 36″ firebox for a fireplace. All dimensions are given for the layout of the firebox on the hearth. Notice that all measurements are taken from the center of the fireplace. The outside of the brickwork is not tooled, because it will be faced later with face brick. This is the reason the firebrick extends to the face of the brick wall in the front.

firebox. Repeat this on the opposite side and the walls of the firebox are laid out.

Note: The front opening width could vary slightly depending on variations in the size of brick being used.

BUILDING THE WALLS OF THE FIREBOX

One course is laid in mortar on the hearth, completely around the firebox to establish the bond. This is then backed up with rough bricks as a filler, allowing approximately 1 inch for expansion.

The purpose of the angled sides of the firebox is to make it narrower than the front opening and to create a reflector shape which radiates heat out into the room. The slope of the back wall also helps the smoke and gases to curl upward and reflect the heat out from the firebox. The angle of the side walls should not exceed 6 inches per foot of depth. After the first course of the firebox wall has been laid out on the hearth, sprinkle a little sand on the hearth so any mortar dropped while building the rest of the firebox does not stain the hearth. Make sure to tool the mortar joints on the hearth before spreading any sand on them.

The firebox walls are then built up plumb 3 courses (approximately 14 inches). At this point the back wall is started in either a slope or roll that brings the firebrick up under the entire near flange of the damper. It is very important to lay the head and bed joints as tight as possible because the fire will burn out large joints in a period of time. Refer back to Figure 34-5 to see how the back of the firebox is rolled in for the damper.

Edges of firebricks that are cut to work bond against other firebricks should either be cut with the masonry saw or the hammer and chisel.

> **Caution:** Eye protection should be worn when cutting any masonry unit.

The cut firebricks are then rubbed with another piece of firebrick until the edges are smooth and straight. This is important because poorly fitted bricks allow fire to enter and may crack the firebox. Properly fitted firebricks also indicate good workmanship.

As the unexposed back of the firebox wall is built, it is a good practice to back parge it with mortar to help hold the bricks more securely in place. This also helps keep the wall from burning out too early.

The slope of the back wall, if built correctly, should not need any propping to prevent it from falling in while it is being built. Each course is built all the way around the firebox before the next course is started. This helps to keep all of the work tied together. Any wall ties projecting from the previously built chimney wall should be built into the wall as it is laid up.

Strike the mortar joints with a convex jointer. This type of joint is recommended because it gives a good seal and a pleasant appearance.

THE SMOKE SHELF

The smoke shelf is built by rolling the back wall of firebrick lining forward until it meets the rear flange of the fireplace damper. The depth of the smoke shelf varies depending on the size of the flue lining and the depth of the firebox.

Use a good, hard fired building brick. Lay it in back of the firebox wall in mortar as the back wall is built. As the height of the wall increases, the brick backing is laid in such a way that it becomes a *clipped header* (a header that has a small portion cut off of it). At the top of the shelf it becomes a full header brick. This is actually corbeling the backing behind the firebox wall until the top of the smoke shelf is reached. A small space should be left between the back of the firebrick wall and the backing material to allow room for expansion. The area at the top of the firebox is the hottest spot. This is where the hot gases and smoke concentrate when passing through the damper and up the chimney flue.

Note: Regardless of the type of masonry unit used to fill in back of the firebox and form the smoke shelf, it is very important that it be about level with the last course of firebrick laid, where the damper rests.

The smoke shelf is one of the more important parts of the fireplace. If it is not installed properly, the fireplace smokes into the room. The smoke shelf turns the wind coming down the flue back up the chimney, carrying the smoke and gases from the fireplace. If the smoke shelf is too high or too low, the smoke eddies around the damper area, resulting in a fireplace that functions incorrectly.

Figure 34-10 shows a cutaway view of the finished firebox with corbeling, smoke shelf, and damper in position on the top of the firebox. Any holes or voids on top of the smoke shelf should be filled with mortar. Troweling a coat of mortar over the entire area of the smoke shelf makes it smooth and waterproof. Be sure the top of the smoke shelf is level with the bottom of the damper.

The smoke shelf can be formed with a slight curve where it meets the back wall of the chimney by coving

(sloping) up the mortar at that point. This helps to promote a smoother flow of air through the smoke chamber. Care should be taken to prevent any buildup of mortar droppings or particles of brick from being left on the top of the smoke shelf. This can be accomplished by laying an empty cement bag or a piece of plastic on the smoke shelf as soon as it is installed. After all construction around and over the smoke shelf is completed, remove the protective cover and debris through the open damper. Be careful not to dump any of the debris

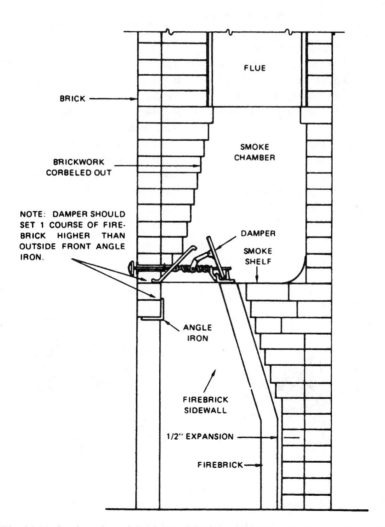

Fig. 34-10 Section view of the finished firebox showing corbeling, smoke shelf, and damper in position.

on the damper rod. Operate the damper a couple of times after removing the cement bag to make sure that it works correctly.

FIREPLACE THROAT AND DAMPER

The throat is the part of the fireplace just above the opening. It connects the firebox with the smoke chamber. Correct shaping of the throat is one of the most critical parts of fireplace design and must be done correctly. A properly designed damper relieves the mason of the task of designing the throat.

Dampers for use in the fireplace have been designed and put on the market by various manufacturing companies. There are two main types of dampers: the poker and rotary control. Figure 34-11 shows the two types.

The poker type is operated by inserting a poker in the hole under the damper, in the ratchet arm, and pushing upwards. This action opens the damper to the desired width.

The rotary control damper is more popular. It is operated by turning a projecting handle that is built into the face of the finished fireplace. This causes the same opening operation as the poker type but a person does not have to work in the hot area of the firebox to open the damper.

The damper allows the draft to be shut off during nonuse periods, conserving other sources of heat or air conditioning. The fire burns best, and the maximum amount of heat enters the room, when the damper is opened just enough so the fire does not smoke but burns freely.

Since the damper is exposed to direct heat from the fire it must be made of a material that withstands a very hot fire over a long period of time. A cast iron damper is recommended. Care should be taken when installing a cast-iron damper; if it drops, it can break.

INSTALLING THE DAMPER

As a rule, the mason must assemble the damper before it is installed. Be sure that it is put together according to the manufacturer's instructions. All moving parts should work freely before the damper is installed in the fireplace. Once the damper is walled in, it is very difficult to make any corrections.

The worm that operates the lid of the damper (on a rotary damper) sometimes has rough metal edges left from the casting process. It is a good idea to file any rough places on the worm until they are smooth. This permits smooth working operation of the damper. The poker type has a ratchet arm instead of a worm.

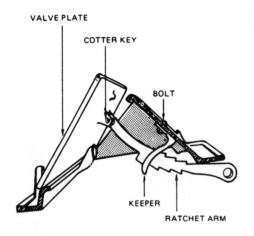

POKER CONTROL
(RATCHET ARM IS REVERSIBLE)

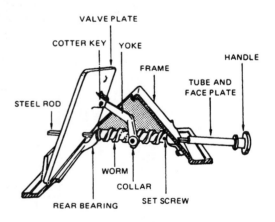

ROTARY CONTROL
(INTERCHANGEABLE WITH POKER CONTROL)

Fig. 34-11 Two types of dampers.

The setscrew holds the rotary worm securely against the handle and prevents it from pulling out the front of the fireplace. It should not be fully tightened until the damper is in position and the facing of the fireplace is completed. This is usually the last operation before the fireplace is finished.

To set the damper in the fireplace, first center the handle of the damper with the center of the fireplace. The handle should be out far enough so the face of the finished fireplace front wall comes in contact with the metal faceplate on the handle.

Set the damper in place in a mortar bed and tap down lightly to settle firmly. Do not tap too hard because the cast iron edge can break. The damper should be at least one shiner course of firebrick higher than the angle iron lintel which covers the opening on the front of the fireplace. Including the mortar joints, this distance is approximately $5\frac{1}{2}$ inches.

With the pointing trowel, smooth out the mortar around the bottom edges of the damper at the point where it sets on the firebox wall. The edge flange of the damper is made so it fits back on the firebox neatly if the firebox is built correctly. Finally, point up the mortar joints on the underside of the damper, fully closing any holes. As a last check, try the damper to see if it opens and closes smoothly before laying any more masonry work.

BUILDING THE SMOKE CHAMBER

The space from the smoke shelf and damper up to the bottom of the flue lining is called the *smoke chamber*. Before any bricks are laid to form the smoke chamber walls, the damper must have a minimum of $\frac{1}{2}$-inch clearance between both ends to allow for expansion.

A simple method of providing for this expansion space is to place fire-resistant materials (such as fiberglass) at the end of the damper. This prevents the mortar from sealing in solid against the ends of the damper and provides for expansion. If this is not done, fireplaces crack in steps at the point where the lintel covers over the opening on the front of the fireplace. Figure 34-12 shows the damper in place with the proper $\frac{1}{2}$ inch expansion space at the ends of the damper.

FORMING THE SMOKE CHAMBER

Good, hard brick should be used to build the smoke chamber, as it will be fireproof. The back wall of the smoke chamber is built straight and parged with a coat of mortar. The distance from the bottom of the smoke shelf to the flue lining is stated for different size fireplaces on the table in Figure 34-6. For a standard 36-inch wide fireplace, the distance is 27 inches. After slushing in (with mortar) all flatwork at the damper height, the brickwork is corbeled in from the front and both sides. It is corbeled until it narrows and meets the area where the flue sets. Care should be taken that the corbeling does not interfere with the opening of the damper.

A metal angle iron can be used to support the interior front brick corbeling that forms the wall of the smoke chamber, Figure 34-13. This is an improvement over the old, established practice of building directly on the front flange of the damper, as this is cast iron and not meant to act as a supporting lintel. Make sure, however, that any area around the damper has a noncombustible insulation such as fiberglass filled in around it to prevent any masonry from touching the metal damper. If the damper cannot expand, it will surely crack the brickwork once the fireplace gets hot.

This brickwork is laid as full headers that bond and transmit the weight better without falling over on the damper. As the work progresses, it should be tied into the main chimney with metal wall ties.

The slope of the sides and front should be approximately 30 degrees. They should be corbeled uniformly to the center of the chimney. This prevents the draft from drawing from one side or the other and allows the fire to burn more evenly in the firebox.

The corbeling out of the brick courses should not exceed $\frac{3}{4}$ inch per course. After all of the corbeling is finished and the flue lining is ready to be set, the underside of the corbeling should be smoothly parged with mortar. This is done so the smoke has an even flow with nothing to obstruct its passage to the flue.

The following is a good method for determining the correct slope to corbel the brickwork for the smoke chamber. Before laying any of the bricks in the smoke chamber area, locate the center of the

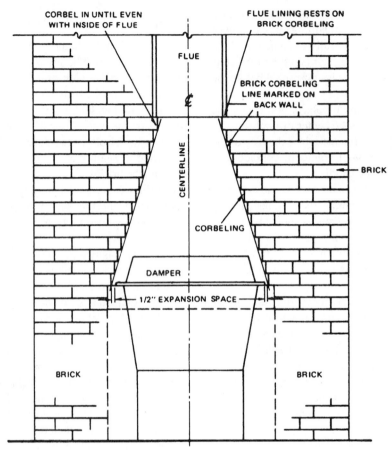

Fig. 34-12 Method of forming the smoke chamber walls. The brick is corbeled to a line drawn on the back wall from the damper to the area where the flue lining sets. Leave ½″ expansion space at each end of the damper.

chimney on the back inside wall. Using a rule, measure up from the top of the smoke shelf to the point where the flue lining is to be set. Then, mark the width of the flue on the wall. Next, draw a line from that point down to where the corbeling will start on the smoke shelf level. This can be marked clearly on the parged wall. The corbeling is laid out to match the line. This makes the task of dividing the corbeling much easier than random measuring or guessing the incline. Figure 34-12 shows how this is done.

Note: In cases in which the chimney has been built first and the fireplace roughed in, the front interior corbeled brick wall over the damper is built up and sealed off under the trimmer arch. The brickwork at the point

where it meets the supporting brick arch must be wedged and filled in tightly with brick and mortar to seal off the smoke chamber from the front of the fireplace.

Figure 34-14 shows the completed firebox and rough work of the chimney around the fireplace area walled up under the trimmer supporting arch. This is an excellent method of finishing off the interior of the fireplace.

DETERMINING THE SIZE OF FLUE LININGS

The size of the flue lining needed for a particular fireplace is determined by using the opening of the firebox

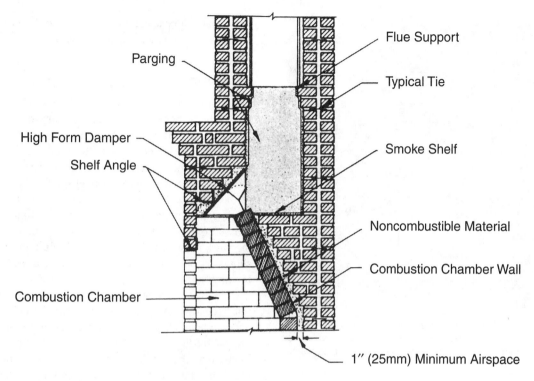

Parging

Flue Support

Typical Tie

High Form Damper

Shelf Angle

Smoke Shelf

Noncombustible Material

Combustion Chamber Wall

Combustion Chamber

1″ (25mm) Minimum Airspace

Fig. 34-13 Damper installed in the fireplace with front interior brickwork being supported by angle iron over damper.

as a guide factor. The total size of the opening is found by multiplying the finished open width by the finished height. The result is the total square inches of the fireplace opening.

The rule for a conventional fireplace (one side open only) is that the cross-sectional area of the flue lining is at least $\frac{1}{12}$ the total area of the fireplace opening. For multiple-opening fireplaces (with two or more open sides) the cross-sectional area of each open side must be figured and added together. The cross-sectional area of the flue lining should be at least $\frac{1}{10}$ the total area of all open sides for this type of fireplace.

The height and width of the opening has a definite function in the operation of a fireplace. If there is a question as to the size flue lining required for a fireplace, or if the determined size is on the borderline, always use the next larger size.

NONMODULAR AND MODULAR SIZE FLUE LININGS

Careful consideration must be taken when determining flue lining sizes because there are two types of dimensioning in use. In the past, flue linings were all of a nonmodular size (not based on the 4-inch modular grid). Today, modular size flue linings are also available from building supply dealers. The modular linings coordinate with dimensions of other masonry units more efficiently.

The table in Figure 34-6 gives the old rectangular flue lining size that is still used in many parts of the country. The new modular size is also given. Do not substitute one style flue for the other without first checking the size of the cross-sectional area inside the flue lining. If working from a set of plans, follow the size flue indicated on the plan.

Fig. 34-14 Completed firebox and rough work of the chimney. Notice the way the brickwork is sealed off under the trimmer arch. The fireplace is ready for facing work.

SETTING THE FLUE LINING OVER THE SMOKE CHAMBER

At the proper height, the flue lining should set in the center and against the back wall of the chimney, minus space for expansion. It should rest on the brick corbeling. The inside of the flue lining should be flush with the edge of the corbeling so it does not obstruct the draft. Refer back to Figure 34-12 to see how this was accomplished.

If the flue lining must be set on an incline or angle, make it a gradual one, no more than 30 degrees from vertical, to prevent it from choking off. The rest of the chimney up to the roof is built as previously described.

BUILDING THE FRONT FACING OF THE FIREPLACE

Workmanship

The facing of the fireplace is very important since it reflects the ability of the mason. It is the part of the fire-place that can be seen; therefore, the work must be of the highest quality and neatness possible.

Dry Bonding

The first course of masonry work is laid off to a line across the front of the fireplace to ensure that it is perfectly straight. If brick is being used for the facing it should be dry bonded across the fireplace from one end to the other. This is done to make sure it works whole bricks when it crosses over the lintel at the top of the opening. This is very important because a half brick can appear in the center of the lintel when crossed over if not bonded properly. If this happens the work must be torn down and rebuilt. Figure 34-15 shows dry bonding of the brickwork across the front of the fireplace.

As a rule, if stone is to be used for the facing, it is not necessary to dry bond. They run different sizes and can be made to work out without problems. Figure 34-16 shows the mason laying out the first course of stone on the fireplace. Notice the wall ties projecting from the rough face of the fireplace. They are used to tie the stonework together with the rough work.

Setting the Lintel over the Opening

The masonry piers are built up to the height where the lintel sets over the opening. This height can be determined by referring to Figure 34-6 for the particular size fireplace being built.

Use a heavy angle iron lintel over the opening. The lintel should be at least $\frac{3}{8}$ inch thick, $3\frac{1}{2}$ inches high, and $3\frac{1}{2}$ inches wide for the average fireplace. If the opening of the fireplace is 48 inches or more, a heavier angle iron must be used to prevent sagging. If the angle iron has been cut off with a torch, hammer the metal beads off the edges of the iron before laying on the piers. This allows the bricks or stone to lay level on the angle iron.

The exterior angle iron should set one course of firebrick lower than where the damper sets. It bears a minimum of 6 inches on each side of the opening for proper support.

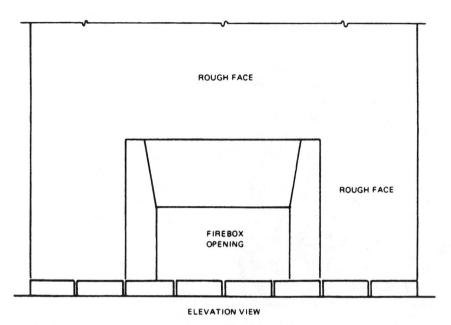

ROUGH FACE

ROUGH FACE

FIREBOX
OPENING

ELEVATION VIEW

Fig. 34-15 Laying out the first course of facing brick dry to determine the bond when crossing over the opening. This must be done or a half brick can be in the center of the opening when built.

Fig. 34-16 A mason lays out the stone on the face of the fireplace. This does not have to be bonded over the lintel because it is random stonework. It is lined up, however, across the fireplace with a line.

Mantels

The design of the mantel, if one is to be used, depends on personal preference. If it is a wooden mantel, the carpenter may supply and install it. In this case, the only thing the mason may have to do is to build wood nailing blocks into the masonry work at the appropriate height, so that the mantel can be nailed or fastened by the carpenter.

A brick or stone mantel is installed by the mason. The standard height from the finished floor to the top of the mantel is 54 inches. The fireplace can drop off at this point and be plastered or paneled, or continued in face brick the rest of the way to the ceiling. Figure 34-17 shows a finished brick fireplace with a raised hearth and a corbeled mantel.

Raised Hearth

The raised hearth, Figure 34-18, should be square with the main fireplace wall and the height of the inside hearth. It can be used as a bench seat around the fireplace.

Fig. 34-17 Finished brick fireplace with raised hearth and brick mantel. The fireplace is not cleaned down yet.

Fig. 34-18 A stone fireplace with a raised hearth built in a basement that is to be developed into a recreation room.

Strike the mortar joints on top of the raised hearth with a flat tool. The slicker striking tool is recommended. Raised hearths are very popular for several reasons: they give a nice appearance because the firebox is raised to a higher level, heat is radiated at a higher level, and the hearth can be used as a seat.

Washing Down

The final task of building the fireplace is the cleaning of the masonry work. Wash down the fireplace with a solution of 1 part hydrochloric acid (muriatic) to 10 parts clean water. Use a nonmetallic container for the solution.

The fireplace should always be washed down and cleaned before the finished floor of the room is laid because the solution can damage the floor. Washing down is important since it is the finishing touch in building a fireplace.

A CHIMNEY WITH MORE THAN ONE FIREPLACE

Many homes have chimneys with two fireplaces and a furnace flue. In this case, the flues must be slanted on angles to divide them up equally in the chimney and to allow for the second firebox to be built. As mentioned previously, do not slant or lay the flue lining on more than a 30-degree angle or it can choke off. It is also very important to remember that only one flue can be used for each heat source.

Figure 34-19 shows how the flue linings are slanted in a chimney that has two fireplaces. This is done to provide space for the second fireplace and to divide them equally in the section above the second fireplace.

Since the flue lining is drawn in, the size of the chimney above the second fireplace can be reduced to save materials. This can be done in several different ways. The two most popular methods are (1) to roll the brickwork in and lay a curved rowlock cap on it, or (2) to cut the chimney back on a rake and add a brick paving to run the water off. The rake can be on one side or both sides depending on the architect's preference or existing windows in the wall. A form of wood can be used for the rake guide or roll pattern and attached to the chimney until it is completed by the mason. Figures 34-20 and 34-21 show a reduced chimney built using

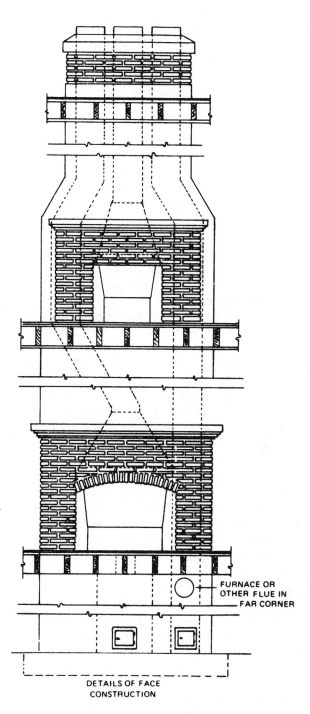

DETAILS OF FACE
CONSTRUCTION

Fig. 34-19 Elevation of a typical chimney that has two fireplaces and a furnace flue. The slanted direction of the flue lining is indicated by the broken lines.

the two methods described. This is done when the chimney is being built and the fireplace is installed later.

How a Rumford Fireplace Is Constructed

In Unit 33, the Rumford fireplace was mentioned. With the return of fireplaces to the home for heating purposes rather than decorative purposes, there has been a renewed interest in the Rumford fireplace. As a review, Rumford designed a fireplace that featured a shallow depth in the firebox with larger than conventional angled sides and back in the firebox and a higher opening than in a regular conventional fireplace.

Figures 34-22 and 34-23 provide views, details and dimensions for building a Rumford fireplace. The numbered list of instructions will provide anyone who

Fig. 34-20 Chimney that has been reduced in size by rolling the brickwork.

Fig. 34-21 Chimney that has been reduced in size by cutting the chimney on a rake and paving the rake area with brick.

wishes to build this type of fireplace guidance on how it is designed and constructed.

In addition, this plan utilizes the idea of using outside air for combustion and draft for maximum efficiency, which was also discussed in Unit 33.

Building a Rumford Fireplace

The Rumford stone fireplace described here is not a true historically designed Rumford fireplace but a very successful variation of the same and is recommended by the Brick Industry Association.

The Rumford stone fireplace described here was built according to the section, elevation plan, and tables illustrated in this unit for a Rumford fireplace. After

the fireplace was completed and had cured for approximately 30 days, it was tested by building a fire in it and it performed excellently. The amount of heat it radiated was very good and even though it has a shallow firebox, there was no smoking of the fireplace in the room.

This fireplace was built in a large vacation lodge and easily heated the entire area. To achieve the best results, however, one must follow the plans and tables as accurately as possible. The Rumford fireplace has become increasingly popular in recent years and is highly recommended as an alternative source of heat in addition to creating a nice appearance. Following is a brief description of how this fireplace and chimney was built.

The Firebox Area

This particular fireplace has a 40″ front firebox opening. According to the Rumford Fireplace Dimensions table, Figure 34-23, you will see that if the front opening is 40″, (A) there are some choices of how high the top of the opening can be, shown in the table. I selected 40″ (B), as this was a floor to ceiling fireplace of stone and I wanted a large fireplace. Since this fireplace has an arch over the opening, the 40″ height is to the center of the bottom of the stone arch. This particular fireplace had a segmental arch over the opening. A segmental arch is one of the strongest arches, in which the abutment or pier is ample to resist the thrust. For openings over windows, doors and fireplaces in home construction, it is the type most often used. Although the rise of a segmental arch will depend somewhat on the architect's design, you should follow a specific minimum height-to-span ratio when designing. A good rule to follow for a segmental arch is to always make the minimum rise at least equal to one-eighth the width of the span or opening. Segmental arch rises of less than this are not recommended as they could fail. Arches less than this can be built, but must be supported by a frame or permanent support of noncombustible materials if over a fireplace opening, such as an angle iron. This is especially true in the case of a Jack Arch which is relatively flat. For additional information on masonry arches, refer to Unit 37, *Construction of Semicircular and Segmental Arches*. The rise of this particular arch was 12 inches, which is easily above the minimum rule.

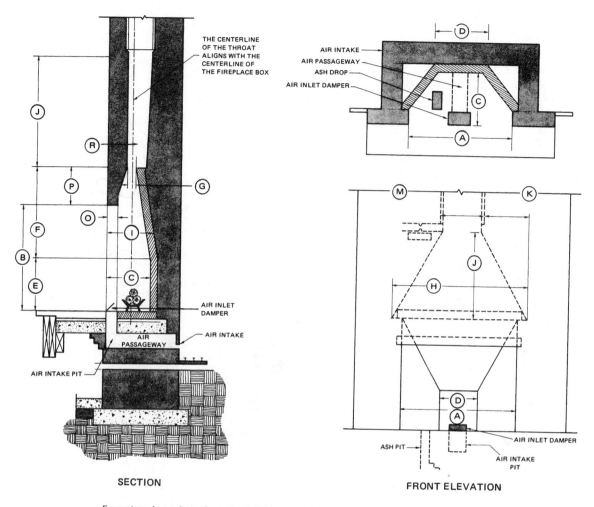

SECTION

FRONT ELEVATION

For maximum heat radiation from a Rumford Fireplace:

1. The width of the firebox (D) must equal the depth (C).

2. The vertical portion of the firebox (E) must equal the width (D).

3. Thickness of the firebox (1 minus C) should be at least 2 1/4 inches.

4. Area of fireplace opening (A x B) must not exceed ten times the flue opening area.

5. The width of the fireplace opening (A), and its height (B) should each be two to three times the depth of the firebox (C).

6. Opening height (B) should not be larger than width (A).

7. The throat (G) should be no less than three or more than four inches.

8. Centerline of the throat must align with the centerline of the firebox base.

9. The smokeshelf (R) should be four inches wide.

10. The width of the lintel (O) should be no less than four nor more than five inches.

11. Vertical distance from lintel to throat (P) must be at least 12 inches.

12. A flat plate damper is required at the throat, and must open towards the smokeshelf.

Fig. 34-22 Rumford fireplace with exterior air supply system. (Courtesy Brick Industry Association)

Rumford Fireplace Dimensions[a,b]

Finished Fireplace Opening						Rough Brick Work					Flue	Angle	Throat and Smoke Shelf		
A	B	C	D	E	F[c]	G	H	I[c]	J	K	L × M	N	O	P	R
36	32	16	16	16	28	4	44	19½	27	14	12 × 16	A-48	4	12	10
40	32	16	16	16	28	4	48	19½	29	16	16 × 16	A-48	4	12	14
40	37	16	16	16	33	4	48	19½	29	16	16 × 16	A-48	4	12	14
40	40	20	20	20	32	4	48	23½	29	16	16 × 16	A-48	4	12	14
48	37	16	16	16	33	4	56	19½	36	18	16 × 20	B-60	4	12	14
48	40	20	20	20	32	4	56	23½	36	18	16 × 20	B-60	4	12	14
48	48	20	20	20	40	4	56	23½	36	18	20 × 20	B-60	4	12	16
54	40	20	20	20	32	4	66	23½	45	23	20 × 20	B-72	4	12	16
54	48	20	20	20	40	4	66	23½	45	23	20 × 20	B-72	4	12	16
54	54	20	20	20	46	4	66	23½	42	21	20 × 24	B-72	4	12	16
60	48	20	20	20	40	4	72	23½	45	24	20 × 24	B-72	4	12	16

aThese are approximate dimensions based on historical data of Rumford fireplace construction. As is true with all fireplaces, successful performance is experimental.
bThese dimensions have been developed from the following formulas. These formulas may also be used for opening dimensions other than those listed. Minimum dimensions are taken from the CABO One and Two-Family Dwelling Code, 1986 Edition.

NOTES TO TABLE 1

A = Fireplace opening width, in.
B = Fireplace opening height, in. where: $2/3\ A < B \leq A$
C = D = E where: $1/3\ B \leq C \leq 1/2\ B$
F = B − E + P where: P = 12 in. minimum
G = 4 in.
H = A + 8 in. for: $A \leq 48$ in.; A + 12 in. for $A > 48$ in.
I = C + 3½ in. minimum when fire brick are laid as shiners or C + 5½ in. when fire brick or common brick are laid as stretchers.
J = K/u, where: $0.50 \leq u \leq 0.58$
K = 1/2 (H − M)
L × M > 0.16 (A × B)

N = A = 3 × 3 × 3/16 in. angle (number denotes length, in.)
B = 3½ × 3 × 1/4 in. angle
O = Nominal brick thickness
P = 12 in. minimum
Q = 8 in. minimum when $A \times B < 864\ in^2$; 12 in. minimum when $A \times B \geq 864\ in^2$
R = Smoke shelf width (flue opening, in.)
S = 8 in. minimum when fire brick lining is used; 10 in. minimum when common brick lining is used.
T = 16 in. minimum when $A \times B < 864\ in^2$; 20 in. minimum when $A \times B \geq 864\ in^2$

cMinimum dimensions

Fig. 34-23 Rumford fireplace dimensions.

A slightly higher rise was desired, as the fireplace was built floor to ceiling with a raised hearth.

The splay (angle) of the sides of the firebox is also determined from the table. The width of the fireplace opening (A), and its height (B) should each be two to three times the depth of the firebox (C).

By studying the plans and tables, one can easily see how these dimensions coordinate together. Space does not allow discussion of all the rules and measurements here. These plans and tables do allow a mason or person building a Rumford fireplace to select different size openings to fit their desires and needs. Once an opening

size is selected, then the matching dimensions must be followed from the tables without making any changes. This is because successful fireplaces, especially Rumford fireplaces, are based on correct proportions.

It is also highly recommended to lay the firebricks in a stretcher position for a Rumford fireplace, as they will last longer, radiate more heat and are more secure than in a Shiner position due to the additional depth of the brick. The back wall of the firebox should be bonded in behind the sides for extra strength. Metal wall ties are built into the mortar joints on the front jamb periodically to tie the facing materials, in this case, stone, to it. Due to the location of the fireplace shown here, an outside hearth vent could not be installed and was not found to be necessary.

Making a Damper for a Rumford Fireplace

A flat plate steel damper sized to fit the splay and required dimensions of the fireplace was custom-made at a metal machine shop, as standard dampers to fit a traditional Rumford fireplace are usually not available from local building supply dealers. A heavy $\frac{1}{2}''$ plate steel frame and lid were used so they would not warp from the heat. At a local masonry building supplier, some left-over parts from standard-type poker dampers were located and welded onto the plate frame. This worked out well, but you could make your own. The damper depth, or throat, for a Rumford is very narrow and should not be more than $3''$ to $4''$ deep. This should be followed accurately, as this narrower opening increases the velocity of the draft, which is very important for the fireplace to operate correctly without smoking. This is shown on the section view as (G). Figure 34-24 shows a completed firebox with plate damper in place.

Building the Rough Work behind the Facing

The rough work or un-tooled brickwork behind the facing is built the same as in a normal fireplace, cor-

Fig. 34-24 Completed firebox with steel plate damper in place.

beling the brickwork out equally to the point where the flue lining will set. These dimensions are shown on the plans and tables. The size of the flue lining is obtained from consulting the table, Figure 34-23. Reading across the table from left to right, you will see that this fireplace requires a $16'' \times 16''$ flue. If you cannot find this exact size in your area, then select a flue that will have approximately the same amount of cross-sectional inches and it will work fine. This is a very large flue, but is also the reason that a Rumford works so well. The narrow damper opening, the big firebox and the large flue working together create a very strong draft, resulting in a fireplace that does not smoke but radiates a lot of heat into the room. The centerline of the throat where the damper sets must align with the centerline of the firebox inner hearth. When the flue lining is set into position, the center of the flue should line up with the center of the damper on top of the throat. This can be seen clearly in Figure 34-22 Section View.

After the rough brickwork has been completed and the chimney has been finished, the facing of the fireplace work can be done. This particular fireplace had a stone arch over the opening which rests on a wooden form as shown in the illustration, Figure 34-25.

Facing the Fireplace with Stone

The stone work was laid in a 1:1:6 portland cement, hydrated lime, to sand mortar. The mortar joints were smoothed and filled in fully with a slicker tool and then raked out with a short rounded-off length of broom handle after they had dried enough as not to smear and then brushed, Figure 34-26.

A wood mantle was installed as the fireplace was built with stone brackets to support it. The fireplace was cleaned by scrubbing with a stiff brush the next day after it was completed and before the mortar had set hard. A wood paddle was used to scrape any particles of mortar off of the surface. Stubborn mortar stains can be removed with a masonry cleaner after it has cured for at least two weeks. As a rule, if the work is brushed at the correct time as it is built, this is not necessary. A completed Rumford fireplace is shown in Figure 34-27.

Fig. 34-25 Rumford fireplace with rough brickwork and arch support in place.

Fig. 34-26 Filling in the mortar joints with a slicker pointing tool.

ESSENTIALS OF FIREPLACE CONSTRUCTION

If the information and techniques given in this unit are carefully followed, the finished fireplace will burn properly and safely, giving off the maximum amount of heat possible according to its particular type or design.

- The fireplace should fit the size of the room.
- Each fireplace or heating device must have its own individual flue.
- Use portland cement-lime type M or equal mortar for foundation and type N or equal for above-grade work.
- A fireplace will operate more efficiently if a source of outside air is introduced directly into the firebox rather than using air from within the room.

- Building a fireplace is a matter of combining correct proportions, as they are stated in the table, with the plan of that particular fireplace. Do not experiment with these.
- Build the firebox with angled (splayed) sides and a sloping back.
- Provide the proper size smoke shelf as shown on the plan or illustration.
- Make sure that the smoke shelf is level with the back of the damper.
- Try the damper to see if it works smoothly and correctly before building it into the fireplace.
- The flue lining that sits at the top of the combustion chamber should be in the center of the fireplace so it does not draw to one side or the other.

Fig. 34-27 Completed Rumford fireplace.

- Slope the sides and front of the smoke chamber brick walls evenly to the point where the flue lining will be set in place.
- Flue lining should be set on corbeled brickwork, not on nails driven in the wall.
- Do not leave pockets or obstructions to interrupt smooth passage of smoke and gases in the smoke chamber.
- Select the size flue lining on the basis of having $\frac{1}{12}$ of the area of the fireplace opening.
- Use a heavy steel angle iron over all openings so it will support the load without sagging.
- Parge (plaster) the area within the smoke chamber to promote the smooth passage of air and smoke.
- Build the chimney high enough above the roof to prevent downdraft or smoking (2 feet above ridge of roof).

- Perform high-quality work as the fireplace is the focal point of most homes and a true reflection of the best work in the mason's craft.

BUILDING THE FIRST FIRE IN THE FIREPLACE

Before the first fire is built in the fireplace, the fireplace should cure at least 30 days after construction. If fires are built before this amount of time passes, the moisture necessary for proper curing of the mortar is driven off and could cause a cracking of the bricks or mortar.

Start the fire slowly with a small amount of kindling wood in the back of the firebox. Gradually increase the fire until a moderate one is built. No fireplace will draw perfectly at first, as the air in the flue must be heated somewhat to get the draft flowing properly. Never start with a big, roaring fire. The heat buildup is too fast, which will crack or damage the fireplace and could set the home on fire.

> **Caution:** Make sure that the damper is open before lighting any fire in the fireplace.

BUILDING AN ADD-ON CHIMNEY AND FIREPLACES

One thing that will enhance the value of a home and add additional comfort is to add a masonry chimney and fireplace. It also provides an emergency heat source in case there is a prolonged electrical service loss due to events such as storms. In the past, the fireplace in a home was valued for its esthetic features, but in recent years, it has once again been regarded as a necessity, as a back-up to provide warmth and even as a place to do emergency cooking.

Because I live in the northern part of the United States, I have had to resort to these uses of the fireplace a number of times. A fireplace will also increase the value of a house from approximately $12,000 to $15,000 and make it much more attractive to prospective home buyers, if it has to be sold. There also is nothing as satisfying as gathering around a fire burning in the fireplace over the holidays for a feeling of warmth and security. For these reasons, building and remodeling contractors have experienced a large up-surge in constructing add-on chimneys and fireplaces. The add-on chimney and fireplaces described here are typical examples of what is required to plan and build them.

Building Permit

You will have to apply for and get a building permit from your local government and have at least a clear drawing or sketch with particulars, before being allowed to start any construction. In most cases, this does not require the services of a registered architect, but just has to be accurate and clear. The local building inspector will advise you on the proper footing to install depending on the weather extremes where you live and in most cases will want to visit the site and inspect the poured footing to make sure it meets local code requirements before allowing any masonry work to be done.

One or Two Fireplaces?

How many fireplaces do you require for your home? Since I was building this chimney for my son and there was no labor cost involved, we decided to go with two fireplaces, one in the basement recreation room and the other in the first-floor family room. The only difference is that two fireplaces will require a larger chimney and, of course, will involve more work. If you are paying for it, it will also be more expensive. Once the question of one or two fireplaces is answered and the permit is granted, construction can start.

Cutting an Opening through the Foundation Wall for the Basement Fireplace

Once the footing is in place and has been inspected and approved, an opening must be cut through the concrete block foundation wall large enough for the firebox of the fireplace that will face the interior of the room. This is determined by consulting the tables of standard fireplaces in this unit for rough openings. The size of the opening will depend on whether you are going to build a standard or a Rumford fireplace. Both of the fireplaces for this chimney were the standard design. However, there is a set of tables for both types of fireplaces in this unit. Be sure to allow extra height

for a raised hearth, if you are going to build one. Mark off the center of the fireplace on the outside wall and also the size of the opening to be cut out on the wall. It can be cut out neatly with a masonry saw or by using a sledge hammer, 2-pound striking hammer, and masonry chisels. In any event, be careful to cut it out as neatly as possible. In the event that the masonry courses above the opening show any signs of cracking or not staying in place, be sure to put a prop under them to support them. Remove all the loose debris from the area. In the event you have a poured concrete basement wall, it will have to be cut out with a masonry saw, Figure 34-28.

Laying the Chimney out with Concrete Blocks

Lay out the concrete block work for the chimney. Keep in mind that it should be laid out for width and length, and that will work full brick when the finished grade height is reached, even if this means making the chimney a little bigger than necessary. The concrete block will also have to be parged with mortar and tarred once it has built up to the height of finished grade. Although I could have used an $8'' \times 8'' \times 16''$ block, I like to use a little wider $8'' \times 10'' \times 16''$ block as it is more stable and allows more room to lay the brickwork out, Figure 34-29.

Fig. 34-28 Cutting the opening in the foundation wall.

Fig. 34-29 Laying out the chimney with concrete block.

Fig. 34-30 Laying the brickwork out at finished grade.

Laying out the Brickwork at Finished Grade

I always lay out the brick three courses lower than the actual finished grade of the earth that will be around it to allow for settling. There is nothing more ugly that having the tarred surface of the foundation showing after the earth has settled around a chimney! If there is a stepped grade, starting at the lowest step and working up, dry bond the brickwork out so it will work for proper bond at finished grade. Metal wall ties need to be attached to the foundation wall and the inside corners of the chimney as the work proceeds to attach it securely to the masonry house wall, using case-hardened or tempered nails, Figure 34-30

**Building the Firebox and Enclosing
the Rough Opening**

After the outside chimney has been built above the interior opening, the firebox is laid up with firebrick, the damper is set in place, and the rough brick built up to the ceiling height of the room following the procedures described in this unit for building fireplaces. Make sure that the rough brickwork in the interior of the chimney area is built up and sealed tightly against the original house wall that you cut through, so it will be secure and strong. Wall ties should be built into the mortar bed

joints of the rough brick every 16″ in height, as they will be utilized later to bond the finished face brick to the rough bricks or block backing, Figure 34-31.

Building the Front Facing of the Fireplace

Since we decided to use a double rowlock segmented arch to support the opening over the firebox, the first course of face brick of the fireplace was dry bonded across the front of the fireplace to make sure it works full brick across the top of the arch. I then laid it up in mortar up to the height of the firebox opening. The arch form was set in place and the double rowlock ring arch laid into position on top of the wood form. See Figure 34-32.

The Completed Fireplace

After a suitable length of time had passed and the mortar joints in the fireplace had cured sufficiently, it was cleaned with a masonry cleaner and the mantel was installed, completing that part of the project, Figure 34-33.

Finishing the Chimney

As the chimney was built, a 1″ air space was maintained free of mortar against the side of the house as

Fig. 34-31 Firebox and rough opening installed.

Fig. 34-32 Laying the brick arch on the wooden form.

required by the building code. Metal wall ties were nailed through the siding into the wood studs at 16″ intervals in height to tie the chimney to the house. The top of the chimney was corbelled out and the last two courses of brick set back to form an ornate cap and give it a little design and character. I used a Type S mortar mix on a bevel for the wash coat to seal the top of the chimney. As the chimney was not right next to the peak or ridge of the roof, I made sure that the top of the chimney was built a good 3 feet higher than the top

of the peak of the roof on a level line to make sure it would draw well, Figure 34-34.

To test the draft in the fireplaces, I took a piece of rolled-up newspaper and lit it with a match in both of the fireplaces with the dampers open to make sure they drew properly. Both fireplaces worked perfectly. Although it is not always possible, for best results, it is recommended to let the masonry cure a week before washing it down with a masonry cleaner. This completed the job. See Figure 34-35.

Fig. 34-33 Completed brick fireplace.

Fig. 34-35 Completed chimney.

Fig. 34-34 Finished chimney top.

ACHIEVEMENT REVIEW

A. Select the best answer from the choices offered to complete each statement. List your choice by letter identification.

1. One of the major problems in building a fireplace that projects into the room is
   a. supporting the hearth.
   b. the damper rod may not be long enough.
   c. the firebox is more difficult to design and build.
   d. there is not enough working room for the mason.

2. The best all-around mortar to use for building a fireplace and chimney is type
   a. O.   c. N.
   b. K.   d. M.

3. The rough hearth, which was built of rock on a wooden form, was in the form of an arch. This type of arch is known as a
   a. jack arch.   c. trimmer arch.
   b. hearth arch.   d. spandel arch.

4. Raised hearths should not be built higher than
   a. 8 inches.   c. 14 inches.
   b. 12 inches.   d. 16 inches.

5. A standard fireplace that measures 36 inches wide by 29 inches high should have a rough opening to allow laying of the firebrick and damper of
   a. 44 inches by 48 inches.   c. 52 inches by 60 inches.
   b. 48 inches by 52 inches.   d. 54 inches by 62 inches.

6. The mortar joints on the top of the hearth should be tooled with a
   a. V-jointer.   c. slicker jointer.
   b. convex jointer.   d. rake-out jointer.

7. The finished fireplace should be laid out from
   a. the outside wall.   c. the center of the chimney.
   b. the edge of the chimney.   d. the inside of the outside wall.

8. The sides of the firebox are built on an angle
   a. to conserve firebrick.
   b. for design and beauty.
   c. to reflect the heat outside.
   d. to enable the smoke and gases to draw up the flue.

9. The roll or slant in the back of the firebox should be started on top of the
   a. second course.   c. fourth course.
   b. third course.   d. fifth course.

10. The part of the fireplace that deflects the air coming down the chimney back up the other side of the flue is called the
    a. damper.   c. smoke shelf.
    b. firebox.   d. hearth.

11. Another name for the damper in a fireplace is the
    a. smoke chamber.   c. firebox.
    b. smoke shelf.   d. throat.

12. The area of the fireplace from the point where the damper sits up to where the flue lining is located is called the
    a. smoke chamber.   c. smoke shelf.
    b. hearth.   d. firebox.

13. Applying the formula discussed in the unit stating that the cross-sectional area of the flue lining is at least $\frac{1}{12}$ the total area of the fireplace opening, which of the following modular-size flues is needed for a fireplace with an opening of 36 inches by 29 inches?
    a. 8 inches by 12 inches   c. 12 inches by 16 inches
    b. 12 inches by 12 inches   d. 16 inches by 16 inches

14. Using the table of standard-size fireplace dimensions in inches given in the unit, which of the following measurements is correct for the height from the damper to where the flue sits in a fireplace with a finished opening of 36 inches by 29 inches?
    a. 20 inches   c. 27 inches
    b. 24 inches   d. 30 inches

15. The standard height of the top of a mantel from the finished floor is
    a. 48 inches.   c. 60 inches.
    b. 54 inches.   d. 62 inches.

B. Identify the parts of the chimney and fireplace that are lettered in Figure 34-36.

MATH CHECKPOINT

1. Eighty firebricks are required to build the firebox in a standard 36″ wide fireplace. Each brick cost $.55. What will the total cost of the firebrick be?

2. The front of a fireplace is laid with standard-size face brick. It measures 6′ wide by 8′ high. Deduct the opening of the firebox, which is 36″ wide by 30″ high. Use the figure of seven brick per square foot of wall area to determine the number of bricks needed. In addition, allow 5% extra for broken or chipped brick. How many bricks will be required to build the front of the fireplace?

3. How many square inches are there in a flue opening that measures $13'' \times 13''$?

4. How many flue linings will be required for a chimney that is 36′ in height? Deduct three flue linings from this figure because the first flue lining will be set into place at the top of the fireplace throat or smoke chamber.

5. The outer hearth of a fireplace is to be laid with $4'' \times 4''$ quarry tile. How many will be required if the hearth measures $16'' \times 16''$ long?

PROJECT 29: BUILDING A BRICK FIREPLACE AND CHIMNEY

OBJECTIVE

- The student will lay out and build a brick fireplace and the section of the chimney directly over the fireplace area. This project is typical of a fireplace open on one side. It is the type found in most homes.

  Note: Two students should work as a team on this project.

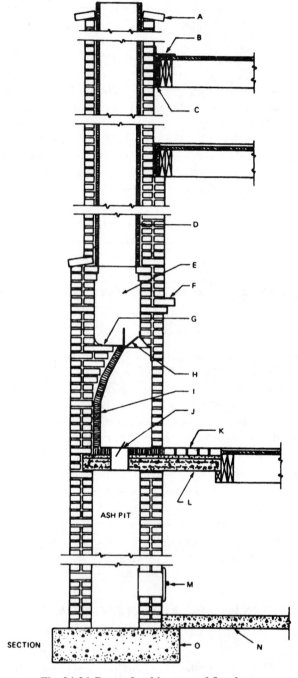

Fig. 34-36 Parts of a chimney and fireplace.

EQUIPMENT, TOOLS, AND SUPPLIES

Mortar mixing tools and box

Basic set of mason's tools

Chalk box

2-foot framing square

Mortar pan or board

Mortar stand

Supply of mortar materials

Water

Brick to be estimated from the plan by student: supply of good face brick for front of fireplace; supply of common brick for unexposed work inside the fireplace and chimney area

80 standard firebricks

Ash dump door

One 48-inch angle iron ($3\frac{1}{2}$ inches x $3\frac{1}{2}$ inches)

One 36-inch rotary control damper (poker type can be used if rotary is not available)

Small amount of insulation to be used at end of damper for expansion

One 13-inch $\times$ 13-inch nonmodular flue lining (modular flue can be used if nonmodular is not available)

Supply of bricks or 4-inch $\times$ 4-inch quarry tile for outer hearth estimated from the plans, Figures 34-37 through 34-39

SUGGESTIONS

- When laying out the first course for the fireplace, be sure to use 2 or 3 different bricks of various lengths if there is going to be a variation in the bricks being used for the fireplace.
- Lime and sand mortar can be used instead of real mortar because it can be reused.
- Form all mortar joints fully since a fireplace is subject to intense heat.
- Keep all of the mortar joints of the firebrick small because heat leaks through a large mortar joint.
- Wear safety glasses whenever cutting any of the masonry units, especially the firebricks since they are hard and brittle.
- Work all measurements from a centerline.
- After cutting any firebrick, rub the cut edge until it fits neatly against the brick previously laid.
- Do not mix more mortar than can be used in one hour.
- Strike the mortar joints as soon as they set enough so as not to cause burn marks from the steel striking tool.
- Practice good workmanship because the fireplace is the focal point of a room and should be of the best quality.
- It is not necessary to fill in on the sides of the firebox for a training project since all of this material must be reclaimed.

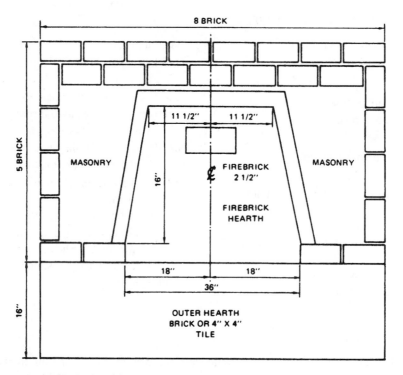

Fig. 34-37 Project 29: Building a brick fireplace and chimney (plan view).

PROCEDURE

Refer to Figures 34-37 through 34-39.

1. Mix the mortar to the stiffness desired.
2. Lay out the first course dry completely around the project according to the plans.
3. Lay the first course in mortar as shown.
4. Lay up the outside walls 3 courses high, leaving room for the opening. Lay all standard courses of brick to number 6 on the modular rule.
5. Working from the center of the fireplace, lay out the hearth to the correct size and height shown on the plans.
6. Install the ash dump in the hearth as shown.
7. Strike the hearth with a flat jointer (slicker).
8. Working from the centerline of the hearth, lay out the firebox walls as shown. Use the framing square to mark off the lines; be sure lines are true.
9. Lay out the first course of the firebox as shown on the plans.
10. Tool the mortar joints on the side and back walls of the firebox with a concave jointer.
11. Sprinkle a light coating of sand on the bottom of the firebox to prevent mortar from sticking.

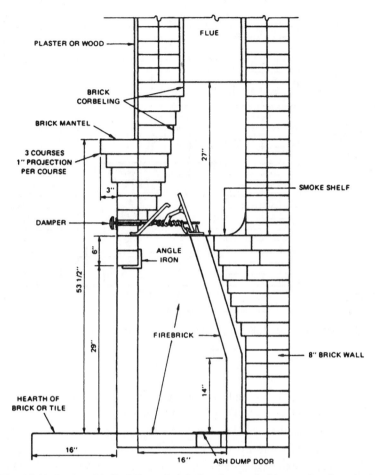

Fig. 34-38 Project 29: Building a brick fireplace and chimney (section view). Notice: once the height of the flue is set, no purpose is served by building higher. This is why a broken line is shown.

12. Build the outside walls of the fireplace 6 more courses of bricks high and fill in on the inside with masonry.

13. Lay up the firebrick in the firebox until it is 3 courses high.

14. Continue building the firebox, slanting or rolling the back to the throat size of the damper being used. See the plans.

15. At the completion of the firebox, parge the unexposed back wall with mortar.

16. Back up brick behind the firebox walls, being sure to leave ½ inch for expansion.

17. Level off the smoke shelf at the back of the fireplace with mortar where the damper will sit. Parge a slight cove of mortar at the point where the back of the smoke shelf meets the back wall. See Figure 34-38.

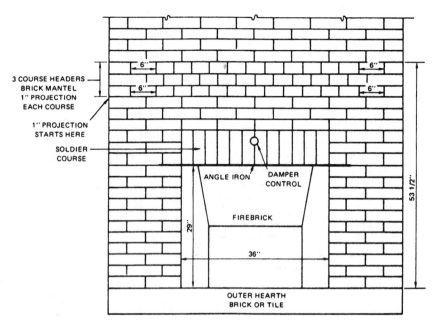

Fig. 34-39 Project 29: Building a brick fireplace and chimney (elevation view).

18. Lay up the outside masonry jambs in the front of the fireplace until they are 29 inches high. This is 11 courses of standard brick. See the plans.

19. Set the angle iron on the piers dividing the bearing up on each side of the opening.

20. Assemble and set the damper in mortar on the top of the firebox so that the flange rests securely on the firebox walls. Lay some insulation for expansion at each end of the damper.

21. Build up the outside of the fireplace walls to the point where the corbeling starts for the smoke chamber. At the same time, build in the handle of the rotary control damper flush with the outside wall. Do not mortar in solid the turning handle rod.

22. Build up the back and side walls of the fireplace and chimney racking away from the front of the fireplace to scaffold height.

23. Use a pencil to mark off on the parged back wall the correct slope of the corbeling for the smoke chamber. (This procedure was described in the unit.)

24. Lay the corbeling course as shown on the plans until it reaches 27 inches above the damper. This is where the flue lining sits on the corbeling. Plaster the projecting edges of the corbeling until smooth, as shown on the plan.

25. Continue building the front of the fireplace up to the bottom of the mantel height as shown.

26. Corbel out the mantel course as shown in Figure 34-38, being careful to note how each course starts as related to the bond. Project each course out 1 inch. (Be careful not to let the projected headers droop down to the front as this will be very noticeable.)

27. Finish the front of the fireplace, tooling the mortar joints with either a convex or V-jointer. Brush all work when it is sufficiently dried so that it does not smear.

28. Set the flue liner in place on the corbeling, and brick up around it as shown on the plans.

29. The smoke chamber should now be completely sealed off up to the damper. There is no point in building the chimney and fireplace any higher. (This depends on the preference of the individual instructor.)

30. Clean the loose sand from the inside hearth, and tighten the setscrew in the damper rod so it cannot pull out. Remove any loose mortar from around the damper or on the smoke shelf by reaching up and inside the damper.

31. Lay the outside hearth to the measurement shown on the plan, either of bricks laid flat or hearth tile. (Check with the instructor on this.)

32. Recheck the front of the fireplace for any holes, brush off all areas, and have the instructor evaluate the project.

UNIT 35
Multiple-Opening and Heat-Circulating Fireplaces

——————————— OBJECTIVES ———————————

After studying this unit, the student will be able to

■ list the different types of multiple-opening fireplaces and state why a special damper improves performance.

■ explain why the metal heat-circulating fireplace was developed and list some of its advantages over the conventional type.

■ describe briefly the steps in constructing a metal heat-circulating fireplace.

The conventional fireplace is open on only one side. It is the most popular type of fireplace because it is the simplest to build. When fireplaces are open on more than one side they are called *multiple-opening fireplaces,* Figure 35-1.

Although the multiple-opening fireplace (also called multiple face) may seem to be a modern design, it is actually quite old in origin. For example, the corner fireplace, which has two adjacent sides open, has been in use for several hundred years in the Scandinavian countries.

Another example is a fireplace in which the two opposite faces are open. It is used sometimes as a room divider or when a fireplace is desired on both sides of two rooms, using only one chimney.

Fireplaces with three or four sides open are usually designed by the architect to be the focal point of the home. When they are located in the center of the room with three or four sides open, they are known as *island fireplaces.* Although they are pleasing to look at, multiple-opening fireplaces present more problems concerning draft and operation than a conventional fireplace opened on one side.

Adequate draft must be obtained by using over-sized flues and controlled face opening sizes. Cross drafts can go through the fireplace when front and rear doors in the home are opened at the same time. This can cause the smoke to blow out of the fireplace into the room.

Glass fire screens on one or more of the face openings help to prevent this.

DAMPERS FOR MULTIPLE-OPENING FIREPLACES

Regular dampers have always been used for multiple-opening fireplaces. Sometimes two dampers are laid back to back with a masonry partition wall between them. This is the case when a fireplace opening fronts in two different rooms but uses the same chimney. Each fireplace requires a separate flue.

Special dampers are now on the market for use in multiple-opening fireplaces that function more efficiently than the standard throat damper. A regular damper has a long, narrow throat designed to make the fire burn evenly along the back wall of the fireplace with only one side open. Fireplaces with more than one open side require a damper shaped to promote even burning of the fire anywhere on the hearth. A high funnel shape and larger throat area help solve this problem.

The conventional fireplace discussed in Unit 34 is closed on three sides and has a fixed area of opening. Fireplaces with more than one side open, therefore, have a larger area of opening, and the size of the flue and the area of the damper throat must also be bigger. A fireplace with more than one side open is affected more by any crosscurrents of air in the room. The multiple-opening fireplace must have a bigger flue and stronger

446

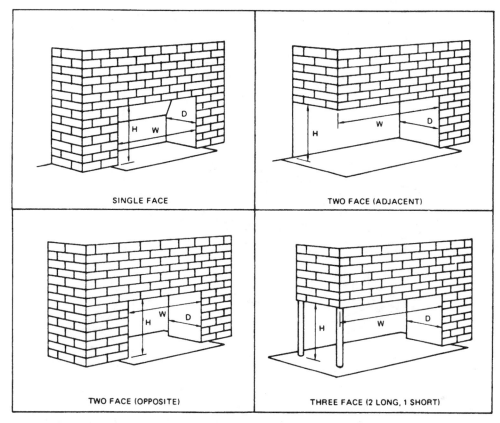

SINGLE FACE

TWO FACE (ADJACENT)

TWO FACE (OPPOSITE)

THREE FACE (2 LONG, 1 SHORT)

Fig. 35-1 Single face and multiple-opening fireplace designs.

drafts to overcome this problem. An example of a special damper for multiple-opening fireplaces is shown in Figure 35-2.

ADVANTAGES OF THE MULTIPLE-OPENING DAMPER

The most important advantage of the multiple-opening damper is that it permits a choice of flue locations as shown in Figure 35-3. When building a multiple-opening fireplace, often the flue cannot be located directly above the fireplace, so the flue lining must be slanted or angled. This obstructs the draft if it is not done correctly. Figure 35-3 shows how the smoke shelf

Fig. 35-2 A multiple-opening fireplace damper.

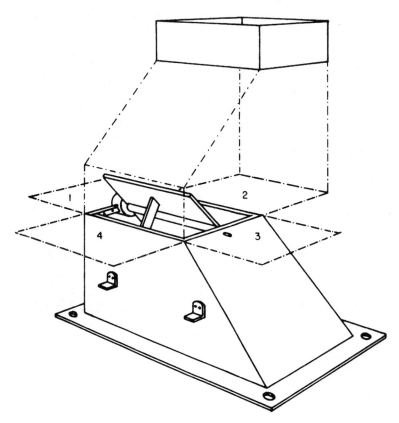

Fig. 35-3 An advantage of the multiple-opening damper is that it allows a choice of four locations for the flue lining.

can be built on any one of four sides when using the special multiple-opening damper. Reversing the damper allows four positions depending on the best location for the flue. It is important that all masonry be kept at least one-half inch away from all metal to allow for expansion and contraction.

Glass wool is supplied by the manufacturer with each damper. The glass wool insulation is placed in 1-inch thick pads along the four corners. This prevents the masonry work from coming in contact with the metal. It also keeps the proper expansion space. A thin mixture of mortar brushed on the steel helps to hold the glass wool in place until walled in by the mason.

Other construction advantages are shown in Figure 35-4, and are explained below. (The numbers correspond to those in the figure.)

1. The damper is designed with a smooth, high funnel throat shape to offer as little obstruction to draft as possible.
2. It is designed to permit rapid laying of masonry work. All four sides have a strong lintel that can carry the load. Corner posts can be obtained to carry the outside corner of the damper when the fireplace is open on two sides.
3. The damper is designed to permit the use of different size flue tile depending on the area opening of the fireplace.
4. It allows for easy construction of the smoke shelf. If a conventional damper (as described in Unit 34) is used in fireplaces with more than one opening, it presents a problem in forming the smoke shelf.

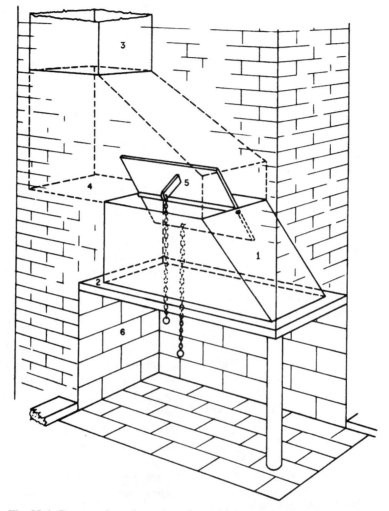

Fig. 35-4 Construction advantages of a multiple-opening fireplace. Numbers correspond to advantages listed in this unit.

5. The removable valve plate on the damper has adjustable tension to permit draft adjustments. Once the adjustments are made it is rarely necessary to change.
6. A durable chain controls the damper opening. The multiple-opening damper is also good for fireplaces with only one side open, but with an opening that is usually high or deep.

DETERMINING THE FLUE SIZE FOR MULTIPLE-OPENING FIREPLACES

Fireplaces with more than one face or opening are not treated the same as those having only one face when determining the correct flue size. Flue size is obtained by adding together the sums of the square-inch areas of each open face of the fireplace. The cross-sectional area of the flue lining should be at least $\frac{1}{10}$ the total area of all open sides for this type of fireplace.

Remember that any fireplace with more than one open side is much more difficult to build than a conventional one-side open fireplace. This is due to draft problems.

Multiple-opening fireplace construction differs mainly in the damper arrangement. The special multiple-opening damper provides more flexibility in constructing fireplaces that have more than one opening. Rules of construction for single-face fireplaces, with the exception of the special damper, also apply to the multiple-opening fireplaces.

DEVELOPMENT OF METAL HEAT-CIRCULATING FIREPLACES

Metal heat-circulating fireplaces have been in use for about 40 years. They were developed to answer the need for a fireplace that was more efficient in heat output. It was felt that the advancement of furnaces and heating plants would outdate the regular fireplace and possibly replace it in the home.

The problem of smoking, loss of heat up the chimney, and a limited amount of radiation from the firebrick caused designers to look for a better method of building a fireplace. The metal heat-circulating fireplace was developed as a solution to the problem.

The unit is designed to eliminate smoking and greatly reduce the loss of heat up the chimney. Two of the main causes of fireplaces not working properly are poor workmanship and not following the correct principles of construction. The chimney must be high enough for a good draft and the flue must be the correct size for the fireplace opening. Key parts such as the firebox, damper, smoke shelf, and smoke chamber also have to be built correctly or the fireplace does not function properly. The metal heat-circulating fireplace unit is scientifically correct in its design and construction. When delivered to the job it relieves the mason of building many of the key parts of the fireplace. Once the hearth is laid and the steel form set on the hearth, all the mason has to do is follow the instructions with the unit and continue to build the chimney. A mason is more likely to build a successful fireplace by using the metal unit.

CIRCULATION OF HEAT

The ability of a metal fireplace shell to circulate and radiate heat more efficiently is the main reason for its increasing popularity in these times of ever-rising heating bills.

The walls of the metal unit are hollow with enough space between them for the passage of cool air. This cool air is taken from the floor level through a vent built into the masonry work, and drawn into this hollow space. The heating chamber extends all the way around the sides and the back of the unit. Fire in the firebox area heats the steel walls. The air drawn in is heated and quickly rises. As the air expands and rises, it is forced out of the heat chamber through vents built into the face of the fireplace. These vents can be built below or above the mantel height or ducted into any adjoining room to convey the heat where it is needed.

THICKNESS OF SIDES OF HEAT CHAMBER

In order for the sides and back of the metal formed unit to withstand the intense heat, it is designed much the same as a furnace wall. The sides are usually $3/16$-inch steel plate. Outer walls that are next to the masonry work are not as thick because they do not have to withstand as much heat. The outer walls merely help to form the passageway for the flow of air and also serve as a guide for the exterior masonry to follow. The lighter weight steel also makes the price lower per unit than if made completely of heavy steel.

ECONOMY

Considering fuel shortages and high prices, the heat-circulating fireplace unit is desirable because it provides an emergency source of heat and helps lower the overall price of heating a home. That feature alone makes this type of fireplace attractive to the homeowner. In the spring and fall months when it is not necessary to fully heat the home, this type of unit can be used to keep the home comfortable.

It also offers long-range savings to the homeowner over a period of years. The output of heat is greater than from a standard fireplace, due to the movement of air through the heat chambers. A fan can be installed inside the venting area near the heat chamber to draw the air through the vent faster. When the sole source of heat is the fireplace (as in a basement fireplace or summer lodge), a heat-circulating unit is much more efficient.

The metal formed unit can also be constructed faster, and requires less materials. The parts that require the most time for the mason to build are prebuilt at the factory. This faster construction results in a savings in labor for the builder or homeowner.

The design and outward appearance is not limited since the unit is enclosed within the masonry work. Vents are finished off in such a way that they do not look offensive nor do they mar the finished job. Because of the strength of the steel and the fact that no welded joints are used across the back or corners, there is little possibility of cracking.

CONSTRUCTION DETAILS

Planning

The size of the unit to be used is determined by the size of the room, amount of heat desired, and location of the unit in the room. The size selected then determines the size of the chimney, foundation, and related details of construction. It is necessary to decide on the size of the unit before building any part of the fireplace or chimney, since it must fit properly into the chimney. As a rule, on new construction the architect specifies the size of the unit. Heat-circulating fireplace forms, Figure 35-5, are available from building supply dealers.

Flues

The same rule applies to heat-circulating fireplaces that applies to regular fireplaces when determining the

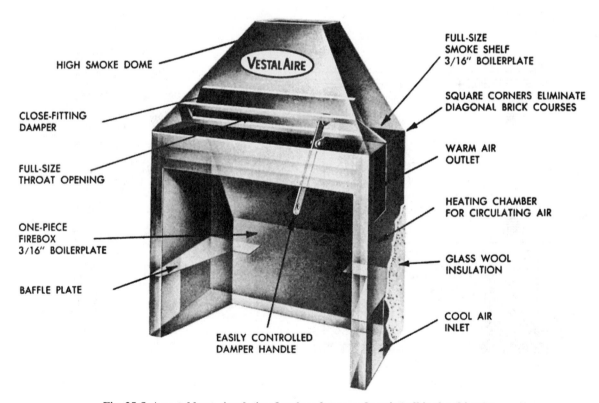

Fig. 35-5 A metal heat-circulating fireplace form ready to install in the chimney.

size flue lining to be used. Each unit must have a separate flue for proper draft. Any other flues built into the chimney for furnaces or heating plants are built the same way as for a regular fireplace and chimney; the only difference is that they must not interfere with the cold and hot air venting of the metal formed unit.

GRILLES AND DUCTS

The location of the grilles for cold and hot air are determined by the location of the rough openings in the metal form. Air passages should be finished smoothly to help speed air and to prevent the loss of heat. The ducts can be made of masonry formed as a chamber to where the air enters the metal form. Ducts can also be metal formed if there is a special arrangement. For best results, sharp turns and corners should be avoided and the duct arrangement should be kept as simple as pos-

sible. See Figures 35-6 and 35-7 for the construction detail of a grille and duct system.

Note: Figure 35-7 is a good example of how brickwork is built around grilles.

INSTALLATION

Hearth

The firebrick hearth for a metal unit must be laid level in all directions. The fireplace form is centered on the hearth, even with the edge of the front course of firebrick. Mortar should not be used to level the unit because it cracks out from the heat.

Insulation

Insulation, usually glass wool, should be applied to the metal back and sides with a thin coating of mortar

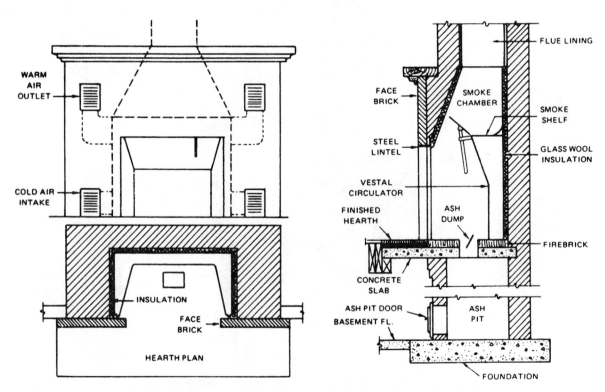

Fig. 35-6 Construction details for a metal heat-circulating fireplace form.

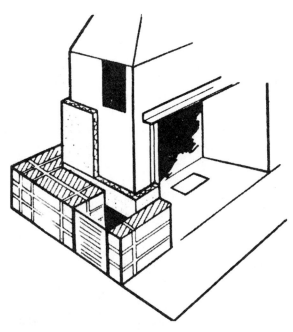

Fig. 35-7 Building the cold air inlet into the metal form grille cover provides a finished appearance.

Fig. 35-8 Workers applying insulation to metal form. Insulation must cover all metal surfaces.

brushed on the unit. The cold and hot air ducts should not be covered with insulation. As the masonry work is built, it must not touch the metal at any place because of the expansion of the metal. Insulation is supplied with the unit for that purpose.

Figure 35-8 shows the metal heat-circulating form resting on the firebrick hearth and two workers applying the insulation around the metal surfaces.

Figure 35-9 shows the metal heat-circulating shell built into the masonry wall. The remainder of the insulation will be applied to cover all of the metal surfaces when the brick facing is done. Notice the metal wall ties projecting from the rough face of the fireplace. They will be built into the facing work to unite the rough work and the facework together.

Laying Out the Opening

Lay the first course in the front of the fireplace, making the front opening 1 inch to 2 inches less than the actual width of the steel opening. This allows the edges of the metal jamb to be hidden from view. Do not fill

Fig. 35-9 Metal fireplace form installed in rough masonry work, ready for the facing of the fireplace.

in any mortar at the point where the metal jamb and masonry work meet, since it can crack out. See Figure 35-6, Hearth Plan.

Cold and Hot Air Ducts

The cold air ducts must be built in such a manner that no air passes directly from the cold air to the hot air ducts. The plans supplied with the unit show exactly how this is done. Air must circulate freely throughout the metal walls and out the hot air duct.

Electric fans that are made by the manufacturer of the unit can be installed in the cold air duct attached to the grille covers. They move the air more efficiently through the unit. All wiring or electrical connections must be built into the masonry away from the heat. Never locate the fan in the hot air duct because it will burn up.

Figure 35-10 illustrates intake and outlet grilles for a metal heat-circulating unit. The frame with the

INTAKE AND OUTLET GRILLES

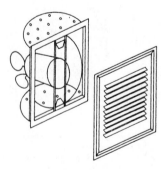

ELECTRIC FAN GRILLES

Fig. 35-10 Grilles.

rounded metal edge is walled into the brickwork and the flat louvered plate is attached to the installed frame with screws.

Electric fans are installed in the cold-air grille to help pull more air through the shell. The fan is sold as an accessory but is considered to be essential if the fireplace is to operate with maximum efficiency.

In the stone fireplace pictured in Figure 35-11, it can be seen how the grilles are built into a stone fireplace that faces a metal heat-circulating unit. The wiring needed to operate the fans must be built into the masonry work as the stonework is being laid. Care should be taken to keep the wiring away from the metal form so it does not damage the wire or insulation on the wire.

Setting the Lintel over the Firebox

The lintel should be set low enough to conceal the metal apron on the unit. This is considered when giving the heights in the installation instructions for the unit. Pack insulation behind and at the ends of the lintel to allow for expansion, Figure 35-12. The lintel is not a part of the unit so it will have to be supplied by the contractor.

Fig. 35-11 Stone fireplace with grilles and wiring for cold air fans built in masonry.

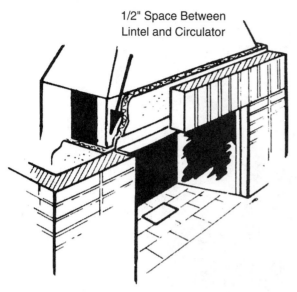

1/2" Space Between Lintel and Circulator

Fig. 35-12 Insulation installed in back of angle iron lintel and air space.

Fig. 35-13 Completed stone fireplace with a metal heat-circulating unit installed.

Setting the Flue Lining

The corbeling of the brickwork in the smoke chamber is done in the same manner as for a conventional fireplace. The most important point to remember, however, is that the flue lining must be set on the brick corbeling and not on the metal unit. This is very important, otherwise the height of the flue will rest on the metal fireplace form.

The connection between the flue lining and metal form is made tight by packing insulation around the flue and the top of the metal form. The chimney is then finished in the same way as for a regular fireplace.

Figure 35-13 shows a completed stone fireplace which encloses a metal heat-circulating unit. Once the wood mantel is installed in position the rest of the room will be finished off with wood paneling.

IMPORTANCE OF FOLLOWING MANUFACTURER'S INSTRUCTIONS

Since several different manufacturers make metal fireplace units, the installation may vary. A complete set of plans accompanies all metal fireplace units. Follow the plans exactly for best results.

The finished fireplace looks like a conventional fireplace, except for the grilles and the metal firebox. Figure 35-14 shows the completed brick metal heat-circulating fireplace.

PROJECTED CORNER FIREPLACE

The projected corner fireplace plan shown in Figure 35-15 is similar to a regular single-face fireplace with one side removed. Steel angle irons supported by a noncombustible steel post with plates on, carry and support the masonry work above the opening. The only real difference between this fireplace design and the conventional single-face fireplace is the shape of the damper with tapered ends. Instead of using a damper with tapered or angled ends, a square-end damper as shown in Figure 35-16 should be used for a projected corner fireplace. The open side of the fireplace should have a short masonry wall to help stop the escape of combustible gases when cross-drafts occur. This protection can also be increased by corbelling the top of the short wall. The flanges of the damper should be supported on masonry work as a protection against intense heat. It must be remembered that all metal dampers, regardless of the

Fig. 35-14 Finished metal heat-circulating fireplace in a home. Notice the grilles in the front of the fireplace for cold and hot air ducts.

design, should never be imbedded solidly in the masonry work, as they must have some room for thermal expansion. Review Unit 34 for the correct procedures to follow to allow dampers to expand. They should always be free at all edges in order to expand without cracking the adjoining masonry work or mortar.

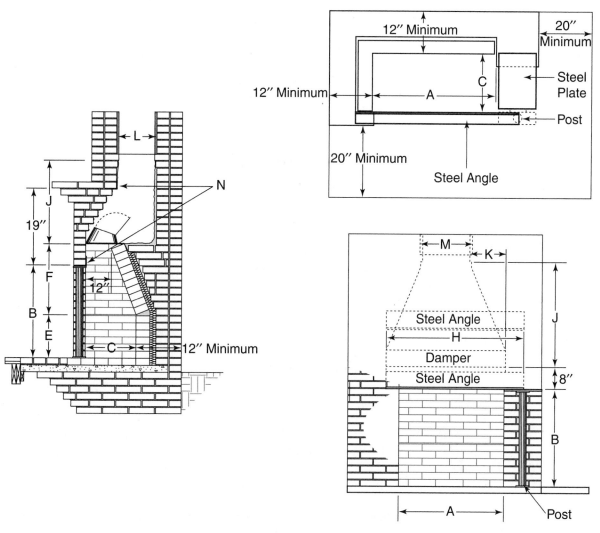

Projected Corner Fireplace Dimensions, Inches[a]

A	B	C	E	F	H	J	K	L	M	N[b + c]	O[d]	P
28	$26\frac{1}{2}$	16	14	20	35	$29\frac{1}{2}$	12	12	12	A-36	11×16	16
32	$26\frac{1}{2}$	16	14	20	40	32	12	12	16	A-42	11×16	16
36	$26\frac{1}{2}$	16	14	20	44	35	14	12	16	A-48	11×16	16
40	30	16	14	20	48	35	16	16	16	B-54	11×16	16
48	30	20	14	24	56	43	20	16	16	B-20	11×16	20
54	30	20	14	23	62	45	23	16	16	B-72	11×16	20
60	30	20	14	23	68	51	24	16	20	B-78	11×16	20

[a] SI conversion: mm = $\times$ 25.4   [b] Two required   [c] Angle sizes A = $3 \times 3 \times \frac{1}{16}''$ B = $3\frac{1}{2} \times 3 \times \frac{1}{4}''$ [d] Plate lintel

Fig. 35-15 Projected corner fireplace. (Courtesy Brick Industry Association)

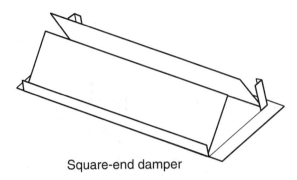

Square-end damper

Fig. 35-16 Square-end damper. (Courtesy Brick Industry Association)

ACHIEVEMENT REVIEW

Select the best answer from the choices offered to complete each statement. List your choice by letter identification.

1. The major problem with a fireplace that is open on two opposite sides and has fronts in two different rooms is that
   a. it is very difficult to build.
   b. crosscurrents of air cause it to smoke.
   c. the cost of building the fireplace is very high.
   d. both sides of the fireplace cannot be used at the same time.

2. The most important advantage of the multiple-opening damper is that it
   a. is made of heavier metal and lasts longer.
   b. is faster to install than a damper in a regular fireplace.
   c. permits a choice of flue locations in the chimney.
   d. is insulated to prevent cracking of the masonry work on the front of the fireplace.

3. The insulation can be attached to the damper during installation by brushing the metal with
   a. a light coating of tar compound.   c. a thin coat of mortar.
   b. epoxy cement glue.   d. contact cement adhesive.

4. When determining the correct size flue lining to use in a multiple-opening fireplace, the net inside sectional area of the flue lining and the total area of all open sides added together should be in the proportion of at least
   a. one to ten.   c. one to fourteen.
   b. one to twelve.   d. one to sixteen.

5. The most important advantage of the metal heat form fireplace is that
   a. the metal form outlasts the masonry fireplace.
   b. it is less expensive to buy and install in the masonry work.
   c. it is more efficient because the heat is circulated through the unit.
   d. it is easier for the mason to build because many of the key parts are prebuilt at the factory.

6. Metal was selected for the construction of the firebox mainly because it
   a. becomes hotter and reflects the heat better.
   b. is fireproof and lasts longer.
   c. is lighter in weight and is easier to handle when installed by the mason.
   d. can be built larger than a regular firebox and contains a larger fire more safely.

7. The metal firebox is completely insulated with
   a. asbestos.
   b. an air space.
   c. firebrick.
   d. glass wool.

8. The flue lining that is set above the smoke chamber over the metal fireplace form should rest on
   a. the top of the metal form.
   b. brick corbeling.
   c. a concrete shelf.
   d. a steel lintel attached to the metal form.

9. For best results, the cold air intake should be located at
   a. the middle of the unit.
   b. the bottom of the unit.
   c. the top of the unit.
   d. about ¾ of the way up the face of the unit.

10. If other flues are built into the area of the fireplace near the metal heat-circulating unit, they should be separated by at least
    a. 4 inches of masonry.
    b. 8 inches of masonry.
    c. 12 inches of masonry.
    d. 14 inches of masonry.

MATH CHECKPOINT

1. You are going to build a fireplace and chimney with a metal heat fireplace form in it. The form costs $325.00 plus 10% tax and there is also a 5% shipping charge. What is the total cost of the unit?

2. Allowing 25 firebricks for the hearth at $.55 each and adding 5% sales tax, what would the total cost be?

3. If 26 bags of masonry cement are needed at a cost of $4.75 a bag plus 5% sales tax, what would the total cost of the cement be?

4. Four tons of building sand are required at $20.00 a ton plus 5% sales tax. What is the total cost of the sand?

5. Nine, 13″ × 13″ flue liners will be needed costing $8.60 each plus 5% sales tax. What is the total cost of the flue liners?

6. The job requires 3,500 bricks at $.28 each plus 5% sales tax. What is the total cost of the brick?

7. The mason estimates it will take him and one laborer 32 hours each to do the work. The mason charges $20.00 per hour for himself and $8.00 per hour for the laborer. What would the total cost be for both of them?

8. What is the total cost of materials and labor for the job?

SUMMARY, SECTION 8

- Early fireplaces were not very efficient because there was no damper to control the draft.
- In the mid1700's, two Americans established successful fireplace principles that have lasted to the present time. They were Benjamin Franklin and Sir Benjamin Thompson, who was also known as Count Rumford.
- Franklin studied and experimented with the action of smoke in a fireplace and what made it draw properly. Rumford advocated a more shallow fireplace, with greater angled sides in the firebox and a higher than usual opening, which in turn, radiated more heat into the room. Both men published important papers on the subject of fireplaces.
- Terra-cotta flue lining, highly improved portland cement base mortars, and firebrick have made the modern fireplace and chimney more durable and fireproof.
- Successful fireplaces operate on the principle of air coming down one side of the flue lining, striking the smoke shelf, and being deflected back up the other side of the flue lining. During this process air must also be sucked in from the room to create this draft.
- Chimney tops must be 2 to 3 feet above the top of the ridge of the roof to eliminate downdrafts.
- If there is more than one flue lining in a chimney, they should be separated by at least a four-inch brick wall.
- There must be a separate flue lining for each heat source, such as a fireplace and furnace.
- A space of at least 2 inches should be allowed between all woodwork and the chimney inside the house and 1 inch outside between framed walls and the chimney.
- The best method to use when building a fireplace and chimney is to build the chimney first, forming only the rough opening of the fireplace. The firebox and facing are built later.
- The base of the chimney should be dry bonded with brick so that it works whole brick above finished grade.
- Type M portland cement mortar is recommended for building chimneys and fireplaces below grade, and type N above grade.
- Hearths can either be flush with the finished floor, or raised to provide a higher level of radiation and a bench seat near the fire.
- One of the major relatively new developments in fireplace construction is to build into the chimney base under the hearth a duct system that supplies fresh outside air to the firebox area. This, in turn, prevents the drawing of heated air from within the house for the purpose of combustion.
- Fireplaces and chimneys should always be laid out from the center of the chimney.
- The angled (splayed) sides of the firebox and the slanted or rolled back help to deflect the heat into the room and make the smoke and gases rise up more smoothly into the smoke chamber.
- The smoke shelf should be level with the bottom of the damper.

- For the fireplace to radiate the most heat into the room, the damper should be regulated so that it does not smoke, but burns freely.
- The damper should be closed when the fire is out and the fireplace is not in use to conserve heat in the house.
- The walls of the smoke chamber should be built on an incline from the damper to the point where the flue sits. This is done by corbeling the brickwork and plastering the underside smooth.
- When selecting a flue lining for a fireplace open on one side, the inside cross-sectional area of the flue lining must be at least $1/12$ the total area of the finished firebox opening.
- Flue linings should always be set on brick corbeling flush with the inside of the flue.
- The mason must be especially neat when building the firebox and front of the fireplace since it is the focal point of a room and reflects the quality of the mason's work.
- If the fireplace requires cleaning, it should always be done before the finished floors are laid.
- Always be sure to open the damper before starting a fire in the fireplace.
- Multiple-opening fireplaces require a special damper because they are open on more than one side.
- One of the most important advantages of the multiple-opening damper is that it permits a choice of flue locations.
- Flue sizes for multiple-opening fireplaces are determined by adding the square inches of all openings. They should be in a proportion of not less than one to ten.
- Metal heat-circulating fireplaces were developed because a more efficient heat-radiating fireplace was needed.
- The metal heat-circulating fireplace operates on the principle of cold air entering the hollow metal wall of the fireplace form at the bottom. The air is heated and forced up and out through vents at the top of the fireplace. Fans can be installed in the ductwork to improve the flow of air.
- The prebuilt metal fireplace form is manufactured scientifically correct.
- The use of metal heat-circulating fireplaces has increased due to the energy crisis and the ability of metal fireplaces to radiate more heat than a conventional fireplace, burning the same amount of fuel.
- The metal shell of a heat-circulating fireplace must be completely insulated in order to allow it to expand when hot.
- A fan placed in the cold-air inlet will improve the operation of a metal heat-circulating fireplace as it moves the air through the hollow walls of the form at a greater rate of speed. It has the same effect as the operation of a forced-air furnace.

SUMMARY ACHIEVEMENT REVIEW, SECTION 8

Select the best answer from the choices offered to complete each statement. Refer back to the material presented in Section 8, if necessary.

1. Which of the following American patriots was also considered to be an expert on the function and design of fireplaces?
   a. Thomas Jefferson
   b. George Washington
   c. Benjamin Franklin
   d. James Monroe

2. The Rumford fireplace was successful because it
   a. had a metal back in the firebox.
   b. was open on two sides and therefore would radiate more heat.
   c. had a shallow firebox and a higher front opening than a regular fireplace.
   d. was much deeper than a regular fireplace and therefore would hold a lot more wood in the firebox.

3. The popularity of the fireplace is increasing today due to
   a. a greater interest in the design and architectural effect offered in the main room of the home.
   b. the concern for energy conservation.
   c. the structural support it offers the rest of the home.
   d. the cheery atmosphere it adds to the home.

4. The movement of draft in a chimney depends on the
   a. type of fuel being burned.
   b. size of the fire built in the firebox.
   c. height of the chimney and the temperature difference.
   d. size of the damper being used.

5. The idea of bringing in outside air into the firebox by using a duct or tube arrangement is an important improvement in fireplace construction because
   a. a larger volume of air can be introduced into the firebox than with a regular fireplace.
   b. the fireplace would be less likely to discharge smoke into the room.
   c. it would greatly reduce the loss of heated air from within the room.
   d. the fireplace would not burn as much wood as a conventional one without an outside air source.

6. The best type of masonry material to use for areas that are subjected to intense heat from the fire is
   a. concrete block.
   b. stone.
   c. tile.
   d. brick.

7. The portland cement mortar that should be used in the general construction of a fireplace is type
   a. M.
   b. N.
   c. O.
   d. K.

8. The purpose of building a fireplace with angled sides is to
   a. add design and character to the fireplace.
   b. save firebrick.
   c. radiate the heat outward better.
   d. make it easier to clean and remove ashes from the firebox.

9. The walls of a fireplace, if built of brick, should never be of a thickness less than
   a. 4 inches.
   c. 12 inches.
   b. 8 inches.
   d. 14 inches.

10. To prevent the firebox from cracking due to expansion and contraction, leave a space between the firebrick wall and the backing materials of
    a. $\frac{1}{4}$ inch.
    c. 1 inch.
    b. $\frac{1}{2}$ inch.
    d. 2 inches.

11. If the smoke shelf is to work correctly, it must be laid
    a. 1 inch lower than the bottom of the damper.
    b. level with the bottom of the damper.
    c. 1 inch above the bottom of the damper.
    d. 2 inches above the bottom of the damper.

12. The slope of the sides of the smoke chamber corbeling should be no more than
    a. 30 degrees.
    c. 60 degrees.
    b. 45 degrees.
    d. 90 degrees.

13. The ratio of the cross-sectional area of the flue lining for a conventional fireplace to the area of the fireplace opening should be
    a. $\frac{1}{6}$.
    c. $\frac{1}{3}$.
    b. $\frac{1}{12}$.
    d. $\frac{1}{2}$.

14. The lintel that supports the facing materials of the front of the fireplace should be installed
    a. 1 course of firebrick below where the damper sits.
    b. 1 course of regular brick below where the damper sits.
    c. level with the damper.
    d. 1 course of firebrick above where the damper sits.

15. Chimneys with more than one flue should have at least a minimum brickwork partition of
    a. 4 inches
    c. 8 inches.
    b. 6 inches.
    d. 12 inches.

16. Fireplaces that have more than one opening are called
    a. island fireplaces.
    c. multiple-opening fireplaces.
    b. conventional fireplaces.
    d. heat-circulating fireplaces.

17. The main advantage of a special damper in a fireplace with more than one opening is that
    a. a bigger fire can be made in the firebox.
    b. the fire can be built anywhere on the hearth.
    c. it allows heavier construction and does not burn out as quickly.
    d. the design of the damper eliminates the building of a smoke shelf.

18. One important benefit of the metal heat-circulating fireplace for the homeowner is that
    a. the metal unit is stronger.
    b. it is more economical to have it installed.
    c. the circulation of heat is more efficient.
    d. the metal unit is easier to keep clean and requires less upkeep.

19. The metal heat-circulating fireplace is more efficient in heat radiation because
    a. it is constructed scientifically correct at the factory.
    b. the air space in the metal walls causes the heat to move through the unit and into the room.

    c. metal radiates heat better than bricks.

    d. it has a smoother finish and there is no place for the heat to be absorbed, such as a firebrick.

20. Insulation is placed around the metal heat-circulating unit mostly to

    a. prevent heat loss.

    b. make a fireproof installation between the unit and the rest of the chimney.

    c. allow for expansion of the metal unit.

    d. close all air holes and to completely seal off the unit from the rest of the fireplace and chimney.

SECTION NINE
ARCHES

UNIT 36
Development of Arches

―――――― OBJECTIVES ――――――

After studying this unit, the student will be able to

- describe the major types of masonry arches used today.
- explain the difference between a minor and major arch.
- describe the theory and function of an arch.
- define the terms associated with arch construction.

An *arch* is a section of masonry work that spans an opening and supports not only its own weight, but also the weight of the masonry wall above. With a curved arch, the more weight bearing on the arch, the more thrust is exerted outward to the sides of the wall.

BRIEF HISTORY

Arches date back many centuries. Chinese and Egyptian civilizations used arches before the Christian era, but it was the Romans who expanded the use of the arch in the construction of canals and waterways called *viaducts* that carried water to the cities. This was the semicircular arch, which is still known as the *Roman arch*.

Figure 36-1 shows a series of semicircular Roman arches in the Coliseum in Rome, Italy, which, after 2000 years, still support the masonry walls above them. Even though the marble facing stones have been removed and reused in many buildings in Rome, the arches still remain strong and support the walls, which is a testimonial to Roman engineering.

Arches have appealed to builders for centuries because of the various shapes, types, and designs that can be built. The beauty of the arch is expressed in the many forms that are used for balance, proportion, scale, rhythm, and character. Figure 36-2 shows a series of semicircular brick arches.

The structural advantage of the arch over a level beam or lintel is that when supporting a load, the pressure is mostly compressive (within) and to the sides, instead of straight down. For this reason, the arch frequently provides a very efficient way of supporting masonry work over an opening.

Arches were also used for building stone bridges, especially in the eastern part of the United States. Figure 36-3 shows a stone bridge which uses the Roman arch for support.

Concrete and steel lintels, which shorten installation time, have now greatly replaced the masonry arch. Arches are still, however, used in many buildings for architectural effect and design. Examples of arches can be found in churches, banks, educational institutions, and many homes.

465

Fig. 36-1 Semicircular Roman arches in the Coliseum, Rome, Italy.

TYPES OF ARCHES

Many arch forms have been developed, ranging from the jack arch, through the circular, elliptical, parabolic, and others, to the high-rise Gothic or pointed arch. Figure 36-4 shows some of the most frequently used masonry arches today.

Looking at Figure 36-4, it can be seen that arches are constructed with various curvatures. One of the most beautiful arches is the Gothic arch. It was used in the past in many European cathedrals. Figure 36-5 shows Gothic arches in a church.

CLASSES OF ARCHES

The two general classes of arches are minor and major. The classifications are based on span and the load the arch can support.

Minor arches have a maximum span of 6 feet and a maximum equivalent load of about 1,000 lbs. per linear square foot of span. Minor arches are usually used in masonry walls as lintels over openings.

Major arches have spans or loadings whose dimensions start where the maximum stops for minor arches. In general, a minor arch is relatively short and a major arch is long.

Figure 36-2 shows what is considered to be a minor arch and Figure 36-3 shows a major arch.

ARCH CONSTRUCTION

An arch is really a beam which is curved in the plane of the loads resting on it. Therefore, any section in an arch may be subjected to movement or shear as any ordinary beam would be under a load.

Due to the nature of its construction, the arch exerts not only a downward pressure but also a strong sideway pressure or a tendency to spread. This is known as the "thrust" of an arch. In order for the arch to remain strong and not "give," the thrust must be resisted by abutments, buttresses, or the strength of the wall itself. How the arch is designed directly relates to the amount of the thrust it will be expected to resist.

Fig. 36-2 These semicircular arches were designed by Thomas Jefferson and constructed during the period in which he lived.

In a fixed arch, which most masonry arches are, there are three conditions that must be maintained to ensure pure arch action.

- The length of the span must be constant.
- The elevations at the ends must remain unchanged.
- The inclination (angle) of the skewback must be fixed.

Fig. 36-3 A stone bridge that used the semicircular Roman arch for support of the masonry over the stream.

If any of these conditions are altered or changed by sliding, settlement, or movement of the abutment or pier, stresses will occur that could cause the arch to fail. There can be no weak points in the arch including the workmanship performed and the use of full mortar joints.

Masonry arches are built with the aid of a temporary wood form braced in position and made to the shape of the arch it will support. The purpose of the form is to carry the dead load of the arch and other weight until it has cured and is capable of carrying all of the weight or load imposed on it. This time allowed for curing cannot be rushed. For plain masonry arches, leave the arch form in place for 7 days after the completion of the arch. In a case in which there is a very light load or the majority of the weight will not be placed over it soon, the arch form could be removed in several days. This decision would have to be made for each particular arch and job condition. The decision is made by the person in charge who would assume responsibility if a problem arose.

As a rule, the wood form is the responsibility of the carpenter on the job. The mason should give the carpenter ample time beforehand to construct the form so the work is not delayed. Once the arch has cured, if there are several arches to build, the form can be removed and reused.

There are two general methods used in building masonry arches. One method requires buying special-shaped bricks that are made at the brick company on a taper or angle. The purpose of this is to keep the mortar head joints the same uniform thickness throughout the arch. Upon request, the brick company will supply these bricks already in a box, the size of the window or opening. They, of course, cost more than regular bricks and the order should be placed well in advance of their use. The other method is to use bricks of regular size and alter the mortar joints so that the bricks fit the wood form. This is the least preferred, as the mortar joints will be wider at the top of the arch than at the bottom due to the angle of turn or radius. Arch bricks can also be sawed at the job site by a trained masonry saw operator, but this is expensive and difficult to do. Preboxed arches should be ordered from the brick company if the cost of the job warrants it, and if they are available.

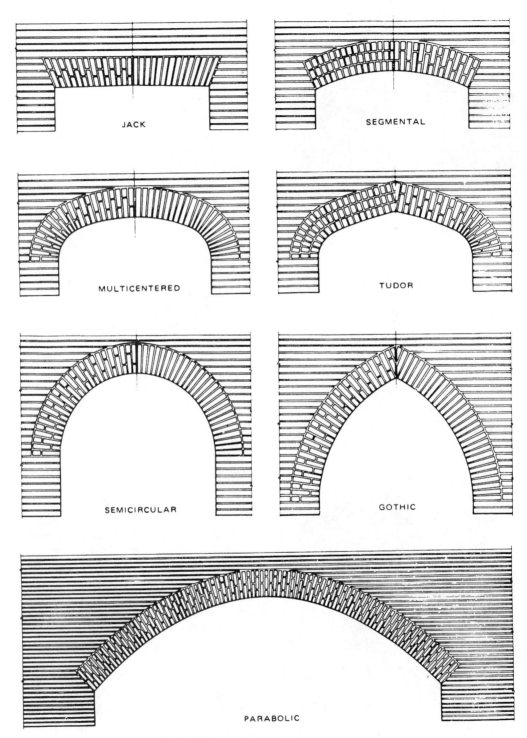

JACK

SEGMENTAL

MULTICENTERED

TUDOR

SEMICIRCULAR

GOTHIC

PARABOLIC

Fig. 36-4 Various types of masonry arches.

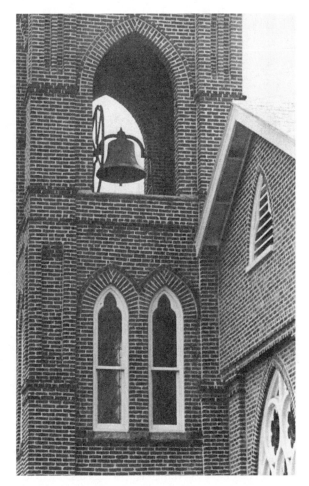

Fig. 36-5 Examples of Gothic arches in a church.

Arches that have a curvature are usually built of soldier, header, or rowlock position brick so they will appear different from the wall and fit the edges of the form better. Where the curvature of the arch is of a short radius, soldier bricks do not work well as the mortar head joints would be too large. In a case like this, use either header or rowlock position. In addition to helping achieve a better curvature and neater head joints, the header or rowlock provides a bond through the wall, making the arch stronger.

After the arch has cured sufficiently and the forms have been removed, the underside or soffit mortar joints between the brick are cut out carefully with a chisel and repointed. This could not be done earlier because they were resting on the wood form during the construction of the arch.

ARCH TERMINOLOGY

To understand how arches are constructed, one must know the various terms associated with arches. Figure 36-6 shows the various parts of an arch. Refer to Figures 36-4 to 36-10 when studying the list of arch terms in this unit.

abutment—the skewback of the arch and the masonry that supports it.

arch axis—the median line of the arch ring.

blind arch—an arch whose opening is filled with masonry.

bullseye arch (see Figure 36-8)—an arch whose intrados is a full circle. Also known as a circular arch.

camber—the relatively small rise of a jack arch.

constant cross-section arch—an arch that has a constant depth and thickness throughout the arch.

creepers—bricks that are cut to fit against the curvature of a semicircular arch.

crown—the apex (center) of the arch ring. In symmetrical arches the crown is at midspan.

depth (see "D" in Figure 36-6)—the dimension that is perpendicular to the tangent of the axis. The depth of a jack arch is its greatest vertical dimension. Depth as shown in the figure refers to height.

elliptical arch—an arch with two centers and continually changing radii.

extrados—the convex curve that bounds the upper limits of the arch.

fixed arch—an arch that has the skewback fixed in position and inclination (slope). Plain masonry arches are, by the nature of their construction, fixed arches.

Gothic or pointed arch—an arch with relatively high rises, the sides of which consist of arcs of circles, the centers of which are at the level of the springing line. The Gothic arch is often referred to as a drop, equilateral, or lancet arch (depending upon whether the spacings of the centers are less than, equal to, or greater than the clear span).

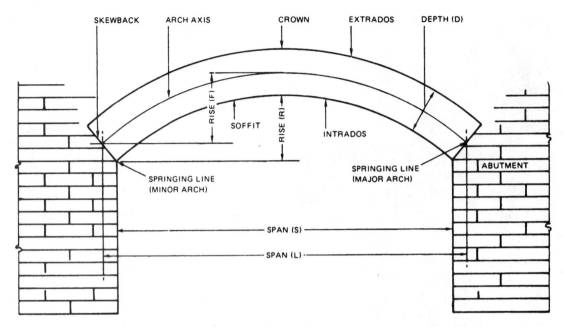

Fig. 36-6 Parts of an arch.

horseshoe arch—an arch whose intrados is greater than a semicircle and less than a full circle. Also known as an Arabic or Moorish arch.

intrados—the concave curve that bounds the lower limit (the entire curvature) of an arch (see Figure 36-6). The difference between the soffit and the intrados is that the intrados is linear (length), while the soffit is a surface, in this case the bottom surface of the arch.

jack arch (see Figures 36-7 and 36-9)—a flat arch that can carry a load but should be supported on steel angle irons if the opening is more than 2 feet long. It is the weakest of all arches.

keystone—the center masonry unit of stone in an arch that is cut to a taper on each side from the top to the bottom of the arch. It locks the other pieces in place and helps prevent the arch from collapsing.

multicentered arch—an arch that has a curve consisting of several arcs of circles that are usually tangent at their intersections.

relieving arch—an arch built over a lintel, jack arch or smaller arch to divert loads, thus relieving the lower arch or lintel from excessive loading. Also known as a discharging or safety arch.

rise (see "R" in Figure 36-6)—the maximum height of the arch soffit above the level of its springing line. The rise of a major parabolic arch (shown as "F" in Figure 36-6) is the maximum height of an arch axis above the springing line.

segmental arch (see Figures 36-7 and 36-10)—an arch that has a circular curve that is less than a semicircle.

skewback—the inclined (sloping) masonry units that the beginning arch bricks rest against at each end of the opening.

slanted arch—a flat arch that is constructed with a keystone or brick whose sides are sloped at the same angles as the skewback and uniform width brick and mortar joints.

soffit—the undersurface or bottom of the arch.

span—the horizontal distance between abutments (masonry jambs on each side of an opening). For minor arch calculations, the clear span of the opening is used (shown as "S" in Figure 36-6). For a major parabolic arch, the span is the distance between the ends of the arch axis at the skewback (shown as "L" in Figure 36-6).

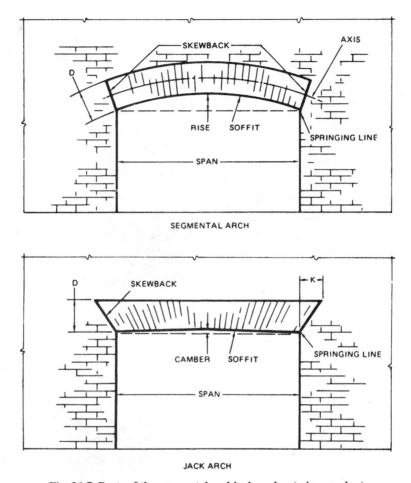

Fig. 36-7 Parts of the segmental and jack arches (minor arches).

springer—the first voussoir (masonry unit) that is laid against the skewback angle of an arch.

springing line—for minor arches, the line where the skewback cuts the soffit. For major parabolic arches, the term commonly refers to the intersection of the arch axis with the skewback (see Figure 36-6).

tudor arch—a pointed four-center arch with medium rise-to-span ratio (shown in Figure 36-4).

Venetian arch—an arch formed by a combination of jack arches at the skewbacks (ends) and semicircular arch at the middle or center of the arch. Also known as a Queen Anne arch.

voussoir—one of the wedge-shaped masonry units that form the arch ring. An example is a brick in a jack arch.

APPLYING ARCH TERMINOLOGY TO THE CONSTRUCTION OF A MINOR ARCH

The rise of a segmental arch should not be less than 1 inch per foot of the span and the skewback should be at right angles to the arch axis. In the jack arch the camber should be ⅛ inch per foot of span and the inclination of the skewback should be ½ inch per foot for

Fig. 36-8 A beautiful example of a bullseye arch in Bruton Parish Church in Colonial Williamsburg, Virginia.

Fig. 36-9 Bonded jack arch over a window in a house.

each 4 inches of arch depth. Both arches, shown in Figure 36-7, are minor arches used over the head of a window or door.

USES OF ARCHES IN STRUCTURES TODAY

The most commonly used arch types today are the semicircular and the jack arch. They are used more for ornamental purposes than to support a load. Minor arches should not be expected to carry heavy, concentrated loads.

Gothic and bullseye arches are still used mainly for religious institutions, Figure 36-8. The segmental arch is used to obtain a slightly arched effect in buildings in which ornate brickwork is desired. Figures 36-9 and 36-10 are examples of three popular arches used in masonry work.

The building of arches is considered one of the more difficult tasks of the mason. Close attention must be given to correct bonding and workmanship. Units 37 and 38 deal with the laying out and building of various arches.

Fig. 36-10 Applications of two popular forms of arches. The arch in the bottom part of the figure is a segmental arch, and the arch in the top part of the figure is a semicircular arch.

ACHIEVEMENT REVIEW

A. Select the best answer from the choices offered to complete each statement. List your choice by letter identification.

1. The small rise found in the bottom of a jack arch is called the
   a. skewback.
   b. crown.
   c. camber.
   d. springing line.

2. An arch that is flat and supported by an angle iron is known as a
   a. Gothic arch.
   b. segmental arch.
   c. jack arch.
   d. parabolic arch.

3. The wedge-shaped masonry units that make up the arch ring are known as the
   a. keystones.
   b. soffits.
   c. skewbacks.
   d. voussoirs.

4. An arch that is shorter in length than 6 feet is classified as a
   a. major arch.
   b. minor arch.
   c. multicentered arch.
   d. constant cross-section arch.

5. The underside of an arch is known as its
   a. crown.
   b. depth.
   c. fall.
   d. soffit.

6. The inclined masonry units that are cut on an angle on each side of a jack arch and that the arch bricks are laid against are called the
   a. axis.
   b. skewbacks.
   c. intrados.
   d. extrados.

7. The most popular arch built in churches is the
   a. semicircular.
   b. Gothic.
   c. tudor.
   d. parabolic.

8. Of all the types of arches constructed, the one considered to be the weakest is the
   a. jack.
   b. semicircular.
   c. tudor.
   d. Gothic.

B. Identify the various parts of the arch indicated by the letters in Figure 36-11. Refer back to the unit if necessary.

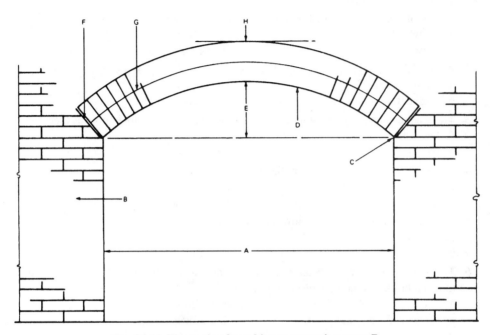

Fig. 36-11 Illustration for achievement review, part B.

UNIT 37
Construction of Semicircular and Segmental Arches

————————————— OBJECTIVES —————————————

After studying this unit, the student will be able to

■ describe how wood arch forms are constructed for semicircular arches.

■ explain how the arch bricks are spaced on the arch form before laying them on mortar.

■ describe the different types of brick positions used in the construction of semicircular arches.

INTRODUCTION

If properly constructed, a brick arch will support a heavy load such as a masonry wall. The arch can support this load mainly because of its curved shape. Although some arches, such as jack arches, have a flat shape, they can also support a load if the arch is designed properly. Semicircular arches are considered to be the strongest of all masonry arches because the pressures and overhead loads applied are divided equally over the entire area of the arch.

Successful masonry work depends on correct layout and following good working practices. This is especially important in the construction of arches. Constructing arches requires careful planning and understanding of the tool skills needed to lay the bricks accurately and neatly.

Many architects specify arches in the construction of certain brick structures such as banks, churches, office buildings, and educational institutions. The arch is mostly used today for design and beauty in a brick building, not as a main supporting member, as in the past, Figure 37-1.

The building of brick arches is a challenge to the mason's ability and is considered to be a difficult, but satisfying task. It is one of the more creative parts of the masonry trade and is required of all journeyman ma-

sons. This unit deals only with the construction of semicircular arches and segmental arches.

WOOD ARCH FORMS

Before a semicircular arch is constructed, a curved wood form must be built to support the arch until the arch has cured. This wood center arch form is usually constructed by the carpenter using information given on the plans.

For arches up to six feet long, a good construction grade of $\frac{3}{4}$-inch plywood is excellent for the front and back of the arch form. The two pieces are cut to the specified curvature. Then, 2-inch × 4-inch wood pieces are cut and spaced in between them as spreaders which serve to hold the form together. A piece of $\frac{1}{4}$-inch plywood is cut and fitted over the top, or over the part of the arch form that the bricks will lay on. The $\frac{1}{4}$-inch plywood bends easily over the curved form without breaking. A sufficient number of 2-inch × 4-inch pieces should be spaced between the front and back to prevent the $\frac{1}{4}$-inch plywood from sagging. Figure 37-2 shows a wood center arch form constructed from plywood.

The purpose of the arch form is to support the dead load of the arch and wall above the arch until the masonry has cured and is able to carry the weight. The length of time to leave the arch form in place depends

475

Fig. 37-1 Unusual variation of a semicircular arch to create a decorative entrance way.

on the size of the arch, the weight it must support, and the curing conditions. For minor arches, three days is usually enough time if the temperature is not freezing and the mortar is setting correctly. Major arches, which are larger, should not be disturbed nor have the forms removed before ten days have passed, if conditions are the same. These allowances are not absolute rules. The most important considerations are the prevailing weather conditions and the setting time of the mortar.

DETERMINING THE CORRECT CURVATURE FOR AN ARCH

Although the carpenter usually constructs the arch form, sometimes the mason is asked to construct it. This is especially likely on a small job or when the repair of old work involves replacing arches that have deteriorated.

In order to lay out the curvature of an arch, the span and rise must be known. These dimensions are usually given on the plan. It is also necessary to find the radius in order to mark the curve of an arch on the wood form. A simple geometric method of finding the radius is shown in Figure 37-3. Multiplying one-half of span A by itself and dividing the result by rise B gives distance C. To obtain the radius add B and C, and then divide

the answer by 2. The following example illustrates this method.

Example: Find the radius of an arch that has a 4-foot (48-inch) span and a 12-inch rise. Solution: Take one-half of the span and multiply it by itself (24 inches times 24 inches, which equals 576 square inches). Divide 576 by the rise (12 inches), and get 48 inches. Then add together 48 inches and the rise (12 inches), to get 60 inches. Divide 60 inches by 2, which equals 30 inches, the radius.

The radius of an arch is $\frac{1}{2}$ of the springing line. (The *springing line* is a line across the span of an arch passing through the points where the arch intersects with the skewbacks.) To correct the optical illusion of flatness in semicircular arches, the *radial center point* (exact center of the circle) is usually raised one or two inches above the springing line.

SETTING THE FORM IN PLACE

After the arch form is constructed and brick piers are built to the height where the arch is to be laid over the opening, the arch form is set in place. A pair of legs is made from 2-inch × 6-inch or 2-inch × 8-inch lumber (depending on the depth of the arch). The legs are

Fig. 37-2 Wood arch form constructed from plywood. A strip of ¼″ plywood is nailed over the top for the arch bricks to lie on. Notice the 2″ × 4″ bracing around the inside of the curvature.

cut about 1 inch shorter than the actual height of the opening.

The legs are set in place, plumb and level. Four wood wedges are then laid on top of the legs, two on each. The arch center form is set on top of the wedges and adjusted into the proper position by tightening up on the wedges. Care must be taken during the process, so that when it is finished, the arch form is at the correct height, level and plumb with the face of the brickwork. Appropriate braces can be fastened to the form to hold the arch center securely in place. Figure 37-4 shows a

wood arch form for a semicircular arch, braced in position, ready for the bricks to be laid on the form.

MARKING OFF THE ARCH BRICKS ON THE FACE OF THE FORM

There are two different methods of allowing for mortar joints in the construction of brick arches. The first involves the use of special arch bricks obtained from the manufacturer. For the second method, the bricks are cut with a masonry saw to a specific angle to achieve

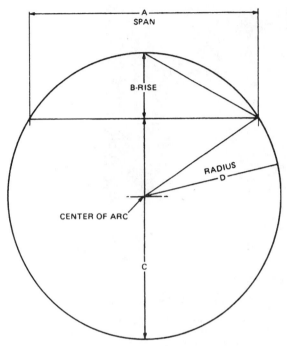

Fig. 37-3 Geometric method of determining the radius. The radius is necessary to scribe the curve of an arch.

uniform mortar joints. In the first method, standard bricks of uniform shape are used and the joint thickness is varied to obtain the desired curvature. The special arch bricks can be cut on the job with a power masonry saw, or they can be ordered from the brick manufacturer as a special order. If ordered from the manufacturer, the arch bricks will be sent separately and carefully marked.

The brick spacing is marked off with a mason's rule. This is done by bending the rule or modular steel tape to fit the curvature of the bottom of the arch form and marking on the face of the form with a sharp pencil. The course counter rule, commonly called a spacing rule, is used most often because it allows the mason to evenly adjust any differences in the joint thickness.

Although the spacing of the brick courses are marked off on the bottom of the arch form, it is difficult to lay and keep the top of the arch bricks at the correct angle. To simplify this step, attach a radius stick to the radial center point with a finishing nail. As the arch is being laid, swing the radius stick to indicate the exact position of the top of the arch brick. This eliminates all of the

guesswork. To be of value, the radius stick must extend to or past the top of the arch. Refer to Figure 37-4 to see the placement of the radius stick.

LAYING THE ARCH BRICKS ON THE WOOD CENTER ARCH FORM

Before laying bricks on the wood arch form, clean out any chips or nails protruding from the construction of the carpentry work. Brush off the form with a fine brush to remove all chips.

Construction of an arch is always begun at the two ends or piers, and laid up to the center or key of the arch. The bottom mortar joint that is set on the arch form should be slightly angled inward. This is because it must be cut out with a chisel to a depth of ½ inch after the arch form is removed and repointed. Whether the key in the arch is brick or stone, it should be laid so that it extends the same distance over the center of the form on each side.

Select the best edge of the brick (the one that is straightest and unchipped) to set on the form in order to obtain a neat appearance. The mortar should be mixed to the proper consistency so that great pressure is not needed to make the joint squeeze out as the bricks are laid. Do not tap too hard on an arch because this will knock the previously laid bricks out of position and weaken the bond.

BONDED ARCH WITH SOLDIER COURSE

Arch bricks can be laid in several positions for semicircular arches. If the arch is a bonded soldier course arch, Figure 37-5, the number of courses needed to build the arch should be counted before any bricks are laid. If the number of courses is even, the sides should start differently (one side with a half brick and the other side with a whole brick). This works out correctly when the center of the arch is reached with the bond. A bonded soldier course is a minimum of 12 inches, or a brick and a half, in height.

ROWLOCK OR HEADER ARCHES

The most popular bond arrangement for semicircular brick arches is either rowlocks, Figure 37-6, or header

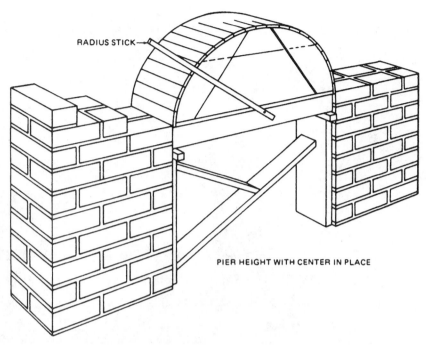

Fig. 37-4 Wood center arch form set in place between two brick piers, ready for the arch bricks to be laid on it.

Fig. 37-5 Bonded soldier course semicircular arch with a stone keystone in the center.

course rings. The use of two or three rings (in height) is recommended for more strength. The shorter height of the rowlock or header bricks allows them to be turned on a sharper radius. This results in better curvature with smaller mortar joints. In addition to the mortar joints being smaller (which is desirable), the rowlock or header gives a through-the-wall tie since each brick is 8 inches in length.

FLASHING AND WEEP HOLES FOR ARCHES

Any type of arch needs to have flashing and weep holes installed to prevent moisture penetrating and staying in the wall. Installing flashing is easiest with jack arches as they are basically flat. To be effective, flashing for a jack arch should be installed below the arch and above the window frame or steel angle lintels. It should extend a minimum of 4 inches past the openings at each end and be turned up to form an end dam, so that the weep holes will drain the water that is trapped against the square end of the flashing. In the trade, this is also called *tray flashing*. To be effective, weep holes have

Fig. 37-6 A series of brick arches with rowlock rings and a brick keystone in the center.

to be installed at both ends of the flashing and spaced 24 inches on center along the arch span or 16 inches on center, if rope wicks are used. Figure 37-7a shows an example of flashing a jack arch in this manner.

Attachment of the flashing to the backing work and forming the square end dams should be done properly to ensure they will stay in place and not allow water to leak through the ends or into the wall. In the event the arch is built of reinforced brick masonry, the flashing and weep holes should be installed in the first brick course above the arch.

Installing flashing with some other arch types, such as segmental and semicircular arches, can be more difficult, as most rigid flashing materials are hard to bend around an arch with a tight curvature. If the span of the arch is less than 3 feet, one section of tray flashing can be placed in the first horizontal mortar bed joint above the keystone brick and to the width of the arch over the opening. Figure 37-7b illustrates how this is done. For arch spans longer than 3 feet, it would be best to bend pieces with overlapping sections to make sure it does

not leak. The alternative method would be to use a combination of stepped and tray flashing as shown in Figure 37-7c. To form a *step* in the flashing, the end nearest the arch should be turned up to form an end dam, while the opposite end is simply laid flat. A minimum of 15-lb weight building paper or an equivalent moisture protection material should be installed on the exterior facing of the backing work over the full arch and abutments as insurance protection so that moisture will not penetrate the area. This protective covering should overlap the arch flashing as shown in Figure 37-7c.

WORKMANSHIP

The mortar joints should all be the same size and as small as possible (consistent with the wall) to give a neat, attractive appearance. A small joint such as $\frac{1}{4}$ inch to $\frac{3}{8}$ inch is less likely than a larger mortar joint to shrink and absorb water. The mortar joints at the intrados (bottom) and the mortar joints at the extrados (top) should be about the same size. This is not difficult to

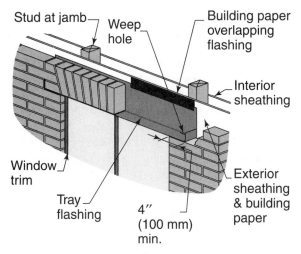

Fig. 37-7a An example of flashing a jack arch. (Courtesy Brick Industry Association)

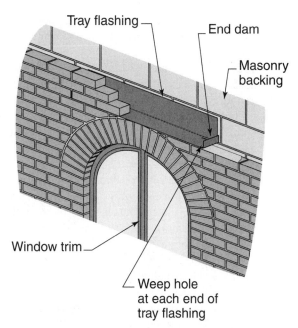

Fig. 37-7b Placing one section of tray flashing. (Courtesy Brick Industry Association)

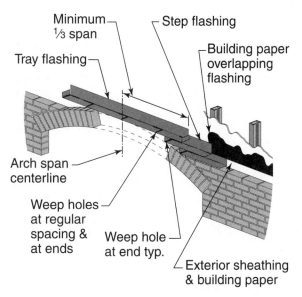

Fig. 37-7c Combination of stepped and tray flashing. (Courtesy Brick Industry Association)

do with the rowlock or header arch, but it is more difficult with the soldier arch because the bricks must be cut to a special tapered shape.

Once the arch bricks are laid and set with the mortar, they should not be moved; otherwise the bond strength of the mortar joint is broken. This causes a loss of adhesion, allowing water to penetrate through the wall. If a brick must be moved after the mortar has set, fresh mortar should be applied and the brick relaid.

The mortar joints should be tooled as soon as they are thumbprint hard and still pliable. The arch should be brushed at the completion of the work. It is also a good practice to have a line attached on the bottom and center of the arch as a guide to make sure the arch does not bulge out of position. A straightedge such as a plumb rule can be used on a small arch to keep the center of the arch from bulging out. This should always be done as a last check after the arch is completed.

Parging the back of the arch with mortar helps to prevent water from leaking through from the face of the wall. It also helps to strengthen the arch and make it more secure.

TYING IN THE ARCH WITH THE BACKING WORK

If an arch does not equal the full width of the masonry wall, it should be tied or bonded into the backing work with metal wall ties and well mortared. The best type to use is the coated wall tie with a galvanized or copper finish. The wall ties should be imbedded at least half the thickness of the width of the brick in the facing of the arch and the backing materials. Every 3 or 4 courses in height should be enough for a good tie. Wire ties allow the mortar to bond more fully. Corrugated ties are used in veneer.

REMOVING THE ARCH FORM AND REPOINTING THE MORTAR JOINTS

After the arch has cured as previously mentioned, the arch form should be carefully removed. This is done by gently driving out the wedges, holding the form in place. When doing this, care should be taken not to chip the bottom of the arch brick.

Since the bottom of the arch rests completely on the wood form, the mortar joints must be cut out with a special chisel, called a plugging chisel, used for repointing joints with mortar. The plugging chisel, Figure 37-8, is made with an angled blade on the end so it does not chip the edge of the bricks, but cuts out only the mortar joint. The special tapered blade cleans mortar from the joints easily without binding.

After removing the form, cut out the bottom of the mortar joints to a depth of $\frac{1}{2}$ inch. Then, with a brush and bucket of water, dampen the mortar joints immediately before repointing with mortar. This is done to prevent the mortar from drying out too rapidly and curing improperly. The same mortar mix that was originally used to build the arch should be used for repointing. Type M or S mortar is usually recommended for building arches.

Some important points to remember when installing a semicircular arch follow:

- Select bricks that are approximately the same size before laying any of the bricks on the form.
- Choose bricks free of chips or cracks.
- Select bricks that match well with the color of the structure.
- Use dry bricks because they are easier to hold in place on the arch form.
- Use well-filled mortar joints. Mix the mortar to a thickness that squeezes out to all edges without the use of undue pressure when laying.
- Mark off all brick spacings on the arch form before laying any of the bricks.
- Work from each edge at the piers up to the crown (top) of the arch.
- If the arch has a keystone (as those shown in Figure 37-9), this should be marked off first. Then the arch bricks must be laid off between the keystone and the bottom of the arch.
- Always try to make the top of the arch work with the stretcher course of bricks in the wall. It is considered better workmanship to have the stretcher

Fig. 37-8 Plugging chisel.

Fig. 37-9 Series of semicircular arches. Notice that each arch has a keystone in the center.

course of brick work over the top of an arch without having any splits of bricks.

- Use a range line across the bottom and center of the arch to make sure the arch has not bulged out of position.
- When finished, straightedge the center of the arch with a plumb rule.
- Provide enough wall ties to tie the arch and backing work together.
- Parge the back of the arch to prevent water from leaking through the wall, and to help strengthen the arch.
- Do not remove the arch forms until the arch has cured enough to carry its own weight without sagging.
- Cut out the mortar joints at the bottom (soffit) of the arch with a plugging chisel and repoint with mortar.

- Observe good safety practices by wearing eye protection when mixing mortar or cutting masonry units for arches.

SEGMENTAL ARCH

A segmental arch is an arch which is not a true half circle but a portion of a circle. The techniques and methods of installation are the same as for a semicircular arch, but it is often a longer or major arch. The segmental arch is used in government or public buildings to accentuate the beauty of the brickwork. Figures 37-10 through 37-13 show various construction procedures for installing semicircular arches in a wall.

Fig. 37-10 Brick piers built to receive the arch forms. Notice the skewback angles on each side of the top of the piers.

Fig. 37-11 Wood arch form in a position ready to lay brick on.

Fig. 37-12 Completed brick arch in place on form. These bricks are laid in soldier position.

Fig. 37-13 Masons building in wall around and over segmental arch.

The building of arches by the mason requires good workmanship in addition to an understanding of the principles of arch construction. The practices and techniques in this unit should be studied and learned to aid the student in the construction of semicircular arches.

Rule for Segmental Arch Construction

The minimum rise should always be equal to one-eighth the width of the opening or span, if the arch is to be self supporting and not on a form.

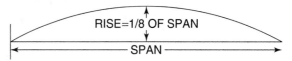

ACHIEVEMENT REVIEW

Select the best answer from the choices offered to complete each statement. List your choice by letter identification.

1. The strongest of all arches is the
   a. jack arch.
   b. Gothic arch.
   c. semicircular arch.
   d. parabolic arch.

2. Arch center forms are usually built of
   a. steel.
   b. plastic.
   c. fiberglass.
   d. wood.

3. The most difficult brick position to build a semicircular arch, retaining all of the head joints the same size, is a
   a. rowlock.
   b. soldier.
   c. header.
   d. stretcher.

4. When building a semicircular arch, the radial center point should be raised above the springing line about
   a. 1 or 2 inches.
   b. 3 or 4 inches.
   c. 5 or 6 inches.
   d. 7 or 8 inches.

5. The top of the arch bricks can be maintained at the proper curvature by using a
   a. mason's rule.
   b. mason's line.
   c. radius stick.
   d. form board.

6. The use of rowlock or header rings rather than soldiers is recommended for semicircular arches. The main reason for this is that
   a. rowlocks or headers present a more pleasing appearance.
   b. rowlock or header bricks are easier to lay on the arch form.
   c. the mortar head joints remain more uniform in size.
   d. it is more economical to lay the rowlock or header bricks than to lay a soldier course.

7. Once the arch bricks have been laid in the mortar joint and have set, they should not be moved. The main reason for this is that if an arch brick is moved,
   a. the design will be changed.
   b. the bond of mortar with the brick is lost.
   c. the arch form may be bumped out of place.
   d. the brick becomes smeared with mortar and requires cleaning before relaying.

8. After the arch has cured and the form is removed, the mortar joints on the bottom of the arch should be cut out with a
   a. hammer.
   b. brick set chisel.
   c. plugging chisel.
   d. blade of a trowel.

9. To make sure the mortar that is pointed in the bottom of the arch bricks stays in place, the joints should be cut to a depth of
   a. $\frac{1}{8}$ inch.
   b. $\frac{1}{4}$ inch.
   c. $\frac{1}{2}$ inch.
   d. $\frac{3}{4}$ inch.

PROJECT 30: LAYING OUT AND CONSTRUCTING A TWO-ROWLOCK SEMICIRCULAR ARCH

OBJECTIVE

- The student will lay out and construct a double-rowlock arch on a previously constructed form. This arch is typical of the type used over single windows.

EQUIPMENT, TOOLS, AND SUPPLIES

Mortar pan or board

Mixing tools

Mason's trowel

Brick hammer

Plumb rule

Ball of nylon line

2-foot framing square

Line pin and nail

Convex jointer

Chalk box

Brush

Plugging chisel

Brick, lime, and sand or Carotex for mortar, to be estimated from the plan by the student

One 2-inch × 8-inch, 4-foot long board, for legs and bracing boards

Arch center form to be constructed by the carpenter or student

SUGGESTIONS

- Use $\frac{3}{8}$-inch mortar head joints spacing for layout.
- Lay the courses of brick to number 6 on the modular rule.
- Stock all materials approximately 2 feet back from the wall line.
- Select dry bricks for the project.
- Choose arch bricks of the same size and free from chips.
- Mix ahead only enough mortar that can be used in one hour because building an arch is slow work.
- Strike the mortar joints as soon as they are thumbprint hard.
- Wear eye protection whenever mixing mortar or cutting bricks.
- Recheck all jambs before attempting to set the arch form in place.

PROCEDURE

1. Dry bond the first course of the project as shown in the plan, Figure 37-14.
2. Set the arch center form in place over the dry bonded bricks to be sure it fits and then remove it.
3. Lay out the first course in mortar.
4. Build the piers to a height of 8 courses, level with each other.
5. Cut a pair of wood legs from the 2-inch × 8-inch board to the height of the pier minus enough room for placement of wood wedges.
6. Level the arch form on the legs with the wood wedges so that the radial center point is 1 inch above the springing point.
7. Brace the form into position.
8. The height of the arch center form plus the 2 rowlocks should be even with the brick coursing.
9. Attach a radius stick to the radial center point with a small nail.
10. With the mason's spacing rule, mark off the coursing around the bottom of the arch form.
11. Lay the rowlock arch bricks from the edges of the piers to the center of the arch.
12. Swing the radius stick in line with the marks on the bottom of the arch form to check the top of the rowlock curvature.
13. As the arch ring is laid, continue building the piers up and cut the bricks on the correct angle to fit against the arch ring.
14. Check the alignment of the arch with the piers using a line attached across the face of the project.
15. Continue building the project until it is completed as shown on the plans in Figure 37-14.
16. After the project has cured, remove the arch center form and point up the bottom (soffit) of the arch.
17. Have the instructor evaluate the project.

PROJECT 31: LAYING OUT AND CONSTRUCTING A DOUBLE OVERLAPPING ROMAN ARCH

OBJECTIVES

- The student will lay out and construct two Roman arches, when they overlap each other at the springing points for a 12-inch wall.
- This will be a challenging project. Careful attention should be given to keeping the brickwork in line by using a line attached to both ends of the wall for every course.

EQUIPMENT, TOOLS, AND SUPPLIES

Mixing pan or board

Mixing tools

2-foot framing square

Line pin and nails

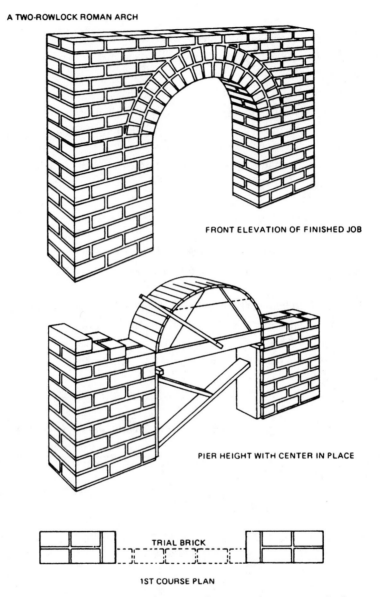

A TWO-ROWLOCK ROMAN ARCH

FRONT ELEVATION OF FINISHED JOB

PIER HEIGHT WITH CENTER IN PLACE

TRIAL BRICK

1ST COURSE PLAN

**Fig. 37-14 Project 30: Laying out and constructing a two-rowlock
semicircular arch.**

Mason's trowel	Convex jointer (small and sled runner type)
Brick hammer	Chalk box
Plumb rule	Brush
Ball of nylon line	Plugging chisel

Brick, lime, and sand for mortar to be estimated from the plan by the student

Two 2-inch × 8-inch, 4-foot long boards for legs and bracing materials

Supply of nails and plywood necessary for construction of forms

Arch center forms to be constructed by the carpenter, wood shop, or masonry students

SUGGESTIONS

- Use $\frac{3}{8}$-inch mortar head joints spacing for layout.
- Lay the brick coursing to number 6 on the modular rule.
- Stock all materials at least 2 feet away from the project to allow working room.
- Select dry brick free of chips for the project.
- Select bricks of approximately the same length.
- Mix only enough mortar ahead that can be used in one hour due to the nature of the work.
- Firmly brace the arch forms in place and make sure they are plumb and level with each other.
- Strike the mortar joints when they are thumbprint hard.
- Wear eye protection at all times when cutting bricks.
- Constantly check with the line to make sure the arch and wall do not get out of line.
- Keep all waste and scraps out from under foot as the work progresses.

PROCEDURE

1. Dry bond the first course of the project, Figure 37-15.
2. Set the arch forms in place over the dry bonding to be sure it fits and then remove it out of the way.
3. From the layout line, lay out the three piers as shown on the first course plan. All piers must be squared, level, and in line with each other.
4. Build each pier to a height of 8 courses, square and plumb. Make sure that each pier is level with the others at the point where the arch forms will be set in place.
5. Install the arch forms in place on the wood supports so the forms are firm and braced. Make sure that the forms all line up across the project, and the tops are level with each other. Attach a radius stick to the center of the form at the bottom (as in project 25) and mark off the course of brick on the form.
6. Begin to lay the rowlock rings and the end leads. Use the radius stick as a check point to keep the rowlock bricks in the proper angle as they are laid around the ring or form.
7. Make a center mark on the eighth course in the middle of the center pier. This will be the vertical head joint for the second ring of the arch.
8. As you build the arches, keep this plumb head joint until full-size headers are obtained for the second ring, Figure 37-15.
9. Cut all of the creepers (brick that touch the arch on its curvature) using the radius stick for the radial joints.

OVERLAPPING ROMAN ARCHES

FINISHED FRONT ELEVATION

FIRST COURSE PLAN

TRIAL BRICK

**Fig. 37-15 Project 31: Laying out and constructing a double overlapping
Roman arch.**

10. Be sure to bond the first ring on the soffit or where it is set on the arch form as shown on the plan.

11. Continue to use the line to keep the wall and arches in line. A good wood straight-edge can be used as a leveling board from one end of the project to the other.

12. Strike the mortar joints as they become thumbprint hard. Use a small convex jointer for the head joints and a sled runner jointer for the longer joints.

13. Continue building the project until it is completed as shown on the plans and in Figure 37-16.

Fig. 37-16 Double Roman arch built by student masons in a masonry shop.

14. After the project has cured (hardened) sufficiently, remove the arch forms carefully and point up the mortar joints on the bottom (soffit) of the arch.

15. Have the instructor grade the project after it has been brushed and wiped off with a burlap cloth.

UNIT 38
Construction of a Jack Arch

—————————————— OBJECTIVES ——————————————

After studying this unit, the student will be able to

■ name the two types of jack arches.

■ describe how to lay out a brick jack arch.

■ explain how a jack arch is installed by the mason.

INTRODUCTION

A jack arch is a flat arch capable of supporting a load imposed on it over an opening. Because of its lack of curvature, it is the weakest of all the arches. It is supported on steel angle irons if the opening is more than two feet long.

Workmanship has a great deal to do with the appearance and durability of the finished jack arch. The arch should be constructed so that the mortar head joints are the same width for the entire height of the arch. To accomplish this, special bricks must be molded at the brick plant according to the plan specification for the inclination of the arch, or the arch bricks must be cut on the job to a pattern and rubbed perfect with a stone.

The top and bottom of the arch bricks or voussoirs must be in line with the angle iron at the bottom and the line at the top of the arch. The bottom and top of the arch bricks must be cut to the correct angle so they are level with the angle iron and with the line since the bricks are laid on a slant. These are cut with the masonry saw.

A jack arch that is built correctly gives the appearance of sagging in the center. This is only an optical illusion but can be corrected by having a slight camber (upwards curvature) in the angle iron. The longer the arch, the more it will appear to be sagging. Three lines are usually used in the construction of a jack arch. Once the wall is built and the skewback of the arch is installed, a line is attached at the bottom of the arch

and across the face. Another line is used at the top edge to keep the top of the arch in line and as a guide for making the angle cut for the top of the bricks. Then, a range line is used across the center of the arch to make sure the arch does not bulge out of position.

TYPES OF JACK ARCHES

There are two main types of jack arches. One is called common, the other, bonded.

A *common jack arch,* Figure 38-1, is composed of only one full brick in height, laid in a soldier position. It is the simplest type of jack arch and requires the least amount of work to install. A common jack arch serves to decorate the house or structure.

A *bonded jack arch,* Figure 38-2, is laid out and built the same way as a common jack arch except that the bonded arch is higher and is usually either a brick and a half or two bricks in height. To make the arch stronger, it is bonded by staggering the mortar joints the same as in a brick wall, only the arch bricks are laid in a vertical position. Since the height of the bonded arch is greater, the brick skewback on each side is five courses in height. The bricks are cut parallel to the angle iron and to the line at the top of the arch.

Bonded jack arches can also have a keystone in the center for architectural effect and design, Figure 38-3. If the arch has a keystone, the stone should be centered in the arch and marked on the frame allowing for the

Fig. 38-1 Common jack arch over a window.

Fig. 38-3 A bonded jack arch with a keystone in the center. This type of jack arch is used on buildings that have limestone trim.

Fig. 38-2 Bonded jack arch over a window.

mortar joints on each side. The bricks are then laid to that point.

LAYING OUT A JACK ARCH FROM THE RADIAL CENTER POINT

How to Determine the Angle or Inclination of a Jack Arch

Before actually laying a jack arch on the angle iron lintel, the inclination (angle) of the skewback or end brick in the jack arch has to be determined. You need to work this out first because the abutment the skewback will rest against has to be cut on the same angle to receive it. Regardless whether the arch is of brick or stone, the same theory has to be applied. The accepted rule to follow, as shown in Figure 38-4, states that "the skewback angle should be ½″ per foot of span of the opening for each 4″ of arch depth."

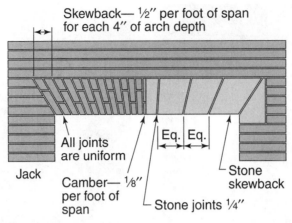

Fig. 38-4 Determining the skewback angle of a jack arch. (Courtesy Brick Industry Association)

Next locate the radial center of the arch, which is the center point at the bottom of the arch opening. Once this point is located, drive a small nail at the center point and attach a line, Figure 38-5.

The line is then held from the radial center to the edge of the opening at the point where the jack arch starts on the angle iron. The line can be attached to a wood board and the brick skewbacks cut to the correct

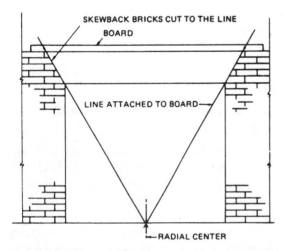

Fig. 38-5 A line is attached from the radial center to the skewback of the jack arch. The line is fastened to a board spanning the opening above the arch and the skewbacks are cut to the line.

angle. The skewbacks are either cut with a sharp brick chisel or cut with the brick saw for accuracy.

MARKING OFF SPACING OF THE ARCH BRICKS

To make sure that the arch bricks work even and full bricks, the spacing of each brick must be marked off on the bottom of the angle iron and the top of the arch. Refer to Figure 38-6 while reading the procedure below.

First, lay a wood board (2 inches × 4 inches or 2 inches × 6 inches) on top of the skewback on each side of the arch so that it spans the opening. This board is used to mark the top of the arch bricks.

Next, attach the line to the radial center. Hold the line tightly from the center of the arch. Mark each course of arch brick on the top of the board, and on the angle iron until the center of the arch is reached. Remember that the center of the arch should be the center of the keystone. Mark in the same manner from opposite sides until the arch is completely marked off. It is important that both marks at the top and bottom of the arch be marked to keep the correct angle.

Jack arches can be built of two different type bricks. The first is an especially larger shape brick made at the brick plant, which is then tapered or sawn to match and fit the required skewback angle, so that the brick can be laid on an incline and the mortar head joints will consistently be the same width the full height of the brick. The second is a fill-size regular uncut rectangular shape

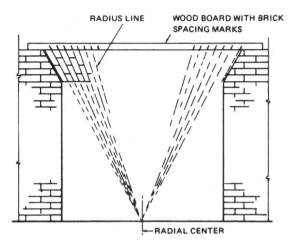

Fig. 38-6 Spacing the arch bricks using the line from the radial center. The line is swung to the center of the arch, marking off the individual bricks.

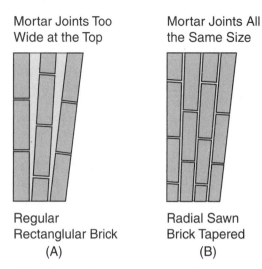

Fig. 38-7 Jack arch examples.

brick. When a regular brick is used, the mortar joints themselves have to be tapered, which makes the mortar head joints larger at the top and smaller at the bottom due to the angle of inclination. It is also unsightly at the bottom edge where the arch brick rests on the lintel, as mortar has to be put under the brick to hold it at an angle. The selection of a tapered or regular rectangular brick can be determined by the arch type, arch dimensions, and the appearance desired. Figure 38-7a shows a regular rectangular brick with a larger mortar joint at the top than at the bottom. The radial or tapered sawn brick in Figure 38-7b shows the same thickness of mortar joint from top to bottom, making a much better appearance.

Usually, on a job that requires a number of jack arches built of tapered radial bricks, the best practice is always to provide the brick manufacturer with the dimensions of the openings and the height of the arches and to place an order well ahead of the time when you will need them. The bricks will be sawn to the correct sizes and angle at the brick plant and will be placed in a wooden box in straw. The mortar joint specified spacers will be placed between each brick to protect it during shipment to the job site. When the time comes to install the bricks on the job, they are set on the scaffolding next to the opening and the mason takes them from the box, one by one, and lays them into position

in mortar. This procedure ensures that the jack arch bricks are going to be in the correct order and will guarantee that you will get the best job possible. You can sometimes also order what is called "arch blanks" or larger special-made brick units and saw them yourself on the job to fit your needs. This is not done very much anymore as it is very labor intensive and costly. If you are going to install arches, the workmanship should be as good as possible!

It is recommended that the thickness of mortar joints between arch bricks be a maximum of $\frac{1}{4}''$ and a minimum of $\frac{1}{8}''$. Generally, the architect on the job will specify the thickness of the mortar joints. When using mortar joints that are this thin, it is very important that the arch bricks be uniform in size. That is why ordering arch brick that is sawn at the brick plant is the best way to go!

LAYING THE JACK ARCH IN MORTAR

After the jack arch has been laid out on the board and angle iron, attach the three lines in the way previously described. The mortar should be mixed so it is stiff enough to squeeze out when laid, without too much tapping on the bricks. Use solid mortar joints on all bricks.

Start laying the arch bricks from the ends to the center of the arch. The spacing marks on the board and on the angle iron must be checked often or the inclination may be lost. Recheck the radius line at intervals as the work progresses. Check the angle of the arch bricks to make sure the correct angle is maintained.

A relatively small mortar joint should be used between the arch bricks because it gives a better appearance and allows the work to set up more quickly. Large mortar joints detract from the overall appearance and therefore should not be used in the construction of arches.

When the last bricks are laid in the center of the arch (key) make sure to butter both ends of the bricks in place. (*Buttering* is using a trowel to apply mortar on a masonry unit, to form a joint.) Also butter both ends of the keystone to make sure the mortar joint is well filled and a complete bond is established.

Tool the mortar joints as soon as they are thumbprint hard and then brush the work. Be sure to look under the bottom of the arch and point up any holes neatly with mortar when the work is finished. A slicker does a nice job of pointing. The jack arch is the easiest of all arches to construct, but good workmanship should be used at all times to obtain the best results.

ACHIEVEMENT REVIEW

Select the best answer from the choices offered to complete each statement. List your choice by letter identification.

1. A jack arch should be supported on angle irons if the span is more than
   a. 2 feet.   c. 4 feet.
   b. 3 feet.   d. 5 feet.

2. The layout line for a jack arch is taken from the
   a. springing line.   c. center of the angle iron.
   b. radial center.   d. top of the arch.

3. The jack arch is laid from
   a. the center to the jambs.   c. one end to the other end of the arch.
   b. each end to the center.   d. the center to the ends.

4. A jack arch is marked off on the top and bottom of the arch in order to
   a. maintain the exact coursing of each brick.
   b. correct the curvature.
   c. find the keystone.
   d. make it easier for the mason to see the marks.

5. The last brick laid in the center of the arch is called the
   a. crown.   c. extrados.
   b. keystone.   d. inclination.

6. The radial center of a jack arch is always taken from
   a. the end of the arch.
   b. the top of the arch.
   c. the bottom of the arch.
   d. halfway up the opening and in the center of the arch.

7. The position that jack arches are laid in is called
   a. rowlock.   c. soldier.
   b. stretcher.   d. header.

PROJECT 32: CONSTRUCTING A COMMON JACK ARCH

OBJECTIVE

- The student will use standard and Norman bricks to lay out and construct a common jack arch from a radial center. This kind of jack arch is typical of the type used in a house where a simple arch is desired for design.

EQUIPMENT, TOOLS, AND SUPPLIES

Mortar pan or board

Mixing tools

Mason's trowel

Plumb rule

Mason's spacing rule and modular rule

Chalk line

Mason's hammer

Ball of nylon line

2 corner blocks

Line pins and nails

Convex sled runner striker

Brick set chisel

Brush

Pencil

Square

Supply of bricks, hydrated lime, and sand or Carotex for the mortar—to be estimated by the student from the project plan, Figure 38-8. Bricks are standard face bricks, the exception is that the arch brick are 12″ long Norman brick.

One 8-inch × 8-inch × 16-inch concrete block to put the level in when it is not in use

Two wood 2-inch × 6-inch boards, approximately 56 inches long

Three $3\frac{1}{2}$-inch × $3\frac{1}{2}$-inch angle irons, each 56 inches long

SUGGESTIONS

- Use solid joints for all of the brickwork.
- Lay out the first course with $\frac{3}{8}$-inch head joints.
- All brick coursing is to be number 6 on the modular rule.
- Do not mix more mortar than can be used in a 1-hour period.
- Keep all materials at least 2 feet away from the wall line for working space.
- Use eye protection whenever cutting bricks or when mixing mortar.
- Strike the mortar joints when they are thumbprint hard.
- Make all cuts with the brick set chisel for accuracy.
- Select straight bricks that are free from chips to be used for the arch.
- **Important.** A standard or regular brick will not work for this single jack arch project, as it will not be long enough when the skewback angle is cut on the bottom and top edge to make it sit on the lintel properly and to be laid level with the line at the top of the arch. Since special tapered jack arch bricks are not available in a shop or training situation, the practical solution to installing this arch is to use *12″ Norman bricks*, which can be sawn to the proper skewback angle, thereby resolving that problem.

They should be available at your local masonry supplier in a variety of colors, preferably in a red or pink color which will usually match somewhat the color of the other bricks in the project.

• Applying the rule discussed in this unit, which states that the angle of the skewback should be equal to $\frac{1}{2}$ inch per foot of span for each 4 inches of arch depth, the opening of this project is 4 feet (48″), therefore $4 \times \frac{1}{2}$″ equals 2 inches. This is the angle of the skewback that needs to be cut on the brick abutment on each side and on the top and bottom of the arch brick, so it will fit correctly. This angle will change somewhat as you proceed toward the center of the arch, where the keystone brick is laid. That is the reason the project utilizes a radial center at the bottom of the project. You have to check the inclination (angle) constantly with a string line from the radial center to the top of the arch as it is installed and make the proper adjustments.

PROCEDURE

1. Mix the mortar for the project.
2. Lay out the project with the chalk line according to the plan in Figure 38-8, and dry bond the first course.

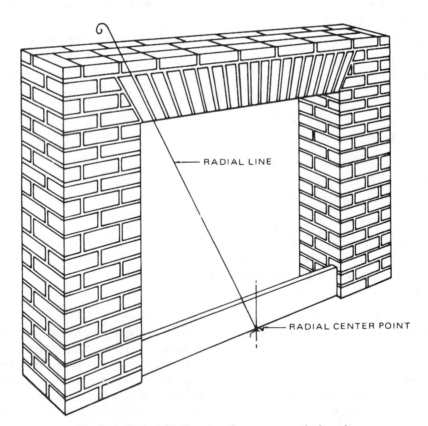

Fig. 38-8 Project 32: Constructing a common jack arch.

3. First build the piers on each side as shown on the plan. Use number 6 on the modular rule.

4. Set the angle iron in place when the height of the piers is reached. Divide the bearing on each side of the pier equally. This project requires 3 angle irons because the wall is 12 inches wide.

5. Rack up a lead on both sides of the project to a height of 3 courses, but do not build all the way out to the jamb.

6. Cut the radius board to fit between the piers at the front bottom of the project. The radius board is a 2-inch × 6-inch piece of framing lumber.

7. Find the center of the radius board and attach a small nail at the bottom center.

8. Attach a line to the radial center point and draw it up to the inside of the piers to determine the angle of skewbacks, which is two inches.

9. To the other 2-inch × 6-inch piece laid on top of the project, attach the skewback line at the top of the arch line with a nail.

10. Cut the skewbacks to the line and lay in position.

11. By moving the radius line across the project to the center and using the spacing rule, mark off the individual bricks, both on the top and on the angle iron with a pencil. Because the Norman arch bricks are not tapered, the mortar joints at the top half edge will be tapered larger. (Use a ¼-inch mortar joint on the spacing rule. This may differ according to the thickness of the face bricks available and the varying angle to the key brick.)

12. Make sure that the keystone (key brick) is laid out so it is in the center of the arch.

13. Attach a line at the top and bottom of the arch and proceed to lay the jack arch from each end to the middle. Cut the top and bottom of all arch bricks so they lay level.

14. Check the inclination of the arch periodically as the arch is laid with the radius line.

15. Continue laying the arch until it is completed.

16. Lay the jack arch on one side of the wall only. Lay stretcher courses in the middle and back of the wall because the project is 12 inches in thickness.

17. Strike the mortar joints and point up the joints on the bottom of the arch.

18. Continue to lay the project as shown on the plan in Figure 38-8, until the course over the arch is laid and the backing is completed.

19. Strike all work and have the instructor evaluate the project. Use a sled runner jointer for best results.

PROJECT 33: CONSTRUCTING A BONDED JACK ARCH

OBJECTIVE

- The student will lay out and construct a bonded jack arch of standard and Norman bricks for the arch. The angles at the top and bottom of the arch should be cut with the masonry saw or brick set chisel.

  This type of bonded jack arch is typical of the kind used on a colonial building such as a bank, school, or educational institution. Bonded jack arches are laid over windows and door openings mainly for design purposes, not to support a load.

EQUIPMENT, TOOLS, AND SUPPLIES

Mixing tools

Mortar pan or board

Mason's trowel

Plumb rule

Mason's spacing rule and modular rule

Chalk line

Mason's line

Ball of nylon line

2 corner blocks

Several line pins and nails

Convex sled runner striking iron

Brick set chisel

Brush

Pencil

Square

Supply of standard bricks, Norman bricks for arch, hydrated lime and sand or Carotex for the mortar to be estimated by the student from the plan in Figure 38-9.

Three $3\frac{1}{2}$-inch $\times$ $3\frac{1}{2}$-inch angle irons, 56 inches long, for the opening

Two wood 2-inch $\times$ 6-inch pieces of framing lumber, 56 inches long

One 8-inch $\times$ 8-inch $\times$ 16-inch concrete block to store the level in when not in use

Masonry saw with a diamond blade (if available)

SUGGESTIONS

- Select good, straight Norman bricks for the arch.
- All bed joint coursing should be number 6 on the modular rule.
- Lay out the head joints for the first course with $\frac{3}{8}$-inch mortar joints.
- Do not mix more mortar than can be used in one hour.
- This project can be built by a team of two students.
- Strike all mortar joints as soon as they are thumbprint hard.
- Observe good safety practices at all times and wear eye protection whenever mixing mortar or cutting bricks.
- Since the opening span is the same (4′) as in Project 32, the skewback angle is also 2″.

PROCEDURE

1. Mix the mortar for the project to the consistency desired.
2. Lay out the project with the chalk line according to the plan in Figure 38-9.
3. Dry bond the first course of bricks as shown on the plan.
4. Build the brick piers up to the angle iron height. Keep the bricks level with each other.
5. Set the angle irons in place dividing the bearing evenly on both sides. (Three angle irons are needed because the wall is 12 inches wide.)
6. Rack up a lead in mortar on each side of the pier 5 courses high above the lintel height.
7. Locate the radial center point. Cut and lay the radius 2-inch $\times$ 6-inch board, in line with the front of the project.
8. Attach a line to the radial center point and draw it up to the top of the arch at the pier line.

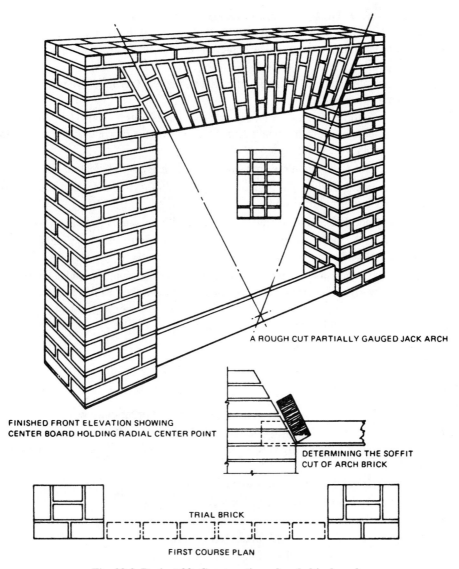

A ROUGH CUT PARTIALLY GAUGED JACK ARCH

FINISHED FRONT ELEVATION SHOWING
CENTER BOARD HOLDING RADIAL CENTER POINT

DETERMINING THE SOFFIT
CUT OF ARCH BRICK

TRIAL BRICK

FIRST COURSE PLAN

Fig. 38-9 Project 33: Constructing a bonded jack arch.

9. Lay the other 2-inch × 6-inch board on top of the project so it spans the arch opening and is in line with the face of the project.

10. Holding the radius line in position, attach it to a nail in the board at the point where the skewback bricks must be cut.

11. Cut the skewback angle (2″) on each side to the line with either the brick set or masonry saw (if available).

12. By moving the radius line toward the center of the arch and using the spacing rule, mark off the individual arch bricks on the angle iron and the edge of the board. (Remember to keep the brick key in the center of the arch while the spacing is being laid out.)

13. Attach a line at the top and bottom of the arch as a guide. One may also be attached in the center as a range line.

14. Set a brick on top of the angle iron dry (without mortar) at the ends and mark the 2″ angle of the bottom cut with a pencil. Cut the angle with the saw or chisel. Rub the bottom cut smooth with a rubbing stone.

15. Repeat step 14 for the top of the arch since it must be cut to fit level with the line.

16. Cutting the bricks as shown in Figure 38-9, proceed to lay the arch on the angle iron toward the center from both sides.

17. Periodically check the inclined angle of the arch with the radius line.

18. Lay the brick key last and double-joint or butter the edges of the bricks and brick key in place for a solid joint.

19. Strike up the mortar joints with a sled runner jointer and point up under the bottom (soffit) of the arch. Mortar joints will be larger at the top half as this is not a tapered arch brick.

20. Continue to build until the project is completed.

21. Recheck the work to be sure it is level and plumb. Have the instructor evaluate the project.

SUMMARY, SECTION 9

- In the past, arches were used for supporting the weight of masonry work above openings; now they are used more for design and architectural effects.
- The semicircular arch is called the Roman arch, named for the builders of the Roman era.
- The strongest of all arches is the semicircular arch.
- The Gothic arch is often used in building churches.
- The two major classifications of arches are minor and major.
- The theory of the arch is that the upward curve resists downward pressure by exerting pressure to the sides rather than straight down.
- The intrados is the bottom curvature and the extrados is the top curvature of an arch.
- A jack arch is a flat arch supported by angle irons.
- The center of an arch is called the key.
- The individual bricks in an arch are called its voussoirs.
- Arch construction is considered to be one of the more difficult tasks of the masonry trade.
- Semicircular arches are constructed on a wood arch form.
- The curvature of a semicircular arch is determined by using geometry.
- The installation of a semicircular arch can be simplified by using a radius stick attached to the center of the arch form.

- Each arch brick should be marked on the arch form before any of the bricks are laid.
- The most popular brick positions for a semicircular arch are rowlocks or headers.
- The mortar joints in an arch should be kept to a minimum width, such as ¼ inch, and the maximum should be ⅜ inch.
- Three lines should be used when installing an arch: one at the bottom, one at the middle, and one at the top.
- Parging the back of an arch makes it stronger and tends to stop moisture from penetrating.
- A masonry arch should be tied into the backing part of the wall using wall or masonry ties.
- Arch forms should not be removed until the masonry work has cured.
- When cutting out the bottom of an arch which is resting on a form for repointing, it is important to dampen the cut out portion with water before repointing.
- Well-filled mortar joints should always be used in the construction of brick arches.
- The building of masonry arches is one of the most difficult tasks for the mason and requires more advanced tool skills.

SUMMARY ACHIEVEMENT REVIEW, SECTION 9

Complete the following statements referring to the material found in Section 9.

1. A masonry arch that is constructed with a half-circle is a _____ arch.
2. The advantage of an arch over a beam or lintel is its ability to withstand a load under downward pressure. This type of pressure is called _____.
3. The simplest type of arch is the _____.
4. An arch whose curve is circular but less than a semicircle is a _____ arch.
5. The two general classifications of arches are _____ and _____.
6. The pressure exerted outward to the sides of a semicircular arch under a load is called its _____.
7. The stretcher bricks cut on an inclined angle on each side of a jack arch are called _____.
8. An arch that has a curve consisting of several arcs of circles, which are usually tangent at their intersection, is a _____ arch.
9. The bottom of an arch is called the _____.
10. The line where the skewback intersects with the soffit is the _____.
11. Bricks that form the arch ring are known as the _____.
12. The upward curvature in the bottom of an angle iron used for installing a jack arch is called the _____.
13. The top center of the curvature of a semicircular arch is the _____.
14. Bricks that are cut to fit against the curvature of a semicircular arch are called _____.

15. A pointed four-center arch with a medium rise-to-span ratio is a _____ arch.

16. Semicircular arches are constructed on forms built of _____.

17. Before a form for an arch can be built, it is necessary to locate a certain point where a line can be attached. This point is known as the _____.

18. To aid in laying out the correct curvature of the arch when marking off the individual courses, a _____ is attached to the arch form.

19. Arches are built from both ends to the _____.

20. Due to its length, the most difficult brick position for the construction of semicircular arches is the _____.

21. When building a jack arch it is important to tie the arch bricks into the backing materials by installing _____.

22. An arch brick must not be moved after it has initially set because the bond strength _____.

23. A special chisel is used for cutting out mortar from the soffit of a semicircular arch after the form has been removed. This chisel is called a _____.

24. The arch that is usually associated with churches is the _____.

25. The distance between piers or abutments the arch crosses is known as the _____.

26. A jack arch, which consists of two bricks in vertical height and in which the mortar joints are staggered for strength, is called a _____.

27. Individual bricks are marked off across the angle iron with the aid of a _____ rule.

28. The top of a jack arch can be regulated for brick spacing by marking each brick with a pencil on a _____.

29. The mortar joints should be struck with a striking tool as soon as they are _____.

30. When a masonry arch has a stone-shaped wedge installed in the center to hold the arch in place, this piece is called a _____.

SECTION TEN
CONCRETE

UNIT 39
Concrete Forms and Placing Footings

OBJECTIVES

After studying this unit, the student will be able to

- determine the size of footings for various widths of concrete block walls.
- estimate concrete for a footing.
- describe the correct concrete for the footing for a house or small commercial building.
- define the two types of footings used for a foundation.
- describe how a footing is placed and cured.
- clean concrete tools properly.

INTRODUCTION

Although concrete work is considered a trade of its own, the mason on occasion will be required to estimate concrete, to build at least some minor forms, and to place concrete on the job. This is especially true with small jobs such as footings, porches and steps, etc., where a regular concrete contractor does not need to be present. Concrete work is another form of masonry work that the brick, block, and stone mason should become familiar with and be able to perform at least on a limited basis. This unit deals with estimating concrete for small jobs, understanding how concrete forms are constructed and installed, and placing concrete in footings, porches, and steps. Commercial concrete form work is also discussed briefly to provide some background information and an understanding of concrete for the mason who

will be involved with or will work in and around larger commercial work.

The footings are the first actual structural work done on a building. A *footing* is the base on which a foundation is built. Footings are usually concrete, which is placed in a trench or form at the bottom of the foundation. Their function is to support the weight of the structure and distribute it over a greater area.

The two key factors that influence the size of a footing are the weight of the structure and the load-bearing strength of the soil it rests upon. The engineer determines the load-bearing strength of the soil. This is considered in writing the building codes that govern the size of footings. The architect shows the size of the footings on the plans for the building. Observe all local building codes when designing and placing footings.

Good footings are important because without them a building can settle, and cracks may develop in the walls. Since the footings are never seen after the floor is in place, their importance is often overlooked.

SIZE OF THE FOOTINGS

The earth the footings rest on should be firm and undisturbed to give maximum support. Concrete is the material used most often for footings, because it can be made to conform to the requirements necessary for strong footings and is economical.

A general rule to follow for residential buildings (houses or apartments) is that the footing should be twice as wide as the masonry wall foundation that will rest on it. The depth of the footing should be one-half to three-fourths of its width. For example, the average footing for an 8-inch wide concrete block should be 16 inches wide and 8 inches deep. Of course, this rule does not fit all conditions; it is only suggested as a general rule when there are no special soil or load-bearing problems. Special cases or conditions are the responsibility of the engineer or architect.

The following are three important rules for placing any footing.

- Never make the footing less than 6 inches in depth, regardless of the size of the masonry wall.
- Make the footing square and the same width for its full depth for maximum strength.
- Always try to place the entire footing at one pour, with no breaks, to insure full bonding of the concrete.

Footings for chimneys, columns, or piers should be larger than for an ordinary masonry wall because they must support a concentrated load. This heavy weight must be supported without settling because not only will the chimney settle, but the wall it is bonded into may be damaged when it pulls away. A good recommendation for chimney footings, if they are poured at the same time as the rest of the footing for the house, is 12 inches deep. If the footing is poured at a later time, as is the case with an add-on chimney, then it should be reinforced with steel rods and can be increased to a depth of 16 inches. The concrete footing for a chimney that will have a fireplace should be approximately 18

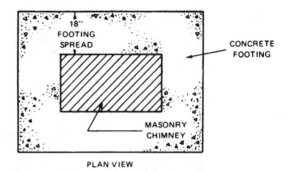

PLAN VIEW

Fig. 39-1 Masonry chimney foundation showing the footing spread for weight distribution. Consult local building codes for exact footing size in a specific area. This is an average footing for a chimney.

inches wider than the actual chimney base to provide a better weight distribution, Figure 39-1. The footing for a chimney is commonly called a chimney pad on the job.

STEEL REINFORCEMENT

Steel reinforcement in the footing is usually not required for houses or small commercial buildings. However, in the case of chimney and fireplace footings, it is a good practice to place a few steel rods or concrete reinforcement wire in the footing for extra strength. If steel is to be used, about 4 inches of concrete is placed, the steel is laid on the concrete, and then the remainder of the concrete is placed. Never lay the steel on the ground and place concrete on top of it. The steel must be encased in the concrete to be of any value. Half-inch steel rods are the size selected for most small jobs. When piers that carry a heavy load are to be built on the footing, the reinforcing rods can be left projecting out of the footing, Figure 39-2. In this manner, masonry or concrete piers can be tied to the footing.

CONCRETE FOR FOOTINGS

Concrete footings for most homes or small commercial buildings should support at least 2500 pounds per square inch (psi). Regular concrete for general use in floors and walks of 3000 to 3500 psi is a little richer (more portland cement in the mix per cubic yard) than

Fig. 39-2 Steel rods projecting out of the footings to reinforce piers that will be built at a later time.

a footing concrete due to the finishing or troweling process.

Concrete for a footing mix can be ordered either by specifying a 5-bag mix, called a prescription mix (which means that it takes 5 bags of portland cement to a cubic yard) or by a psi rating of 2500. Both mean the same thing; it is merely two different ways to order. Concrete prices vary depending on how far it has to be delivered from the batching plant, so this should be taken into consideration when figuring concrete.

If a small footing is being installed, it may not be economical to order concrete from a plant due to the minimum charge most concrete companies have. In this case a utility mixer can be used on the job to mix the concrete. A good, average footing mix proportion for an on-the-job mix is 1 part portland cement to 2 parts

sand, to 4 parts crushed stone or gravel. Add enough water to make the concrete workable but not runny, or it will be too weak.

If you order concrete from a ready-mix batching plant delivered to the job site by a mixer truck, the ingredients are measured accurately by weight utilizing computerized equipment. This will produce the identical quality concrete each time. An interesting statistic concerning the materials needed for a cubic yard of concrete is that it will contain approximately 1800 lbs. of gravel or crushed stone (coarse graded aggregate); 1300 lbs. of sand, which is considered to be a fine aggregate; 500 lbs. of Type I or II portland cement; and about 35 gallons of water. In addition, about 15 ounces of water-reducing agents and between 3 to 4 ounces of an air-entrainment agent to resist freeze-thaw cycles are added. This concrete mix will have a compressive strength up to 3000 lbs. per square inch (psi) after it has cured for 28 days and is strong enough for most residential or normal uses.

There is a lot of pre-packaged mixed concrete used; it is convenient when you only need small amounts and do not want to take the time to combine all of the ingredients. Ingredients in pre-packaged concrete such as in the Sakrete™ and Quikrete™ brands are precisely formulated and conform to the ASTM C-387 Standard. When you read that designation on the bag, you can be confident that after the proper amount of mixing water has been added, the concrete will test out to a minimum of 3500 psi after a 28-day curing cycle. One difference between pre-packaged or pre-mixed bagged concrete and ready-mixed concrete delivered by a concrete mixer truck is the size of the aggregates they contain. Stone or gravel in ready-mix concrete typically ranges from $\frac{1}{4}''$ to $1''$ in size unless specified otherwise. Coarse aggregates in bagged pre-packaged or pre-blended concrete tends to be a little smaller, ranging from $\frac{3}{8}''$ to $\frac{9}{16}''$ in size. For small jobs this should not be much of a concern.

HOW TO ESTIMATE CONCRETE FOR A FOOTING

The unit of measure for concrete is the cubic yard, which contains 27 cubic feet. To figure the amount of concrete needed for a footing, first find the total number of lineal (running) feet around the walls of a building. Include the lengths of walls of any porches, retaining walls, etc. that will be a part of the foundation footings. Multiply the lengths of the footings in feet by the thickness in feet by the width in feet. This will be the number of cubic feet in the footing. Now divide the number of cubic feet by 27 to find the number of cubic yards.

The formula is expressed as follows:

$$\frac{\text{Length in feet} \times \text{thickness in feet} \times \text{width in feet}}{27} = \text{cubic yards}$$

The amount of concrete figured by this method does not allow for any waste or slight differences in thickness due to uneven fill or base. Therefore, in estimating the amount of concrete needed, an extra 5 percent should be added in order to allow for these factors.

The simplest method of estimating the concrete is to express the thickness dimensions and width in feet or parts of a foot rather than use the fractions. Figure 39-3 gives both the fractional and decimal parts of a foot for a number of common widths and thicknesses.

Following is a sample job which applies this formula. The wall lengths are 40 feet by 60 feet. Adding all of these together shows a total of 200 lineal feet around the house. The front porch is 6 feet by 10 feet and has three walls because the porch butts up against the house. Adding these together the sum is 22 feet. The rear porch measures 4 feet by 5 feet and again only has three walls. The sum of the rear porch is 13 feet. Adding all of these together the sum is 235 lineal feet of footing to pour. The addition of the lineal feet of footings is as follows:

Step 1. 200′ Total of lineal feet around the house
22′ Total of lineal feet in front porch
13′ Total of lineal feet in rear porch

235′ Total of lineal feet in footings

The footings are 8 inches thick (deep) by 16 inches wide, as they are for an $8 \times 8 \times 16$ inch concrete block wall. To find out how much concrete will be needed, use the following formula:

(Lineal feet)   (thickness)   (width)
$$\frac{235 \quad \times \quad 2/3 \quad \times \quad 1\,1/3}{27} = \text{cubic yards}$$

INCHES	FRACTIONAL PART OF A FOOT	DECIMAL PART OF A FOOT
4	1/3	0.33
5	5/12	0.42
6	1/2	0.50
7	7/12	0.58
8	2/3	0.67
10	5/6	0.83
12	1	1.00

Fig. 39-3 Expressing inches as fractions and decimals of a foot.

Step 2. Multiply the lineal feet by $2/3$ (expressed 0.67 as a decimal).

235 Lineal feet
$\times\ 0.67$ Thickness

157.45 Total concrete

Step 3. Multiply total concrete by $1\,1/3$ (expressed 1.33 as a decimal).

157.45 Total concrete
$\times\ 1.33$ Width

209.41 Cubic feet

Step 4. Multiply this figure by the waste (5%).

209.41 Cubic feet
$\times\ 0.05$ Waste allowance

10.47 Waste

Step 5. Add the cubic feet and waste allowance together.

209.41 Cubic feet
$\times\ 10.47$ Waste allowance of 5%

219.88 Total cubic feet of concrete

Round this figure to the nearer whole number which is 220 cubic feet.

Step 6. Divide 220 cubic feet by 27 cubic feet to find the number of cubic yards.

$$\frac{220}{27} = 8.15 \text{ cubic yards, rounded to } 8\ 1/4 \text{ cubic yards}$$

TYPES OF FOOTINGS

When laying out footings, the building lines are first established using nylon line. A *plumb bob* (a small, pointed weight attached to a line) is suspended from the point where the building lines cross, Figure 39-4. A stake is driven in the ground at this point to locate the exact outside corner of the building.

There are two major methods of placing the footings for a structure. *Trench footings* are constructed by digging out the soil and placing the concrete in the trench. *Formed footings* are constructed in forms that are built before the concrete is placed. Either method is an acceptable way of placing footings, depending upon the job conditions or presence of rock.

Trench Footings

One of the most important items to consider in digging the trench for a footing is the shape of the sides and bottom of the trench, Figure 39-5. Often the trenches are dug carelessly and the footings do not have the strength which is expected of them. The trenches should have straight sides and should be dug to the full thickness and width indicated on the plans. The soil should be shoveled back far enough out of the way to provide working room. This also helps prevent soil or rock from sliding back into the trench before the concrete is placed.

After the trench is dug, level stakes are placed in the center of the footing. These are stakes that show where the top surface of the footing is to be. Usually, wood stakes or pointed lengths of steel rod are used, depending upon the hardness of the ground. These stakes or rods are placed about 5 feet apart in the trench and are leveled with each other. The concrete is then placed to the top of the stakes, Figure 39-6.

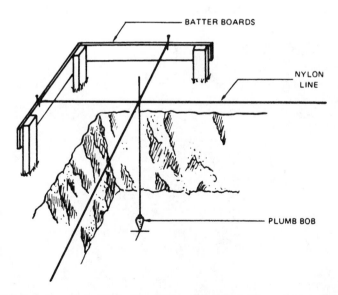

Fig. 39-4 Locating the building lines with a plumb bob. The point located by the plumb bob is the exact corner of the building. This method is used to transfer the corner of the building from batter board level to the bottom of the excavation, determining the location of the footings.

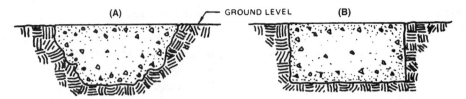

Fig. 39-5 Trench footings. Improperly shaped footing (A) will not support the
load of the structure as the sides are not straight. Properly shaped footing (B)
will support the load of the structure as it completely fills the footing area.

If the soil in the trench is very dry, it should be dampened with a garden hose before placing the concrete. This prevents too rapid drying of the concrete.

Formed Footings

When the foundation is being dug, rock is sometimes found. If rock is present, it may be necessary to form the footing with wood or steel forms, Figure 39-7, and place the concrete into the forms. This is sometimes done because digging out the rock is expensive and delays the job.

On large buildings the footing forms are constructed by carpenters. However, it is common practice on small buildings, such as houses and garages, for the mason to form and place the footings. Lumber used to form footings must be rigid and braced securely to prevent the concrete from pushing the forms out of position. Once this happens it is nearly impossible to correct.

An economical way of forming footings on a small job is to order the first floor joists (the framing lumber for the first floor) as soon as the foundation is excavated. This lumber is used to construct the forms for the footings so that no special lumber must be ordered. After the concrete has set, remove the forms, pull out the nails, and clean the lumber with a steel brush. The same lumber can then be used for floor joists when needed.

When setting the forms, it is very important that the top of the boards or forms are level with each other and

Fig. 39-6 Trench footing ready for concrete.

Fig. 39-7 Formed footing. Notice the bracing that holds the forms in place.

Fig. 39-8 Stepped footing that has been placed and the forms removed. Steps are 8″ in height so they will work the full height of the concrete block without any cutting required.

at the height specified on the plans. The footing must be placed exactly as specified.

Stepped Footings

When a building is constructed on sloping ground it is not always possible to place the footing around the foundation at the same level. In this case the footings must conform to the existing grade and be placed in steps, Figure 39-8. Each step must be level and positioned in divisions of eight inches from the top of the next step. (Eight inches is the height of a course of concrete blocks.) The steps should be no more than 2 feet high and no less than 2 feet long, to help prevent settling and cracking. Where there is a sharp drop in the grade of the excavation, it may be necessary to use a number of steps. This is done by using form boards cut to the size of the steps, Figure 39-9. The boards are braced securely in position. Steel rods can be inserted

across the steps to increase the strength of the footing. The width and thickness of a stepped footing are the same as for a regular footing.

PLACING CONCRETE FOR FOOTINGS

Note: Recheck all forms for accuracy before placing the concrete.

Position the truck carrying the concrete relatively close to the trench or form (inside the building lines if possible), so the concrete can be transferred directly from the truck by means of the metal chute, Figure 39-10. This saves time and labor that would be wasted in wheeling the concrete in a wheelbarrow from the truck to the footing.

The tools used in placing concrete for a footing are very simple, but it is important to have them ready when the concrete is delivered to the job site. Shovels, rakes, mortar hoes, and a few brick trowels are usually adequate.

Fig. 39-9 Stepped footing forms. Boards are in place, ready to receive the concrete. Notice the steel rods laid from the level top of the footing into the step for strength.

Fig. 39-10 Placing the concrete into the formed footings

Part of the water should be left out of the concrete mix until the truck arrives on the job site. The builder or mason can then tell the driver of the truck what consistency (stiffness or slump) mix is desired. This is very important because just enough water is needed to allow the concrete to pour from the truck and to allow the mason to work it. Too much water in the mix greatly decreases the strength of the concrete.

Immediately after placing the concrete in the forms, it should be compacted (settled) so that it fills out completely against the forms. This causes the footing to have a neat, smooth finish, free from holes or voids when the forms are removed or stripped.

Compacting can be done easily on footings for a house by tapping the sides of the forms with a hammer. It also helps to insert a trowel blade in the concrete along the side of the form and work it up and down with a slicing motion. The tapping and slicing motion causes the concrete to compact and the cement paste in the mix to fill out the form again, closing all voids and holes. On large jobs, gasoline-powered vibrators are used.

The top of the footing should be left relatively rough. This provides for good bonding to the mortar joint when the first course of concrete blocks is laid. Leveling with an ordinary garden rake and tamping with the rake tines leaves an excellent surface for the first course. Remember, it is very important to place the entire footing in one day, with no breaks, so that it cures as one mass, Figure 39-11.

CARE OF TOOLS

As soon as tools and equipment are no longer being used, they should be washed off with water to remove all concrete. If left to harden, the concrete is difficult to remove and can rust or pit the metal parts.

It is advisable to have a supply of water, preferably a hose with a nozzle, to clean the tools. If a water supply is not available, have the driver of the concrete truck rinse the tools when washing out the truck before leaving the job. All concrete trucks are supplied with a water tank and hose.

Fig. 39-11 Concrete trench footing in place ready for block work

CURING TIME

In warm weather, concrete blocks should not be laid on the footing until three days have passed. The forms should not be removed until four days have passed. In cold weather, the footing should not be loaded or forms removed for at least eight days. Freezing can be prevented by covering the footing with straw. The testing strength rating for concrete is usually based on twenty-eight days curing time. However, concrete actually takes longer to reach full strength. The ideal temperature for curing concrete is about 72 degrees Fahrenheit. Common sense should always be used when starting concrete block work on a green (not cured) footing. If masonry materials are being stocked in the foundation area before the footing has cured completely, care should be taken not to damage the footing.

OTHER USES OF FORMS FOR CONCRETE

Placing concrete in forms for footings is probably the simplest and most basic use of concrete in construction work. Since concrete is fluid in nature, it can mold into almost any form desired. At the same time, it is tremendously strong and meets many building requirements. Because of its strength, concrete is being specified for a variety of building purposes.

When there is repeated use of the same forms, such as found on a large housing or apartment project where the buildings are all the same, commercial-built metal forms can be used or rented at a considerable cost savings for the builder. However, in buildings that are one of a kind, the carpenter usually builds the forms from plywood and framing lumber. The forms must be very strong and braced so they will not move or leak while the placing of concrete is taking place. Metal reinforcement rods are inserted into the forms and wired together to make the concrete stronger. The construction and erection of steel or wood forms is not the job of the mason but the carpenter, or in some cases, the cement worker, Figure 39-12.

Fig. 39-12 Carpenters constructing plywood forms for poured concrete walls.

Fig. 39-13 Wood-form work ready to receive concrete on a large job.

Once the forms are constructed, braced, and coated inside with form oil to help create a smooth surface and prevent the concrete from sticking to the plywood, the concrete can be poured into them, Figure 39-13.

Pan Joists for Roof and Floor Systems

Construction using pan joists provides a very strong structural and ceiling floor system that uses a ribbed slab formed with pans, Figure 39-14. The reason that architects like this type of system on large buildings with concrete floors is that it is fast and strong and the pans may be reused many times over. The pans are almost always rented from a company that specializes in not only supplying them, but also installing them in place on the job. After the concrete has cured sufficiently, this crew returns, removes the pans, and hauls them away to be used elsewhere. Standard pan forms produce inside dimensions of 20″ to 30″ and depths from 6″ to 20″.

Concrete Curb and Gutter

The use of concrete to form curbs and gutters provides the most lasting of any method for containing a roadbed. Steel forms are put into place and leveled before any concrete is placed in them.

Fig. 39-14 Workman installing pan forms for suspended concrete floor and ceiling.

Fig. 39-15 Placing concrete in steel curb and gutter forms.

The concrete is delivered from the batching plant in ready-mix trucks and placed in the forms, Figure 39-15.

Once the concrete is worked fully in the forms, it is smoothed off with the trowel and rounded off on the edges with an edger, Figure 39-16.

Concrete Porches and Steps

Another very popular use of concrete is in constructing porches and steps on most residential projects, such as a home or apartment complexes. A wood form is built out of either plywood or framing lumber to hold the concrete in place until it has cured. The earth under it should be settled firmly or compacted, or the steps will sink. This is a common complaint of persons who buy new houses, as the fill earth sinks as time goes by.

Figure 39-17A and 39-17B show typical concrete steps and porch platform. The top of the porch and the steps should have a little fall to drain any water that may collect.

Steps do not always have to be in a straight line, but can be curved as desired, Figure 39-18. The best choice of framing materials for this is $3/4''$ marine plywood, as it can be curved as required. The forms must be braced very well so they do not move when the concrete is

placed in them, as shown in the photo. The steps in this particular photo were to an in-ground swimming pool at the rear of the house. They were curved due to the shortness of the distance from the back door sill to the pool area.

SUMMARY

This unit has discussed footings and some of the many uses of forms used to place concrete. Although the construction of footings is considered to be one of the simpler tasks performed by the mason, it is one of the most important parts of the structure. The entire building rests on the footings. For this reason, it is important to be familiar with all aspects of footings and how to properly construct them.

Good footings make it a lot easier for the mason to build walls on them that are level and to the correct height. There is nothing more frustrating for the mason than to have to either cut block down or bed it up with a lot of mortar to get that all-important first course of block level. A large part of the contractor's and mason's profit can be lost at the beginning of a job if the footings are not properly constructed.

Fig. 39-16 Finishing concrete curb and gutter with a floating tool.

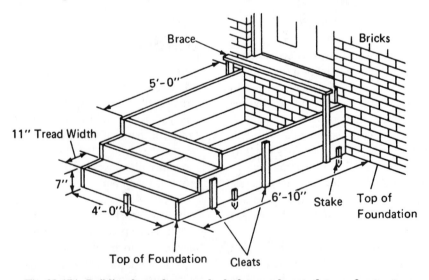

Fig. 39-17A Building forms for a porch platform and a set of steps of concrete. Forms have to be braced strongly and coated on the inside with form oil to prevent the concrete from sticking and create a smooth surface.

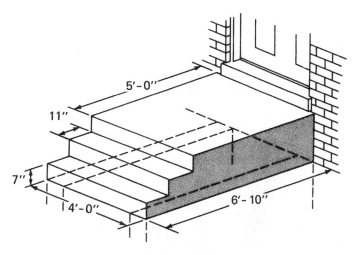

Fig. 39-17B After the steps have cured properly, the forms are removed and the finished product is shown in the illustration.

Fig. 39-18 Concrete steps formed in a curved plywood form.

ACHIEVEMENT REVIEW

The column on the left contains a statement associated with footings. The column on the right lists terms. Select the correct term from the right-hand list and match it with the correct statement on the left.

1. Two key factors that influence the size of footings
2. Primary function of a footing
3. The width of the footing for a ten-inch concrete block foundation

4. The minimum depth of the footing for a fireplace and chimney

5. A specification for concrete for footings in light construction

6. The exact location of the footing is determined from the batter boards by using this

7. The best type of footing to use when rock is encountered in the excavation of a basement

8. Footing used when it is not possible to place the footing around the foundation at the same height, due to a drop in the grade

9. The top of the footing should be in this condition to provide for good bonding to the mortar joint

10. Greatly improves the finished sides of a formed footing and fills voids or holes

11. The final amount of water should be added to the concrete after the concrete is on this

12. Added to increase the strength of a concrete footing

13. In warm weather, the mason should wait this number of days before laying concrete blocks on the footing

14. Used for cleaning tools and equipment immediately after placing concrete

15. In cold weather, the mason should wait this number of days before laying concrete blocks on the footing

a. Stepped footing
b. Steel reinforcement
c. Rough
d. Support the building
e. 3 days
f. 8 days
g. 2500 psi
h. Water
i. The job site
j. A plumb bob
k. Weight of structure and load-bearing strength of soil
l. 16 inches
m. Materials and finish desired
n. Formed footing
o. Tapping with a hammer
p. 20 inches

MATH CHECKPOINT

1. How many cubic yards of concrete are needed to pour a wall footing 18″ wide by 8″ deep and 32′ long?

2. How many cubic yards of concrete are needed to pour a footing for a retaining wall that is 8″ deep by 16″ wide by 60′ long?

3. How many cubic yards of concrete would be required to pour eight separate footings to support round steel lally columns? Each footing should be 12″ deep by 24″ wide by 24″ long.

4. How many cubic yards of concrete would be required to pour a fireplace footing that is 12″ deep by 36″ wide and 9′ long?

5. If one cubic yard of concrete weighs 4000 lbs and a concrete truck will haul nine cubic yards in one load, how many tons will this be?

UNIT 40
Essentials of
Concrete Work

OBJECTIVES

After studying this unit, the student will be able to

- describe the composition of concrete.
- explain what air-entrained concrete is and its advantages.
- describe how concrete is tested for strength.
- list and explain the steps in preparing, placing, and finishing concrete flat work.

Unit 39 described the use of concrete for footings, foundations, and curb and gutter work. This unit deals with background information that is important to know in doing various types of concrete work, in particular, what is known as *flat work*. It also discusses how to form, place, and finish concrete flat work such as walks, slabs, etc.

Concrete is very important to the mason because without it, masonry work could not be done. Concrete supports the masonry walls and in many cases reinforces the masonry wall to make it stronger. This unit will provide basic information on the composition of concrete, what ready-mix concrete is, what air-entrained concrete is, the water-cement ratio law, additives of concrete, how concrete is tested for strength, tools used for concrete work, preparing for placement of concrete, placing and finishing of concrete, curing of concrete, and how to treat concrete in hot and cold weather. All of this information is necessary if you are to understand concrete and be successful with it.

Concrete is one of the most versatile of all structural building materials and one of the most useful because it is durable, strong, fire resistant, and economical. Due to its plastic fluid nature before it cures, it can be molded to almost any type of form and shape. Concrete depends more on being proportioned and mixed correctly for its success than any other building material. Carelessness and not paying attention to the ingredients, mixing, and

curing procedures can be devastating and will surely cause the concrete to fail if proper directions are not followed.

INGREDIENTS OF CONCRETE

Concrete is a mixture of five basic materials used in proper proportions. They are:

1. portland cement
2. fine aggregates (sand or crushed stone)
3. coarse aggregate (gravel, crushed stone, or manufactured aggregate)
4. water
5. admixtures (optional)

On occasion, certain additives are added to concrete to achieve special properties or to meet job specifications and requirements. These additives are called *admixtures*. Let us take a closer look at the materials that make up concrete.

Portland Cement

Portland cement causes the concrete to become hard. It is a finely ground material that consists mainly of compounds of lime, silica, alumina, and iron. When mixed with water, it forms a paste that hardens and binds the aggregates together to form what we know as concrete.

519

This setting and hardening is due to the reaction between the cement and water and is known as *hydration.* Hydration will continue for a period of years if the concrete is kept moist.

Portland cement is manufactured by blending the above materials together and heating them in a rotary kiln to a temperature of approximately 2600° to 3000° F to form a clinker. This clinker is cooled and pulverized resulting in a very fine gray powder that is portland cement. It is either bagged in 94-pound bags or sold in bulk, which is hauled by tank-type trucks to concrete batching plants. Each bag holds one cubic foot. Portland cement will last indefinitely if kept dry. When stored for a long period of time, it will sometimes develop what is known as *warehouse pack.* This means that the cement has taken a certain firmness or pack. Rolling or adjusting the bag will correct this problem. This packing effect does not affect the strength of portland cement. Generally speaking, Type 1 portland cement is used for most concrete. There are four other types made for special purposes. This can be reviewed by referring to Unit 10.

Aggregates

The sand and crushed stones or gravel used in concrete are known as the aggregates or the binding materials. Aggregates make up about 60% to 80% of the volume of concrete. The selection of good clean aggregates is very important for strong concrete. This is true because the cement paste has to bind all of the ingredients together into one common mass.

Fine aggregates are composed of sand or other acceptable materials that can be man-made. A good concrete sand should contain grains or particles that vary uniformly in size up to no larger than $\frac{1}{4}''$. These particles help fill the spaces between the other larger aggregates. However, too fine a sand particle will end up requiring more cement paste to attain the proper strength and will result in making the concrete too expensive.

Coarse aggregates are made up of crushed stone, gravel, or other man-made materials that are larger than $\frac{1}{4}''$ in diameter. The maximum size of coarse aggregates is usually $1\frac{1}{2}''$ in diameter. The larger aggregates are used in concrete for foundations, footings, or thick walls. Less cement paste is needed when using larger

aggregates, and as a result more concrete is obtained per bag of portland cement used. This is the reason footing concrete is several dollars cheaper than finish concrete, which has a richer mixture for troweling. Angular shaped stones, such as crushed limestone, are better for concrete than gravel, which is more rounded in shape. The deciding factor concerning the size of aggregates to be used is the shape of the form it is being placed in and the reinforcing steel being used. This is true because the concrete must fill completely in the form and around the reinforcement steel. Causing the concrete to fill around the steel rods in forms is done by using a vibrator that has a long, metal, flexible tube with a metal cylinder-shaped head attached. Job specifications dictate the amount of vibration required to achieve certain results. The maximum size of aggregates should not exceed the following recommendations:

a. One-fifth the dimensions of nonreinforced members.
b. Three-fourth the clear spacing between reinforcing bars or between reinforcing bars and the form.
c. One-third the depth of nonreinforced slabs on grade.

Aggregates both fine and coarse should be clean and free of loam, clay, or any vegetable matter that would prevent cement paste from binding to the aggregate, as this would reduce the strength of the finished concrete.

Water

Generally speaking, any natural water that is drinkable is fine for concrete. It should be free from any measurable amounts of oils, acids, or alkalis. However, water that is suitable for mixing concrete may not always be fit for human consumption as drinking water. The intent is to use good clean water for concrete if possible.

The purpose of water in concrete is to cause it to blend together and hydrate. As a rule water makes up approximately 14% to 21% of the total volume of concrete.

The relationship between the amount of water and portland cement used to make it into a paste is known as the *water-cement ratio.* The quality of the paste is determined by the amount of water used with the cement. This can be compared to mixing up wood glue from a powder. If the glue is too runny or thin it will be weak. If it does not have enough water in it, it will be unworkable and too dry. One important problem over the years in concrete work has been too much thinning of concrete to make it run into forms or become easier to handle. This is the major reason many large jobs conduct concrete tests on the site to prevent this from happening.

Admixtures

Admixtures are considered to be any materials that are added to concrete during the mixing process other than portland cement, water, and aggregates. There are a variety of admixtures that can be added to achieve special properties. They are retarders (slow down the setting time), accelerators (speed up the setting time), water-reducing agents (increase the flow), workability agents (increase the workability of the concrete), pozzolans (help to control high internal temperatures that occur in some types of concrete), and air-entraining admixtures.

The most important by far is the use of air-entraining agents, which because of their ability to withstand freezing and thawing cycles have become very popular for use in concrete. The development in the late 1930s of air-entrainment agents is considered to be the greatest advance in modern concrete. Air-entrained concrete is made by using an air-entraining cement or by adding an air-entraining admixture into the batch as it is being mixed. This additive traps bubbles of air that are extremely small. The diameter of these bubbles range from about a thousandth of an inch to a hundredth of an inch. These bubbles are not interconnected, and therefore work better. As many as 400 to 600 billion bubbles may be present in a cubic yard of air-entrained concrete. This type of concrete has a creamier composition and improves workability. The addition of air-entrained agents also greatly reduces segregation and bleeding, which means separation of the water from the paste in the mix. All of these are very desirable features of a good concrete. It also gives a water-tightness that is vastly superior to regular concrete. The advantage of being able to place concrete that has air-entrained agents in the cold winter months has greatly increased the use

of concrete as a structural material. The strength of air-entrained concrete is also equal to regular concrete. The great majority of concrete sold now is of this type. When one considers the little difference in cost over regular concrete, it is well worth the money.

TESTING CONCRETE ON THE JOB

Many job specifications require that fresh concrete be tested on the job to make sure that it meets with the specified requirements. There are two main types of test used, a *slump test* and a *cylinder test.*

Slump Test

To do a slump test, the concrete is placed in a slump cone of 16-gauge galvanized metal, Figure 40-1. The cone is 8″ in diameter at the base and 4″ in diameter at the top; its height is 12″. The base and top are open. The mold is placed on a firm, flat surface and filled to about one-third of its capacity. The concrete is then rodded with 25 strokes of a ⅝″ steel rod about 24″ long and having a hemispherical tip. The filling of the cone is completed in two or more layers. Each layer is rodded 25 times and each rod stroke must penetrate into the underlying layer of concrete. After the top layer is rodded it is struck off using a screeded and rolling motion of the tamping rod. The mold is then lifted off very carefully. The mold is placed next to the concrete and the slump is measured. For example, assume that the center of the top of the slumped concrete is 5″ below the center of the top of the cone. The slump of this sample concrete would be rated as 5″. From this test the inspector can immediately determine if the concrete has the correct amount of water in the mix without waiting for a longer technical test. Slump tests for concrete are recognized throughout the concrete industry and considered to be very accurate for a quick, on-site inspection.

Cylinder Test

These tests are performed to determine if the concrete has the correct compressive strength. These take much longer, as the test ratings are assigned after the concrete

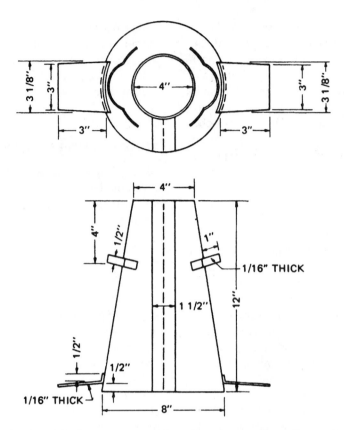

Fig. 40-1 Metal slump cone mold used to measure slump of concrete.

has cured for a determined length of time. Following are basic instructions for performing a test.

Most job requirements state that a sample of the concrete be taken at three or more intervals while the batch is being discharged from the truck. However, the test samples cannot be taken at the beginning or very end of the discharge. The test specimen is made into a watertight cylinder mold that is 6″ in diameter by 12″ in length, providing the coarse aggregate does not exceed 2″ in nominal size, Figure 40-2. The mold is filled in three equal layers. Each layer is rodded with 25 strokes of a ⅝″ round steel rod about 24″ long. When rodding the second and third layers, the rod should just break through into the layer below. Reinforcing rods or other tools should not be used for tamping, as there is a special test rod used for this. After rodding the top layer, the mason should use a trowel to scrape off this layer

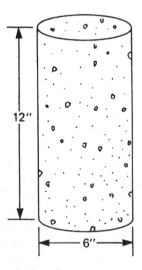

Fig. 40-2 Test cylinder of concrete.

so it is even with the top of the mold. The surface is then immediately covered with glass, metal, or plastic plates to stop evaporation. The cylinder is placed on a rigid, level, flat surface free from vibration or any disturbances for 24 hours. It should also be stored under conditions that have a temperature balance within $60°$ F to $80°F$. After the 24 hours have passed the test is sent to a test center for standard curing. As a rule, after the test cylinder has been cured for 28 days it is tested for compressive strength and results are published or made known to the builder or owner on the job site.

PREPARATION TO PLACE CONCRETE

The first step in placing concrete is preparing the area to receive it by removing all sod, large rocks or stones, or trash. This is done by using a common garden shovel, flat shovel, pick, etc. Next, the ground should be tamped with a tamper to settle any loose earth. Never place concrete on loose fill earth if possible, as it will settle. This is especially true in the case of a porch and steps, which will sink as much as a foot on fill earth. After the ground or base has been prepared, a form of some type must be built or set up to restrain the concrete after it has been placed. This form must be firm and cannot be allowed to give or the concrete will follow the form and the job will not be acceptable. This is generally why framing lumber or ¾″ plywood or metal forms, such as those used in curb and gutter work, are used. Since the last unit covered curb and gutter work, and footings, the reference here will be to what is known as *flat work,* which is floors, slabs, walks, etc. If a concrete floor is being placed in the basement of a house, the basement walls of the house act as the form and all that is necessary is nailing expansion strips around the edges to prevent cracking of the floor. The top edge of the expansion strip will be the top edge of the concrete finished floor when it is poured.

Concrete sidewalks and patios are good examples of exterior flat work. As previously stated, after the earth has been leveled a wood form should be built and braced into position, Figure 40-3. Using wood $2'' \times 4''$ boards for the sides of the form, wood $2'' \times 4''$ stakes should be driven along the sides periodically and nailed into the sides of the forms. It is a good practice to use a double-headed nail, which is called a form nail, be-

Fig. 40-3 $2'' \times 4''$ **wood forms in position to place concrete in.**

cause it can be pulled out rather easily without disturbing the concrete after the concrete has cured. Divider strips of $2'' \times 4''$ wood are spaced approximately every 10 feet of walk and a piece of tar paper or an expansion strip set next to it when the concrete is poured in the form. These divider strips should be nailed in lightly with a small nail as they must be removed as the concrete is placed in the form. Care must be taken not to disturb the main outside forms when the concrete is placed in them. Placing a sheet of plastic vapor barrier on the ground is a good recommended practice for preventing the water in the concrete from being sucked up too rapidly, which would not allow it to cure properly. Last, some reinforcement wire should be laid on top of the vapor barrier. When the concrete is placed, the reinforcement wire is held up with the end of a rake so it will be fully integrated in the concrete. It is a good practice to brush some form oil on the sides of the forms to prevent the concrete from sticking. Any clean motor oil will work, but there is a special form oil made for this, Figure 40-4.

Fig. 40-4 Steel reinforcement rods and concrete wire in place for a patio slab.

Ready-mixed concrete should be ordered well in advance. The order should state the time of delivery and psi strength desired. Generally, a 4000-psi concrete is ordered for floors and walks. This means in layman terms that $6\frac{1}{2}$ bags of portland cement are mixed to the cubic yard of concrete. This is known as the *prescription method* of ordering concrete and is recommended by architects and engineers. Ready-mix concrete should be placed in the forms as soon as possible after arriving on the job. Excessive mixing in the drum on the truck may cause the concrete to become too stiff and difficult to finish properly. The rule of thumb is not to wait any longer than $1\frac{1}{2}$ hours before placing. Hot weather will require placing the concrete even sooner than this because heat will cause concrete to dry out faster. Delays are very costly once the concrete arrives, as the drivers only have a certain length of time to unload before charging waiting time. Generally, this time limit is $\frac{1}{2}$ hour to 45 minutes for most concrete suppliers.

Estimating concrete for the job was covered in Unit 39. Review this information as needed when you figure concrete for a job.

TOOLS NEEDED TO FINISH CONCRETE

Figure 40-5 shows the basic tools needed to place and finish concrete. They are as follows:

a. screed (can be a manufactured one or a good straight wood $2'' \times 4''$ board)
b. concrete rake (garden rake will do in a pinch)
c. bricklayer's trowel
d. magnesium float (magnesium is preferred due to its lighter weight)
e. wood float (works best when floating a rough place to bring up the cement paste to the top)
f. bull float (comes with threaded extension handles that can be screwed together to allow the finisher to reach farther)
g. finishing trowel (for smooth finish)
h. sidewalk edgers (used to round off the edge of concrete to prevent flaking or chipping)
i. cement groover (used to cut a groove at expansion or control joints)
j. power trowel (used to trowel the finish smooth when there is a large floor or slab)
k. exposed aggregate broom (specialty tool often used now to wash off excess materials and expose the aggregate; hooks up to a garden hose)

Although there are a number of specialty tools used to achieve special effects in concrete, these are considered to be the basic ones. They can be purchased from most building suppliers who stock trade tools and hardware stores, or can be rented from rental companies. It is highly recommended when buying tools for professional use to only purchase brand names. Your local dealer can recommend the best brand names available!

One very excellent source of professional masonry tools is Bon Tool Company (formerly Masonry Specialty Co.), 4430 Gibsonia Road, (Rt. 910), Gibsonia, PA 15044. Free catalogs are available on request.

PLACING THE CONCRETE

When the concrete truck arrives on the job site, it should be located as close to the forms as possible. This is particularly true in the event that a large slab or floor has to be placed. The truck has a sectional metal chute on the rear that can be positioned in place and the concrete poured directly from the mixing drum, down the chute, and into the forms. This, of course, saves a lot of labor and time. This is not always possible, and in many cases where walks, patios, etc. are being placed the

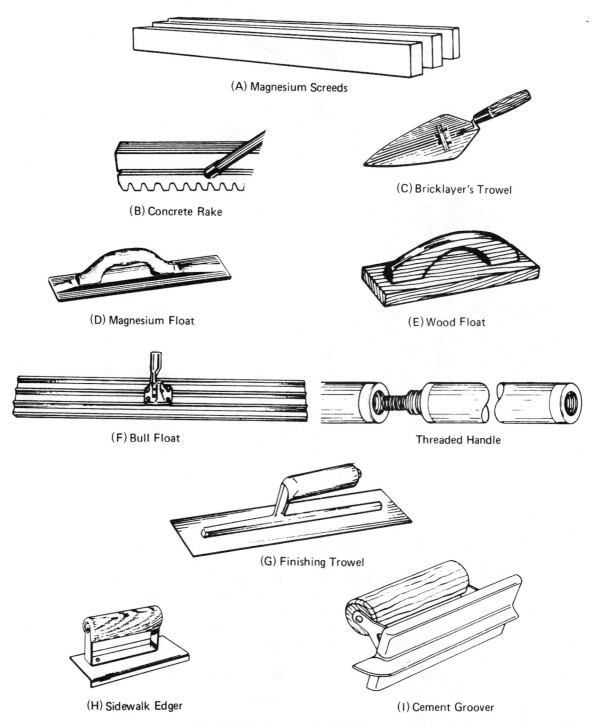

(A) Magnesium Screeds

(B) Concrete Rake

(C) Bricklayer's Trowel

(D) Magnesium Float

(E) Wood Float

(F) Bull Float

Threaded Handle

(G) Finishing Trowel

(H) Sidewalk Edger

(I) Cement Groover

Fig. 40-5 Concrete tools and equipment.

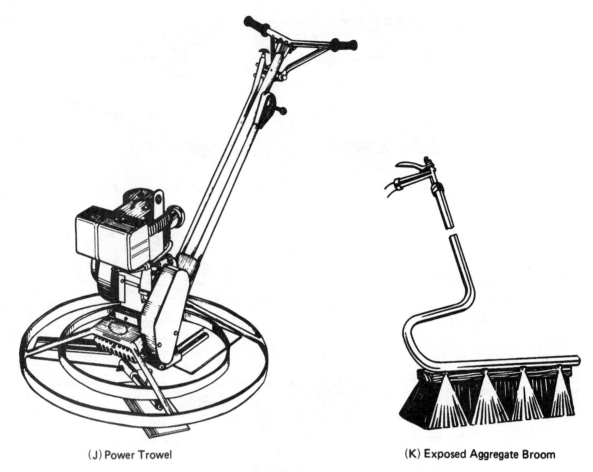

(J) Power Trowel

(K) Exposed Aggregate Broom

Fig. 40-5 Concrete tools and equipment (continued).

concrete is discharged into a wheelbarrow. The wheelbarrow is then used to transport the concrete from the truck to the formed area. The concrete placement procedures that are shown here are for a typical walk, Figure 40-6.

Start at one end of the form, being careful not to rest the end of the wheelbarrow on the form. Dump the concrete into the form, working it well with the shovel and concrete rake into the corners and along the edges to fill out all areas. Lift the reinforcement wire with the end of the rake to make sure it is embedded into the mix and not lying under it. Tap along the edges of the forms with a hammer to help settle the concrete against them. This will help to produce a smoother, well-filled edge

when the forms are removed later. A bricklayer's trowel can be inserted between the edge of the form and the concrete and worked up and down with a slicing motion. This tends to promote the cement paste in the mix to fill in along the side of the form and create a smoother finish free of any holes. Add additional concrete with the shovel if there is any settlement. Excess concrete is removed and is added to the next area to be placed.

FINISHING CONCRETE

Screeding

Screeding (leveling the concrete to the top of the forms) is more easily done with two persons, but depends on

Fig. 40-6 Placing the concrete in the forms.

Fig. 40-7 Screeding the top of a concrete walk with a wood $2'' \times 4''$ screed.

the width of the form, Figure 40-7. A manufactured metal screed can be used, or a good straight wood $2'' \times 4''$ board will usually work fine. The screed should be long enough to extend approximately $6''$ to $10''$ beyond the edges of the form boards. The screed is then worked back and forth in a sawing motion toward the end of the form, letting it rest at all times on the top of the form board. This will force all of the aggregates down into the mix and cause the cement paste to rise to the top, which is the desired result. In the process all of the high spots will be removed by the screeding action and excess concrete will be dragged into any of the low places to fill them. Any concrete that piles up in front of the screed as it is being worked should be removed either with the shovel or trowel and returned to the wheelbarrow for the next section of the form not yet placed.

Note: It is very important that the screeding operation is done immediately after placing the concrete in the form to obtain good results. One very important fact

about placing concrete is that you must be ready at all times to work it as the situation dictates and never leave for any period of time, as the concrete will take an initial set before you are ready for it. Once this happens it cannot be worked out as it should and the cement paste will not rise to the top for finishing later.

Immediately after screeding, the screed should be rinsed off with water from a garden hose with a nozzle so the screed will be ready for use next time. If a water supply is not available, the concrete truck has a rubber hose attached to the rear and connected to a supply of water on board. All you have to do is ask the driver if you can use it, and he or she will be happy to oblige.

If there are $2'' \times 4''$ form boards in the walk or slab, gently remove them as they are reached in the screeding process so concrete can be placed there, Figure 40-8. Make sure that the tar paper or control joint previously discussed is in position correctly and is level with the top of the concrete.

Fig. 40-8 Removing the wood $2'' \times 4''$ at the expansion joint location.

Fig. 40-9 Floating the concrete to work up the cement paste for finish troweling.

Floating

As soon as the screeding process is completed, the concrete should be *floated* (to rub or trowel the surface of the wet concrete with either a wood or metal tool that is flat on the edge touching the concrete, Figure 40-9). This action or procedure causes the cement paste in the mix to reactivate and to come to the top of the concrete. A good layer of cement paste has to be on top in order to have a smooth finish and is very necessary to completely seal in all of the aggregates in the concrete. In the trade this is called the *cream* in the concrete. For small jobs such as walks or a slab that you can reach the middle of, a metal float can be used. For larger floors or slabs where you cannot reach out far enough, a metal *bullfloat* is used very successfully. This metal bullfloat was shown in the suggested tools needed and has a number of extension handles that can be screwed together to allow the finisher to reach out for long distances. The preferred favorite of concrete finishers is a bullfloat made of magnesium because it is very light and strong. As was the case in screeding, the bullfloat or regular float should be rinsed off immediately with water after use to prevent the concrete from sticking to it. When floating concrete, care should be taken not to overwork the concrete or it will cause some of the ag-

gregates to work up to the top of the finish and become sandy. This knowledge only comes with experience!

Edging

Edging is done to round off the edges of the concrete and helps to prevent chipping or breaking of exposed edges. It also creates a more pleasing and finished appearance. The proper time to do the edging is directly after the concrete has been floated, as there is plenty of cement paste present on the edge of the concrete. Run the edger around all edges with a steady movement and try not to move backwards over the edges that have just been done, as it tends to pull the aggregate loose, Figure 40-10. If there is a place that is not filled out well, add a little cement paste from the surface of the slab and re-edge it. Care should be taken not to let the edge dig in and work the aggregate loose. After some experience you will discover that if the front edge of the edger is held up slightly it will not dig in and a more slick smooth edge will result. Usually after the concrete sets a little and some of the surface water evaporates, edging a second time around will do a better job.

Grooving

Basically the same procedure used in edging will be repeated at the point where expansion joints are located.

Fig. 40-10 Rounding off the edges with the edger tool.

Fig. 40-11 Finish troweling of concrete.

However, a groover tool is used here. The groover has a straight, projecting length of steel located in the center of the bottom of the tool that is approximately $\frac{3}{8}''$ to $\frac{1}{2}''$ deep and approximately $\frac{1}{2}''$ wide. To form a straight groove, place a piece or strip of wood that runs the width of the concrete. Hold the groover level and run it along the strip from one end to the other, using the same technique as for the edger. Press just hard enough to form a sharp, distinct groove in the concrete. After the first troweling, as was the case with edging, it is a good idea to repeat the grooving for best results.

Troweling

Troweling with a steel trowel will produce a smooth finish and help to avoid any problems later from water entering and being absorbed into the concrete, Figure 40-11. This is one of the most important stages in finishing a concrete slab. The trick is to start troweling the first time at the right moment. The troweling should be done when the surface water just starts to dry off. Weather conditions play a big factor in determining when this is done. In cool weather you will have to wait longer, as the concrete will dry more slowly. In hot weather, the concrete will dry more quickly. The key is to wait there, observing to determine when the surface

is dry enough. If the trowel is not sinking in or leaving a lot of deep trowel marks as you trowel, the time is probably right. Only experience will teach you that!

If the area you are troweling is too large to reach across, a wide board can be laid on the concrete, and you can kneel on that without sinking in. These minor impressions from the board can be troweled out as you back your way out of the slab to the edges. If the cement paste has to be worked up on the surface a little to promote a smooth finish, then it is time to use a wood float. The wood float tends to draw the cement paste up on top. Smaller knee boards can be used in place of a wood plank after the concrete has dried enough to support them.

Use an arc-type sweeping motion when troweling, raising the front edge of the finishing trowel slightly so it will not dig in. Overlap the previous trowel strokes about an inch and try to use smooth trowel strokes without an undue amount of pressure. It will take two or three trowelings to achieve a smooth surface. More pressure will be required as the surface dries out.

Once a smooth surface is obtained, the mason should not repeat trowelings. Excess troweling will cause black marks on the surface from the metal blade of the trowel. This is very similar to overstriking mortar joints in brickwork. Excessive troweling will also cause the finish of the concrete to become sandy or gritty. As with other techniques of finishing concrete, experience is the best teacher.

FINISHES FOR CONCRETE

Broom Finish

Sometimes a rougher finish is desired so the concrete will not be slippery. If this is the case, the concrete can be brushed after the first troweling with either a stiff bristle brush or an old broom. A regular corn straw house broom or a wide, floor broom work fine. Use even strokes to create the gritty surface but be careful not to cause the broom to dig into the concrete and leave dips or large rough places. Be sure to wash the broom out after using as the cement paste will harden on the end of the broom and ruin it. If the concrete is too dry, wet the broom a little and this should do the job.

Exposed Aggregate Finish

Another method of finishing concrete is by creating an exposed aggregate finish. Briefly, this is achieved by following all of the procedures described, but instead of finish troweling, spray water in a fine mist from a garden hose on the surface; at the same time use a broom or brush to expose the aggregates. Care should be taken not to wash out the cement paste between the aggregates, but to just expose it. As a rule, gravel is used in the concrete for this type of finish. Refer back to the tools needed and you will see a manufactured, exposed aggregate broom that has a length of pipe attached to it that a hose fits onto. Water sprays through four plastic ports to wash off the cement paste and expose the aggregate. This special tool does an excellent job and is highly recommended.

CURING

In order for concrete to attain its maximum strength it has to be *cured* (hardened) properly. This is done by not letting the concrete dry out too fast and keeping it at the right temperature. Soon after the concrete has been placed, its increase in strength is very rapid for a period of three to seven days. After that time the cur-

ing process is very gradual, and in about 28 days the concrete is considered to be cured enough to meet standard compressive strength requirements.

Concrete can be cured by a number of methods depending on job specifications and requirements. These can be divided into two groups, those that provide extra moisture (water) to the concrete and those that prevent the moisture present from escaping. Spraying water onto the concrete or flooding the surface with water is one method. It is the most effective method of preventing water loss. Generally, sprinkling water from a hose in a fine spray is the method most often used on the job, along with laying burlap down on the surface of the concrete and keeping it wet for a predetermined period of time. This constant supply of water keeps the concrete from drying out too fast and promotes excellent curing. Engineers say that if concrete is kept damp for a month, the strength is about twice that of concrete cured in dry air. There are other, more technical methods of curing concrete, such as barriers of waterproof paper or plastic film seals, which prevent the water from evaporating. Some also use certain chemicals that are sprayed on the surface.

Fresh concrete should not be walked on for a period of at least three days. The curing time depends on the temperature of the concrete and weather conditions. When the temperature is 70° F or above, concrete is cured for at least three days. When it is 50° F or above, it is cured for at least five days. The most favorable range of temperature for curing concrete is from 55° to 73° F. At higher temperatures, the hydration occurs faster and the concrete does not attain its full strength. The process of hydration occurs much slower when it is colder; almost no chemical reaction occurs when the temperature is near freezing. If concrete is allowed to freeze in the first twenty-four hours after it has been placed, it is almost certain to be permanently damaged. This is the reason that great care must be taken when placing and curing concrete in the wintertime. The practice of covering the area with plastic tents and providing heat through blowers is required on most cold weather construction. Careful checks are made to make sure that this is done correctly and that the concrete does not dry out too quickly. Large commercial jobs that use concrete as a part of the supporting structure are monitored very carefully by inspectors to make sure that the proper practice is followed. The same is true in stripping forms from this type of construction.

Wood forms for flat work such as walks, floors, slabs, etc. should be removed as soon as possible after the concrete has hardened enough so that they will not sag or move. This promotes the curing of the edges along with the surface and allows the mason to fill in any voids or rough spots on the edges before the concrete has cured. Filling in holes is accomplished by troweling in some cement paste in these areas and either troweling them smooth or rubbing them out with a rubbing stone.

All forms should be scrubbed clean as soon as possible so they will be able to be used again.

PROBLEM AREAS IN CONCRETE

Following are a few of the most troublesome problems that occur in concrete if not placed, finished, and cured correctly:

Crazing. This is the formation of many fine hairline cracks on the surface of new concrete. It is caused by shrinkage. The cracks form a pattern that looks like crushed eggshells. They are caused by the surface of the concrete drying out too fast and are easily avoided by sprinkling water on the surface as previously mentioned or keeping the surface wet by laying wet burlap on it.

Scaling. Scaling is when the surface of concrete chips or flakes away to a depth of about $\frac{1}{16}$ of an inch to $\frac{1}{4}$ of an inch. It is caused when the fine aggregates and cement paste separate and do not reunite. A layer of fine silt, cement, and clay rises to the surface. This thinned, weak top coat will easily scale after exposure to the elements.

This occurs early in the curing of the concrete and as a rule is caused when the concrete has frozen or suffered frost damage, or there has been poor workmanship. As previously discussed, using air-entrained concrete will greatly avoid freezing and thawing. Poor workmanship practices such as screeding, floating, or troweling too soon while too much free water is on the surface will in some cases cause scaling. Adding too much water to the top of the concrete to make it trowel smoother once it has already dried will cause concrete to scale.

Dusting. The formation of dry, powdery materials on the surface of newly cured concrete is called *dusting.* Too much clay or silt in the sand or stones is one of the principal causes of dusting. Only clean sand and aggregates should be used in the mix. It is a poor practice to store sand or aggregates on the ground or in soft areas or mud.

Another cause of dusting is condensation appearing on the surface of the concrete. This happens most of the time due to cold nights and warm days. In the spring and fall seasons this is most likely to occur. The way to combat this problem is to keep the temperature fairly constant by using a heat blower or building a plastic tent around the concrete if there is a drastic change in temperatures.

Floating or troweling excess collected moisture into the concrete can also cause it to dust. The spreading of powdered, dry portland cement on top of a concrete slab to speed up the drying procedure is another cause of dusting. Letting concrete set too long before finish troweling and sprinkling water on it to make it trowel up smooth promotes the dusting process. The same is true when excessive troweling is done once the concrete has taken its initial set. As stated before, concrete must be finished at the right time or stages of hydration to achieve the best results.

Placing Concrete Too Thinly. Placing concrete too thinly can result in cracking on the surface. Unless the concrete is a special mixture or is reinforced, it is risky to place concrete less than 4 inches in depth. This is especially true when placing concrete on a base that soaks up the moisture too quickly. Placing a plastic vapor barrier on top of the ground or using crushed stones and placing the concrete on top of them is highly recommended. A good motto is to use good, sound, proven practices because good results will follow.

PROPORTIONS TO MIX YOUR OWN CONCRETE

There are two commonly used formulas for mixing concrete that will serve most needs. When mixing footing concrete, use a 1:2:4 proportion with enough water to blend the materials together to make them workable. This means one part portland cement to two parts sand or fine aggregate, and four parts crushed stone or gravel. This is equal to approximately a 3000-psi concrete and can be used for most footing mixes for residential construction.

Secondly, for concrete slabs, floors, walks, etc., use a 1:2:3 mixture, which means one part portland cement to two parts sand or fine aggregate, and three parts crushed stone, and of course, enough water to make it workable. This is equal to approximately a 3500-psi concrete and is standard for industry. There will be occasions when only a small amount of concrete is needed, so use these formulas as guidelines.

It is especially important to float concrete as soon as possible on steps or walks, after it has been screeded off to force the cement paste to rise to the top of the mix. You must have a thin layer of cement paste on the surface to be able to trowel a smooth surface and seal all of the aggregates in the concrete. It will take several repetitions with a cement finishing trowel to obtain a good finish. Figure 40-12 shows a concrete finisher floating the surface of a set of concrete steps to level off and force the concrete paste to the surface for finish troweling.

Fig. 40-12 Floating the surface of a set of concrete steps.

ACHIEVEMENT REVIEW

Select the best answer from the choices offered to complete each statement. List your choice by letter identification.

1. Additives combined with concrete to achieve certain properties are called
   a. additions.
   b. admixtures.
   c. chemicals.
   d. formulas.

2. The hardening of concrete due to the adding of water is known as
   a. water-cement ratio.
   b. setting.
   c. hydration.
   d. chemical change.

3. Sand and crushed stone used in concrete are known as
   a. aggregates.
   b. blenders.
   c. accelerators.
   d. fillers.

4. The best shape of stone to make concrete with is
   a. round.
   b. rectangular.
   c. square.
   d. angular.

5. Concrete that is placed to form a ceiling or floor in a building can be made much stronger by adding
   a. more portland cement.
   b. steel reinforcement.
   c. larger stones.
   d. lime.

6. Water in concrete makes up approximately
   a. 14% to 21% of its volume.
   b. 18% to 25% of its volume.
   c. 20% to 30% of its volume.
   d. 30% to 40% of its volume.

7. The relationship between the amount of water and portland cement in concrete is known as
   a. its separation point.
   b. its saturation point.
   c. chemical balance.
   d. water-cement ratio.

8. The most important property of air-entrained concrete is that it
   a. improves the appearance.
   b. makes it easier to finish.
   c. will absorb coloring agents more readily.
   d. withstands freezing and thawing better.

9. The purpose of a slump test for concrete is to
   a. check the amount of stone in the mix.
   b. determine if the correct amount of water is in the mix.
   c. determine if the correct amount of portland cement is in the mix.
   d. measure the fineness of the sand in the mix.

10. Plastic is laid beneath concrete that is on bare ground to
    a. create a smoother finish.
    b. speed up the process of placing the concrete.
    c. stop the ground from soaking up water from concrete.
    d. prevent the concrete from settling unevenly.

11. Floats are used to
    a. make the surface smooth.
    b. remove the water from the surface.
    c. bring the cement paste to the top.
    d. cause any objects such as chips of wood or nails to surface.

12. Leveling the concrete even with the top of the forms by dragging a wood board across it is called
    a. screeding.
    b. floating.
    c. jointing.
    d. edging.

13. Curing means
    a. making the mix richer.
    b. adding color to the concrete.
    c. letting the concrete get hard.
    d. adding additives to the mix.

14. The term *crazing* in concrete means
    a. the surface turns to dust.
    b. when the surface settles.
    c. hairline cracks appearing on the surface.
    d. a defect due to expansion.

15. It is not generally recommended to place concrete to a depth of less than
    a. 4 inches.
    b. 6 inches.
    c. 9 inches.
    d. 12 inches.

MATH CHECKPOINT

1. How many cubic yards of concrete will it take to place a floor that is 4″ deep by 20′ wide and 40′ long?

2. How many cubic yards of concrete will it take to place a garage floor that is 4″ deep by 20′ wide by 24′ long?

3. If you were mixing two cubic yards of concrete and it takes six bags to the cubic yard, what would the total cost of the portland cement be if each bag costs $6.00?

4. A cement finisher charges 25 cents per square foot of concrete finished. How much would be charged to finish a concrete slab that is 20′ wide by 60′ long?

5. If one cubic yard of concrete weighs 4000 lbs, how many tons would 19 cubic yards weigh?

SUMMARY, SECTION 10

- Concrete flat work is defined as any type of work that is flat or on a horizontal plane.
- Concrete is very versatile because it can be placed or poured in forms or molds to take almost any shape.
- The four ingredients of concrete are portland cement, fine aggregates, coarse aggregates, and water.
- Any additive added to concrete is called an admixture.
- The hardening of concrete due to the reaction between portland cement and water is known as hydration.
- Concrete gets harder over a period of years if kept moist.
- One standard bag of portland cement holds one cubic foot.
- Aggregates make up 60% to 80% of the volume of concrete.
- Coarse aggregates are larger than ¼″ and do not exceed 1½″ in diameter.
- Generally speaking, water for drinking purposes is acceptable for concrete mixing.
- The relationship between the amount of water and portland cement to make concrete is known as the water-cement ratio.
- The most important feature of air-entraining cement in concrete is that it allows the concrete to expand and contract from freezing and thawing.
- Two main types of tests are used to determine the proper amount of cement and water, and strength requirements. These are the slump and cylinder tests.
- Framing lumber and plywood are the two most commonly used building materials for forming concrete.
- Expansion strips are inserted into concrete walks or flat work to allow expansion.
- Concrete should not be placed on fill earth without being compacted.
- Concrete forms must be braced firmly to prevent any movement.
- Flat work such as walks and floors should have a vapor barrier installed beneath them and reinforcement wire in the mix.
- Ready-mix concrete should be placed as soon as possible after arriving on the job and never more than 1½ hours later.
- Good-quality brand tools should be used for concrete work.
- A filled, smoother edge can be obtained on concrete by working the trowel up and down along the edges when placing in the forms.

- Screeding is the process of leveling the concrete to the top of the forms and causing the cement paste to rise to the surface by dragging a board across the surface.
- Floating is done after the screeding to rework the cement paste to the surface of the concrete.
- The edger is used to round off the top edges of the concrete.
- Grooving is done to form a neat appearance where expansion joints are located.
- Steel troweling is done last to create a smooth finish.
- Broom finishing causes the concrete to have a slightly rough finish to prevent it from being slippery.
- The use of exposed aggregate finish is becoming very popular for a rustic appearance in concrete.
- Standard curing time for testing concrete is approximately 28 days.
- The keys to curing concrete properly are proper temperature and moisture levels.
- Some of the major problem areas in concrete are crazing, scaling, and dusting. Poor workmanship and incorrect finishing techniques along with freezing are the causes most of the time.

SUMMARY ACHIEVEMENT REVIEW, SECTION 10

Complete each of the following statements referring to materials found in Section 10.

1. The main material in concrete that causes it to get hard is _____.
2. Sand and stones in concrete are known as _____.
3. Any other materials that are added to concrete to achieve special results are called _____.
4. When portland cement packs in the bag from storage it is called _____.
5. Generally, of all the types of portland cement used for concrete, Type _____ is used the most.
6. A good concrete sand should contain particles no larger than _____.
7. Maximum size of coarse aggregates is _____.
8. Concrete can be caused to compact or fill in around reinforcement steel more efficiently by using a _____.
9. Concrete that has any amount of clay or vegetable matter in it will prevent the cement paste from _____.
10. The purpose of water in concrete is to cause _____.
11. The relationship between the amount of water and portland cement used in concrete is known as _____.
12. Adding accelerators to concrete causes it to _____.
13. Additives in concrete that help control high internal temperatures in the mix are called _____.
14. The most important advantage of air-entrained concrete is _____.
15. The slump test is performed by using a steel _____.

16. The usual assigned curing time to perform a cylinder test for concrete is _____.

17. The first step in placing concrete is to _____.

18. Divider strips are placed in concrete walks to allow expansion every _____ apart.

19. It helps to prevent concrete from sticking to the edge of form boards if _____ is applied.

20. Generally, concrete for walks and floors has a _____ psi compressive strength rating.

21. Concrete is finished smooth by using a steel _____.

22. The edges of a slab are made round by using a tool known as a _____.

23. A concrete finish that has the stone showing on the surface for appearance is called _____.

24. Dragging a wood $2'' \times 4''$ board across the surface of concrete to level it up and bring the cement paste to the top is called _____.

25. Larger floors or slabs can be floated with a tool that has metal-threaded extension handles attached and is known as a _____.

26. Using a tool to round off the edges of concrete at the points where the expansion joints are located is called _____.

27. The process of controlling temperatures and moisture to season the concrete is called _____.

28. The most favorable range of temperature to season concrete is from _____ ° F to _____ ° F.

29. The appearance of hairline cracks on the surface of concrete looking like crushed eggshell patterns is called _____.

30. When concrete has frozen or suffered frost damage the top may flake away. This is known as _____.

SECTION ELEVEN
UNDERSTANDING AND READING CONSTRUCTION DRAWINGS

UNIT 41
Line and Symbol Identification

―――――――――――― OBJECTIVES ――――――――――――

After studying this unit, the student will be able to

■ explain the use of lines in a set of plans.

■ recognize the most commonly used symbols and explain their meaning.

■ define certain abbreviations found in plans.

■ explain what comprises a schedule and its importance to the mason.

Working drawings or plans consist of a group of drawings whose sole purpose is to inform various craftspersons what their specific jobs are and how they are to be completed. All phases of construction require specific information from plans. Since plans can be lengthy and complicated, architects developed a language of their own involving various types of symbols, conventions, and abbreviations. (A *convention* is a pictorial representation of something that cannot be identified on a plan by a symbol or abbreviation.) By studying and learning these, the mason can read and interpret the architect's drawings and specifications and proceed with the work as indicated.

It is extremely important that the workers understand exactly what the architect wants accomplished on the structure. It is equally important that the architect not omit details of construction. Constant communication between the architect's office and the jobsite is very time consuming and can be avoided if the plans are written and interpreted properly. Symbols and abbreviations help make plans concise and readable.

THE USE OF LINES IN PLANS

Lines as used in building plans comprise a major trade language. The thickness of the line and the way it is drawn have specific meanings, Figure 41-1.

Main Object Line

The *main object line* is a heavy continuous line that shows visible outlines or edges of what a person sees when looking at the structure.

Dimension and Extension Line

A *dimension line* shows the actual distance between two points. The line is lighter in weight than the main object line and has an arrowhead, slash, or dot at each end. The arrowhead, slash, or dot, which touches the extension lines, represents the edges of the main object or structure. The numbers in the break of the dimension line give the exact distance between the two points.

Dimension lines are used in conjunction with extension lines. An *extension line* is of the same weight (thickness) as the dimension line and extends from the edge of the main object.

Cutting Plane Line for Section Indicator

Cutting plane lines are solid lines indicating that a wall, building, or object has been cut at some point along the line. The cross sections of areas indicated by cutting plane lines are shown in greater detail elsewhere in the plans. The arrows indicate the direction in which you are looking as you view the cross section. The particular cross sections may be indicated by letters such as A, B, or C.

When the cross section to which the cutting plane lines refer is not shown on the same page of the drawing, symbols such as F/24 or G/24 are used. These symbols inform the reader that the particular cross section will be found in the plans in Section F on page 24 or in Section G on page 24.

Broken Line

The *broken line* indicates that parts of the structure or object are omitted from the plan or that the full length of the object is not drawn on the plan. This is done because an item on a plan that is shown in full by a long, constant line would require a great amount of space. One broken line allows the item to be shown in shorter form, and vital information is not omitted.

Invisible Line

The *invisible line,* consisting of a series of short dashes, indicates edges of objects that are hidden from view or that are under another part of the building. When the line continues through the actual structure, the worker can be sure that there are parts of the structure below the surface. In this case, the worker must locate another view in the drawings to determine at what level the hidden item occurs. Often, these hidden parts are explained in an elevation view or a cross section view of that area to which the hidden line refers.

For example, consider a first floor plan. Assume that the mason is viewing the plan to examine the layout of the fireplace. The foundation of the fireplace is not shown on the first floor plan, but on the basement plan. If the base of the foundation of the fireplace is larger than the fireplace itself, the size is shown by an invisible line. Steel beams are also indicated by invisible lines to help the carpenter locate the beams after the subfloor is laid.

Centerline

A *centerline* (often called a *dot and dash line*) is composed of alternating long and short dashes. This line is relatively light in weight and is used to indicate or locate center points. It is a common practice in the building trades to locate an object by measuring from the center of the object to the center of another object.

The mason uses this system of measurement when laying out masonry walls. For example, 4″ concrete

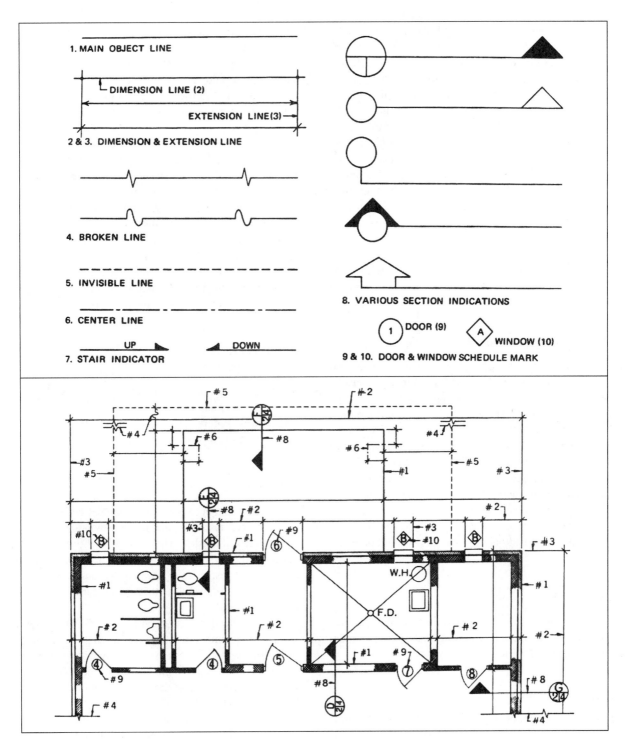

Fig. 41-1 Typical working drawing floor plan (bottom) with various lines used (top).

block does not really measure a full 4″, but is actually 3⅝″ in width. If, when laying out a long space, the mason assumes that all walls are 4″ in width and measures the space between walls from the outside of one wall to the outside of the next wall, the total measurement will be inaccurate. Depending on the length of the space, the difference in measurement can be great. If, however, all measurements are taken from the center of one wall to the center of the next wall, any difference in measurement is undetectable. Therefore, it is recommended that masonry units be measured from center to center whenever possible.

Stair Indicator

The *stair indicator* is a short, unbroken line with an arrowhead at one end indicating whether the stairs are ascending or descending.

Door and Window Indicators

Many times, doors and windows of various types and sizes are used on the same job. Therefore, they must be identified correctly before they are installed.

There are different ways to designate doors and windows. One of the most common methods is shown in Figure 41-1. The door type is indicated by a number inside a circle; the window is shown by a number inside a diamond. To determine the sizes and descriptions of the windows and doors as shown on the plans, the window or door schedule must be studied. As a rule, the schedule appears on a separate sheet of plans and can be found by consulting the index or table of contents.

SYMBOLS

Without the use of symbols, architects cannot show all necessary information regarding materials, methods, and location of components.

Types of Symbols

The types of symbols used include those used in elevation views and those in sectional views. Elevation symbols are easily recognized, as they look very much like the actual material or object, Figure 41-2. An *elevation view* is a vertical picture of an object showing the front, side, or rear view of an object, room, or structure as one would view it while facing it.

The materials shown in an elevation view appear differently in a sectional view. A *sectional view* shows the object as if it were sliced vertically, showing of what the object would be composed, Figure 41-3. For example, a sectional view of a masonry wall would show the thickness of the joints and the units, how the wall ties are installed, and, many times, the exact height of the wall.

The mason should be familiar with some of the more common symbols for the mechanical trades, as they may affect the work when building in or around certain equipment. For example, it may be necessary for the mason to provide for electrical switch boxes, heating units, or built-in plumbing fixtures. Figure 41-4 shows typical electrical and plumbing symbols.

ABBREVIATIONS

As masons view a set of plans, they should notice the use of abbreviations. Since they are used so often, the more important ones affecting the mason should be memorized. Figure 41-5 shows abbreviations commonly used in construction work.

SCHEDULES

A *schedule,* in a set of plans, is a list added separately to the plans that describes such items as windows, doors, floors, and wall finishes. The mason must be able to recognize and interpret schedules correctly, or great expense could result from necessary changes made at the conclusion of the job so that the job conforms to the finished schedule.

As mentioned before, windows and doors are designated by a number or letter on plans. The same letter or number is duplicated in the schedule, with a brief description of the item, Figure 41-6. Typical door and window schedules must show their relationship to the floor plan, Figure 41-7.

It should be remembered that schedules consist of brief descriptions; workers must also consult the drawings and specifications to obtain all the necessary infor-

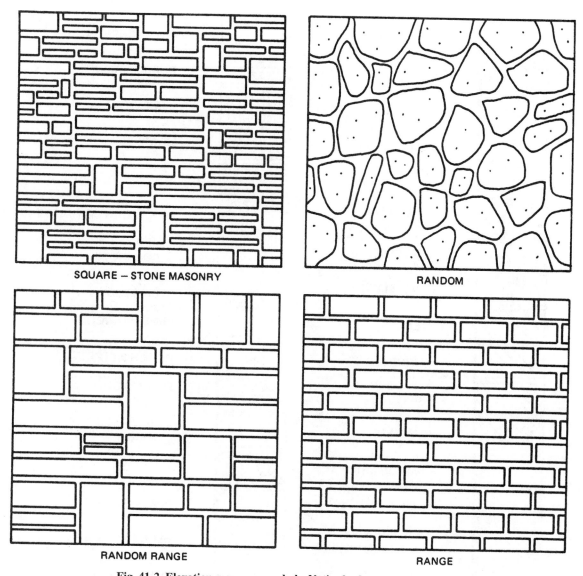

SQUARE – STONE MASONRY

RANDOM

RANDOM RANGE

RANGE

Fig. 41-2 Elevation masonry symbols. Notice both materials used and construction details.

mation. Schedules help greatly in estimating the cost of a job. For example, by studying the floor plan and schedule in Figure 41-7, it can readily be seen that 3 Type 2 windows are required. It can also be determined that the living room has oak floors, plaster walls, a plaster ceiling, and a wooden base at the floor line.

The schedule is very important to the mason when working on commercial buildings. For example, it would be very easy to build a masonry wall and strike the joints, when plaster may have been specified on the plans. This type of misunderstanding may be avoided by carefully studying the schedule.

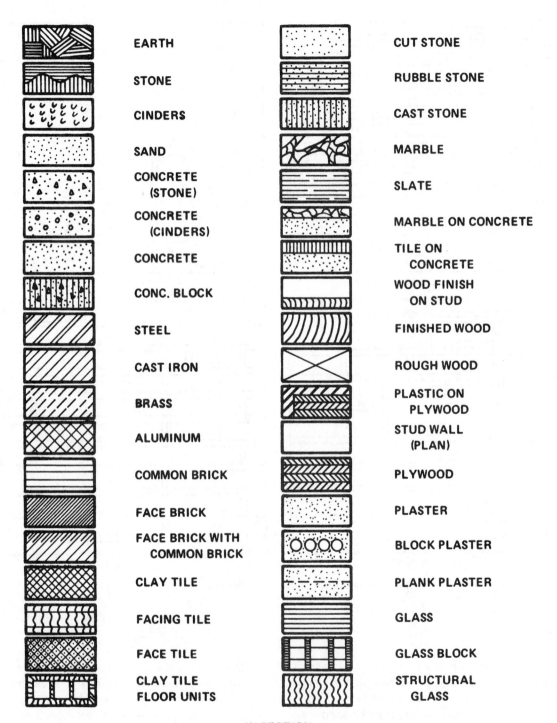

IN SECTION

Fig. 41-3 Architectural symbols showing materials in the sectional view.

	RECESSED FLUORESCENT FIXTURE
	STEM MOUNTED FLUORESCENT FIXTURE
	PANELBOARD
	INCANDESCENT FIXTURE
	WALL MOUNTED INCANDESCENT FIXTURE
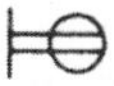	DUPLEX CONVENIENCE OUTLET
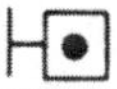	PUSH BUTTON
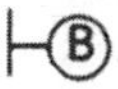	BELL
	FLOOR OUTLET
	CLOCK OUTLET
	MOTOR -- FAN
	TELEPHONE OUTLET
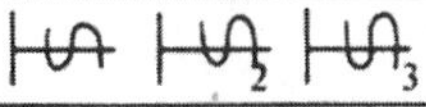	ONE POLE SWITCH / TWO POLE SWITCH / THREE WAY SWITCH
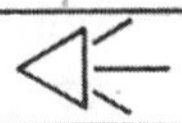	EXTERIOR LIGHT FIXTURE
PLUMBING SYMBOLS	
	BATHTUB
	SHOWER STALL
	COMMODES
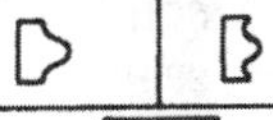	URINALS
	LAVATORY
	SINK
	SINK WITH DRAINBOARD
 F.D.	FLOOR DRAIN
 WH	WATER HEATER
 HB	HOSE BIBB

Fig. 41-4 Typical electrical and plumbing symbols.

access door	AD	exterior	EXT
access panel	AP	feet	FT
acoustical tile	AT	finished floor	FIN. FL
aggregate	AGGR	firebrick	FBRK
anchor bolt	AB	fireplace	FP
angle	L	flashing	FL
approximate	APPROX	floor	FL
barrel	bbl or BBL	footing	FTG
basement	BSMT	foundation	FND
beam or bench mark	BM	glass block	GL BL
blocking	BLKG	grade	GR
blueprint	BP	height	HB
board	BD	hot water	HW
brick	BRK	I beam	I
building	BLDG	inch	in or "
building line	BL	insulation	INS
cast iron	CI	level	LEV
cast stone	CS	lineal feet	lin ft
catch basin	CB	louver	LV
caulking	CLKG	manhole	MH
ceiling height	CLG HT	marble	MR
cement mortar	CEM MORT	masonry opening	MO
centerline	C or CL	modular	MOD
center to center	C to C	not to scale	NT
channel	CHAN	on center	OC
cinder block	CB	opening	OPG
cleanout door	COD	partition	PTN
concrete	CONC	plaster	PLAS
concrete masonry unit	CMU	plate glass	PL GL
column	COL	precast	PRCST
common	COM	reinforced	REINF
contractor	CONTR	rough opening	RGH OPNG
courses	C	schedule	SCH
cross section	X-SECT	specifications	SPEC
dampproofing	DP	wall vent	WV
detail	DET	weatherproofing	WP
elevation	el or EL	weephole	WH
excavate	EXC	window	WDW
expansion joint	EXP JT		

Fig. 41-5 Architectural abbreviations.

WINDOW	SCHEDULE		MATERIAL	LEGEND
NO.	SIZE	TYPE		
①	3'-9" x 2'-9"	STEEL SECURITY SASH		CONCRETE
②	1'-9" x 2'-9"	STEEL SECURITY SASH		BRICK
③	2-2'-8" x 4'-6"	WOOD DOUBLE HUNG		CINDER BLOCK
				WOOD STUD PART.

DOOR	SCHEDULE		LINTEL	SCHEDULE	
NO.	SIZE	TYPE	NO.	TYPE	
Ⓐ	3'-0" x 6'-8"	KALAMEIN - SELF CLOSING WITH FIRE CODE LABEL	L-1	4 - 3/4" φ RODS	
Ⓑ	3'-0" x 7'-0" x 1 3/4"	WOOD - PL. GLASS PANEL	L-2	3 Ls - 4" x 3 1/2" x 5/16"	
Ⓒ	3'-0" x 7'-0" x 1 3/4"	WOOD - OBSCURE WIRE GL. PNL	L-3	1 L - 4" x 4" x 5/16" / 2 Ls - 4" x 3" x 5/16"	
Ⓓ	3'-0" x 6'-8" x 1 3/4"	WOOD - OBSCURE GL. PNL.	L-4	3 Ls - 3" x 3" x 1/4"	
Ⓔ	2'-8" x 6'-8" x 1 3/8"	WOOD			
Ⓕ	2'-4" x 6'-8" x 1 3/8"	WOOD			
Ⓖ	2'-4" x 6'-8" x 1 3/8"	WOOD - LOUVERED LOWER PNL			

Fig. 41-6 Schedule for a small job, including information on windows, doors, and lintels.

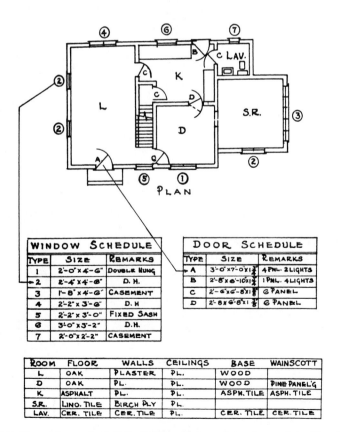

Fig. 41-7 Floor plan and related schedule. Note: arrows show where the schedule items should be applied to the floor plan.

ACHIEVEMENT REVIEW

The column on the left contains lines, symbols, and abbreviations that appear in building plans and schedules. The column on the right (Figures 41-8 and 41-9) contains the actual item. Select the correct item from the right-hand list and match it with the proper term on the left.

Part A. Line Identification

1. Invisible edge line
2. Stair indicator
3. Door and window (schedule mark)
4. Dimension line
5. Centerline
6. Section indicators
7. Broken line
8. Main object line

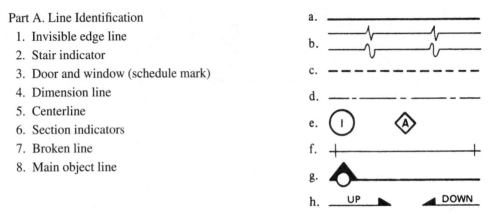

Fig. 41-8 Illustrations for achievement review, part A.

Part B. Symbols (Sectional)

1. Plaster
2. Concrete
3. Rough wood
4. Marble
5. Face brick
6. Earth
7. Steel
8. Cast stone
9. Finished wood

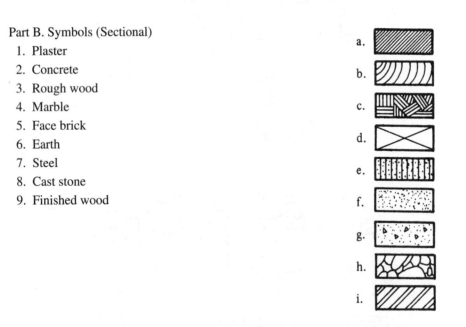

Fig. 41-9 Illustrations for achievement review, part B.

Part C. Abbreviations

1. Basement
2. Concrete masonry unit
3. Center to center
4. Cinder block
5. Plaster
6. Specifications
7. Modular
8. Brick
9. Contractor
10. Grade
11. Weep hole
12. Glass block
13. Aggregate
14. Barrel
15. Column
16. Bench mark
17. Masonry opening
18. Building
19. Louver
20. Channel

a. BBL
b. MO
c. SPEC
d. CHAN
e. GL BL
f. AGGR
g. COL
h. CMU
i. CB
j. BM
k. PLAS
l. BSMT
m. C TO C
n. BRK
o. MOD
p. GR
q. WH
r. BLDG
s. CONTR
t. LV

UNIT 42
The Working Drawing

OBJECTIVES

After studying this unit, the student will be able to

- describe the major sections of a set of working drawings.
- explain the relationships between various parts of a drawing.

Every construction job, regardless of the size and design, must have a set of working drawings. *Working drawings* are completely dimensioned views accompanied by all necessary notes. Working drawings are supplied to all of the trades involved in that particular job.

A complete set of drawings for a structure usually includes three major sections: architectural, structural, and mechanical and electrical. If the project is large or complicated, there may be a set for each major section to reduce the time necessary for the various tradespersons to locate information.

Within the three major sections are subdivisions, usually assembled in the following order: plot plan, foundation plan and floor plan, elevation drawing, sectional drawing, and detailed drawing.

When studying plans, the mason must be especially careful to read all notes included in the drawings, since they supply a great amount of information supplementing the drawing. It is important to consult the specifications for any information not shown on the plans.

SPECIFICATIONS

Specifications are defined as a written or printed description for contract purposes for a structure or equipment, and the materials and workmanship standards required on a particular job by the architect. The specifications should agree with the working drawings and contain standard provisions and any special provisions, as may be deemed necessary by the architect, pertaining to the quantities and qualities of materials and the scope of work by the various trades working on the job. They

form a part of the contract between the builder and the architect. Since they are considered to be a legal and binding part of the contract, in the event of a dispute, they can be used in court of law as evidence!

When conflicts arise on the job concerning a difference between the working drawing and the specifications, the job supervisor or foreman and the architect or his representative should meet and settle the issue. In most cases, the specifications will overrule or take precedence over the working drawing, unless there is an obvious mistake that can be corrected on the job through a mutual agreement between the two parties. In any case, the mason's employer should always be a part of any agreement and a letter or change order be instituted and signed by both parties, to avoid a misunderstanding at the completion of the job. If there is going to be additional cost to the mason contractor, this should be determined and agreed to before proceeding. Both the builder's superintendent and the masonry foreman should initial on the working drawing their names and the date the agreement was made, so there will be no argument later. In addition, any changes or inconsistencies on the drawing should be indicated in red pen and initialed by all parties. These are referred to as "as-builts." *Never orally agree to a change that is going to cost the masonry contractor more money, as you may be left holding the bag at the end of the job!* The mason or the foreman who is responsible for the job should always study the plans and read the specifications before starting any work on a new job. That is the time to have any questions resolved, not later when it may involve tearing down and rebuilding masonry work. It is important to remember that basically the main difference between the specifications and the working drawings is that the working drawings show the location, height, length, and thickness of masonry walls, along with the sizes of openings and overall illustrations, etc., as the building will look when completed. The specifications state the other information needed to construct the structure, such as kinds of brick or masonry units, methods of bonding, type of mortar to use, etc. The specifications and the working drawings go hand in hand and both are necessary to complete or renovate any structure.

PLOT PLAN

The architect usually draws a plot plan for a building of any size or importance. The *plot plan* (also known as a site plan), Figure 42-1, supplies information on the following:

- Property line
- Overall building lines for the proposed structure
- Contour lines showing the rise and fall of the ground

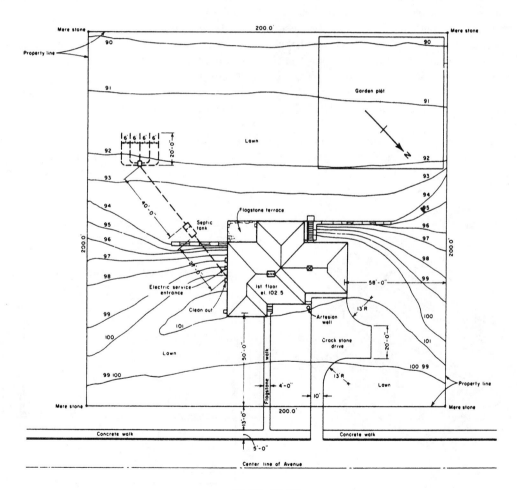

PLOT PLAN
Scale $\frac{3}{4}$ = 20'

Fig. 42-1 Typical plot plan.

- Indications of the location of trees, shrubs, sidewalks, driveways, septic systems, and well
- Utilities—electric, water, etc.

Also included on the plot plan is a reference point for determining elevations, commonly called a *bench mark,* and the *grade line* or *grade level,* which shows the rough and finished levels of the ground surrounding the structure. The elevation is calculated from coastal sea level points established by the National Oceanographic and Atmospheric Administration. Information on the grade line is very important to the masons on the job because they must be able to determine at which level to stop laying concrete block for the foundation and where to begin laying the exposed masonry such as brick or stone. The location of the land on which the job is to be built is determined by the surveyor from city or county land records, deeds, or known reference points.

A complete plot plan is necessary before local planning and zoning authorities will issue a building permit. Many banks and mortgage companies require that a survey and plot plan be submitted before loans on the property can be made.

The plot plan usually indicates the location of the building on the plans in reference to the points of a compass. This greatly simplifies locating a particular view on a plan. It is a good practice to spread the plot plan out on the worktable in the same position as the building is to actually be built on the lot to eliminate possible errors.

FLOOR AND FOUNDATION PLAN

Floor plans show the measurement and location of such items as walls, windows, doors, chimneys, and electrical devices, Figure 42-2.

Most sets of plans also contain plan views for the foundation and second floor if they exist, Figure 42-3. The information found on the foundation plan is of extreme importance since all of the upper parts of the structure depend upon accurate layout and construction of the footing and foundation. All measurements and dimensions should be checked at least twice with those on the floor plans. In cases where an overall measurement is shown on the drawing of a wall, with many smaller measurements of windows and doors shown on

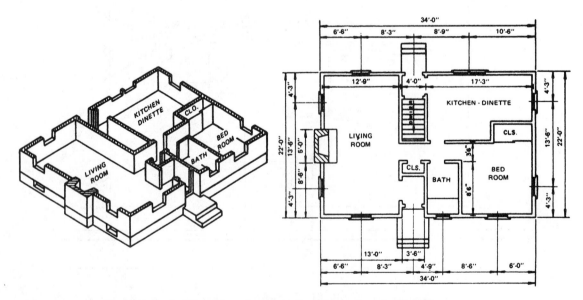

Fig. 42-2 Typical floor plan. To obtain the proper view of a floor plan, assume that the top half of the structure has been removed and that the viewer is examining the exposed section from above.

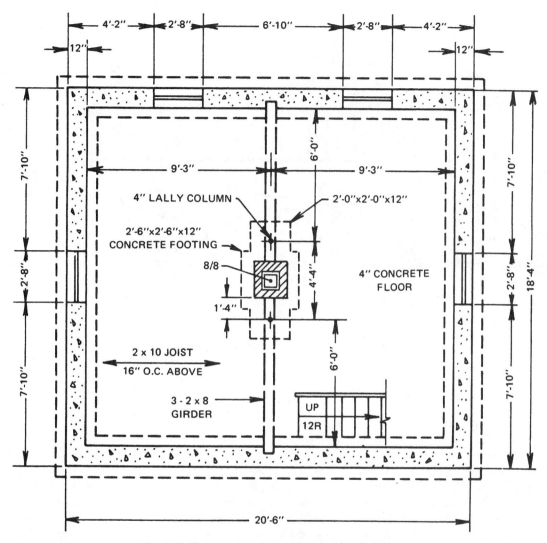

Fig. 42-3 Concrete foundation plan for a typical home.

the same wall, the mason should add all of the smaller measurements together to be sure that the total is the same as the overall measurement. Simple mathematical errors should be caught before they become problems later. This process is known as *cross-checking*.

ELEVATION DRAWING

An *elevation* is the view of the structure that the worker has from a normal standing position, Figure 42-4. Elevation views indicate certain exterior features with

which the worker must be familiar, such as the floor heights, finish grades, foundation and footings, and exterior and interior walls.

A typical house elevation shows a right elevation, left elevation, front elevation, and rear elevation. On many drawings, they are indicated as north, east, south, and west elevations. All of the elevation views of the building must be studied before it is possible to visualize the structure as it will look when completed.

Floor plans and elevations are usually drawn to the same scale. Windows and doors on the plan view are

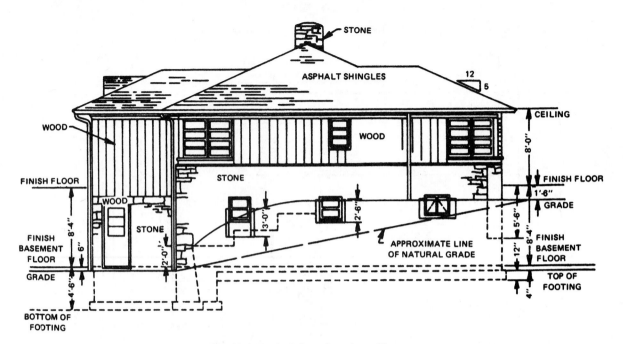

Fig. 42-4 Typical elevation view of house.

the same size and the same distance from the corners of the building on both views. By checking and comparing the plan view with the elevation view, the mason should be able to visualize approximately how the building will appear when it is constructed.

It is important to remember that the major difference between a plan view and an elevation view is that the plan view is always shown in a horizontal plane and the elevation view is shown from a vertical position. The information concerning measurements should be the same for each. Another important difference between the plan view and the elevation view concerns windows and doors. Measurements of windows and doors are given on the plan view but only the appearance of the window and door and the type are pictured on the elevation views.

Details not shown on the elevation view can usually be found by using the notes or references on the elevation plan which refer to another sheet on the drawings where they may be drawn to a larger scale.

SECTIONAL DRAWINGS

Features of construction, such as stairs, chimneys, wall thickness, placement of wall ties, and flashing installation cannot be shown clearly by an elevation view. *Sectional drawings* are views of specific portions of the structure. To understand the view given by a sectional drawing, compare it with a grapefruit cut in half. The exposed area of the fruit after cutting shows the inside of the fruit vertically from top to bottom. This is the same idea which the architect is trying to project to the builder through a sectional drawing.

There are three different definitions of a sectional view: a *longitudinal sectional view,* which is defined as a vertical cut through the long dimension of a building or roof; a *transverse sectional view,* which is a vertical cut through the short dimension of a building or room; and a *cross sectional view,* showing the composition of the wall, floor, or roof. All sectional views show how the separate parts of the structure are to be assembled or incorporated into the total structure.

A typical sectional view of a wall shows specific information and measurements, Figure 42-5. As an elevation view shows the outside of a structure, the sectional view allows the viewer to look inside the wall and see items that cannot be seen on other views of the plans.

Sectional plans are also used in the construction of fireplaces, Figure 42-6. Information such as the height of the damper, location of the ash dump, height of the mantel, and materials used in the hearth are shown in detail.

DETAILED DRAWING

As the name implies, a *detailed drawing* is an enlargement of a drawing on a smaller scale such as a sectional or elevation view. The detailed drawing, Figure 42-7, is used when the working drawing cannot show the desired information clearly or in enough detail.

Some of the many items generally shown in detail are front entrances, specific wall sections, complicated bond patterns in brick or stone, millwork, fireplace sections, and window and door installation details such as lock arrangements or sash mechanisms.

Since these details of construction are larger than the normal drawing, the mason must check the scale under each detailed drawing, as they may change on the same page of the plans. Details may vary from a scale of $1\frac{1}{2}'' = 1'0''$ up to actual-size drawings.

Shop Drawing

In recent years, many manufacturers have developed building products that require specific instructions for their installation. Since installation instructions for specific brands of the same item may vary, architects sometimes require that a shop drawing be supplied by the company. A *shop drawing* is a drawing provided by a manufacturer to explain the installation of a product. Many times, this is the case with doors and windows, Figure 42-8. After the architect approves the shop drawing, a copy is sent to the contractor on the job. Another is given to the particular supervisor to whom this drawing pertains, who uses it as a working drawing.

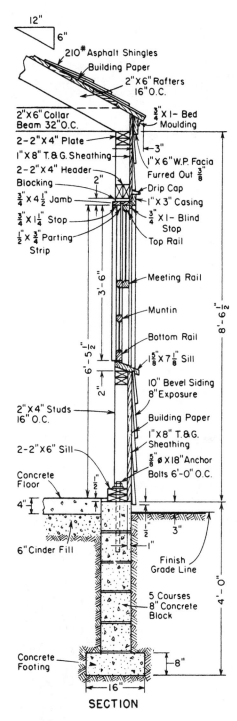

Fig. 42-5 Sectional view of wall.

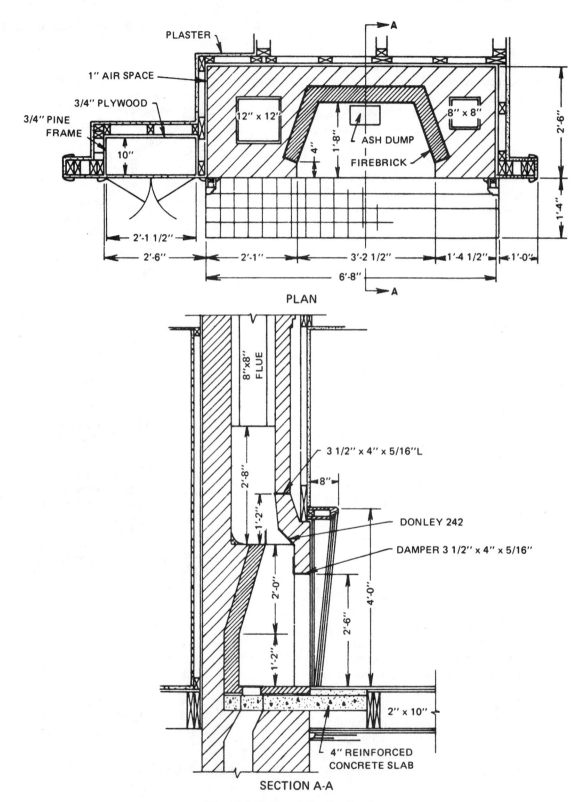

Fig. 42-6 Sectional plan for fireplace.

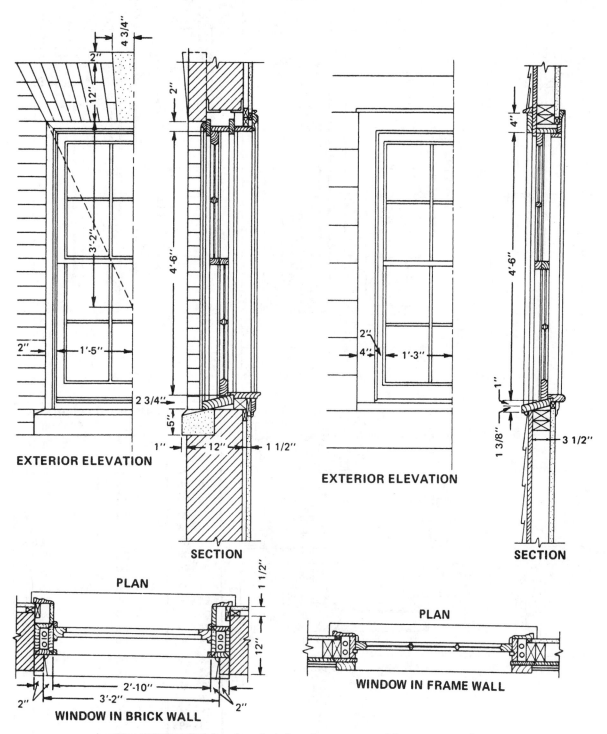

EXTERIOR ELEVATION

SECTION

EXTERIOR ELEVATION

SECTION

PLAN

WINDOW IN BRICK WALL

PLAN

WINDOW IN FRAME WALL

Fig. 42-7 Detailed drawing of windows in masonry and frame construction.

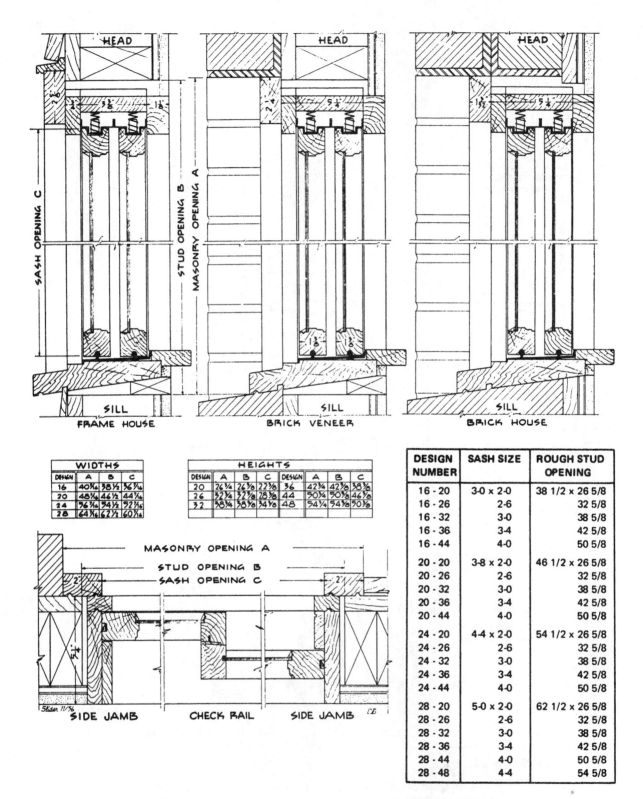

DESIGN NUMBER	SASH SIZE	ROUGH STUD OPENING
16 - 20	3-0 x 2-0	38 1/2 x 26 5/8
16 - 26	2-6	32 5/8
16 - 32	3-0	38 5/8
16 - 36	3-4	42 5/8
16 - 44	4-0	50 5/8
20 - 20	3-8 x 2-0	46 1/2 x 26 5/8
20 - 26	2-6	32 5/8
20 - 32	3-0	38 5/8
20 - 36	3-4	42 5/8
20 - 44	4-0	50 5/8
24 - 20	4-4 x 2-0	54 1/2 x 26 5/8
24 - 26	2-6	32 5/8
24 - 32	3-0	38 5/8
24 - 36	3-4	42 5/8
24 - 44	4-0	50 5/8
28 - 20	5-0 x 2-0	62 1/2 x 26 5/8
28 - 26	2-6	32 5/8
28 - 32	3-0	38 5/8
28 - 36	3-4	42 5/8
28 - 44	4-0	50 5/8
28 - 48	4-4	54 5/8

Fig. 42-8 Shop drawing of sliding window from the Woodco catalog.

REVISIONS TO THE WORKING DRAWING

Sometimes after the specifications have been written and the plans are drawn for a project, changes are made by the owner or architect for various reasons. These changes, known as *revisions,* must accompany the plans so that the contractor is informed of them before work is started. Revisions should be held to a minimum, but are a part of most plans.

As an example of a necessary revision, consider a door to the main entrance of a house. The size of the masonry opening is given on the floor plan, but the architect discovers after the floor plan is drawn that a door in the size specified is not available. To change the door size so that one may be obtained, a revision to the plans must be made. If the work has already been laid out by the mason, a *change order* is then issued by the architect.

CARE OF THE WORKING DRAWING

Proper care must be taken of the drawings if they are to last the entire job. The drawings should not be allowed to become wet and they should not be left in the sun for any length of time, as they will bleach out and be unreadable. The plans should be kept away from mortar or concrete mixtures and caution should be exercised when handling the drawings in the welding area. The plans should be collected at the close of the workday and returned to the job office or masons' shanty. Plans are expensive to duplicate and should be treated with great care.

ACHIEVEMENT REVIEW

A. Indicate in which of the following views the items below would be found: plot plan, floor and foundation plan, elevation view, sectional plan, and detailed drawings. If an item is shown in more than one plan, select the plan that gives the best view.

Sample: Chimney base _____ foundation plan _____

1. Masonry opening for windows _____
2. View of chimney above the roof _____
3. Location of the driveway _____
4. Layout of the partition walls on the first floor _____
5. Slope or pitch of the roof _____
6. Layout of the smoke chamber in fireplace _____
7. Pier to support first floor beam _____
8. View of materials that make up the roof _____
9. Height of fireplace mantel _____
10. Outside walls as they would appear when completed _____

B. Answer each of the following questions.

1. What does the term *bench mark* mean?
2. What general information is shown on a plot plan?
3. What type of information is shown on a typical floor plan for a house?
4. How many elevation drawings are shown on a typical house plan?
5. What is meant by the term *shop drawings?*
6. What does the term *revision* mean in a set of plans?

7. What is a longitudinal section view?
8. What is a transverse sectional view?
9. What happens to a set of plans if they are left out in bright sunlight for a long period of time?
10. Which of the working drawings discussed could also be called a site plan?

UNIT 43
Dimensions and Scales

—————————————— OBJECTIVES ——————————————

After studying this unit, the student will be able to

■ describe two methods of dimensioning when laying out masonry walls.

■ define scale in relation to working drawings.

■ identify the various scales used on a set of plans.

■ explain the use and limitations of the mason's rule when scaling a drawing to determine a measurement.

Dimensions and scales are the architect's way of stating critical measurements to the persons carrying out the plans. It is essential for the mason to understand scales and the various methods of dimensioning if a project is to be laid out and constructed correctly, Figure 43-1. A plan without accurate measurements is a useless picture of the proposed structure. When studying working drawings, notice that most dimensions (more commonly called measurements) appear on the floor plans. Dimensions indicate the size of all important parts of the structure, drawn to a specified, predetermined scale.

The mason should not assume that all figures on a drawing are correct. All individual measurements on each wall in the plan should be rechecked. This is done by adding all of the separate dimensions and checking this figure with the overall dimension. If there is any difference in the two figures, the supervisor should be consulted. The mason should never change the original drawing without first consulting the supervisor.

DIMENSIONING

Methods

There are two general methods of dimensioning commonly used. One consists of measuring the overall dimension of the structure. The other is to measure the dimensions just to the outside of the unfinished wall. An overall dimension is taken from one extreme point to the other extreme point of the object or wall, Figure 43-2.

Both methods of measuring dimensions are shown in Figure 43-3. The dimension shown is $22'-0''$ to the exterior face of the sheathing or plywood. This indicates that the sheathing is flush with the foundation wall. The dimension to the outside face of the wood studs is $21'-10\frac{1}{2}''$.

Some architects prefer to set the outside studs even with the outside face of the foundation wall. The sheathing would, in this case, overlap on the face top of the foundation wall. This method simplifies the procedure since the overall dimensions and the framing measurements are the same. The mason must take careful note of points of dimension location before laying out any work.

Dimensions as Stated on Plans

Dimensions are shown on plans in feet and inches, with a hyphen (-) between. For example, various measurements may be stated as $5'-4''$, $1'-4''$, $6'-6''$, or $10'-5\frac{1}{2}''$. Even measurements are shown with a hyphen and a 0 with inch marks ($''$) following the foot measurement, such as $4'-0''$. This method reduces the possibility of any misunderstanding concerning the measurement.

The exception to this practice is dimensions, which are recognized standards in the modular system of construction work, such as $16''$ center-to-center measurement of framing studs. This measurement could be shown simply as $16''$ C to C.

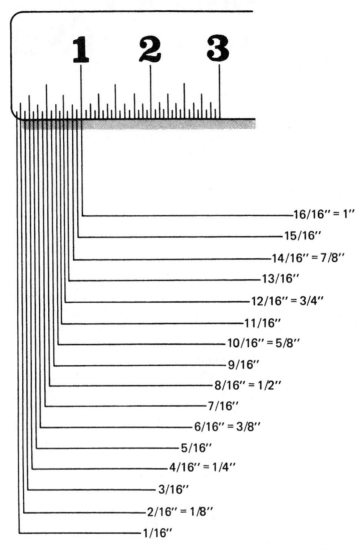

Fig. 43-1 Before scaling any drawings with the rule, review the breakdown of one inch in fractions as shown in the illustration.

Generally, the majority of dimensions needed by the mason are located on floor plans. However, sectional views of walls should be examined especially carefully.

Practical Application

Dimensioning Partition Walls. Masonry partition walls are used to divide a certain amount of space into various rooms. They may or may not be load bearing. If they are not load bearing, they are known as *partitions.*

There are two general methods of laying out masonry partition walls. The first is to measure from the outside of the exterior wall to the center of the partition wall, Figure 43-4.

The second method is to measure from the outside of the exterior masonry wall to the inside face of the partition wall, Figure 43-5.

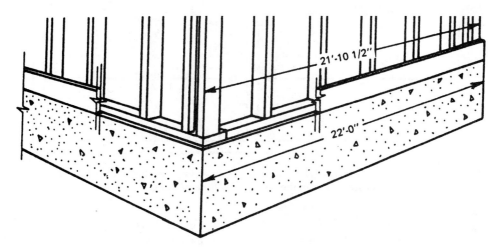

Fig. 43-2 Framed wall construction on concrete foundation. Notice that the overall measurement of the foundation is 22'-0".

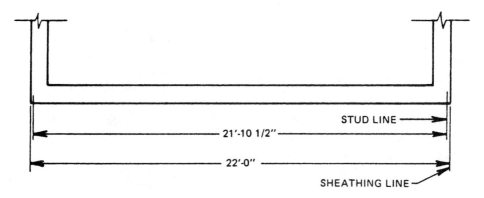

Fig. 43-3 Comparison of the dimension measured to the face of the studs and the dimension measured to the face of the sheathing. The difference between the two dimensions is the thickness of the sheathing.

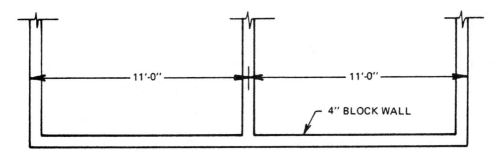

Fig. 43-4 An accurate method of dimensioning a partition wall involves measuring from the outside face of the wall to the center of the partition wall.

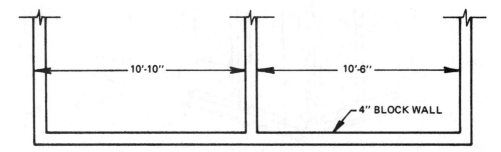

Fig. 43-5 Dimensioning a partition wall by measuring from the outside face of the main wall to the inside face of the partition wall

Dimensioning Openings in Masonry Walls. In dimensioning openings in a masonry wall, the measurements may be taken from either the center of the opening, Figure 43-6, or the side of the opening, Figure 43-7. Taking the measurement from the center of the opening is considered more accurate.

Points to Remember

When studying dimensions, the student should remember the following:

- Overall measurements are taken from one extreme end of the wall or object to the other extreme.
- Dimensions from the outside of the studding to the outside of the studding in framework must include sheathing thickness to be considered an overall measurement.

- The dimensions of masonry partition walls can be taken from the outside face of the exterior wall to the center of the partition wall, or from the outside face of the exterior wall to the inside face of the partition wall.
- All dimensions are shown with the foot measurement first, followed by a hyphen and a 0 or the remaining number of inches (for example, 4'-0" or 4'-6").
- Although it is general practice to take measurements from the outside wall to the edge of openings, it is considered more accurate to measure to the center of openings than to the edge of the jambs. Always double-check all measurements of openings before proceeding with the work.
- The overall dimensions and all dimensions in between must always be checked against each other to be sure they are the same length.

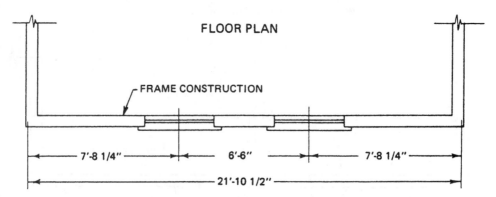

Fig. 43-6 Measuring from the outside face of wood studs to the center of window openings. This is customary for framing work.

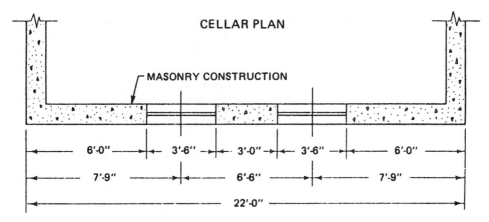

CELLAR PLAN

Fig. 43-7 Measuring from the outside face of the masonry wall to the edges of openings

- Dimensions should be shown on drawings according to their importance. The overall dimension is always given first, with the dimensions to center of openings next. The sizes of the openings are then shown. All of the smaller dimensions are shown inside the complete overall measurement.
- Dimensions serve two specific purposes; one, to indicate the location of a construction feature, and two, to indicate the size of this object or feature.

SCALES AND THE WORKING DRAWING

It stands to reason that working drawings cannot usually be shown in the actual size of the specific projects. The obvious solution is to reduce the plans, or draw the project to *scale*. On such drawings, all dimensions are reduced proportionately to reflect a certain fixed ratio between the size of the drawing and the actual finished structure. The size of the scale is determined by how complicated or important the information is for the viewer to understand.

Where to Find the Scale on a Drawing

To obtain dimensions from plans, the scale to which the plan is drawn must first be determined. The scale is usually shown on the drawings in the title plate. The scale is usually found in the lower right-hand corner or under a particular detail or sectional view. The *title plate*

gives necessary information about the drawing not given in the drawing itself or in accompanying notes. It is possible that the scale will change several times on the same sheet. When studying working drawings, be sure to check all views for possible changes in scale.

Floor plans for various floors in a set of drawings are drawn to the same scale.

Large commercial drawings usually have an index of the various sheets contained in a set of plans and the page number where they can be found. The index is on the first page following the title sheet. After locating the correct sheet, determine the scale before making any measurements.

Other Scales Used in Construction Drawings

There are 10 basic scales used in dimensioning plans and drawings besides the full scale. These scales are known as the architect's scales, Figure 43-8.

Commonly Used Scales

The most commonly used scale is $\frac{1}{4}'' = 1'\text{-}0''$. This indicates to the mason that every $\frac{1}{4}''$ on the plans represents $1'\text{-}0''$ of actual structure area. For example, in this scale, $1''$ represents $4'\text{-}0''$ on the job; $5''$ represents $20'\text{-}0''$. The $\frac{1}{4}''$ scale is usually used on housing or small commercial drawings. This is done because all of the information can be drawn on a small sheet of paper and

Designated Scale	*Designated Scale*
$3/32'' = 1'\text{-}0''$	$1/2'' = 1'\text{-}0''$
$1/8'' = 1'\text{-}0''$	$3/4'' = 1'\text{-}0''$
$3/16'' = 1'\text{-}0''$	$1'' = 1'\text{-}0''$
$1/4'' = 1'\text{-}0''$	$1\ 1/2'' = 1'\text{-}0''$
$3/8'' = 1'\text{-}0''$	$3'' = 1'\text{-}0''$

Note: At times, a full scale ($1'' = 1''$) is used.

Fig. 43-8 The architect's scales

still be seen clearly. Buildings constructed from a $1/4''$ scale are actually 48 times larger than the actual drawing when constructed.

On larger structures, the scale of $1/8'' = 1'\text{-}0''$ is most often used. Scales any smaller than $1/8''$ would be very difficult for the worker to read and the possibility of error would be greatly increased. It should not be necessary for the mason to use a magnifying glass to read the plans. Buildings constructed to the $1/8''$ scale are actually 96 times larger than the actual drawing.

The smaller the scale is, the more careful the mason must be when dimensioning. A mistake of $1/16''$ on the $1/8''$ scale would account for $6''$ on the job. A line $2\frac{1}{2}''$ long drawn to the $1/8''$ scale would actually measure $20'\text{-}0''$ on the job.

THE USE OF RULES IN DIMENSIONING DRAWINGS

The Architect's Scale

The *architect's scale* is a special instrument or ruler that the architect uses to dimension drawings which contain the above mentioned scales. A triangular-shaped architect's scale ruler is used in the architect's office or by drafting students when preparing a drawing. This instrument is usually not practical on the job because of its length. A straight scale, Figure 43-9, is made that will easily fit into the shirt pocket. This type of scale can usually be obtained from any reliable store which stocks drafting supplies.

How to Read the Architect's Scale

Notice the scale in Figure 43-10. The left and right edges of the scale are laid out so that they contain two different scales with a common denominator. The $1/8''$ scale (left side) is marked off with short lines, each representing $1'\text{-}0''$. The scale is read from left to right. The $1/4''$ scale (right side) is marked with longer lines, each representing $1'\text{-}0''$. This scale is read from right to left. Each long line represents $1'\text{-}0''$ on the $1/4''$ scale or $2'\text{-}0''$ on the $1/8''$ scale. Each short line represents $1'\text{-}0''$ on the $1/8''$ scale and $6''$ on the $1/4''$ scale.

Each scale has a short section at the end with several lines. The fine lines represent inches or fractions of inches when the measurement exceeds an even foot dimension, such as $10'\text{-}4''$. The $10'\text{-}0''$ measurement is read on the scale and the remaining $4''$ are found in the finely calibrated lines.

In the $3/4''$ scale, notice that there are 12, $1/16''$ marks in $3/4''$. Since $3/4''$ equals $1'\text{-}0''$ according to the scale, each of the marks represents $1''$ since there are 12 of them. Therefore, $1/16''$ in a $3/4''$ scale always represents $1''$ on the job.

In the $1''$ and $1/2''$ scales, every $1/8''$ represents $1''$ since there are 12, $1/8$s in $1\frac{1}{2}''$.

To determine what fraction is equal to $1''$ on the job, divide the fraction which equals $1'\text{-}0''$ on the job

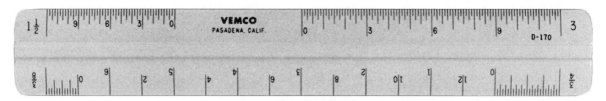

Fig. 43-9 Architect's scale rule. This rule contains the eight scales used most often in construction and fits easily into the worker's pocket. Four of the scales are shown in the illustration; the other four are on the reverse side.

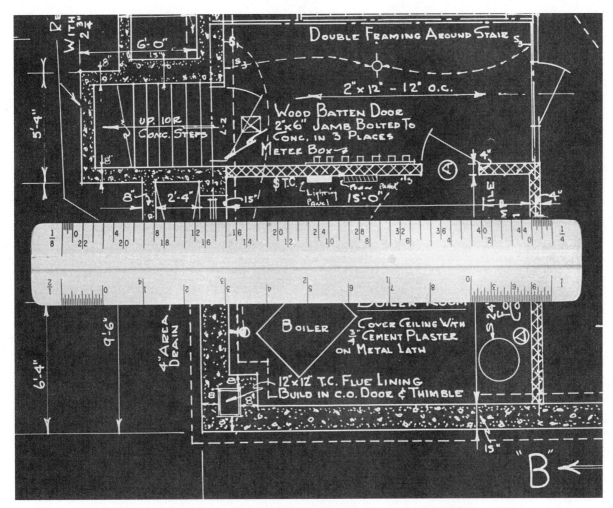

Fig. 43-10 **Determining a measurement with the architect's scale rule. Using a ¼″ scale, notice that the length of the boiler room is 15′, just as the plan indicates.**

by 12, since there are 12″ per foot. With a scale of ½″ = 1′, divide the ½″ by 12; $\frac{1}{24}$″ on this scale equals 1″ on the job.

If the scale used is $\frac{3}{8}$″ = 1′-0″, $\frac{3}{8}$″ is divided by 12 to determine the amount equal to 1″, which would be $\frac{1}{32}$″ on the scale.

The Mason's Rule

Since the mason's rule is divided into sixteenths, it cannot be used easily for all scales. Where the scale is such that $\frac{1}{16}$″ equals even inches on the job, the rule can be

used with no problem. In the scale $\frac{3}{8}$″ = 1′-0″, $\frac{1}{16}$ equals 2″ because $\frac{1}{32}$″ is equal to 1′-0″ on the job. If possible, the architect's scale rule should be used to scale an unknown dimension from a set of plans. The possibility of error is decreased with the use of this rule, as opposed to the mason's rule.

Determining a Measurement with the Mason's Folding Rule

Scaling an unknown dimension from a plan with a folding rule should be done only as a last resort. The small

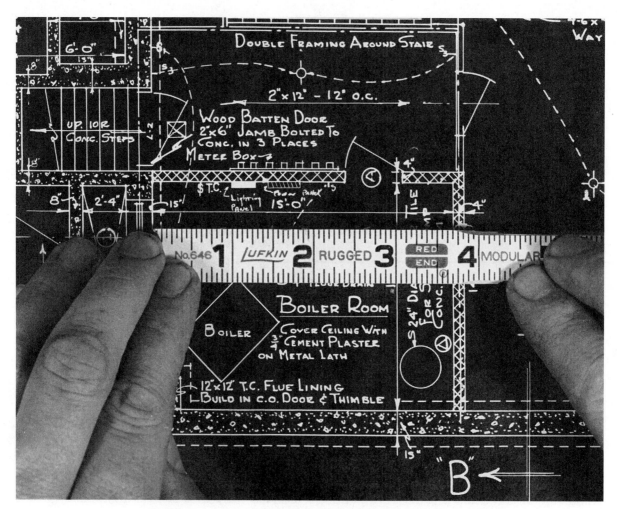

Fig. 43-11 Determining a measurement with the mason's folding rule. The plan, drawn to the ¼″ scale, shows that the length of the boiler room is 15′. Using the mason's rule, the measurement falls short of the correct measurement by $\frac{1}{16}$″. (Compare Figure 43-10, in which the measurement was very accurate.)

divisions that are shown on the architect's rule are not shown on the mason's folding rule, Figure 43-11. Also it may be difficult to remember the various scales discussed and apply them correctly when measuring with the folding rule. On the architect's scale rule, the mason merely matches the scale rule with the item on the plans.

There are times, however, when the mason's rule is very useful in working with plans and drawings. For example, in cases in which the mason must determine a missing dimension from a drawing, or sketch a particular feature to scale for clarification of an important point, the mason's rule is invaluable.

Blueprints are reproduced from original drawings by a process that may cause a small amount of shrinkage. Because of this, and for various other reasons, it is always good practice for the mason to double-check dimensions and scales. However, if an error is found, the supervisor should be consulted before any changes are made.

ACHIEVEMENT REVIEW

A. Select the best answer from the choices offered to complete each statement. List your choice by letter identification.

1. An overall dimension is taken from
   a. one extreme point on a drawing to the other extreme point.
   b. center-to-center of the opening.
   c. the outside face of the wall to the center of the opening.
   d. the total length of all outside walls.

2. The most accurate method of laying out a partition wall is from the
   a. exterior face of the wall to the inside face of the partition wall.
   b. interior face of the exterior wall to the inside face of the partition wall.
   c. outside face of the exterior wall to the center of the partition wall.
   d. center of the outside wall to the center of the partition wall.

3. A measurement of 5 feet and 4 inches is shown on a plan as
   a. $5'-0''-4''$.   c. 5-4.
   b. $5'-4''$.   d. $5''-4'$.

4. The most commonly used scale in house construction is
   a. $\frac{1}{8}'' = 1'-0''$.   c. $\frac{1}{2}'' = 1'-0''$.
   b. $\frac{1}{4}'' = 1'-0''$.   d. $\frac{3}{4}'' = 1'-0''$.

5. The most commonly used scale for large commercial buildings is
   a. $\frac{1}{4}'' = 1'-0''$.   c. $\frac{1}{2}'' = 1'-0''$.
   b. $\frac{3}{8}'' = 1'-0''$.   d. $\frac{1}{8}'' = 1'-0''$.

6. The specified scale for a drawing is usually found
   a. above the view.   c. under the view.
   b. to the side of the view.   d. in the middle of the view.

7. Not counting the full scale, the total number of scales found on an architect's scale rule is
   a. 10.   c. 6.
   b. 8.   d. 4.

B. Answer each of the following questions. Express your answer in inches.

1. In a $\frac{1}{4}''$ scale, what length of line in a plan would represent an actual measurement of $20'-0''$?

2. In a $\frac{1}{4}''$ scale, what length of line in a plan would represent an actual measurement of $40'-6''$?

3. In a $\frac{1}{8}''$ scale, what length of line in a plan would represent an actual measurement of $50'-0''$?

4. In a $\frac{1}{8}''$ scale, what length of line in a plan would represent an actual measurement of $30'-6''$?

5. In a $\frac{1}{2}''$ scale, what length of line in a plan would represent an actual measurement of $15'-0''$?

6. In a $\frac{3}{4}''$ scale, what length of line in a plan would represent an actual measurement of $10'-0''$?

C. The following lines drawn to a $\frac{1}{4}''$ scale in a drawing represent what lengths in an actual building?

1. $\frac{3}{16}''$
2. $\frac{1}{8}''$
3. $\frac{7}{8}''$
4. $4\frac{1}{2}''$
5. $10''$
6. $15\frac{1}{8}''$

D. Draw a line on a sheet of paper for the following:

1. In a $\frac{1}{8}''$ scale, draw a line equal to $12'$.
2. In a $\frac{1}{8}''$ scale, draw a line equal to $20'$.
3. In a $\frac{1}{4}''$ scale, draw a line equal to $15'$.
4. In a $\frac{1}{4}''$ scale, draw a line equal to $25'$.
5. In a $\frac{3}{4}''$ scale, draw a line equal to $8'$.
6. In a $\frac{3}{4}''$ scale, draw a line equal to $14'$.

SUMMARY, SECTION 11

- Specifications are the written or printed instructions that accompany building plans and describe information not already shown on the plans.
- To construct a building, building plans and specifications must be studied together before any work is done.
- Specification points of instruction should be listed in the order the work is to be accomplished.
- When there is a conflict between specifications and plans, the specifications take precedence.
- Points of good workmanship should be a part of all masonry specifications.
- Lines of different thicknesses and design are used on building plans to denote various meanings to the viewer.
- Symbols are used to save space on plans and to identify various materials to be used in the construction of the building.
- Symbols for the same materials are shown differently in elevations and section views.
- Schedules are lists contained in the building plans that describe such specific items as sizes of windows and doors and finishes for floors and ceilings.
- Plans and specifications should be protected on the job to ensure that they last for the duration of the job.
- Generally, a set of working drawings consists of a plot plan, foundation and floor plans, elevation view, sectional view, and detailed drawing.
- The plot plan shows the location of the structure on the lot and important features such as the sewer, driveway, and elevation of land.
- A floor plan is a horizontal view of the floor level of a structure, showing important information such as the layout of windows, doors, partitions, and stairways.
- An elevation drawing is a vertical view as one would see it while standing and facing the structure.

- A sectional drawing shows how the separate parts of the structure are to be assembled or incorporated into the total structure. It is a vertical slice through the structure showing all objects that would normally be hidden from view.
- Details on a working drawing show objects on a larger scale than they appear on other drawings. They provide a better understanding of the construction of the building.
- All of the working drawings for one project are directly related to one another and should be studied carefully to assure complete understanding of the job to be done.
- All dimensions of the length of a wall should be added together and the total figure checked against the overall dimensions to be sure they agree.
- The most accurate method of laying out masonry partitions or openings is to measure from the edge of the exterior face to the center of the opening or wall.
- Dimensions serve two basic purposes: to indicate the location of the object, and to indicate the size of the object.
- Drawing to scale involves reducing all views of a structure to a predetermined, fixed ratio that will allow the drawing to fit on a relatively small sheet of paper.
- The two most frequently used scales are the $\frac{1}{4}''$ (for home and small commercial construction) and the $\frac{1}{8}''$ (for larger commercial buildings).
- There are 10 basic scales used by architects in dimensioning plans.
- Plans should be scaled with a mason's rule only as a last resort.
- The supervisor should be consulted before any measurements on a plan are changed.

SUMMARY ACHIEVEMENT REVIEW, SECTION 11

Complete each of the following statements referring to material found in Section 11.

1. A printed or written description of work to be done that accompanies building plans is called the _____.
2. A typical set of general specifications has a standard number of sections or divisions totaling _____.
3. When a question arises on the job as to whether the specifications or plans are correct, the _____ are considered correct.
4. A heavy, continuous line used to indicate visible outlines or edges of objects is called a (an) _____.
5. A wavy line indicating that parts of a structure have been left out or that the full length of the object has not been drawn on building plans is called a (an) _____.
6. A line composed of short dashes used to show that there are parts hidden from view on a plan is called a (an) _____.
7. A list or drawing separate from the plans which generally describes windows, doors, and wall finishes is called a (an) _____.
8. The building plans that the workers read on the job are more commonly called the _____.
9. The part of the plans that shows the structure situated on the lot with service utilities and the contour of the land is called the _____.

10. The drawing giving a vertical view of the wall or object as it should look when completed is called the _____.

11. The part of the building plans that shows the layout of windows, doors, and openings with dimensions is called the _____.

12. The drawing in which a part of the building or object has been cut away, giving a vertical view and exposing the interior of the structure or object, is called the _____.

13. In building plans, drawings made to a larger scale to show important information are called _____.

14. Changes that are made after the plans have been drawn are called _____.

15. A dimension measured from one extreme point to another extreme point is called a (an) _____.

16. The most accurate method of laying out a partition wall is to measure from the exterior face of the outside wall to the _____.

17. A measurement of 5'-0" should be shown on a set of plans as _____.

18. The scale usually used for small buildings and homes is _____.

19. The most commonly used scale for commercial buildings is _____.

20. There are _____ scales shown on the architect's scale rule.

21. A set of plans should not be scaled with a mason's folding rule because _____.

SECTION TWELVE
MASONRY AS A CAREER
AND MASONRY PROJECTS

UNIT 44
Masonry as a Career

OBJECTIVES

After studying this unit, the student will be able to

- describe some of the types of work a mason does.
- list some of the benefits derived from a career in masonry work.
- discuss and explain a masonry apprenticeship program and how one enters the program.

Masonry work offers a rewarding and productive career for persons who like to work with their hands and are interested in construction work. It is a trade that rewards ambition, initiative, and hard work and can put you on the road to being independent. The future of the construction business is solid and there will always be a demand for craftspersons who excel at their chosen trade.

Masonry work is one of the oldest and most time-honored crafts, and has always attracted skilled workers who find enjoyment, contentment, and worthwhile careers. They not only advance their vocational goals, but also gain personal satisfaction in being able to create a building from a mass of materials and mortar. The financial reward is excellent, as masons earn very good wages, receive good health benefits as a rule from their employers, and can build a good retirement security by wise management of their resources.

One should always consider the masonry trade as an art, not just a job, and try to do the class of work that

will earn one the reputation of being called a "First-Class Mechanic!"

WHAT WORK DOES THE MASON DO?

The term *masonry* refers to the craft or trade of building with mortar, brick, stone, tile, etc. Today it has a broader meaning in that we also have cement masons who place and finish concrete. Even though a mason or bricklayer does at times need to pour some concrete for a footing or small job, a cement mason has to learn a separate trade and training is quite different than that for a regular bricklayer or stone mason.

The term *bricklayer* means a person who lays brick and blocks to form walls and structures. Some years ago there were block masons in some of the larger cities who only dealt with concrete blocks. Nowadays, this is seldom true as the modern mason is expected to perform both tasks because of the construction of solid walls of brick and concrete blocks and the increased use of concrete masonry units of different types. Some of the many tasks a bricklayer performs include building brick and block walls, porches and steps, fireplaces, window sills, arches, window and door openings, piers, pilasters, chases, chimneys, etc., Figure 44-1.

Masons are also required to operate masonry saws, clean and wash down masonry work with cleaners, and perform a variety of related jobs that are consistent or applicable to the craft. In addition, all masons are expected to know the composition of various types of mortars and how to mix them. Although this is normally the job of the laborer or tender, on many small jobs the luxury of providing one laborer per mason is not affordable.

PHYSICAL QUALIFICATIONS

Masonry work is strenuous, and a mason must be in good health to be productive. A mason must stand for long periods of time, be able to withstand heat and cold caused by seasonal weather conditions, be able to kneel, bend, and reach, and possess good balance. Masons also have to be able to work on scaffolds high in the air

and in tight or closed places such as closets, those areas around pipes, elevator shafts, etc. Good eyesight is necessary to perform masonry work, and what is called a "keen eye" can be developed as the trade is learned by being taught to sight walls for level and plumb.

BENEFITS OF A CAREER IN MASONRY

There are many benefits derived from being a professional mason. Some of the most important are being able to earn a living with good pay, having a trade of your own that no one can take away from you, working outside in a healthy environment, receiving self and job satisfaction in working with your hands, and being creative and independent. Your work is long-lasting and probably will be around when you are long gone. Also, masons have opportunities to build their own homes and save a lot of money. They gain a basic knowledge of real estate and often have the opportunity to progress and be in business for themselves one day with a minimum of expense. In addition there is a peace of mind in bricklaying that is not found in many office-type jobs. One rarely sees a bricklayer who has had a nervous breakdown from job-related stress.

Fig. 44-1 A journeyman mason building a brick fireplace.

APPRENTICESHIP

What is an apprentice? An apprentice is a worker who learns a craft through planned, supervised work on the job in conjunction with receiving technical instruction. They are taught the proper use, care, and safe handling of tools and equipment used in their work while on the job.

How does one get started in masonry? One way is to work part time in the summer as a helper, watch the skilled masons, and experience the working conditions. If seriously interested in learning the trade, one can get facts and information from a high-school counselor, vocational teachers in the school system, union offices, a state employment office, and masons working in the trade.

A vocational high school background in construction trades or bricklaying courses can qualify a potential apprentice for advanced standing or credit in an apprenticeship program, Figure 44-2. This is usually left to the determination of the employer but is a practice generally followed. Other courses taken in high school that would be very beneficial are drafting, architectural drawing (Figure 44-3), and math classes. Leadership training offered through youth organizations such as VICA (Vocational Industrial Clubs of America) is invaluable for the apprentice who wishes to progress in the trade. The skill contests of VICA also help to sharpen students' skills and train them to be competitive.

Entering an Apprenticeship

The best way to enter into an apprenticeship program is to sign up under the terms of a written agreement. These agreements are made with a joint apprenticeship committee for the full term of the apprenticeship and should follow United States Department of Labor guidelines. The apprenticeship committees represent both management and labor and are composed of persons with experience in the industry. Committee members select applicants in accordance with an approved, unbiased procedure. The committees also determine the need for the number of new apprentices and set the minimum standards of education, experience, and training.

Fig. 44-2 Student masons building brick fireplaces in a vocational masonry class.

Fig. 44-3 Apprentice studying plans in the classroom. This is part of the 144-hour technical program.

An apprentice can be assured of proper training through regulations adopted by either the Bureau of Apprenticeship and Training (BAT) or a state apprenticeship agency. Sponsors and apprentices both must adhere to the requirements of the program, Figures 44-4 and 44-5. Apprentices who successfully complete their training programs are issued certificates of completion and are registered with the Department of Labor.

An apprentice is usually a person at least 17 years of age and not over 24 years of age. This requirement is flexible, however, in relationship to the limit of 24 years of age, especially in the case of veterans of military service.

Responsibilities of Apprentices

The following terms are set forth by the Manpower Administration, Bureau of Apprenticeship and Training, United States Government.

1. To perform diligently and faithfully the work of the trade and other pertinent duties as assigned by the contractor in accordance with the provisions of the standard.
2. To respect the property of the contractor and abide by the working rules and regulations of the contractor and the local joint committee.
3. To attend regularly and complete satisfactorily the required hours of instruction in subjects related to the trade, as provided under the local standards.
4. To maintain such records of work experience and training received on the job and in related instruction, as may be required by local joint committee.
5. To develop safe working habits and conduct themselves in such a manner as to assure their own safety and that of other workers.
6. To work for the contractor to whom one is assigned to the completion of their apprenticeship, unless they are reassigned to another contractor or their agreement is terminated by the local joint committee.
7. To conduct themselves at all times in a creditable, ethical, and moral manner, realizing that much time, money, and effort are spent to afford them an opportunity to become skilled craftspersons.

Contractors' Responsibilities (and Joint Apprenticeship Committee)

1. To train apprentices a minimum of 144 hours per year of technical instruction in subjects related to their trade. Attendance at related instruction classes shall not be considered as hours worked when given outside of regular working hours, and the apprentice shall not be paid for attending related instructional classes.
2. Only one apprentice may be trained for every three journeymen regularly employed. Journeymen are experienced craftspersons who have successfully completed their apprenticeship and now are employed on the job as masons. In most cases, each apprentice is assigned to one or more journeyman to watch over and help him or her in the performance of his or her work while on the job.

U.S. DEPARTMENT OF LABOR • MANPOWER ADMINISTRATION
Bureau of Apprenticeship and Training

APPRENTICESHIP AGREEMENT
Between
Apprentice and Joint Apprenticeship Committee

CHECK APPROPRIATE BOX
☐ Vietnam Era Veteran
☐ Other Veteran
☐ Non-Veteran

SOCIAL SECURITY NUMBER

THIS AGREEMENT, entered into this ....day of ....19 ....

between the parties to *(Name of local apprenticeship standards)* ....

....

represented by the Joint Apprenticeship Committee, hereinafter referred to as the COMMITTEE, and

(Name of apprentice) ...., born *(Month, Day, Year)* ....,

hereinafter referred to as the APPRENTICE, and *(if a minor) (Name of parent or guardian* ....

hereinafter referred to as the GUARDIAN.

WITNESSETH THAT:

The Committee agrees to be responsible for the selection, placement and training of said apprentice in
the trade of .... as work is available, and in consideration said apprentice agrees diligently and faithfully to perform the work of said trade during the
period of apprenticeship, in accordance with the regulations of the Committee. The apprenticeship
standards referred to herein are hereby incorporated in and made a part of this agreement.

TERM OF APPRENTICESHIP	PROBATIONARY PERIOD	CREDIT FOR PREVIOUS TRADE EXPERIENCE	TERM REMAINING

OTHER CONDITIONS

This agreement may be terminated by mutual consent of the signatory parties, upon proper notification to the registration agency.

SIGNATURE OF APPRENTICE

TO BE COMPLETED BY THE APPRENTICE
Check One: ☐ MALE ☐ FEMALE

ADDRESS *(Number, Street, City, State, Zip Code)*

RACE/ETHNIC GROUP: *(Check one)*
☐ CAUCASIAN   ☐ INFORMATION NOT AVAILABLE
☐ AFRICAN-AMERICAN ☐ NOT ELSEWHERE CLASSIFIED

PARENT OR GUARDIAN *(Signature)*

☐ ASIAN AMERICAN ☐ LATINO
☐ NATIVE AMERICAN

SIGNATURE *(Joint Apprenticeship Committee, Chairman)*

HIGHEST EDUCATION LEVEL: *(Check one)*
☐ 8th GRADE OR LESS ☐ 12th GRADE OR MORE
☐ 9th GRADE OR MORE

SIGNATURE *(Joint Apprenticeship Committee, Secretary)*

NAME OF REGISTRATION AGENCY

DATE *(Month, Day, Year)* | SIGNATURE AND TITLE OF AUTHORIZED OFFICIAL

**Fig. 44-4 Apprenticeship agreement between apprentice and joint
apprenticeship committee.**

U.S. DEPARTMENT OF LABOR • MANPOWER ADMINISTRATION
Bureau of Apprenticeship and Training

APPRENTICESHIP AGREEMENT
Between Apprentice and Employer

CHECK APPROPRIATE BOX
☐ Vietnam Era Veteran
☐ Other Veteran
☐ Non-Veteran

SOCIAL SECURITY NUMBER

The employer and apprentice whose signatures appear below agree to these terms of apprenticeship:

The employer agrees to the nondiscriminatory selection and training of apprentices in accordance with the Equal Opportunity Standards stated in Section 30.3 of Title 29, Code of Federal Regulations, Part 30; and in accordance with the terms and conditions of the *(Name of Apprenticeship Standards)* .... which are made a part of this agreement.

The apprentice agrees to apply himself diligently and faithfully to learning the trade in accordance with this agreement.

TRADE	TERM OF APPRENTICESHIP *(Hours or Years)*	PROBATIONARY PERIOD
CREDIT FOR PREVIOUS EXPERIENCE	TERM REMAINING	DATE THE APPRENTICESHIP BEGINS

This agreement may be terminated by mutual consent of the parties, citing cause(s), with notification to the Registration Agency.

NAME OF APPRENTICE *(Type or Print)*	TO BE COMPLETED BY THE APPRENTICE
SIGNATURE OF APPRENTICE	DATE OF BIRTH *(Month, Day, Year)*
ADDRESS	CHECK APPROPRIATE BOX ☐ MALE ☐ FEMALE
PARENT OR GUARDIAN	RACE/ETHNIC GROUP: *(Check one)* ☐ CAUCASIAN ☐ INFORMATION NOT AVAILABLE
NAME OF EMPLOYER *(Company)*	☐ AFRICAN-AMERICAN ☐ NOT ELSEWHERE CLASSIFIED ☐ ASIAN AMERICAN ☐ LATINO ☐ NATIVE AMERICAN
ADDRESS	HIGHEST EDUCATION LEVEL: *(Check one)* ☐ 8th GRADE OR LESS ☐ 9th GRADE OR MORE ☐ 12th GRADE OR MORE
SIGNATURE OF AUTHORIZED OFFICIAL	
APPROVED BY *(Joint Apprenticeship Committee)*	
SIGNATURE OF CHAIRMAN OR SECRETARY	DATE
REGISTERED BY *(Name of Registration Agency)*	
SIGNATURE OF AUTHORIZED OFFICIAL	DATE

Fig. 44-5 Apprenticeship agreement between apprentice and employer.

3. The hours of work for apprentices are the same as the journeyman based on the standard 40-hour work week.
4. Apprentices shall be paid a progressively increasing schedule of wages based on a percentage of wages paid journeymen. As a rule the apprentice is started off at 50% of the journeymen wages with subsequent raises each six months until the end of the apprenticeship, when they will be paid full wages as journeymen.
5. The International Union of Bricklayers & Allied Craftworkers require masonry apprentices to complete in addition to the 144 hours of technical classroom yearly, a three- or four-year work period experience of 6000 hours on the job under the supervision of journeyman masons before they can be certified.
6. The employer shall at all times exercise reasonable precautions for the health and safety of all apprentices engaged in the performance of their work in compliance with all Federal, State and local codes and statutes.
7. When necessary to lay off any apprentices, it shall be in accordance with seniority, provided any apprentice laid off shall be given the opportunity of reinstatement before any new apprentices may be employed.
8. At the completion of the apprenticeship, the employer shall see that a certificate of completion is awarded to the apprentice, Figure 44-6.

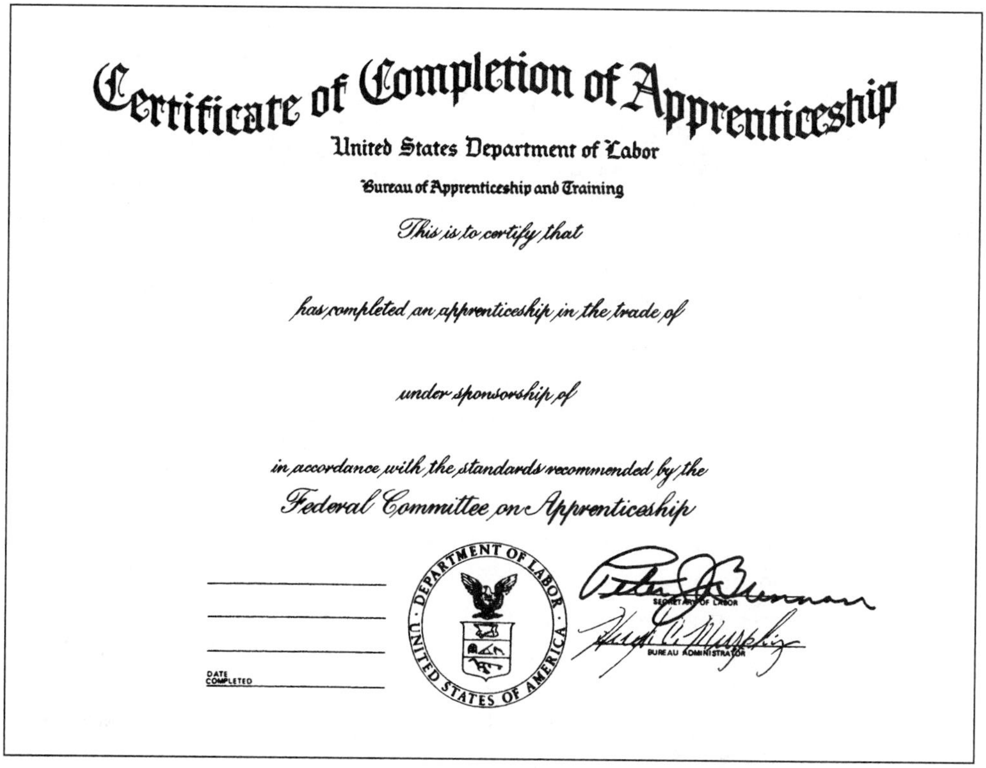
Certificate of Completion of Apprenticeship
United States Department of Labor
Bureau of Apprenticeship and Training
This is to certify that
has completed an apprenticeship in the trade of
under sponsorship of
in accordance with the standards recommended by the
Federal Committee on Apprenticeship
DATE COMPLETED
DEPARTMENT OF LABOR · UNITED STATES OF AMERICA
SECRETARY OF LABOR
BUREAU ADMINISTRATOR

Fig. 44-6 Certificate of completion of apprenticeship.

EMPLOYMENT OUTLOOK AND OPPORTUNITIES

The employment outlook for masons is excellent in today's work force, as many of the journeyman masons are reaching retirement age and are not being replaced. Fewer young people are seeking masonry as a career partly because high school guidance counselors are steering most students toward college. The trades in the United States do not command the same respect they do in foreign countries such as Germany and in most of Europe, where masonry work is considered to be more of a respected craft and in many cases an art, that has been practiced for over several thousand years. Skilled masons take artistic pride in their work and earn very good wages. They get to work outdoors in the fresh air instead of in a small office cubicle all day. They tend to be physically fit and in good shape. Once they have mastered the trade, they can be more independent than in many other types of jobs. Once you have learned a trade, no one can take it away from you. There are many opportunities for the mason who is willing to pursue them.

Many masonry organizations have realized this for some time and now are taking positive steps to increase enrollment in masonry courses. Years ago, people who wanted to learn the masonry trade usually had to come from a family of journeyman masons who took them under their wing and taught them their trade on the job-site. This is rarely the case anymore! The best way to learn the masonry trade now is to apply to enter a high school or career technology center that has a structured approach to teaching a student job entry level skills. In essence, it costs the student nothing but his or her time to take advantage of this. In many local areas, masonry contractor's associations have been formed to promote their craft and expand their business. High schools and career centers should contact these organizations and ask them to serve as advisory committees and to come into the schools and share their experiences with the students. They will be happy to do this, as one of their main purposes is to encourage the training of students as future masons. Most of these organizations will also offer employment to the masonry student who is doing well in the class, possibly in a work-study program or after graduation.

Students who are pursuing careers in engineering, architecture, and the like can benefit from taking a masonry course, as they may utilize it to move into other related construction-based occupations. It is an excellent stepping stone to success in building and owning your own business in the future. Another very valuable side benefit for the student is to become a member of a school skill and leadership organization such as the SKILLS USA-VICA Club. These groups not only promote and conduct skill contests, but also provide leadership development that is not generally taught in any other places in the school system. Many times, the skills and benefits that are learned in these types of programs are the key to a bright future in not only masonry or construction work, but also in the business world in general.

ACHIEVEMENT REVIEW

1. List five types of tasks a bricklayer would perform.
2. List three benefits that a mason would develop from his or her work.
3. List four physical qualifications a mason should possess.
4. If you desire to become a mason and sign up for a program, what would you be called?
5. List three courses in high school that would help a beginning mason.
6. What United States government department handles persons who want to enter into training for a trade?
7. What is the minimum age of a person who wants to sign up for a registered program to learn masonry work?

8. How many hours are required each year for related technical classroom study to be a mason?

9. At the end of a person's training or apprenticeship, if successful, he or she is awarded a certificate of completion. What is the name given to the person at that time?

10. When an apprentice starts his or her career, what would the starting wage rate be in relation to a journeyman mason?

UNIT 45
Selected Masonry Projects

The following projects are combinations of brick and concrete block and offer some challenges to students in a masonry and bricklaying class. All of these were built in masonry classes taught by the author in a public high school. They provide an opportunity for individuals or for a group of students working together to build a common project. Some of them do involve a small amount of basic carpentry work to build the forms or backdrop and can be achieved easily in the shop with some basic tools. If there is a carpentry or woodworking shop available, this is a good opportunity to combine their skills with the masonry class. I have found that the students enjoy these projects very much, and they are excellent if a display or career show is ever planned. A variety of competencies and skills can be tested using projects such as these!

PROJECT 1: STRETCHER AND BASKET WEAVE PANEL WALL

This project is designed to be built by two students working together as a team. I have used it in a number of bricklaying contests and find it offers a wide range of basic bricklaying skills. There are two photos (Figures 45-1 and 45-2) that illustrate how the front and back of the project is built. The front is brick and the back is 4″ block combining to make an 8″ composite wall. It is recommended to lay the brickwork up to number 6 on the bricklaying spacing rule instead of a modular scale. This is done to allow the stretcher brick to be laid next to the soldier brick without showing big mortar bed joints. Notice on the back of the project that the block and brickwork are bonded together by being built over each other. Last, wall ties are used every 16″ in height to bond the two tiers of work together. Another excellent competition project.

PROJECT 2: SPIRAL PIER

This is one of the favorite brick projects in my classes. Simply stated, it is a brick pier that is 2½ brick by 2½ brick with a standard ⅜″ mortar bed and head joint, Figure 45-3. The first course is laid square. From that point on each succeeding course is set back on one end of the pier ½″ and projected out on the opposite end ½″. This requires accurate leveling and plumbing of each course. As the pier is built, it spirals or turns and, if everything is done correctly, in a height of approximately 10′, it will make one full turn. We struck the head joints with a round joint and used a flat slicker to fill out and smooth all bed joints. When completed, the project was wiped and rubbed off with some light oil to give a clean, washed-down appearance.

580

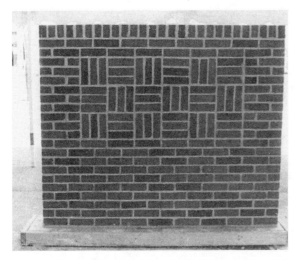

Fig. 45-1 Front view: stretcher and basket weave panel wall.

Fig. 45-2 Rear view: stretcher and basket weave project.

Fig. 45-3 Brick spiral pier.

PROJECT 3: BRICK WALL WITH A HALF MOON ARCH

This is a very popular project with my students, as it offers an opportunity to build an arch within a wall project, Figure 45-4. I used a simple semicircular arch form that was 24″ long by 24″ high to build the arch against. A brick header sill was added last to dress it up. The spacing of the mortar bed joints was basically determined by the height of the arch. Try to make the stretcher courses of brick work out as closely as possible to the top of the arch. It will be close but not exactly even. This project is 8″ in width and is backed up with brick. Four-inch concrete block can be substituted as back-up if desired.

PROJECT 4: BRICK RETAINING WALL

This project is typical of a brick retaining wall built to hold back earth at a driveway, hill, bank, etc., Figure 45-5. The wall is 12″ wide and eight brick long. Small lengths of pipe are built into the wall approximately every 24″ in length to allow ground water to seep through and relieve pressure on the wall. Eight-inch concrete blocks were used as back-up material for the wall. The top of the wall has an unusual design in that a bonded header is used alternated with a bonded rowlock cap. This creates a very attractive appearance; it is also a good idea for retaining walls in that it discourages persons from sitting on top of the wall.

Fig. 45-4 Brick wall with a half moon (semicircular) arch.

Fig. 45-5 Brick retaining wall.

PROJECT 5: BRICK CHRISTMAS TREE

This project is one of the most fun brick projects that can be built in the shop. It involves utilizing a lot of different skills for the students, Figure 45-6. It works as a motivator, especially at the holiday season of the year. The design is relatively simple. Four brick wing walls evolving from the center are built on 90° angles to each other. I found that each needs to be about eight brick long in order to build the tree up to approximately 5'. Four more wing walls are added between the 90° ones at 45°, making a total of eight separate wing walls in the project. On the top we cut out a simple brick star with the brick saw to complete the job. The completed job was rubbed down with a light motor oil to clean it. Students from the school whom we normally never see came to the shop to view it.

PROJECT 6: GARDEN WALK SECTION OF BRICK

This project was a favorite of mine, involving two students working together as a team. It is a typical section of simple brick garden walls with basket weave paving between the walls, Figure 45-7. It is a project that can be built in a relatively short amount of time and provides experience in not only laying the side walls, but in doing a little brick paving work. The width of the walk between the two walls should be laid out so that the paving uses full bricks without any cutting. The same is true for the length of the wall. When the project is completed, wipe off the face of the bricks with a little thin, clean automotive motor oil on a cloth to remove any stains. This adds the final touch.

Fig. 45-6 Brick Christmas tree.

PROJECT 7: BRICK WALK SECTION AND PLANTER

This project is a challenging one and allows several students to work on it at the same time. The end bricks must be solid so there are no holes that show. The brick wall is 8 inches wide with a rowlock on top. Care should be taken when laying out the first course to make sure that the walls are the same length as the rowlock brick cap. The paving is laid in a herring-bone pattern with a brick header at the end of each entrance to the walk. For the project shown in Figure 45-8, a smooth surface brick was used, as was training mortar; and we cleaned the completed project with a thin, light motor oil and cloth. We borrowed a small shrub from the agriculture classes to plant in the planter for the photograph.

Fig. 45-7 Section of a brick garden wall.

Fig. 45-8 Brick walk section and planter.

PROJECT 8: BRICK STEPS AND PORCH

This project represents a typical set of brick steps and porch (Figure 45-9) that masons would be required to build on the job. The rise of each step is 7″ and the depth of the tread is 12″. A ⅝″ projection was formed with the brick rowlock on the top front of each step to allow water to drip off. It is important that each step be the same height and depth so the person using the step does not trip or fall. Wall ties were used to join the steps to the porch. The top of the porch platform has a brick rowlock across the sides to form a border around the top. The hidden interior of the porch was filled with concrete blocks. Brick paving was laid inside the border area on top of the platform.

PROJECT 9: DIAMOND DESIGN GARDEN WALL PROJECT

This project demands that the student pay strict attention to the brick pattern and the placement of dark and light bricks. Brick headers are used to form the bottom and top of the diamond design, Figure 45-10. Wall ties were used to tie the 8″ brick wall together. I

Fig. 45-9 Brick steps and porch.

Fig. 45-10 Diamond design garden wall panel.

usually assign this project to students in the last year of their training, as it is more difficult to build. This would make a very good competition project for advanced students. As with most of the other projects in this unit, we cleaned the completed project with a lightweight, clean motor oil and a cloth to give it the washed-down appearance.

PROJECT 10: SPLIT-FACE CONCRETE BLOCK RETAINING WALL PROJECT

This is a neat project built of split-face concrete blocks that resemble stone, Figure 45-11. This particular project was built next to the front doorway of a building supply dealer and flowers were planted in the top of the combined retaining wall and planter. There is no mortar used in the project. Instead each unit is laid approximately $1''$ back from the face

Fig. 45-11 Split-face concrete block retaining wall project.

of the block beneath it, forming a strong, battered wall. After each course is laid, earth is filled in behind it to strengthen and support the next course of split-face block. Split-face block comes in either 4″ or 6″ depth. The 4″ blocks are recommended for a wall not exceeding 3′ high and the 6″ blocks are used for walls not to exceed 4′ high.

PROJECT 11: TYPICAL WOOD FRAME PLAN FOR BRICK VENEER SHOP PROJECT

An entire masonry class can be involved in a brick veneer project and it provides the invaluable experience of working together on a common project. If your school has a carpentry shop, the building of a framed wall can be simplified by following the plan shown in Figure 45-12. The framed 2″ × 4″ walls were drilled and bolted together in 8′ long sections to form a 24′ wall. You can adapt the wall lengths to fit your needs or situation. We installed electrical boxes for the students to wall in at various locations, made the window sills at varying heights so the students would have to use the scale rule to work them out, and built the door height a little higher than the window heads, as these things would appear on a real job. Angle irons were installed over the windows and door openings. After we had topped the wall out approximately 4′ from the face of the wall, we built a low retaining wall the entire length of the project. This short wall was built 8″ (two bricks) wide, three courses high, and capped off with a brick header course. Sand was placed on the floor and the area was paved with brick in running bond. The students exhibited a lot of pride and were very motivated to build and complete the project. They invited many of their friends who were not in the masonry class to stop by the shop to see their work. The wall section shown in the plan can be sheathed with any suitable inexpensive material or, if money is a problem, the framing can be left exposed. After the project is completed, the wall can be taken down and re-used many times in a number of configurations. The wall does require a lot of materials, but the beauty of this project is that it can be adapted to use the existing supplies of masonry materials in individual shops. Figure 45-13 shows a completed, brick-veneered shop project with an offset wall.

CONTESTS ARE A WAY OF SHARPENING SKILLS

An excellent way of sharpening and improving masonry student's skills is by encouraging them to compete in bricklaying contests in the school masonry shop and pitting their skills against their fellow classmates. I strongly recommend that masonry

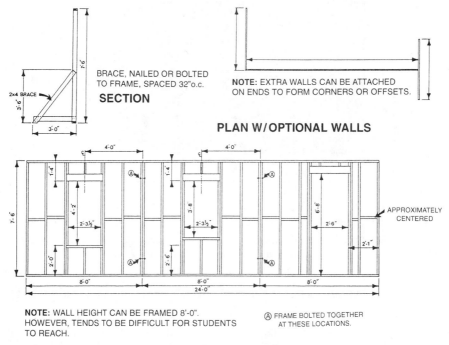

BRACE, NAILED OR BOLTED
TO FRAME, SPACED 32"o.c.
SECTION

2×4 BRACE

NOTE: EXTRA WALLS CAN BE ATTACHED
ON ENDS TO FORM CORNERS OR OFFSETS.

PLAN W/OPTIONAL WALLS

APPROXIMATELY
CENTERED

NOTE: WALL HEIGHT CAN BE FRAMED 8'-0".
HOWEVER, TENDS TO BE DIFFICULT FOR STUDENTS
TO REACH.

Ⓐ FRAME BOLTED TOGETHER
AT THESE LOCATIONS.

ELEVATION

Fig. 45-12 Plan of wall framing.

Fig. 45-13 Completed brick veneer shop project.

students join the SKILLS USA-VICA State and National Organization, which in addition to teaching leadership development, sponsors regional, state and national skill contests. It works like this. The student competes in a school contest; the winner is then qualified to move on to the regional contest and from there the winner represents his or her area in their state contest. The winner of the state contest then competes in the National Skills USA-VICA Contest. These contests are a great learning experience for all of the students who participate, regardless of whether they win or not! Many of the projects shown in this unit can be utilized for contests in the shop. If your school is not a member of VICA and wants to join, contact the National Office at 1-703-777-8810 or write them at P.O. Box 3000, Leesburg, VA 20177-0300 for additional information.

Figure 45-14 shows a masonry contestant building an offset brick pier project at the National Bricklaying Contest.

BRICK SCULPTURES

Brick sculptures are not new and go back to ancient times. Remains are still standing of brick sculpturing on monuments that are part of Babylonian and Mesopotamian structures dating back to 600 B.C. Today this artistic use of brick sculpturing is being revived on a number of brick structures such as churches, corporate offices, restaurants and other commercial type buildings. Sculptured brickwork incorporates artwork into the actual design and fabric of the structure and takes the skills of the mason who installs it to a higher level. Once the builder and the architect agree on design, it is given to a sculptor artist who creates and carves the

Fig. 45-14 Building an offset brick pier project.

artwork from the moist pliable brick clay into the art form design. Cut brick sculptures may have the actual cutting done before or after the clay has been fired and burned. The brick are made without cores to provide more strength or depth to accommodate the design.

When the carving of the design is complete, the sculptures are cut into various working size units and numbered by course, size and specific location according to a carefully laid out plan, so they may be installed correctly after the firing process is completed. The shrinkage that usually occurs from the burning of the brick will as a rule allow for the mortar joints between the units when they are laid up in mortar. After the brick have cooled enough to handle, they are packed in straw with protective spacers between them and shipped to the jobsite for installation. If the design is intricate or complicated, the artist will generally go to the jobsite and supervise the masonry contractor during the unpacking of the sculpture and oversee the laying out and minor adjusting of the design's shape and color and the thickness of the mortar joints, to ensure the proper finished appearance of the design.

The mason needs to take great care when laying the brick units into their proper position in mortar, to ensure that the artist's sculpture is completed and appears as the artist intended it. The artist and the mason form a team working together to blend and install the brick artwork into the structure of the brick wall to create a beautiful lasting design. If the surface of the sculptured brickwork design needs to be cleaned after it has cured, the mason needs to use a relatively mild masonry cleaner that will not damage the bricks or mortar in the finished sculptured design. A directory of recommended brick sculptors artists can be obtained by contacting the Brick Industry of America at 1-703-620-0010.

Figure 45-15 shows a beautiful example of brick sculpture on the wall of a church near Baltimore, Maryland.

Fig. 45-15 Brick sculpture on the wall of a church.

APPENDIX
Additional Projects

PROJECT 1: REHABILITATING AN OLD MASONRY PORCH AND STEPS

As a house gets older, many times the masonry porch and steps tend to settle and not only create a problem for the person trying to get into the doorway, but also detract a lot from the appearance of the home and decrease its value. Since jacking up and raising the porch and steps seldom works, I have found that if the porch itself is in a good shape, the most economical solution is to just reface or brick veneer it! The following project is a good example of this.

Before I started to lay any brick, I demolished and removed the old brick steps, as they usually are not worth saving. After I excavated the soil around the perimeter of the porch down to the safe frost depth, I poured a concrete footing 8″ deep and twice as wide as the brick to be laid on it. Next, I measured down from the top of the existing door sill of the house and worked out, with the aid of a masonry spacing rule, the brick coursing so that it would consist of even courses showing above finished grade with no splitting or cutting of the brick for height. It is also a good idea to lay three courses of brick below the finished grade line in case the topsoil settles some, which is almost certain to happen.

I then laid the brickwork in mortar up to the grade line and in this particular case installed some scrap pieces of flagstone in mortar as a base to start the finished brickwork on. The brickwork was then built up even with the top of the existing concrete slab of the porch. Figure P1-1 shows the brickwork being laid up to the top of the concrete porch slab.

I selected a basket weave pattern design for the brick paving on top of the porch. I strongly recommend that you use a solid paving brick that is classified SW (severe weathering); it

**Fig. P1-1 Brickwork being laid up to the top
of the concrete porch slab.**

591

Fig. P1-2 Laying the brick paving on top of the porch.

Fig. P1-3 Finished brick porch and steps.

is available from your brick supplier and is more durable than just a solid brick. It is not a good practice to use old salvaged brick as it is almost certain that they will not hold up due to the dampness and moisture that a porch and steps are normally subjected to. I projected the border edge bricks about $5/8''$ to form a drip edge and laid the paving brick in mortar to a stretched line from one end to the other. I picked out and used a nice, straight piece of wooden $2'' \times 4''$ to level across the top periodically to adjust any high brick.

You can use a flat or a concave (rounded) joint finish. Personally, I prefer the concave joint finish as it looks neater, doesn't smear the adjoining brick, and seals the joints just as well. The best technique to use in tooling the mortar joints is to check them first to determine if they are dry enough to tool without smearing by pressing your thumb lightly into the mortar joint. If it leaves a light impression of your thumb print, it is time to tool them. I also suggest that you use a sled runner jointer that is a little wider than the width of the joint to prevent depressing it too far below the surface of the adjoining brick, creating a place for water to lay. A sled runner jointer will also keep the mortar joints in more of a straight line in addition to being much faster to use. Be sure that you don't wait too long to tool the mortar joints, or you won't get that nice slick smooth finish that is the mark of a good job. (See Figure P1-2.)

After the paving on the porch was finished, I built the brick steps, projecting the front edges also about $5/8''$ and giving them a little fall so that any water would run off. I replaced the brick paving that existed on the walk and washed and cleaned the brickwork down with SURE-KLEAN 600 Masonry Detergent solution. By building on top of the old porch instead of totally tearing it down, the final result was a beautiful porch and set of brick steps at a minimum cost. Figure P1-3 shows the finished brick porch and steps.

PROJECT 2: BRICK CIRCULAR RETAINING TREE WELL

When I built a brick patio for my father-in-law in back of his house, there was a beautiful dogwood tree growing in the center of the area where he wanted the patio. Since we were both tree lovers, we did not have the heart to cut it down, so we tried to figure out how we could both build the patio and keep the tree. The solution was to build a well around the tree and to design it so it would add to the functional beauty and use of the patio. We tinkered with the design and finally agreed to build a circular retaining wall tree well that would also serve as a bench. See Figure P2-1.

Luckily, when we did the brickwork on the house, we ordered some extra brick in the event of a situation just like this. Ordering extra is always a good idea, because even though the same number and type of brick may still be available, many times the color changes a little and this color difference becomes very noticeable when adding on to an existing house. Since the brick was to be laid in mortar, I had to excavate down below the frost depth and pour a concrete footing. I marked the circular area to be excavated with a garden hose and sprinkled some lime on top of the ground to use as a guide for the excavation. I mixed the concrete to a ratio of 1 part portland cement to 2 parts sand to 4 parts crushed stone, and poured the footing 6″ deep and twice the width of the brick wall, which in this case was 8″. After the concrete had cured about 24 hours, I laid a garden hose in a circle on top of the footing and used it as a guide to follow when drawing a circle with a piece of white chalk around the footing where the wall was to be built. A good way to save on costs is to use some old chipped brick for the work below grade, as they will not be exposed to view.

You could also use concrete block in place of the brickwork below grade, but I wanted to form the brick circle as close as I could in the beginning, so that when it reached the finished grade level, there would be a minimum of adjustments to be made to the finished brickwork that would be exposed above grade.

I dry bonded the brickwork out on the wall at finished grade to establish the bond. The circle was large enough that I could use all full stretcher bricks instead of headers, which would be required on a tighter curvature. I then laid the brick wall, which was two brick wide (8″), around the circle in a mortar bed joint but did not apply any head joints between them. I did this to allow any moisture that would build up inside the tree well to weep out through the open joints and prevent the wall from cracking. Since you cannot lay the brick to a line because it is in a circle, I established plumb points about four brick apart and made a circular wood template out of plywood that I could fit against the brick plumb points and retain the circular pattern. This is a lot more accurate and easier than plumbing every brick on the

Fig. P2-1 Circular brick tree well that encloses a beautiful dogwood tree. It not only preserves the living tree but also serves as a focal point and adds a bench seat to the patio area.

course. After each course of brick was laid, I cleaned out any mortar that had squeezed between the ends of the brick with a steel line pin so that it did not build up as the brickwork was built.

I tooled the mortar bed joints as they became thumbprint hard and brushed them lightly. Every three course of brick in height, I laid corrugated metal wall ties about 24 inches apart across the course to help bond the wall together. I established the same plumb points on the inside of the wall as I did on the outside and maintained the correct curvature with the wood template board. I did not fill any mortar in the collar joint between the two wythe of bricks, as it could cause them to push out. The critical thing is to maintain the $8''$ width of the wall at all times up to the top course so the flagstone cap will fit with a projected drip edge on both sides.

I capped the brick wall off with a course of $1'' \times 12''$ flagstone to seal it against water penetration and form a bench seat. Since the brick wall is $8''$ wide, the flagstone will project approximately $2''$ on each side, which provides a nice water drip on the back and front of the stone cap. In order for the flagstone to fit the curvature of the circle, they will have to be cut on a slight angle from the front to the inside of the wall. You have to work this out once the correct projection is established in the front. The easiest way to cut them is to mark the angle on the stone with a square and pencil and lay it over the edge of a piece of wood $2'' \times 4''$ or a piece of pipe to support it. Then cut it with a brick chisel and hammer, unless you are fortunate enough to be blessed with a masonry saw. In a pinch, you can use a carpenter's portable power circular saw with a masonry blade for the cutting. Be sure to wear safety glasses or goggles to protect your eyes when sawing. I used Vermont slate flagstone for this cap as they have nice shades of red, green, and blue in them.

After all of the masonry work has been completed and before the mortar has gotten hard, be sure to clean out any mortar that remains in the head joints so they will drain properly. Leave it to cure for a couple of days and wash it down with a masonry cleaning solution to remove any mortar stains. My father-in-law filled the tree well with topsoil, finishing off with a layer of peat moss. He plants flowers in it during the warm months of the year. Tree wells do require periodic watering as they are elevated and will not retain moisture over a long period of time. This is a practical project that will add a lot to any patio area, and it is relatively simple to build.

PROJECT 3: LAYING PAVING BRICK WITHOUT MORTAR

The beauty of laying paving brick without mortar is that there are no mortar joints to crack from the frost and freeze cycle that is common to most places in the United States. In addition, it takes a lot less time and labor than installing them in mortar. Basically any one with a little patience and perseverance can handle laying brick paving with a minimum of problems and attain good results. A good example of laying brick paving without using any mortar is in a walk. It is an ideal project for the handy person or homeowner to try.

The first step in installing a walk, regardless of whether it is brick or concrete, is to give some serious thought to the width of the walk before you start excavating any soil for the base. I have learned from experience that, as a rule, a brick path which will allow two persons to walk abreast or to pass each other comfortably needs to have a minimum width of 3 feet. A front walk, however, should be about 4 feet wide or better, which allows for more traffic and moving furniture or larger things through the front door.

The next important thing to do is to make sure that you buy only brick that are made specifically for paving; not just any solid brick will do. Ask your local brick supplier for a Pedestrian and Light Traffic Paving Brick which conforms to ASTM C902 and is classified as a SW Severe Weathering brick. Don't worry, they will know what you are talking about! I never recommend the use of old salvaged brick. To do so is to invite problems down the road, because chances are they will deteriorate a lot sooner than regular paving brick.

The brick I selected for the walk which is shown here were standard size paving brick which are 8″ long by 4″ wide and $2\frac{3}{8}$″ thick. I also decided to lay the walk in a basket weave pattern, which is two brick laid flat (shiner position) in one direction and two brick laid in the same way in the opposite direction. For the sake of estimating, it will take four and a half brick per square foot in this pattern. If you want a brick border around the walk in a different position, you will need a few extra to compensate for that. You can also figure that you will need 5% more brick than estimated to cover cracked, broken or chipped brick.

For the base, I use finely crushed stone, which is called stone screenings or stone dust, depending on the area of the country you live in, with a layer of building sand on top of it as a cushion for the brick to lay on. For most climates, figure on 3″ of stone screenings and 1″ of sand for the setting bed. The amount you need obviously depends on the size of the walk you are laying. If you tell the masonry building supplier the size of your intended walk, it can estimate very closely the amount of materials you will need.

I lay the walk out with a mason's line attached to wood stakes around the perimeter of the walk. To figure the depth to which to remove the soil, add together the height of the brick and allow 4″ for the stone screening and sand. I also give the walk some pitch or fall to allow water to run off. Usually allow $\frac{1}{4}$″ per foot of width to run the water off.

After the forms are all in position and the soil has been excavated fairly level, spread the 3″ layer of stone screenings followed by the 1″ layer of sand on top of it. You can save a lot of time and effort by making a screed board out of a piece of wooden 2″ × 4″ (leveling board) by cutting an L-shaped notch out on the ends the same as the thickness of the brick you are using. Make the screed board a little longer than the wood forms so that it will ride on top of the form. I push the screed board along the form and it spreads and levels out the setting bed to the correct height to lay the brick on, Figure P3-1.

Fig. P3-1 Screeding the setting bed to the correct height.

Fig. P3-2 Laying the border and basket weave pattern brick using the forms as a guide.

Fig. P3-3 Sweeping the sand in the joints between the paving brick.

The rest is rather easy, except of course on your knees! Lay the border brick first well ahead of the paving, so you have something to work against. If the border brick are laid on their edge, you have to dig out a little extra depth to accommodate for this. A border brick that is a little higher will help to restrain the edge of the walk from shifting or moving. Set each brick in the sand bed and tap it firmly into place with the handle of a brick hammer. A rubber hammer also works well to settle a brick that is a little high into place. If I break or crack one, no problem; I just remove it and put a new one in its place. Laying the border and basket weave pattern brick using the forms as a guide is shown in Figure P3-2.

When all of the brick are in place to my satisfaction, I shovel some sand on the walk and sweep it into any voids between the bricks. Later on after a period of time passes, if rain washes any sand out, just sweep some more sand into the voids, Figure P3-3. The last thing I do is to remove the wood form gently and fill in around the edges of the walk with top-soil, tamping it tightly against all of the edges. The only thing left to do is to plant a little grass seed around the edges and the job is completed. Any weeds that crop up can be taken care of with a little weed killer.

PROJECT 4: FIELDSTONE RETAINING WALL FLOWER BEDS

A good friend of mine had a perplexing problem to solve. There was a steep, falling grade from the back of his yard down to his house so that whenever there was a heavy rain storm, the water continually eroded and washed the topsoil down toward the house and onto the sidewalks. He wanted to have a garden at the back of his yard where it did level off pretty well, but he realized that he had to come up with a plan to control the steep grade and prevent the topsoil from washing away. Since he had access to an unlimited supply of fieldstone, I recommended that we build terraced fieldstone walls laid in mortar at spaced intervals up the slope of the hill. We also decided to build them in irregular curved patterns, so they would act as a reinforcement to retain the earth and at the same time provide a creative series of flower beds that would greatly enhance the beauty of the yard.

We sketched out a rough drawing of the hill and marked approximately where we thought the walls should be built. The next step was to lay a long garden hose on the ground across the hill in an undulating circular pattern. We kept adjusting the garden hose in a snake-like design until we were satisfied with what we thought the walls should look like when completed. The earth was excavated where the stone walls would be built to a depth that would be below the frost or freeze line and we poured footings that were 6″ deep and approximately twice as wide as the width of the stone wall. We did not excavate all of the walls in the yard at one time, as we wanted the liberty and flexibility to change our mind as we went along as to where the walls would intersect or join each other to create the most pleasing appearance. My friend was not a professional mason by trade but was very creative and learned the basics by the time we had built a couple of the walls. The mortar mix that I recommend for building stone walls of this type is 1 part portland cement to 1 part mason's hydrated lime to 6 parts washed building sand. The addition of the lime in the mortar causes it to stick or adhere to the stone much better than masonry cement and will decrease the shrinkage and hairline cracks that may occur in stone mortar joints. Since stone is a very dense and hard material, it does not absorb much water from the mortar; therefore, it has to be mixed stiffer than regular brick mortar to prevent bleeding and to support the weight of the stone.

The larger stones were laid on the bottom courses to strengthen the wall more and the smaller ones were used above the finished grade line. Since the walls were not very high and were built on curvatures, we mostly eye-balled the work instead of plumbing it. We did, however, have a level control line stretched tightly from one end to the other, so that the wall would be reasonably close to level at the top. We selected the thinner flat stones for the cap on top of the wall and projected them out about 1″ on each side to form a water drip. The cap course was leveled as best we could, considering the irregularities of the stone. We decided that a dark mortar joint would show up better against the lighter color of the stone, so we raked out the mortar joints about ¾″ deep and would repoint them later when all of the stone work was completed. A variety of special colored masonry cement mortars was available from the masonry materials supply house that we could pick from when we were ready to do the repointing of the joints. The pattern design of the stone work was random, as we mostly laid them in the mortar as we picked them off the piles. When it came time to repoint the mortar joints with the darker mortar, we moistened the raked-out area with a little water and a paint brush to prevent the repointing mortar from drying out too fast. The mortar joints were repointed using a small pointing trowel and a flat slicker jointer, letting the rounded edges of the stone show to create a more natural appearance. The mortar was smoothed out

Fig. P4-1 The completed stone retaining walls with
flower beds.

on top of the cap stones to fill any irregularities and make it as level as possible. After the mortar joints had set enough so that they would not smear, the walls were brushed down with a medium bristle brush to smooth out the mortar joints and remove any particles of mortar remaining. When the mortar joints had cured for about 24 hours, we scrubbed them down with a brush and running water from a garden hose.

The flower beds were filled with topsoil and 4″ of peat moss was added to top it off. My friend and his wife planted a variety of flowers in the beds to complete the job. It took a lot of time and hard work to build the retaining walls, but his yard is now the envy and pride of the neighborhood. Figure P4-1 shows the completed stone retaining walls with flower beds.

PROJECT 5: BRICK PLANTER BASKET

This brick basket planter project is one of my favorite projects. It looks like it would be difficult to build at first glance, but in reality it is only a simple application of basic bricklaying techniques that have been combined! It will attract a lot of second looks from people passing by, as it is a real eye-catcher when built in the front yard of your home with beautiful spring or summer flowers in it. The construction of this project does require some extra special care in building due to the design of the circular wall and the brick handle on the basket. It will test your bricklaying skills, as the curved walls have to be leveled and plumbed every course to maintain its shape, but when completed it is a project that is not only a lot of fun to build, but will also display your bricklaying skills to all of your friends and neighbors.

You should start by driving a center stake in the ground at the location you have selected and then drive a nail in the top of the stake. I decided to build the basket with a 6′ outside diameter because standard 8″ long brick will fit this circle without having to be cut. Since the footing should extended 6″ past the outside wall, I attached a length of mason's line 42″ long to the nail and swung it in a full circle. At the same time you are swinging the line, mark the surface of ground by spraying some orange paint from a pressurized spray can to mark the edge of the circle. Be sure to excavate for the footing to the required freeze depth in your area. It did not take me long to figure out that it was a lot simpler to excavate the entire area rather than trying to just dig a ditch for the wall, due to the small overall size of the project. I formed the edges with some plywood strips, mixed and poured a concrete footing of 1 part portland cement to 2 parts sand to 4 parts stone or gravel.

Because the circular brick wall is only a single brick in width (4″), I started laying brick right on the footing to establish the bond, so that it would work full brick all the way around the circle, making any adjustments where necessary to achieve this. This way, when I built it up to the finished grade, I would not have to make any additional adjustments to the bond.

Notice in Figure P5-1 that there is a double brick pier on each end of the center of the project. This pier will support the basket handle when that point is reached later. Build the brickwork plumb and level up to the finished grade line and tool the mortar joints of the last three courses of brick that will be under the finish grade, in case the soil settles later. On the second course above finish grade line, it is a good idea to insert a piece of $\frac{3}{8}''$ diameter plastic tubing in the mortar head joints every three brick to allow excess water to drain. I tooled the mortar joints with a rounded convex sled joint runner.

Continue laying brick to the height where the brick piers start; form these and build the piers up to where the angle irons will sit to support the bottom of the brick handle of the basket. (See Figure P5-1.) I set the two 4″ wide angle irons back to back and let them rest on the ends of each brick pier 4″ for support. I strongly recommend the use of heavy steel angle irons that are $\frac{3}{8}''$ thick, as they will not sag or bend, instead of the lighter pressed steel angles. Apply a coat of rust preventive paint on the irons before installing and laying any brick on them.

I cut out a pattern from some stiff cardboard for the curvature of the handle on the ends. The curvature is rolled in at the top from the existing wall face at the bottom, approximately 4″. (See Figure P5-1.) Determine where to cut the ends of the adjoining brick that fit against the curved edge of the handle using the cardboard pattern as a template and cut the ends to fit. Lay the cut brick in place and then with the aid of a stretched mason's line fill in the wall that forms the handle. Next, lay the rowlock brick on the ends to fit the curve made by the

Fig. P5-1 Brick planter basket.

pattern. Then, lay the top rowlock brick course to the line to complete the handle. Tool the mortar joints and brush the work when it is dry enough that it will not smear.

I completed the masonry work by parging (plastering with mortar) the inside of the brickwork that was below the finished grade line to make it more waterproof. I allowed the brickwork to cure for at least three days and washed it down with a masonry detergent cleaning agent. After it has been washed down I recommend that the masonry work be allowed to dry for another 24 hours before filling the planter with soil. In conclusion, outside of watering the flowers periodically, you can now sit back and enjoy the fruits of your work.

Glossary

absorption—the weight of water a masonry unit will absorb when immersed in either cold or boiling water for a stated length of time. This weight is expressed as a percentage of the weight of the dry unit.

abutment—the skewback of an arch and the masonry that supports it

adhesion—the characteristic of mortar that enables it to cling to a masonry unit

admixture—a substance or chemical added to mortar to achieve a special condition such as workability, air entrainment, water retention, acceleration of setting time, or retardation of setting time

adobe brick—a brick made in early times from clay and placed in a mold in the sun to dry and cure

aggregate—a material such as sand or gravel that is mixed with portland cement to form a concrete block

anchor bolts—steel bolts installed in a masonry wall to secure a structural member, such as wooden or steel plates

apprentice—one who enters into an agreement to serve an employer for a stated length of time to learn a trade

arch—a section of masonry work that spans an opening and supports not only its own weight, but also the weight of the masonry work above it

arch axis—the median line of the arch ring

architect's scale—ten different scales placed on a rule used to measure dimensions of drawings and plans; scale ranges from $3/32''$ to $3''$

ash dump—flat, metal door with a flange built flush with the top of the inner hearth of the fireplace. Used for dumping ashes into the ashpit below

ash pit—area directly below the inner hearth of a fireplace; used for receiving ashes from the fireplace above

ASTM (American Society for Testing and Materials)—an organization that sets standards for building materials

autoclave—a special metal cylindrical chamber used to rapidly cure concrete block; heat is approximately 360° F

autogenous healing—the ability of mortar with a high lime content to reknot or heal itself if the masonry cracks. Moisture is necessary for this healing to occur.

back filling—the process of filling dirt around a foundation after it has been built to reach grade level

backing up—the process of laying the inside masonry units of a composite wall after the exterior has been laid

banding—the placing of metal straps on cubes of up to 500 bricks by machine as the bricks leave the tunnel kiln

bat—the trade term for broken brick

batch—the combination of materials used to produce concrete block

batter—the recession of masonry in successive courses on a gradual incline; opposite of corbeling

beam—a piece of timber, steel, or other material that is supported at each end by walls, columns, or posts and is used to support heavy flooring or weight over an opening

bearing plate—a steel plate laid in mortar on a masonry wall to distribute the weight of beams that are placed over it

bearing wall—a masonry wall that supports a load other than its own weight

bed joint—a horizontal bed of mortar on which a brick is laid

beehive kiln—a semipermanent circular structure constructed of brick and wrapped with steel bands to control the expansion of the kiln when the bricks are being burned

blowout—the swelling or rupturing of a cavity wall caused by too much pressure from pouring liquid grout in a wall

blueprints or building plans—detailed drawings of a structure showing measurements and the various views that are necessary to build the structure. The term *blueprint* commonly refers to the reproduction of plans with white lines on a blue background. Persons in the masonry trade commonly call all plans blueprints.

bond—the process of (1) tying together various parts of a masonry wall by lapping units one over another or by connecting with metal ties; (2) the pattern formed by exposed face of the units; (3) the adhesion between mortar or grout and masonry units or reinforcement devices

bond beam—a course of masonry units having steel reinforcing rods inserted and filled solidly with mortar; may serve as a lintel or reinforcement beam to strengthen the wall

bonded arch—an arch that is composed of alternating bricks that are bonded together, such as a jack arch that is $1\frac{1}{2}$ bricks high

bonded jack arch—a jack arch that is higher than one brick and is bonded by alternating the mortar joints so they are staggered in appearance. It is the strongest type of jack arch.

box anchor—a square metal tie used to tie two masonry units together

breaking joints—any arrangement of masonry units that prevents continuous vertical joints from occurring in succeeding courses

brick—a solid unit of clay or shale that has been burned in a kiln; usually rectangular in shape

brick kiln—a brick structure used to burn brick at a controlled heat

brick nogging—brick masonry that is built between wood posts or framing

British system of measurement—system of measurement used in the United States and Britain

building code—the legal requirement established by different governing agencies covering minimum construction practices

burning the joint—tooling mortar joints after they are too hard, shown by black marks. The black marks are caused by a reaction between the steel and the mortar joints.

burr—the ragged edge of metal on a chisel head resulting from the strike of a hammer

burred—the condition of a steel chisel when it is struck too frequently with a hammer; the chisel becomes frayed and ragged

buttering—the process of applying mortar on a masonry unit with a trowel to form a joint

buttress—projecting pilaster built against another wall to give the wall greater strength and stability; usually has a battered or sloping top

calcium chloride—a chemical used to accelerate the setting time of mortar

camber—the relatively small rise of a jack arch

carrying the trig—the process of setting the trig brick in mortar on the wall

caulking or calking—a material used to seal cracks around such things as windows, doors, and floor lines; the material usually comes in a tube or a cartridge and is applied with a caulking gun

cavity wall—a wall consisting of 2 tiers of wythes of masonry units separated by a continuous air space not less than 2″ wide. The space may be retained as insulation or filled with grout and steel reinforcements.

chase—a continuous recess built into a wall to receive pipes, wires, or heating ducts

checkpoints—mason's folding rule marks placed every 16″ on the footing or base; used as a reference when laying masonry units

chimney pad—concrete footing for a chimney

cinder block—a concrete block in which cinders are used as the aggregate

cinder brick—a brick made from cement and cinders

cleanout door—a metal frame and door built into a chimney that allows entrance to the chimney so that soot or ashes may be removed

clinker—a very hard, burned brick whose shape is distorted or bloated due to nearly complete burning in the kiln

clock method—the method of cutting and picking up mortar from a pan by using numbers on the face of an imaginary clock as a guide

closure brick—the last brick that is laid in the wall, usually located in the center of the wall

CMU—the abbreviation for Concrete Masonry Units

collar joint—the vertical joint between 2 tiers of masonry

colonial ring-around fireplace—fireplace that has only the facing of the bricks stacked up the jambs with an exposed soldier course over the top of the opening

common or American bond—an arrangement of bricks in stretcher position with a header course on the fifth, sixth, or seventh course of bricks

common jack arch—a jack arch that is only one soldier brick in height

composite wall—any bonded wall consisting of wythes or tiers of different masonry units, such as a brick and concrete wall

compressive strength—the measured amount of downward pressure (load) that can be applied to a masonry material before it ruptures or breaks

concrete block—a hollow or solid block made from portland cement and aggregate

constant cross-section arch—an arch that has a constant depth and thickness throughout the span

contractor—the person(s) who undertakes a job to construct a structure under a contract or agreement

control or expansion joint—a vertical joint built into masonry work in places where the forces acting on the wall cause pressure, possibly creating cracks. The joint is filled with a material that will give flexibility, thereby relieving some of the pressure.

coping—the masonry covering laid on top of a wall. Coping is usually projected from both sides of the wall to provide a protective covering as well as an ornamental design.

corbeling—the projection of masonry units to form a shelf or ledge

crazing—the formation of hairline cracks in the surface of concrete caused mostly by shrinkage

cream—when referring to concrete, this is the cement paste on the top of finished, floated concrete that seals the aggregate in.

creepers—brick in the wall next to the arch that are cut to fit the curvature of the outer edges of the arch ring

crew—a group of persons who work together in constructing a building or project under the direction of a foreman

cross joint—the mortar head joint applied across the long side of a header brick

crowding the line—causing the line to swing or bounce with the brick or the fingers when laying brick; also called putting the thumb in the line

crown—the apex (center) of the arch ring. In symmetrical arches, the crown is at midspan.

cubic foot measurement—the method of measuring masonry materials that are $1'-0''$ long, $1'-0''$ deep, and $1'-0''$ wide

cupping the mortar—a method of cutting and rolling mortar onto a trowel in a cupping motion from the mortar board

cure—a process of hardening, in which concrete attains its maximum strength. It is done by not letting the concrete dry out too quickly and keeping it at the right temperature.

curtain wall—a wall independent of the wall below; supported by structural steel or concrete frame of building

cylinder test—test performed by filling a cylinder full of concrete under specific directions and later testing it for compressive strength after a pre-determined length of time. Usually, concrete has to meet a 28 day psi curing criterion to meet performance standards.

dead load—an inert, inactive load such as in structures due to weight of the members that support the structure and permanent attachments; a constant weight or pressure used in computing strength of beams, floors or roof surfaces.

deadman—a post or prop to which the line on a wall is attached

depth—(of an arch)—the dimension that is perpendicular to the tangent of the axis. The depth of a jack arch is its greatest vertical dimension.

details—in masonry, specific drawings of elements of construction such as lintel layout, flashing details, and installation of bolts in the wall. These are shown on a larger scale to simplify necessary procedures.

double jointing—the process of applying mortar to both ends of the closure brick as well as to ends of the bricks surrounding the area where the closure brick will be placed

dovetail anchor—a metal tie in the shape of a dove's tail that is used to tie masonry units to concrete walls

draft—the current of air created by the variation in pressure resulting from differences in weight between hot gases in the flue and the cooler air outside the chimney

drawing—the process of unloading the brick from the kiln after cooling

drawn to scale—plan drawn to a fixed ratio of measurement between the drawing and the actual completed structure. For example, a $1'-0''$ measurement on the structure usually equals a $\frac{1}{4}''$ measurement of the plan.

drip strip—a strip of wood usually measuring $3'' \times \frac{3}{4}''$ laid in the middle of a cavity wall to catch mortar droppings

drowning the mortar—the trade expression describing a condition of mortar resulting from too much water added to the mix

dry bonding—the laying out of bricks without mortar to establish the bond for the wall

dry-press process—a process in which a mixture of 90% clay and 10% water is formed into bricks in steel molds under high pressure; used for clays with very low plasticity

dummy flue—flue lining installed at the top of the chimney that is not hooked up to any heat source. It is used only to appear as a flue and balance out a chimney that has only two flues and a large empty space between them. Dummy flues should be filled with mortar on the inside to prevent birds or insects from building in them.

dusting—formation of dry powdery materials on the surface of newly cured concrete, usually caused by too much clay or silt in the mix or condensation.

Dutch corner—the method of starting off a corner by using a $6''$ piece of brick

ears (of a block)—the ends of a stretcher block where the head joint is formed; found only on stretcher block that do not have square ends

economy brick—a brick with nominal dimensions of $4'' \times 4'' \times 8''$

edging—using a rounded edge metal tool to round off the edge of concrete. This is done to prevent chipping and breaking of exposed edges of concrete.

efflorescence—a whitish deposit on the exposed surface of masonry units caused by the leaching of soluble salts from within the wall

elasticity—the ability of mortar to stretch and give

elevation—the vertical view of a structure or feature of a structure showing external, upright parts

engineered brick—a brick with nominal dimensions of $3\frac{1}{5}'' \times 4'' \times 8''$

English bond—a bond in which there are alternate courses of headers and stretchers

English corner—the method of starting a corner by laying a $2''$ piece of brick off the return corner brick

exposed aggregate finish—concrete finish achieved by applying a fine water spray to the concrete surface before it hardens. This reveals the aggregates on the surface of the mix and presents a rustic appearance.

extrados—the convex curve that bounds the upper limits of an arch

face brick—a brick used in the front or face side of a wall; usually a better grade of brick

face shell bedding—a method of applying mortar to only the outside webs of a concrete block

face wall—a masonry wall that is tooled and exposed to view

fat mortar—a mortar that contains a high percentage of cementitious materials; sticky and hard to dislodge from the trowel

firebox—area of the fireplace built of firebricks or bricks where the fire is built

firebrick—a brick made from a highly fire-resistant clay found at a greater depth in the ground

fixed arch—an arch that has a skewback that has a fixed position and inclination. Plain masonry arches are fixed arches.

flashing—(1) a special process used to add color to bricks by reducing the temperature in the kiln (reducing the amount of oxygen burning); (2) a type of metal such as lead, aluminum, or tin that is laid in the bed joint of a chimney at roof level to prevent the entrance of moisture and/or provide water drainage

flash set—the premature setting or hardening of cement in a mix due to excessive heat

Flemish bond—the brick bond in which a stretcher and a header alternate on the same course. The headers on every other course should be centered over the stretcher below.

floating—a finishing process on poured concrete that involves rubbing or troweling the surface of the wet concrete with either a wood or metal tool that is flat on the edge touching the concrete. This action causes the cement paste in the mix to reactivate and to come to the surface of the concrete. This paste has to be on top in order to have a smooth finish and is very necessary to completely seal in all of the aggregate, in the concrete.

flue—a built-in passageway that carries gases or fumes from a chimney

flue lining—the fired clay or fireproof terra-cotta pipe used to line the flue of a chimney. They may be round or rectangular in shape.

flue ring—another name for thimble. See thimble.

flush joint—a mortar joint that is even with the unit and in which there is no tooling done

fluted block—concrete block made with projected vertical ribs on face of block; used for textured walls

foot boards (also called blocking or hopping boards)—wooden planks laid on top of bricks or concrete block to allow the mason a higher reach when laying masonry units

footing—the enlarged base, usually made of concrete, that the foundation rests on. Its function is to support the building and distribute the weight over a greater area.

formed footing—concrete footings that are placed into a form on the job

Franklin stove—cast iron stove resembling a fireplace, invented by Benjamin Franklin

frog—a depression in the bed surface of a brick; used for design and as an aid for locking the mortar in place

fullback—concrete block that has all of the cell webs cut out except the last one that holds the block together

full header—a header course that consists completely of headers, except for the corner starting piece

furring—the method of finishing the interior face of a masonry wall to provide space for insulation, prevent moisture transmittance, or to provide a place for application of some type of building material

furrowing—a slight indentation made in the mortar bed joint with the point of a trowel to prepare for laying brick

garden wall bond—the bond in which there are 3 stretchers alternating with a header on the same course

general contractor—the main or prime contractor on a job, who has the responsibility of coordinating all of the subcontractors' work to complete the structure according to the terms of the contract.

gingerbread work—the trade term given very decorative masonry work, especially that which is located on the cornice trim of homes

girth hitch—a form of knot used in masonry work to fasten a nail to a line

Gothic arch (or pointed arch)—an arch with a rather high rise, with sides consisting of arcs of circles, the center of which are at the level of the springing line. The Gothic arch is often referred to as a crop, equilateral, or lancet arch, depending upon whether the spacings of the centers are less than, equal to, or more than the clear span.

gram—the standard unit of weight measure in the metric system. One gram is equal to $\frac{1}{28}$ oz in the English system of measurement.

green brick—brick in its soft, pliable state before burning in the kiln

green staining—a reaction on masonry work caused by the salts contained in vanadium

grooving—using a metal tool that has a projecting edge on the bottom in the center of the tool. Grooving is performed to make a straight indentation where expansion or control joints are in concrete.

grout—(1) a very thin mortar that is poured between two walls for reinforcement; (2) a liquid concrete that is poured in the center of a reinforced masonry wall. Consists of a portland cement, lime, and aggregates.

guardrails—metal rails that fit onto the scaffolding to prevent the workers from falling. Guardrails are required by OSHA federal regulations on any scaffolding exceeding 10′ in height.

guild—an organization of masons formed to guard and protect brickmaking processes and trade secrets

hacking—the procedure of stacking bricks on a pile with the use of brick tongs or carriers

halfback—concrete block that has the insides of two cells cut out

half-high block—a rip block that is cut in half lengthwise

half lap (of unit)—the lapping of one masonry unit halfway over another

hard brick—a brick with a very dense composition and a very low water absorption rate

hard to the line—the term that describes a masonry unit when it is pushing against the line

header—a masonry unit that is laid over two individual walls, thereby tying them together

header block—a special concrete block made to receive a brick header, also called a shoe block

header high—the trade term for the height of the wall where the headers are laid

head joint—the mortar that is placed vertically between the ends of the bricks

high-suction brick—a porous brick that absorbs moisture rapidly

hog in the wall—the trade term describing a masonry wall that is built to the same height on both ends but that has one end of the wall containing more courses than the other. This is usually caused by one mason laying bigger mortar joints under the bricks than are laid by another mason.

humored—the trade term indicating a gentle adjustment of masonry work so that differences are not evident

hydrated or slaked lime—the remaining materials after quicklime has been treated with water. This is the type of lime used in masonry mortar.

hydrochloric acid—see muriatic acid

initial set—the initial drying of mortar that bonds it to the masonry unit

inner hearth—the flat, bottom part of the firebox that the fire is built on; usually built of special firebrick

intrados—the concave curve that bounds the lower limits of an arch. The difference between the soffit and the intrados is that the intrados is linear, while the soffit is a surface, in this case the bottom of the arch.

island fireplace—fireplace built in the center of the room that is open on more than one side

jack arch—a flat arch capable of carrying a load. It should be supported on steel angle irons if the openings are over two feet long. It is the weakest of all arches.

jack-over-jack—the laying of one masonry unit over another without any overlapping; not a good practice unless a stack bond is being formed

jamb—vertical side of an opening, such as the side of a window or door

joints—in masonry work, the mortar between the units

joist—a type of beam used to support floor or ceiling loads, which in turn is supported by a larger beam, girder, or bearing wall

journeyman—a skilled worker who has served as an apprentice in a trade or profession and is now fully recognized as competent in that trade

jumbo or oversized brick—the term used by masons and brick manufacturers to describe a brick larger in size than the standard brick

keystone—the wedge-shaped piece (stone or brick) at the top center of an arch that locks together the other pieces that form the arch ring

kiln run bricks—bricks that meet no special standards as far as color, texture, or design are concerned

lead—the part of a wall that is built as a guide for attaching a line to lay the rest of the wall

lean mortar—a mortar that is lacking in cementitious materials and does not stick to the trowel or masonry unit

lintel—a horizontal member or beam placed over a wall or opening to carry or support the weight of masonry work

litre—the standard unit of measure in the metric system of capacity (more commonly called volume). A litre is primarily used to measure liquids and is equal to 1.0567 qt in the British system of measure.

live load—the non-permanent load to which a structure is subjected or imposed on in addition to its own weight. It includes the weight of persons in a building and freestanding materials, but does not include wind loads, earthquake loads or constant dead loads.

load bearing—the term referring to a wall or other masonry work that supports a load

loam—a type of soil usually composed of clay, silt, sand, or other organic matter

London pattern—a pattern of trowel having a diamond-shaped heel

mantlepiece—the projection of wood or masonry materials that form a shelf on the front of a fireplace; usually very ornate

masonry—a material such as concrete block, bricks, or stone bonded together with mortar to form a wall or structure

masonry cement—prepared cement that contains all of the necessary cementitious ingredients, including any admixtures, so that only sand and water need to be added to form mortar. Masonry cement should meet standards for Type N mortar.

masonry saw—saw powered by an electrical motor used to cut masonry units, either wet or dry

metal heat-circulating fireplace—prebuilt metal fabricated form fireplace unit, with hollow walls and vented openings to allow passage of heat

Material Safety Data Sheet—An information sheet required by OSHA for any hazardous materials that are used in the work place. This sheet must be in English, available to all workers, and contain all important safety information on that particular product.

metre—the standard unit of length measure in the metric system. The metre is equal to 39.37″ in the British system.

Metric system—the decimal system of weights and measures in which the metre, gram, and litre are the basic units of measure. The metric system is used in most of the industrial nations of the world.

modular masonry—a masonry construction in which the size of the building material is based upon the modular unit of 4″

module—the unit of measure used as a standard for construction; a module is 4″

mud—the trade term for masonry mortar or plaster

multicentered arch—an arch whose curve consists of several arcs of circles that are usually tangent at their intersections

multiple-opening damper—type of damper used in a multiple-opening fireplace, much larger than a standard damper.

multiple-opening fireplace—fireplace that is open on more than one side

muriatic acid—an acid used to clean masonry work by mixing a percentage of water to acid. The acid solution works to break down stains and particles on the work.

nailing block—a wooden block inserted into a masonry wall to which the frame is secured by the use of nails

nipples—steel pins used to connect 2 sections of steel scaffolding

nomenclature—description and names used for parts of a brick.

nominal—a dimension greater than a specified masonry dimension because of an amount allowed for the thickness of a mortar joint not exceeding $\frac{1}{2}$″

non-load bearing—a term referring to a wall or other masonry work that supports only its own weight

Norman—a brick with nominal dimensions of $2\frac{2}{3}$″ × 4″ × 12″

OSHA (Occupational Safety and Health Act Public Law 91-596)—the law passed by Congress in 1970 establishing safety requirements for all places of employment in the United States

outer hearth—the part of the hearth that is built out in front of the fireplace; can be flush or raised

overall dimension—a measurement from one extreme point to another to obtain the total distance of the structure, such as a foundation wall

overhand work—the laying of a masonry wall from the inside, with all work done from only one side of the wall

parging—the application of a thin coat of mortar to the back of a wall to waterproof the wall

partition—an interior wall, usually 1 story or less in height, used to separate one room from another

peg end—that end of the line that contains the nail

Philadelphia pattern—a pattern of trowel having a square-shaped heel

pier—an unattached vertical column of masonry work that is not bonded to a masonry wall. The horizontal length of a pier should not be over four times its thickness.

pilaster—a wall portion projecting from wall faces and serving as a vertical column and/or beam

pin end—that end of the line that contains the line pin and to which the line is pulled

plot plan—a plan showing the complete piece of property and important information such as boundary lines, driveways, contour lines, septic systems, and wells

plugging chisel—a chisel with a tapered blade used for removing mortar from joints

plumb bond—the alignment of all head joints in a vertical position

portland cement—the fine, grayish powder formed by burning limestone, clay, or shale and then grinding the resulting clinkers. The result is a cement that hardens under water and that is used as a base for all mortar. Portland cement is a grade of cement, not a brand.

poultice—a paste made with a solvent and an inert material that is used to remove objectionable stains on brickwork

prefixes—abbreviations used to identify the varying degrees of measure in the metric system; the prefixes indicate the relationship of the units to the metre

prescription method—a method of ordering ready-mixed concrete in which the order specifies the time of delivery and the psi strength of concrete desired

presetting—the application of heat at a moderately low temperature (about 72° F) to set cement in concrete block

proprietary compound—a chemical compound protected by a patent or copyright; used to clean masonry work

psi—pounds per square inch

puddling—the process of settling or consolidating grout in a masonry reinforced wall with a stick to prevent the formation of voids in reinforced sections of the wall

pug mill—a machine that extrudes clay in the form of a ribbon that is then cut into brick lengths

pumice—a light, porous volcanic rock used in a finely crushed powdered form as an aggregate of concrete block

putlog—a piece of wood or steel left in the masonry wall with one end resting in a hole and the other end lying on solid ground or on a suitable base. The scaffold frame rests on the putlog, making a strong base on which to start the scaffolding.

quarterback—concrete block that has the inside of two cells cut out

quicklime—a white powder that remains after limestone has been burned at a high temperature in a kiln

racking—the process of laying back each course of masonry work so that it is shorter than the course below it

radial center point—the point at which the radial line begins; used to lay out a jack arch

raggle—a groove in a joint or masonry unit to receive roofing or flashing at the roof line

raised hearth—masonry hearth built above the finished floor level

range number—the identifying number or letter assigned to each product of a brick manufacturer to designate the color, texture, or blend of the brick

ranging the corner—the process of aligning one corner with another by using a line stretched from one point to another

ready-mix concrete—concrete that has already been mixed and is delivered to the job in drum-type concrete trucks

reinforced masonry—a type of masonry work consisting of 2 tiers of masonry units with reinforcements of steel and grout in the center for extra strength

reinforcement rod—a steel rod or bar placed in walls and concrete work to lend strength

return—the turn and continuation of a masonry wall in another direction from which it was started

reveal—that portion of a masonry jamb or recess that is visible from the face of the wall back to the frame; it is placed between the jambs, such as a brick window jamb

rip block—a concrete block less than full size in height that is generally used as a starting piece

rise (of an arch)—the rise of a minor arch is the maximum height of arch *soffit* above the level of its springing line. The rise of a major parabolic arch is the maximum height of the arch *axis* above its springing line.

rock face block—a concrete block made in a mold that resembles a stone wall

rolling scaffolding—a scaffolding that is mounted on wheels; used on relatively small jobs

Roman—a brick with nominal dimensions of $2'' \times 4'' \times 12''$

Roman arch—the name given a semicircular arch because of its use by the Romans in buildings and bridges

rough wall—a masonry wall that is not tooled; must still be plumb and level

round joint—the trade name for a half-circular indentation formed in a mortar joint by use of a convex jointer

rowlock (also spelled rolok)—a brick laid on its face edge so that the normal bedding area is facing the bed joint; commonly used on windowsills

rubbed joint—a flush mortar joint that has been rubbed flat with some type of material to receive paint or to simply appear flat

Rumford Fireplace—fireplace designed by Count Rumford in the 18th century that consisted of a higher, more shallow firebox, smaller damper and a larger flue than normal for greater draft. This type of fireplace radiates a greater amount of heat than a normal masonry fireplace.

running bond or all stretcher bond—a bond in which all of the masonry units are laid in a stretcher position

rustic—a type of brickwork intended to resemble American colonial masonry or to present a roughly finished appearance

sailor—a brick that is laid in a vertical position with the largest side facing the front of the masonry wall

salmon brick—a brick that is relatively soft, pinkish, and underburned; not recommended for outside walls

salt and pepper brick—a brick that has black marks on the finish caused by iron spots in the clay

sandblasting—the process of spraying sand, which is under great air pressure, onto a masonry wall to remove old paint or stains that cannot be removed any other way

sand lime brick—a brick made from sand, lime, and cement

sand-struck brick—bricks that have a sand finish because the mold is lubricated with sand to keep the brick from sticking to the mold

sash block—a specially made concrete block that has a slot in the end to receive a metal window frame

scaffold high—the height to which masonry walls are built when scaffolding is needed to complete the structure; usually $4'$ to $5'$ high

scaffold sill—wooden scaffold frames resting on ground level; helps to prevent the scaffold from settling after heavy rains and distributes the weight of the scaffold and materials

scaling—the flaking away of a concrete surface to a depth of about $\frac{1}{16}''$ to $\frac{1}{4}''$, usually caused by frost damage or freezing of concrete before it has cured.

schedule—a list added separately to plans that describes such items as windows, doors, floors, and wall finishes

scove kiln—an outdated type of brick kiln, rectangular in shape, with holes or tunnels at the bottom in which fires were built to supply the heat to burn the bricks

screeding—the act of using a straight-edge board or metal tool called a *screed* to level off concrete with the top of the finished forms, done to bring to the surface of the mix enough cement paste to allow finish troweling or broom finish.

section view—a drawing that details a vertical slice through a structure, showing all important information such as the composition of the wall, the width and height of the wall, wall ties, collar joints, and, in many cases, the various heights of openings in the wall

segmental arch—an arch that has a circular curve that is less than a full semicircle

semicircular arch—an arch that has a full semicircular curve. It is commonly called a Roman arch.

set—the length of time required for mortar to pass from the soft state to the hard state that will sustain the weight of the masonry placed on it

set of the trowel—the angle of the rise of the trowel handle in relation to the blade when lying in a flat position

setting up—the action of the mortar hardening

sheathing—a facing material such as plywood or insulated siding nailed to the outside of a wood-studded wall

shiner—a brick that is laid in a horizontal position with the largest side facing the front of the wall

shop drawing—a drawing supplied by the manufacturer to explain the installation of a product

shoring—a bracing placed against a wall or under a beam to provide temporary support

shove joint—the procedure of buttering a mortar joint on a brick and shoving it into place against another brick

sieve test—a method of measuring sand by passing it through a specific size of mesh and grading it to ASTM specifications

sill high—the height at which windowsills are installed in a masonry wall

siltation test—a method of determining the amount of loam and impurities in sand by placing the sand in water and measuring the loam that settles on top of the sand

skewback—inclined (sloping) masonry units built into the wall. The beginning arch bricks rest against it at each end of an opening, such as over a window or door.

slack to the line—the term that describes a masonry unit laid too far from the line

slaking—the process of changing quicklime into hydrated lime through a chemical reaction by adding water

slump test—a test done on fresh concrete that is used to determine if the concrete has the correct amount of water in the mix

slushing (a wall)—the process of using a trowel to fill collar joints with mortar

smoke chamber—the area of a fireplace that extends from the level of the damper up to the first flue lining. The slopes of all sides should be equal.

smoke shelf—brick ledge or shelf, laid level with the bottom edge of the damper and back of the firebox that serves to deflect the air currents that come down the chimney back up the opposite side of the flue.

smoke test—test conducted by the mason after the fireplace is completed, consisting of building a fire at the bottom of the flue and then covering the top of the flue tightly. The escape of smoke into any other flue or into the building indicates a leak that must be made tight.

snap header—a half brick laid in a wall to resemble a header

soffit—the undersurface or bottom of an arch

soft-mud process—a process in which units are formed in wooden molds and removed to be burned in the kiln; used for clays that contain too much water for the stiff-mud process

soldier—a brick laid in a vertical position with its longest, narrowest side out, facing the front of the wall

solid masonry wall—a wall built of masonry units laid with full mortar joints between them and with no type of framing present

spall—a small chip or piece of block or brick that has become loosened from the unit

span—the horizontal distance between abutments (masonry jambs on each side of an opening). For minor arches, the clear span of the opening is used. For a major parabolic arch, the span is the distance between the ends of the arch axis.

specifications—the detailed written description of the work to be accomplished in a building. Specifications accompany the plans and describe such things as quality of materials used, workmanship, and methods of construction.

splay—an inclined surface or angle such as the angled sides of a firebox in a fireplace

splitrock—a solid block made of a coarse stone aggregate in which the split face is used as the surface of the finished wall

spotting a brick—the process of laying a brick to establish a specific point without the aid of the line

springing line—for minor arches, the line in which the skewback cuts or intersects the soffit. For major parabolic arches, the intersection of the arch axis with the skewback

stack—the trade term for a chimney that contains a flue for the passage of gases or smoke

stack bond—a bond in which masonry units are laid directly over one another without breaking the bond. All of the head joints should line up in a plumb vertical position.

standard brick—the brick most often used in masonry, with nominal dimensions of $8'' \times 4'' \times 2\frac{2}{3}''$

stiff-mud process—the process in which bricks are made by extruding clay through a pug mill, cutting the column with wire cutters, and burning the units in a kiln

story rod—a wooden pole or rod upon which are indicated all of the masonry courses and openings, such as windows or doors, required in the job

stretcher—a brick laid in a horizontal position with the longest, narrowest side facing the front of the wall. This is the most common position in which bricks are laid.

stringing (mortar)—the process of spreading mortar with a trowel on a masonry wall

structural load—loads or forces that impact on the design and engineering of a structure

subcontractor—a person who performs work negotiated with the prime contractor on a job or project, such as a masonry contractor

suspended or swinging scaffolding—scaffolding with heavy-duty suspended frames on which planking fits

tailing the lead—the process of aligning the ends of the courses on a corner

T-bar tie—a heavy-gauge steel strap tie used to tie intersecting bearing walls to main walls. The tie bar is built into the opposite bed joints of intersecting walls.

tempering—the process of thinning masonry mortar with water to provide workability

tender—a laborer who tends or brings materials to the mason

tensile strength—resistance to lengthwise stress or force, measured by the greatest load in weight per unit area pulling in the direction of length that a given material can stand without tearing apart

texture—the arrangement of particles in masonry materials that accounts for the brick's appearance. The various effects created by tooling mortar joints are also considered part of the texture.

thimble—a round, terra-cotta or fired clay insert that fits into a chimney to receive the furnace pipe, forming a fireproof connection to the masonry wall

throat—the part of the fireplace that narrows at the top of the firebox and through which the gas smoke passes. If a damper is set, it becomes the throat.

throwing a mortar joint—the trade term for the process in which mortar is applied on a brick with a swiping motion to form a joint

thrust (of an arch)—the tendency of an arch to spread to the sides under a load

tier—a single vertical thickness or row of masonry units; also called a wythe

tight away—a term used to signal to the mason on the pulling end of the line that it is all right to pull the line tight against the line pin

toe boards—wooden boards that are attached to the bottom outside edge of scaffolding to prevent objects from falling off. Toe boards are an OSHA requirement on all scaffolds exceeding 10′ in height.

tooling or striking (of joints)—the trade term that describes the finishing of mortar joints by passing a striking tool over the joints

toothing—a toothlike projection of one masonry unit over another that forms a temporary end for a structure

tower scaffolding—a steel-framed scaffolding erected in a towerlike position, with adjustable height

trench footing—concrete footings that are placed in an earthen trench

trig brick—a brick on which a trig is placed to prevent the line from sagging

trimmer arch—masonry arch used in fireplace construction to support the rough hearth or the rough opening over the firebox

tuckpointing—the process of replacing the old mortar in joints with new mortar

Tudor arch—a pointed, four-center arch of medium rise-to-span ration

tunnel kiln—a long, rectangular brick structure in which the process of burning bricks is controlled by computers and is completely automatic. This kiln is the most modern of all kilns used for brick manufacturing.

veneered wall—a masonry wall with a facing, which is attached, but not bonded, to the backing to act as a load-bearing wall

veneer tie—a corrugated metal wall tie used to tie masonry units to a backing, usually of a frame construction, by nailing

viaduct—a long bridge built of a series of masonry spans supported on piers or towers. The Romans used viaducts with arches over the piers to transport water from the mountains to the cities.

void—an open space in an area, such as the space between two masonry walls

voussoir—one of the wedge-shaped masonry units that form the arch ring, such as the brick in a jack arch

walling the brick—the process of stacking brick on the back section of a wall that has been previously built. This process eliminates the need for the mason to reach into the brick pile to pick up individual bricks when building the front wall.

wall plate—a horizontal wooden member anchored to a masonry wall, to which other structural elements may be fastened, usually attached to the top of a masonry wall.

warehouse pack—when referring to cement, the term refers to the settling (called firmness or pack) that the dry cement has undergone after having been stored for a long period of time. Rolling or adjusting the bag will correct this problem.

wash—sloping application of mortar or cement that prevents the entrance of water into masonry work

water-cement ratio—the relationship between the amount of water and portland cement used to make it into a paste

water retentivity—a quality of mortar that prevents the rapid loss of water to masonry units through absorption; enables mortar to retain moisture

water-struck brick—bricks that have a very smooth finish because the molds are lubricated with water to keep the bricks from sticking to the mold

web—the cross walls connecting the face shells of hollow concrete masonry units

weep holes—small holes left in mortar joints for drainage of masonry walls. Weep holes are usually spaced 24″ apart and are placed over the first course of flashing.

welding splatter—the metal beads splashed on the surface of the masonry wall that soak into the surface of the masonry unit

winning—the removal of raw material (clay or shale) from the ground

wire mesh or hardware cloth—metal wire, usually formed in ¼″ squares, used to tie walls, which are not load bearing, to masonry walls

wire track—the trade term for metal wire joint reinforcement used to strengthen or tie 2 tiers of masonry units together to form one composite wall

Z tie—a metal tie used to tie 2 tiers of masonry units together; may have a drip crimp in the center

Index

617